9th International SOLARIS Conference 2018

IOP Conference Series: Materials Science and Engineering
Volume 556

Chengdu, China
30 – 31 August 2018

ISBN: 978-1-5108-9387-0
ISSN: 1757-8981

Printed from e-media with permission by:

Curran Associates, Inc.
57 Morehouse Lane
Red Hook, NY 12571

Some format issues inherent in the e-media version may also appear in this print version.

Printed by Curran Associates, Inc. (2019)

For permission requests, please contact the Institute of Physics at the address below.

Institute of Physics
Dirac House, Temple Back
Bristol BS1 6BE UK

Phone: 44 1 17 929 7481
Fax: 44 1 17 920 0979

techtracking@iop.org

Additional copies of this publication are available from:

Curran Associates, Inc.
57 Morehouse Lane
Red Hook, NY 12571 USA
Phone: 845-758-0400
Fax: 845-758-2633
Email: curran@proceedings.com
Web: www.proceedings.com

Table of contents

Volume 556

9th Edition of the International SOLARIS Conference

30–31 August 2018, Chengdu, China

Accepted papers received: 17 May 2019
Published online: 19 August 2019

Preface

Papers

IOP Conf. Series: Materials Science and Engineering **556** (2019) 011001 doi:10.1088/1757-899X/556/1/011001

Organising Committee

Lingzhi Xie – Sichuan University, China (Chair)
Yupeng Jack Wu – University of Nottingham, UK (Chair)
Wei Zhang, Sichuan University, China
Xia Hao, Sichuan University, China
John Calautit, University of Nottingham, UK
Robin Wilson, University of Nottingham, UK
Yanyi Sun, University of Nottingham, UK
Jing Li, University of Nottingham, UK
Tapas Mallick, University of Exeter, UK
Senthilarasu Sundaram, University of Exeter, UK
Miaoyi Qi, Sichuan University, China (Secretary)
Hao Gao, University of Nottingham, UK (Secretary)

SOLARIS Board members

Professor G N Tiwari, Jamia Millia University, Delhi, India
Associate Professor E Jadraque Gago, Granada University, Granada, Spain
Professor Tariq Muneer, Napier University, UK
Dr Harjit Singh, Brunel University London, UK
Dr H Kambezidis, NOA, Athens, Greece
Dr Danny Li, City University, Hong Kong
Professor J Mohlenikova, Brno University, Czech
Dr M Knez, Maribor University, Slovenia
Professor A Kudish, Ben Gurion University, Israel

International Scientific Committee

Hao Liu, University of Nottingham, UK
Peter Rutherford, University of Nottingham, UK
Mervyn Smyth, University of Ulster, UK
Tariq Muneer, Edinburgh Napier University, UK
Ing-Liang Wong, University of Glasgow Caledonian, UK
Yaodong Wang, University of Newcastle, UK
Ben Hughes, University of Sheffield, UK
Gianfranco Claudio, Loughborough University, UK
Harjit Singh, Brunel University London, UK
Yuan Tian, University of Hertfordshire, UK
Xiaoyu Yan, University of Exeter, UK
Xingxing Zhang, Dalarna University, Sweden
Veronica Soebarto, University of Adelaide, Australia
Tim Zhao, University of Adelaide, Australia; Leibniz Institute of Photonic Technology, Germany
Mohamed M Sabry, Umm Al Qura University, Saudi Arabia
Enshen Long, Sichuan University, China
Jingquan Zhang, Sichuan University, China
Lianhuan Feng, Sichuan University, China
Hassam Chaudhry, Heriot Watt University, UAE
Hasim Altan, University of Sharjah, UAE

IOP Conf. Series: Materials Science and Engineering **556** (2019) 011001 doi:10.1088/1757-899X/556/1/011001

Masa Nagouchi, University of Melbourne, Australia
Zheming Tong, Harvard University, US
Maria Wall, Lund University, Sweden
Filippo Calcerano, Italian National Research Council, Italy
Suresh Sivan , National Institute of Technology Trichy, India
Francis Agyenim, Council for Scientific and Industrial Research, Ghana
Francis Momade, Kwame Nkrumah University of Science and Technology, Ghana

Acknowledgements
We would like to thank the hosting Sichuan University for providing their support for the Conference. We thank our Keynote speaker, scientific committee members, participants, reviewers, session chairs, organizing committee members and all those who have helped us to organize 9th edition of international SOLARIS conference to make it a success.

IOP Conf. Series: Materials Science and Engineering **556** (2019) 011001 doi:10.1088/1757-899X/556/1/011001

Sponsors

Brinell Vision
Protecting and Enhancing Vision

couchperrywilkes
engineering change

EVOENERGY

TALESUN 中利腾晖光伏
ZHONGLI TALESUN SOLAR

成都国检
Chengdu Quality Inspection

YPT
Yorkshire Photonic Technology Ltd

Supporting Institutions

The University of Nottingham
UNITED KINGDOM · CHINA · MALAYSIA

SICHUAN UNIVERSITY 1896

IOP Conf. Series: Materials Science and Engineering **556** (2019) 011001 doi:10.1088/1757-899X/556/1/011001

Photograph

IOP Conf. Series: Materials Science and Engineering **556** (2019) 011002 doi:10.1088/1757-899X/556/1/011002

Peer review statement

All papers published in this volume of *IOP Conference Series: Materials Science and Engineering* have been peer reviewed through processes administered by the proceedings Editors. Reviews were conducted by expert referees to the professional and scientific standards expected of a proceedings journal published by IOP Publishing.

IOP Conf. Series: Materials Science and Engineering **556** (2019) 012001 doi:10.1088/1757-899X/556/1/012001

Analysis of a rotary passive heat recovery device for natural ventilation windcatcher

D O'Connor[1], J K Calautit[2], K Calautit[2], S Shazad[3], B R Hughes[4], C Pantua[2]
[1]Free Running Buildings Limited, UK
[2]Department of Architecture and Built Environment, University of Nottingham, UK
[3] Department of Architecture, University of Sheffield, UK
[4]Department of Mechanical and Aerospace, Engineering, University of Strathclyde, UK
E-mail: dom.oconnor@freerunningbuildings.com

Abstract
Based on the design of traditional architecture, windcatchers are devices which provides passive ventilation by manipulating pressure differentials around buildings induced by the movement of wind and difference in temperature. In temperature climates, it is effective in providing passive cooling during summer months. However, during winter months, the low air temperature supplied to the space can cause further thermal discomfort and increase heat loss which lead to higher energy consumption. This limits the capabilities of windcatchers to provide ventilation all year round. To address this issue, the present study proposes incorporating a rotary thermal heat recovery device into the windcatcher channel and investigate its performance using numerical modelling and experimental tests. The study focused on characterizing the design and performance of the copper radial blades of the proposed heat recovery device. The predicted results of the airflow and temperature showed good agreement with the experimental tests. Two types of radial blade designs were assessed in terms of the airflow velocity and distribution, the pressure drop and the heat recovery performance.

Keywords: built environment, CFD, heat recovery, natural ventilation

1. Introduction

The commitment of UK government in minimizing greenhouse gas emission by at least eighty percent of 1990 levels in the year 2050 has been the greatest drive in reduction of energy use within the built environment [1]. A huge portion of energy consumption comes from space heating which records forty percent of the overall energy use in service and residential sector [2]. Therefore, minimizing the energy demand to heat commercial and domestic buildings can contribute in providing solutions to achieve the aim of reducing carbon emissions. Nowadays, the use of natural ventilation methods is now being recognized and installed in buildings for reducing the mechanical ventilation use and increasing the supply of fresh air [3]. Windcatchers are normally used in Middle Eastern countries and currently increasing its presence in the market [4,5].

Figure 1a shows the structure and function of a windcatcher. This type of system provides fresh air driven by the air pressure on the wind ward area, while exhausting stale air assisted with suction pressure on the leeward area. Moreover, it can also function using the stack effect method; the air mass reduces when the temperature rises, resulting warmer air to go up and depart the windcatcher's openings. In countries with temperate climate like the United Kingdom, their operation is limited during summer. This is due to the low temperature of the incoming or supply air that can lead to thermal discomfort within the occupied space and increase heat loss [6]. While when the use of natural ventilation is restricted, the mixture of contaminants has been observed to increase more than the recognised guidance levels, that can result to illness and poor mental performance [5,6].

IOP Conf. Series: Materials Science and Engineering **556** (2019) 012001 doi:10.1088/1757-899X/556/1/012001

Several researches were already carried out on wind catchers. The main goal is to improve the understanding on the effect of airflow over and through wind catchers and the ventilation level that devices can offer [7]. Moreover, these studies have been conducted to enhance the thermal comfort which can be offered to occupiers [8]. This study will use scaled wind tunnel testing, CFD modelling, and also field or situ testing to recognize the effects in a real-world setting [9].

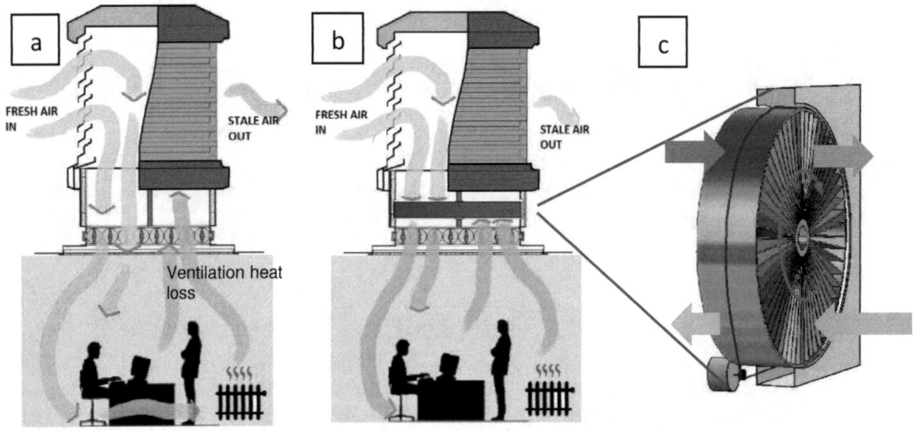

Figure 1. a) standard windcatcher b) wind catcher with heat recovery wheel c) rotary wheel with radial blades

It has been observed that there are several studies on wind catcher cooling, but few on heating incoming air or heat recovery in wind catchers [10-12]. In Shao et al. [13] study, it was revealed that passive stack system design without heat recovery resulted in a huge amount of wasted heat. Woods et al. [14] investigated the application of wind catchers during the winter with the assistance of air mixing methods to dilute the incoming cool air with the indoor air. The wind catcher increased the incoming air temperature, thus reducing the heat demand and at the same time keeping enough level of contaminants, however it required strict control methods. Even though this topic of research is growing with more researchers investigating the performance of heat recovery in passive ventilation techniques, there are still few works carried out in the application of the rotary thermal wheels in natural ventilation. This is due to the high pressure drop across the rotary wheel that can hinder the air flow through the natural ventilation device. This type of device offers heat recovery at a high efficiency, even if compared to other heat recovery device. The energy saving potential of utilizing rotary wheel incorporated with a passive ventilation device like wind tower are high.

In order to enhance the performance of wind catcher devices to allow continuous usage in heating season, integration of a passive heat recovery device is required. This research will assess the performance of the proposed device using numerical modelling and experimental tests for validation. A redesigned rotary heat recovery wheel device will be installed below the wind catcher's channel. Utilizing the components of the thermal wheel as a heat exchanger, thermal energy in the indoor exhaust air is recovered to the incoming air. This technique increases the incoming air temperature. Through increasing the incoming air temperature from the wind catcher, sufficient year around ventilation is maintained during winter, the demand of energy for heating devices will be minimized.

2. Methodology

2.1. Numerical modelling

This study used numerical analysis and experimental testing to examine the potential of the heat recovery and validated the results, as demonstrated in Figure 2. Fluent CFD will be used for numerical modelling, while a combination of wind tunnel testing and full-scale laboratory examination for experimental testing.

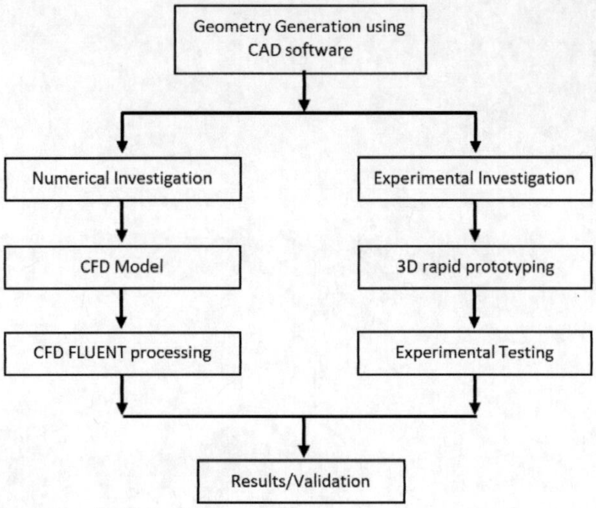

Figure 2. Numerical and experimental methodology

The ANSYS Fluent software was used to investigate the numerical modelling of the passive heat recovery system. It should be noted that the governing equations are not altered in this study. The mass and momentum equations are solved for all flow types. Due to the heat transfer involved in the study, the energy equation is required. As the model flow will be turbulent, the Reynolds-averaged Navier-Stokes and the k-epsilon equations are essential which are well-established methods in the research on natural ventilation flows [15-18]. Second-order upwind scheme is employed to discretised all the transport equations. The most commonly used Semi-implicit method for pressure-linked equations segregated pressure-based algorithm solver is used in this study [19-20].

By using ANSYS CFD, geometry was formed in SolidWorks and imported for flow examination. The geometry created in CAD was transferred to ANSYS Design Modeler that needs further adjustment to identify the domain needed for the CFD modelling. There are two kinds of radial blade rotary wheel arrangements created for this study, as shown in Figure 3. The angular separation among radial blades differ between 15° (20 radial blades) and 10° (32 radial blades).

In order to ensure that the numerical results compared to the experimental data are as close as possible, the simulation should reflect similar conditions as in the experimental tests. For the computational demands to be reduced, the geometry for the CFD models was simplified. In order for the computation to be faster, the intricacy of the heat recovery device and ductwork were decreased, as demonstrated in Figure 3. The 2 mm rotary blades used in this study had 0.100 x 0.100m^2 dimensions and these are made of copper. Every channel has a fluid volume of 0.160 x 0.325 x 0.490$m^{3.}$ The channel consisted an inlet at one end and two circular outlets on the opposite area. During the test, fans were installed at the outlets to direct the flow to the passageways. Additional volumes are also imported and added to the solid geometry to model the volumes of copper in the device, located between the radial blades for simplification.

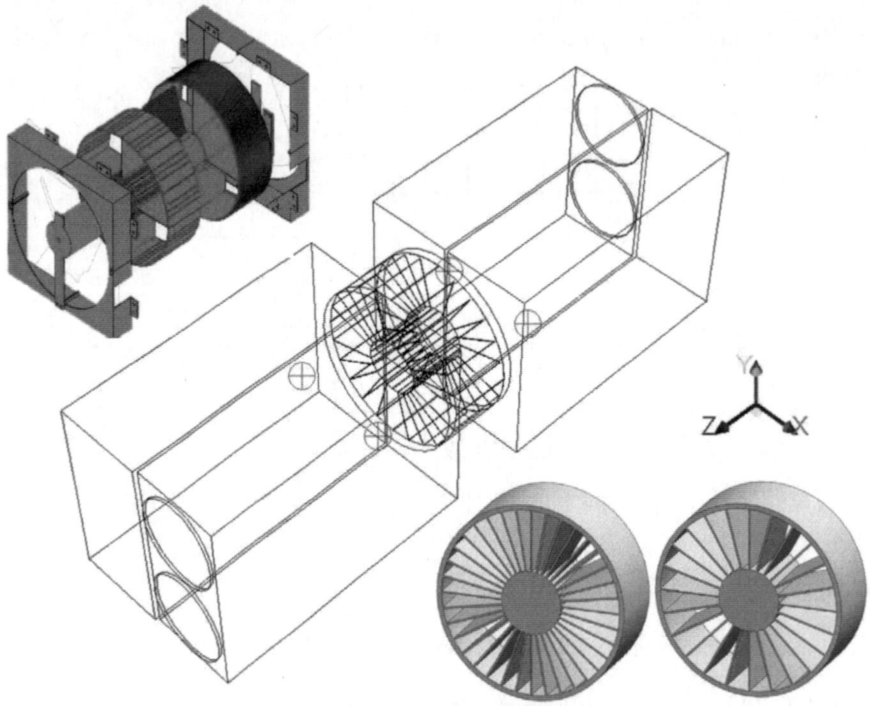

Figure 3. CFD model of the two way cross-flow duct utilised for evaluation.

Since there are variations with the geometry of different arrangements, each needed a single mesh to be created. Mesh sizing was used for the entire geometry surfaces, with element size of 0.0075m for all the arrangements. The volume of mesh elements were approximately 526,000 to 872,000, and this is because of variations in geometry. To validate the CFD grid independence solutions, sensitivity assessment was used. This is done through generating several different mesh sizes with geometries configuration and simulating the model using similar boundary conditions. Table 1 shows the values for the boundary conditions gathered from the calculations during the examinations.

Table 1. Overview of boundary conditions.

Input Condition	Value	
	Supply Air	Exhaust Air
Pressure Inlet (Pa)	15	15
Exhaust Fan Outlet (Pa)	2.5	2.5
Inlet Air Temperature (°C)	21.4- 23.9	24.7-39.
Wheel Rotation Speed (rpm)	6	6
Source (W/m^3)	100,000-750,000	-
Sink (W/m^3)	-	200,000-4,800,000

2.2. Experimental testing

The concept of the rotary wheel with radial blades is shown in Figure 4. To assess the performance of the new design, models of the rotary wheel and suitable covering are created and integrated to the duct work with two counter current passageways (exhaust and supply air flow) with varying material properties to evaluate the thermal performance of the system. The flow properties were measured downstream and upstream of the rotary wheel for each passageway then its potential was evaluated. The measurement points were positioned at the center of every passageway at 0.1 m intervals. The supply air flow settings were ordered through the situations of the lab indoor space (temperature around 21.4-23.9 °C) whereas the departing air stream temperature was conditioned using a heat source (24.7-39.8°C) [21]. The purpose of the experiment was to examine the potential of recovery and thus sufficient continuous supply air flow stream and controlled exhaust air flow stream temperature was useful. The wheel was set to continuously rotate at 6rpm, assisted with motor and gear device integrated inside the case of the system, as demonstrated in Figure 4. With the available motor and given wheel size, 6rpm was decided as the appropriate rotation speed.

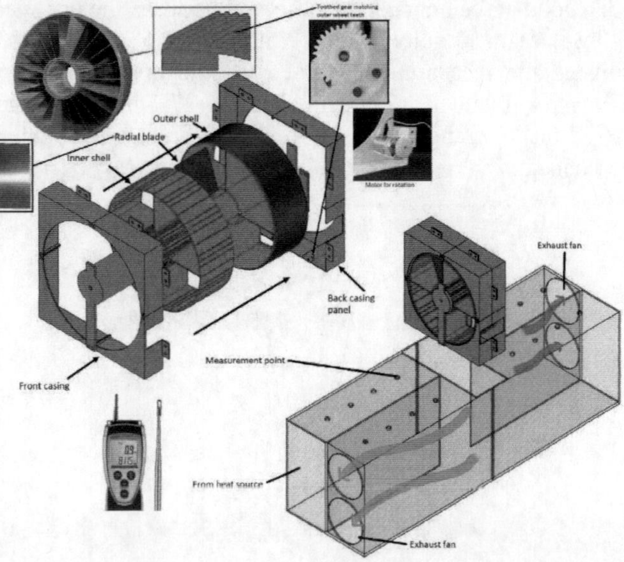

Figure 4. Arrangement of the passive heat recovery wheel experiment.

Four regulated similar axial fans were incorporated at every channel to drive the air flow. Air is drawn from the duct inlets through the channel by the axial fans (suction driven flow). This arrangement showed the greatest uniform velocity profile in the channels. Many of the elements such as casing and rotary wheel were nonstandard and needed the use of 3D printing. Figure 4 demonstrates the total dimension of the ductwork with 5 mm acrylic Perspex material. Casing of the heat recovery system sit flush in the ductwork and any air gaps were closed at the time of experimental testing such as measurement points that are unused. The copper plates were applied for the heat recovery radial blades having 100 mm x 100 mm x 1mm dimensions. The rotary wheel's inner shell with diameter 0.145 m was built to house the radial blades. For the system to rotate, a toothed gear incorporated to a small motor was used. While for the outer shell, it was built to match the teeth on the gear to let the wheel rotate smoothly. In order to build the shell accurately, 3D printing was utilized. To examine the response of the passive rotary wheel on the supply air ventilation level, air flow velocity measurements were conducted. As mentioned in the previous section, low pressure drop over the system was required to make sure that the needed ventilation levels will be reached even at low wind speeds. Velocity measurements were conducted through utilising Testo 425 hot wire anemometer with resolution of 0.01m/s and an accuracy of 0.03 m/s +5%rdg. In every calculation area, the air velocity was calculated for twenty seconds and the mean velocity rate was recorded. The measurement was conducted three times for every area. Air temperature was taken at similar areas with the use of Testo 176 P1 logger, resolution of 0.1°C, and accuracy of 0.03°C. Simultaneous measurements were used for the entire data collection points, every five seconds for one hour. Monitoring was carried out in a long period of time with slow alterations in the air flow setting.

3. Results and Discussion

The numerical simulation is validated to ensure the confidence in the predicted result. The velocity and temperature experimental data are utilised for the validation of each of the configurations of the rotary heat recovery wheel. The contours of the air velocity distribution in the supply or incominh air channel for the 32-blade rotary wheel is shown in Figure 5a. It can be observed that the airflow upstream of the rotary wheel had a uniform air velocity profile and speeding up from 2.15m/s to 3.76m/s as it approaches the surface of the wheel. The airflow speed further accelerates (up to 5m/s) as it enters the wheel and flows between the narrow spaces between the radial blades. As the airflow exits the device, it is still moving at a higher speed than the upstream side. Flow separations are observed downstream of the channel which is expected given the channel shape and the presence of the passive recovery wheel. The airflow velocity measurements 0.1m upstream and downstream of the rotary wheel are shown in Figure 5b and c. As observed, a good agreement can be observed between the two methods. The average error for the upstream and downstream air velocity are 12.5% and 24.6%. For the 20-blade arrangement, the CFD model overestimated the measurement values for the upstream airflow velocity (36% error average) while the downstream values are underestimated (27% error average). Although the error is higher than that of the 32-blade arrangement, the trends of the data are similar enough to compensate for the higher error percentage.

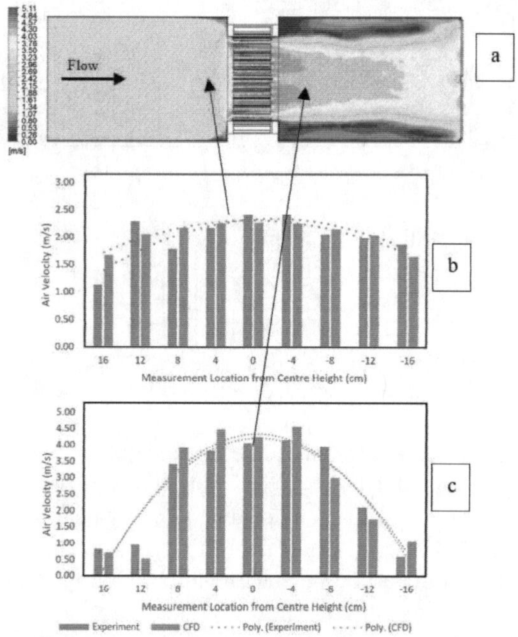

Figure 5. (a) Cross sectional contours of the airflow velocity. Comparison between experiment and CFD air velocity in the supply airstream (b) before and (c) after the 32 blade rotary wheel.

The airflow temperature before and after the device is another parameter assessed in both the experiments and numerical modelling. Validation of the numerical models using airflow temperature is noteworthy given the device purpose is to increase the supply airstream temperature, accurateness of the numerical modelling will support the further development of the proposed concept. Figure 6 and 7 shows a comparison of the air temperature results from the CFD model and experimental tests. As observed, a good agreement is observed between both methods, with the average error at 0.16% and 0.07% for the supply and exhaust airstreams. As the exhaust airstream inlet temperature increased, an increased in temperature change is observed for both airflow streams and configurations. However, the temperature change is not equal between the two airstreams. The temperature change in the exhaust airstream (0.2-2.3°C for HR32-1 to HR32-4) is higher than the supply airstream (014-0.7°C for HR32-1 to HR32-4) which shows that the design of radial blade is effective at conducting heat from the exhaust airflow but not as effective at releasing this energy to the supply airstream. This suggests that significant thermal energy is stored within the radial blades. Increasing the ability of the blades to dissipate the stored thermal energy would increase the performance of the device. It should also be noted that the inlet temperature for the supply airflow is between 23.1-23.9°C which is not the temperature range the device is designed for. However due to the limitations of the experimental tests, the supply air is not conditioned and dependent on the lab room temperature. Given the relatively high supply airflow stream inlet temperature, further investigation would be advised. As the heat recovery device is designed for passive ventilation integration, the supplyg air temperature should represent the outdoor air conditions. Common outdoor air temperatures in the UK, areas where device would be most useful in winter months, would be lower (8-11°C in the UK) than the temperatures measured in here.

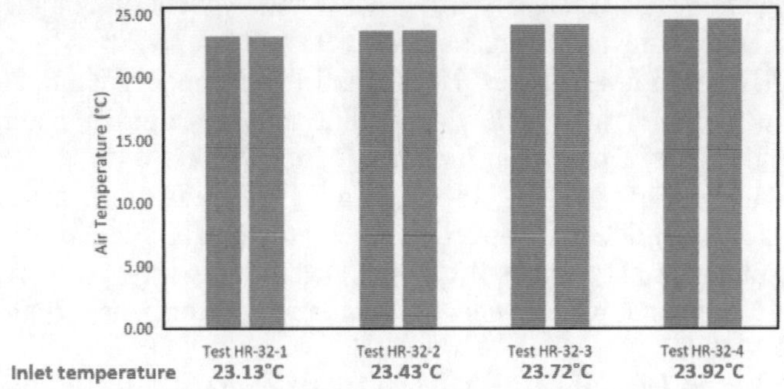

Figure 6. Comparison between experimental and numerical results of air temperature in the supply airstream after the rotary wheel with 32-blades.

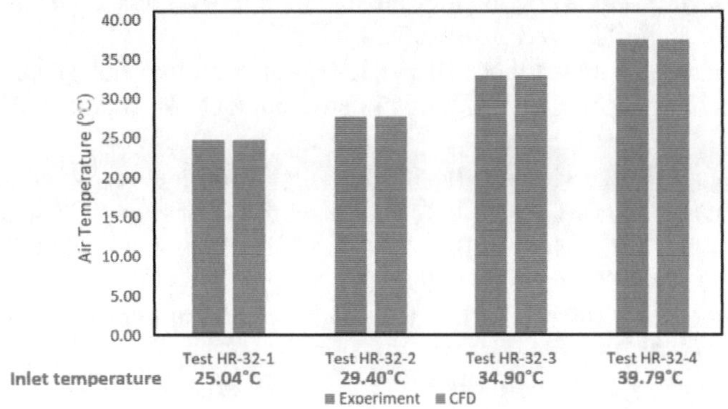

Figure 7. Comparison between experimental and numerical results of air temperature in the exhaust airstream after the rotary wheel with 32-blades.

4. Conclusions

The present study used numerical modelling and experimental tests to investigate the performance of a rotary heat recovery device for a natural ventilation windcatcher. The study focused on characterizing the design and performance of the copper radial blades of the proposed heat recovery device. The predicted results of the airflow and temperature showed good agreement with the experimental tests. Two types of radial blade designs were assessed in terms of the airflow velocity and distribution, the pressure drop and the heat recovery performance. The results showed that although the required pressure drop of 2Pa was not reached, the analysis of the air velocity and pressure upstream and downstream of the device suggested that the it had minimal effect on the airflow rate. The supply rate was 141L/s through the wheel would be suitable to provide fresh air up to 17 occupants. Initial CFD and experimental results showed that the radial blade configurations were able to recover and transfer the heat from the exhaust to the supply airstream. Although, the reduction in temperature of the exhaust air was much higher than the temperature increases of the supply airflow which indicated that the heat conducted by the radial blades is stored and not effectively released to the supply stream. Hence, future studies should focus on further developing the design of the radial blades and increase the heat transfer efficiency considering its potential not to impede the natural ventilation airflow.

References

[1] DECC. Climate Change Act. London, 2008. 2. WBCSD. Energy Efficiency in Buildings: Facts & Trends.
[2] World Business Council for Sustainable Development 2008.

[3] Calautit JK, Hughes BR and Nasir DSNM, 2017, Climatic analysis of a passive cooling technology for the built environment in hot countries, *Applied Energy* **186** 321-335

[4] Hughes BR, Calautit JK and Ghani SA, 2012, The Development of Commercial Wind Towers for Natural Ventilation: a review. *Applied Energy* **92** 606-27

[5] Nejat P, Calautit JK, Majid M, Hughes BR, Zeynali I and Jomehzadeh F, 2016, Evaluation of a two-sided windcatcher integrated with wing wall (as a new design) and comparison with a conventional windcatcher, *Energy and Buildings* **126** 287-300

[6] Calautit JK and Hughes BR, 2014, Wind tunnel and CFD study of the natural ventilation performance of a commercial multi-directional wind tower. *Building and Environment* **80** 71-83

[7] Calautit JK, O'Connor D and Hughes BR, 2014, Determining the optimum spacing and arrangement for commercial wind towers for ventilation performance. *Building and Environment* **82** 274-287

[8] Hosseini SH, Shokry E, Hosseini AJ, Ahmadi G and Calautit JK, 2016, Evaluation of airflow and thermal comfort in buildings ventilated with wind catchers: Simulation of conditions in Yazd City, Iran, *Energy for Sustainable Development* **35** 7-24

[9] Nejat P, Calautit JK, Majid MZ, Hughes BR and Jomehzadeh F, 2016, Anti-short-circuit device: A new solution for short-circuiting in windcatcher and improvement of natural ventilation performance, *Building and Environment* **105** 24-39

[10] Chaudhry HN, Calautit JK and Hughes BR, 2015, Computational analysis of a wind tower assisted passive cooling technology for the built environment, *Journal of Building Engineering* **1** 63-71

[11] O'Connor D, Calautit JK and Hughes BR, 2016, A review of heat recovery technology for passive ventilation applications. *Renewable and Sustainable Energy Reviews* **54** 1481-1493

[12] O'Connor D, Calautit JK and Hughes BR, 2014, A study of passive ventilation integrated with heat recovery. *Energy and Buildings* 82 799-811

[13] Shao L, Riffat SB, Gan G, 1998, Heat recovery with low pressure loss for natural ventilation. *Energy and Buildings* **28** 179-184

[14] Woods AW, Fitzgerald S and Livermore S, 2009, A comparison of winter pre-heating requirements for natural displacement and natural mixing ventilation. *Energy and Buildings* **41**(12) 1306-1312

[15] Hughes BR, Chaudhry HN and Calautit JK, 2014, Passive energy recovery from natural ventilation air streams. *Applied Energy* **113** 127-140

[16] Jomehzadeh F, Nejat P, Calautit JK, Yusof MB, Zaki SA, Hughes BR and Yazid M, 2017,A review on windcatcher for passive cooling and natural ventilation in buildings, Part 1: indoor air quality and thermal comfort assessment. *Renew. Sustain. Energy Rev* **70** 736-756

[17] Calautit JK and Hughes BR, 2014, Measurement and prediction of the indoor airflow in a room ventilated with a commercial wind tower. *Energy and Buildings* **84** 367-377

[18] Calautit JK, Chaudhry HN, Hughes BR and Ghani SA, 2013, Comparison between evaporative cooling and heat pipe assisted thermal loop for a commercial wind tower in hot and dry climatic conditions, *Applied Energy* **101** 740-755

[19] Calautit JK, Hughes BR and Shahzad SS, 2015, CFD and wind tunnel study of the performance of a uni-directional wind catcher with heat transfer devices. *Renewable Energy* **83** 85-99

[20] O'Connor D, Calautit JK and Hughes BR, 2016, Ben Richard Hughes, A novel design of a desiccant rotary wheel for passive ventilation applications. *Applied Energy* 179 99-109

[21] Calautit JK, Chaudhry HN, Hughes BR and Sim LF, 2014, A validated design methodology for a closed-loop subsonic wind tunnel. *Journal of Wind Engineering and Industrial Aerodynamic* **125** 180-94

An economic analysis of the solar photovoltaic/thermal (PV/T) technologies in Sweden: a case study

Xingxing Zhang[1], Jinshun Wu[2], Song Pan[3] and Mengjie Han[1]
[1]School of Industrial Technology and Business Studies, Dalarna University, Falun, Sweden
[2]Architectural Engineering College, North China Institute of Science & Technology, Hebei, China
[3]College of Architecture and Civil Engineering, Beijing University of Technology, Beijing, China

E-mail: xza@du.se

Abstract. The solar energy share in Sweden will soar over the next decades. Such transition offers not great opportunity but uncertainties for the emerging solar PV/thermal (PV/T) technologies. There are still critical challenges for PV/T development in Swedish scenario. This paper aims to investigate the local policies and market structures in Sweden by a set of analytical economic frameworks and a case study. A macro-economic exposure analysis in Swedish context is conducted and a PEST model is then applied for analysis of external environment of PV/T industry in Sweden. The PV/T market is further analysed based on a case product by means of SWOT and Porter's five forces. The whole study concludes that the market of solar PV/T is expanding mostly towards small-and medium-sized energy plants. The market position of the case company meets the trend of PV/T market, which has an overwhelming competence against its competitor in the domestic market. This paper is expected to clarify the economic situations of PV/T in Sweden and promote its further development towards sustainable and low-carbon economics.

Keywords: PV/T; Policy; Market; Strategy

1. Introduction

The energy goals for Sweden are stated in the Government's integrated climate and energy policy bills, which include energy policy targets by 2020 of a 40% reduction in greenhouse gas emissions, at least 50% use of renewable energy, and a 20% more efficient energy end use. In addition, the use of fossil fuels for space and tap water heating is to be phased out by 2020 [1]. These facts are significantly motivating the development of energy efficiency solar gains and decarbonized economy in Sweden for the aid in transition to future sustainable economic growth and social welfare. Such transition offers not only great opportunity but also uncertainties for the emerging solar PV/thermal (PV/T) technologies. Currently, there is no research specifically addressing the economic evaluation of PV/T technologies in Swedish scenario. There are still critical challenges for PV/T development in Sweden, such as unclear market size/position and weak development strategy. It is likely that the economic results would be different if the study is performed in another country or even in the same country today, as costs, tariffs, and policies are changing. It is thus necessary to clarify the local policies, and market structures for the PV/T technology in Sweden. This paper aims to investigate the local policies and market structures in Sweden by a set of analytical economic frameworks and a case study. A macro-economic analysis in Swedish context is conducted in section 2 along with a PEST analysis. The PV/T market is further analysed in section 3 based on a case product in section 4 by means of SWOT and Porter's five forces. This paper is expected to clarify the economic situations of PV/T in Sweden and promote its further development towards sustainable and low-carbon economics.

IOP Conf. Series: Materials Science and Engineering **556** (2019) 012002 doi:10.1088/1757-899X/556/1/012002

2. Macroeconomics of Sweden and PEST analysis related to PV/T industry

2.1. Macro-economic analysis of Sweden

According to the data from World Bank [2], the annual GDP growth rate in Sweden has been rather stable during the last decades, with an average value of about 2.0%, reaching maximum of nearly 6.0% in 2010 and minimum of -5.18% in 2009 owing to the economic crisis. The Sweden GDP growth rate is projected to trend around 2.60% in 2020, mainly depending on net exports and government spending [3]. The Energy use in Sweden varies with the GDP growth rate, which dropped slightly in the last 25 years from a high of 5800 kg of oil equivalent per capita to around 5000 kg of oil equivalent per capita, but the overall volume of energy use is quite stable. The share of renewables increased a lot, ranging from 38% to 64% during the period from 1990 to 2016, which is projected to grow more in near future. These data indicates that Sweden's GDP is growing, and it will consume much more renewable energy, rather than the traditional fuels, during its further development, which opens large opportunities for energy generation firms. The inflation rate in Sweden decreased considerably in general from 2008 until 2017, with an averaged value of 1.20%, achieving peak value of 4.40% in first quarter of 2009 and lowest value of -1.60% in third quarter of 2009. Specifically, the consumer prices increased for housing and utilities (3.1% and 2.8%); meanwhile, cost declined for furniture and household equipment (-0.5% vs 0.4%). The inflation rate in Sweden is modelled at 2.40% by the end of 2017, and in the long-term, it is projected to trend around 2.60% in 2020 [3]. The interest rate in Sweden displays a mean value of 1.2% from 2008 until 2017, with highest value of 4.60% in 2009 and lowest value of -0.50% in 2016. It shows a significant drop in this period. In order to maintain a strong economic activity, the central bank of Sweden is expected to keep the repo rate steady at -0.50% until the year of 2020 [3]. The deposit interest rate is much smaller than the inflation rate, which aims to stimulate the economic activities and call up new business or banking models to return customers' investment at higher rate. The overall wages, and consumer spending in Sweden are keep increasing significantly from 2008 to 2017 [3]. A great increase in the household saving rate is observed from around 5% in 2005 to less than 16% in 2018 [4]. All these figures demonstrate an increasing buying power (potential investment capacity) of consumers.

2.2. PEST analysis for general business environment in Sweden

Political factor: Sweden has committed to 100% renewable energy by 2040, in which solar energy is planning to contribute to 5-10% to electricity generation compared to today's marginal level of less than 0.1% [5]. The government promotes renewable electricity through a quota system, tax regulation mechanisms and a subsidy scheme. In Sweden, tax exemptions are the main incentives to support renewable heating [5]. In Sweden, the grid operator is obliged to connect electricity generation systems to the grid, transmit electricity and expand the grid. But renewable energy is not yet given priority. Companies operating a district heating network are obliged to negotiate terms to connect a heating plant. Since 1 January 2008, a new law on energy declaration of buildings has been in force in Sweden. The government issued a national programme for energy efficiency and energy-smart construction. This energy policy established the conditions for efficient and renewable energy use and cost-effective energy supply in Sweden with low negative impacts on environment and climate. Sweden introduced green electricity certification in 2003. Green electricity must come from renewable recourses, such as wind, solar, geothermal or wave power, biofuels or small-scale hydroelectric plants. Electricity retailers are required to buy a proportion of 'green electricity' as part of their normal supply, while power producers receive certification for the renewable electricity they generate.

Economic factor: The Swedish economy is thriving, which contracted during the global financial crisis and Eurozone crisis, but it has rebounded since then, showing economic growth at about 2.80% in 2017 when consumption and investments persisted, and depreciation of the Swedish Krona boosted exports. Public finances are healthy and the banking sector is well capitalised. Sweden has invested heavily in alternative energy sources since the oil crisis of the early 1970s. Sweden has found a way to reduce emissions while the economy is growing. Swedish businesses investing in renewable energy are increasing, and other sectors are putting greater emphasis on green energy and energy sacing. In addition, the government invests heavily in information and advice for households on how to incorporate renewable energy systems. The Swedish solar market is still limited, but it has begun to grow with the aid of government funding. For 2013–2016, the Swedish Energy Agency is investing SEK 123 million in research into solar cells, thermal solar power and solar fuels [6].

Social factor: The whole society of Sweden indeed shows a great awareness and concern about environmental dilemmas. Swedish carbon emissions are low compared with those of other countries. Sustainable development throughout society is the objective of Sweden's environmental work both inside and outside its borders, although this requires major behavioural changes at all levels and in all areas of society and a new environmental policy. Sweden is playing a leading role in corporate social responsibility , which, also known as sustainable business practice, is used to describe the work done by companies has a positive impact on society, environment or economy.

Technical factor: New environmental technologies have emerged as a new economic sector in Sweden. An essential part of the increase is due to renewable energy technologies that doubled since the oil crisis of the 1970's. Bioenergy and hydroelectric power contribute more than half the increase, but the country has also invested in wind, solar and geothermal energy. Renewable technologies are being applied in Sweden across all sectors, and there is a strong correlation between economic growth and reductions in waste and greenhouse gases, the optimum use of natural and human resources, and the production and conservation of energy through ongoing innovation. Sweden is the world leaders in bioenergy, and is playing the leading role in other renewable sectors, such as concentrated solar technology, thin film solar cells, energy storage and smart gird. However, Sweden's position in the innovation league is declining because its industrial structure is outdated. Many large Swedish companies relied too much on technological breakthroughs a long time ago, which is profitable but not sustainable as they usually increase production capacity for higher revenue.

3. Market analysis of PV/T industry in Sweden

The Sweden's market for solar photovoltaic (PV) energy is very small. Nevertheless, the installation rate of PV continues to increase. In 2016, a total of 79.2 MWp were installed, showing the annual PV market grew with 63 % as compared to the 48.4 MWp installed in 2015. For the PV market, the off-grid PV application accounts for a very small share, with only 1.5 MWp installed in 2016; while the market for grid-connected PV systems has grown rapidly in Sweden, with 77.7 MWp systems installed in 2016 [6]. There has been a clear shift from a market dominated by off-grid systems to a grid-connected market, in which the grid-connected distributed PV systems dominates the market, mainly due to the tax law for large centralized PV systems. Solar energy in Sweden is expected to contribute 5-10% to electricity generation in 2040, which means the share of solar electricity production within the total power generation mix will rise from 0.1% in 2016 to 5-10% in 2040, which is up to 100 times of current solar power capacity if the total electricity generation remain stable as past decades. However, the PV market in Sweden is still relying much on subsidies, and the PV system prices must continue to go down, or the electricity prices to go up if PV needs to contribute to an appreciable part of the Swedish energy mix. Moreover, the Swedish PV market is becoming broader as more and more players with other core businesses, such as utilities and real estate owners, are taking an increasing interest in the PV market. A high competition on PV market will be expected in the near future.

The Swedish solar thermal market is decreasing significantly from 2011. The total number of the installed solar thermal collectors in 2016 was only about 2,823, nearly half of the installed capacity in 2015. The solar thermal market faces increasing competition with other energy technologies, such as biomass boilers, district heating, heat pumps, PV. Typical solar thermal products in Sweden are flat plate and vacuum tubes collectors, while the concentrated solar thermal plants and PV/T products are very rare. Their application includes the fields of hot water and space heating, in operation with other energy system in the meanwhile. The market price of flat plate collectors has dropped in line with the overall development of collector prices across Europe, while vacuum tube collectors have become more expensive owing to less and less players in the area. IEA SHC [1] point out that the main market drivers are the national building code and indirect tax deduction for repair, conversion and extension work can be made for installing solar heating systems. In contrast to PV, there is no direct financial support to solar thermal market. Other barriers are the lower prices of alternative energy technologies and the lack of information on solar heating systems among actors and consumers. The Swedish solar thermal market is disruptive. It has to face high competition from alternative energy systems, but with limited indirect policy support. Great barriers are found for its further development.

Technology drivers offer opportunities for disruptive market innovation. The innovative PV/T technology cogenerates of electricity and heat from the same area, but with less installation cost than installing both individual PV and solar thermals collector, resulting in the lower cost-to-performance ratio (e.g. 1.27 SEK/kWh of levelized cost of energy, 18,812.55 SEK of net present value and 10 years of payback period during its 25 years lifespan [7]) and enabling the possible development of solar thermal market along with the growing PV market. As a result, it offers opportunities for innovation in Swedish solar thermal market. Meanwhile, PV/T technology lowers the energy price per square meter, reducing the dependence degree of PV on governmental subsidy and thus making the target set by the government achievable [8]. Because PV market is shifting to the grid-connected distributed systems, and solar thermal market is indirectly driven by national codes for low-energy buildings and nearly zero-energy buildings, it is expected the Swedish PV/T market will expand mostly in building sector.

4. A Case study of a Swedish company (Absolicon) with PV/T product (X10PVT)

4.1. Market position

In the past, Absolicon company's marketing strategy was to cultivate the market by building up pilot plants with customers to demonstrate how to reduce their energy costs with solar energy [9]. Absolicon's current marketing strategy is to establish a distributor network in export markets and make the sales volumes more efficient. They are looking for a distributor or customer to expand the sales. In the long run, Absolicon recognizes that partnerships are needed to scale up production and reach customers. To increase its market competitiveness, Absolicon develops a production line in the same time that can radically reduce manufacturing costs. Fig. 1 shows the cost comparison between Absolicon's X10PVT and other commercially available PV/T products.

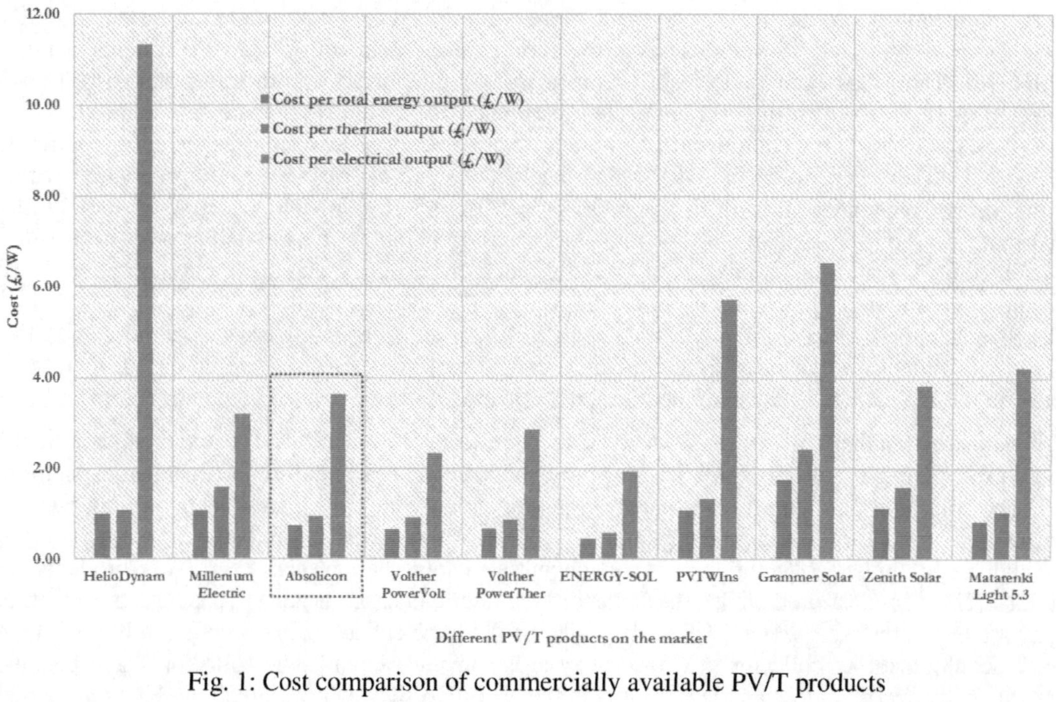

Fig. 1: Cost comparison of commercially available PV/T products

At the similar energy output level, i.e. thermal output of about 400 W/m^2, electrical output of 100 W/m^2, the cost for Absolicon's X10PVT product is around 400 £/m^2, positioning it in the middle class of the cost category. But considering its thermal output is at higher temperature (higher exergy) than the flat-plate ones (e.g. Volther), the cost per exergy of X10PVT is the lowest in this case. When comparing to the concentrated types, such as HelioDynamics of 1.1 £/W and Zenith Solar of 1.2 £/W (belongs to 'Suncore' company), the cost of Absolicon's X10PVT is nearly the lowest at 0.75 £/W. The three types of concentrated PV/T products are similar cost, but targeting at different energy output and applications. So in short, Absolicon successfully keeps its cost competence at the market and has clear costumer group who needs medium-sized plants at 500 W/m^2 in total. This market position meets the trend of solar market in Sweden, which shift towards small and medium-sized energy plants.

4.2. SWOT analysis

Strength: Absolicon is an existing leading player in the Swedish market. It has two production lines with considerable manufacturing capacity for low cost products. It is strong in R&D and innovativeness by organic/sustainable and collaboration with research institutions. Its multiple business streams strengthen the company's competitive advantages and reduces its vulnerability. **Weakness**: Absolicon has a slow grow rate when its revenue was nearly the same in the past three years. It has negative operating profit due to the low return of its investment in pilot projects and poor distribution and sales networks. Its operation efficiency is overall low. It lack a flexible business model and multiple business streams. **Opportunities**: there is a growing demand especially for the PV, and most of the governmental policies support the deployment of Absolicon's products. The modelling results shown in Chapter 6 demonstrate the potential application areas and ways for the company's product. The company has a clear vision of solar market and its customer group. It attracts a lot of attention by doing a few pilot projects, and this would further help the company to attract more investment at current laxer economic environment. This is also desired for the retrieve of solar thermal market by inputting innovative products. **Threats**: Absolicon relied heavily on liability and capital rising while it spent a large amount of investment for the pilot projects in the past two years, which however, has a low return and may increase its future financial risk. The company only has three products and will face competition when the market grows. The Swedish PV industry is becoming broader as more players with other core businesses, such as utilities and real estate owners, are taking an increasing interest in the PV. A high competition in PV industry will be expected in the near future.

4.3. Porter's 5 forces

Fig. 2 illustrates an overview of competitive pressure of Absolicon. The high degree of competitive pressures lie in power of customers and threat of substitution, while the competitive rivalry and power of suppliers stay low and the threats of new entry is medium.

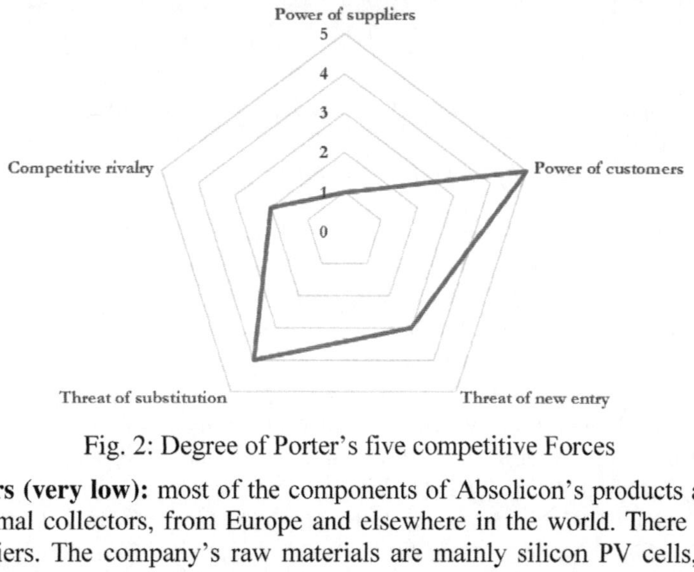

Fig. 2: Degree of Porter's five competitive Forces

Power of suppliers (very low): most of the components of Absolicon's products are similar to other PV and solar thermal collectors, from Europe and elsewhere in the world. There is a wide range of selection of suppliers. The company's raw materials are mainly silicon PV cells, glass and metals, such as aluminium and copper. Other material expenses related to the suppliers, such as plastic, chemicals, thermal oil, follow the general price trends in the markets. **Power of customers (very high):** Absolicon's sale relies heavily on a limited distributors and customers. They are still working on demonstration of their products through pilot projects. In addition, the potential customers have several alterative products, such as standard PV, wind turbine, heat pump, industry waste heat, and other renewable or energy-efficient recourse products. The company has a business only in a small part of Europe and China, where customers' choice is very sensitive to the performance of whole company. **Threat of new entry (medium):** It is hard to establish the concentrated PV/T production line and this is why Absolicon received an order of production line from a Chinese company. There is very limited players on the market who can produce the concentrated PV/T systems as reviewed in section 2. However, it is much easier for interesting players to enter into the flat-plate and flexible PV/T market, since the major manufacturing of these two PV/T products is very similar to that of solar thermal collectors. Moreover, there were many resources and suppliers available for components and raw materials, while the distributors could sell different products at the same time. There is even no need for high capital costs of the production lines of flat-plate and flexible PV/T systems since OEM is very common. So the entry level was considered as medium and it is hard for new players to compete with the Absolicon by offering low-price concentrated PV/T products, but it is easier for new players to start the flat-plate and flexible PV/T products. **Threat of substitution (high):** Although Absolicon provides competitive products at small-and medium plants for combined solar power and heat generation, the threat of substitution stays high since the solar power and heat market are offered with a variety products, such as flat-plate and flexible PV/T, standard PV, and solar thermal collectors. The competition against other types of power and heating products was also high, such as wind turbine, heat pump, bio-boilers industry waste heat, which are different in costs, efficiency, scale/capacity, categories, or even business models. **Competitive rivalry (low):** there are few manufacturers for PV/T products globally, especially for the concentrated PV/T. As a result, the competence for Absolicon remains advanced. On the other hand, the competition is growing with the market, especially due to the gradual participation of large international solar companies.

5. Conclusion

PV/T technologies offer opportunities for the Swedish solar market, which are projected to increase at small-and-medium scales. A high competition in PV/T industry will be expected in the near future since it has to face high competition from alternative energy systems. The market position of the case company meets the trend of PV/T market in Sweden. The high degree of competitive pressures lie in power of customers and threat of substitution, while the competitive rivalry and power of suppliers stay low and the threats of new entry is medium.

Reference

[1]. IEA SHC, 2017. Status of Solar Heating/Cooling and Solar Buildings - 2017. http://www.iea-shc.org/country-report-sweden Accessed on 22/08/2017.
[2]. World bank, 2017. https://data.worldbank.org/country/sweden?view=chart Accessed on 02/10/2017
[3]. Trading economics, 2017. https://tradingeconomics.com/sweden Accessed on 30/09/2017
[4]. OECD, 2017. https://data.oecd.org Accessed on 30/09/2017
[5]. Swedish energy agency, 2017. Proposals for strategy for increased use of solar. https://energimyndigheten.a-w2m.se/Home.mvc?ResourceId=5599 Accessed on 19/08/2017.
[6]. J Lindahl. National Survey Report of PV Power Applications in Sweden – 2016. Swedish Energy Agency.
[7]. Y Gu, X Zhang, J A Myhren et al., Techno-economic analysis of a solar photovoltaic/thermal (PV/T) concentrator for building application in Sweden using Monte Carlo method. Energy Conversion and Management 165 (2018) 8-24.
[8]. Y. Tripanagnostopoulos, M. Souliotis, R. Battisti, A. Corrado. Energy, cost and LCA results of PV and hybrid PV/T solar systems. Progress in Photovoltaics 13 (2005) 235–250.
[9]. Absolicon, 2017. http://www.absolicon.se/product/absolicon-x10-pvt/ Accessed on 30/09/2017.

IOP Conf. Series: Materials Science and Engineering **556** (2019) 012003 doi:10.1088/1757-899X/556/1/012003

Comprehensive assessment of advanced solar facade: thermal, optical and economic assessment

R Liang, Y Sun, Y Wu, R Wilson
University of Nottingham, Nottingham, UK
E-mail: Runqi.Liang@nottingham.ac.uk

Abstract. In order to improve the enegy saving in building, development of glazing system has been widely conducted. CdTe PV window system has the feature of energy generation and adjustment on thermal and visual comfort, in this paper, the thermal performance of a glazing system integrated with semi-transparent CdTe solar cells with different transparency (10% and 50%) under different temperature conditions were investigated by experimental measurement (undertaken in a large climate chamber). The U-value of each PV window has been calculated. Moreover, the spectral transmittance within visible has been measured by spectrometer. This study provided comprehensive fundamental information of different CdTe PV window system for their further investigation on energy and visual comfort.

Keywords: CdTe semi-transparent PV window; Climatic Chamber; Thermal Performance

1. Introduction

For the purpose of improving the performance of glazing systems, generating power on-site and achieving building energy conservation, photovoltaic (PV) glazing has been widely investigated and proposed to be integrated into buildings [1]. Semi-transparent PV (STPV) as one of the most promising PV types has the feature of power generation and daylight adjustment [refs]. Solar cells on a semi-transparent PV glazing partially absorb solar radiation incident on the window surface to generate electrical power, meanwhile block the oversupplied solar energy penetrated into interior space. Therefore, the overall thermal performance of the window system has been affected. This study aims to explore the thermal characteristics of Cadmium Telluride (CdTe) PV windows with different transparency[2, 3]. The U-value of CdTe PV glazing systems under a standard condition was explored through experiments, as well as the thermal resistance under conditions over a range of different mean temperatures of two glazing panes and a range of temperature differences between the surfaces, respectively, to allow a comprehensive picture of heat transfer through STPV window unit. Additionally, the optical properties were measured as well.

2. Methodology

Experimental studies of the selected glazing systems were carried out at the Laboratory at the Energy Technology Building, University of Nottingham. The characteristics (i.e., surface temperatures, heat flux, and spectral transmittance) of semi-transparent CdTe PV window with 10% transparency and 50% transparency, respectively, were measured under a series of controlled temperature conditions.

2.1. Test samples

As illustrated in Figure 1, the construction of these two PV window units consist of five layers in total, from outside to inside, listed as 3.2mm laminated glazing, 0.4 mm conventional series ethylene vinyl acetate (EVA) and CdTe film, 3.2 mm laminated glazing, 20mm Argon-filled gap, and 4mm toughened soft coat low-e glazing. 10% or 50% transparency means that there is 10%/50% percent of total window area that was not covered by solar cells.

IOP Conf. Series: Materials Science and Engineering **556** (2019) 012003 doi:10.1088/1757-899X/556/1/012003

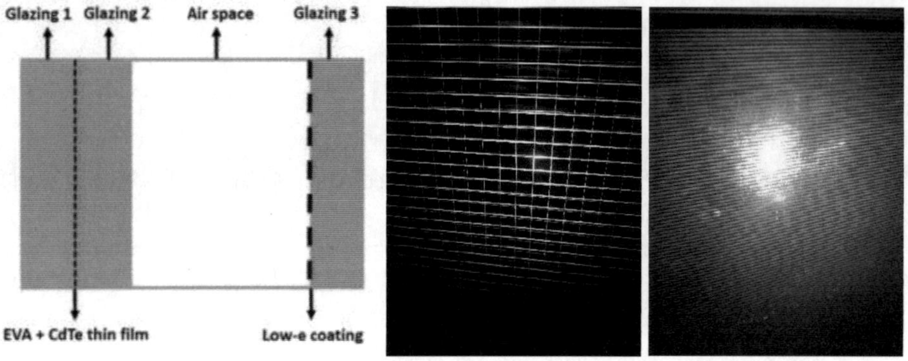

(a) Layers of PV glazing Unit (b) 10% transparency CdTC PV (c) 50% transparency CdTe PV

Figure 1: Structure of the CdTe PV window unit

2.2. Apparatus

Figure 2 (a) shows a TAS series 2 LTCL600 climatic chamber that was used to measure the thermal performance of the glazing units in this study, which comprises two well-insulated walk-in rooms, providing a steady and cyclic simulation of the climatic environment. Each enclosure can be individually controlled. Therefore, it is possible to simulate external climate conditions in one room, whilst mimic internal conditions in the other one. One of the rooms is physically fixed on the ground, while the other can be wheeled to one side to allow the construction of a 'Test Wall'. Once the wall constructed, the two rooms were enclosed together to sandwich the 'Test Wall'. During the test, integral air conditioning units were used to control the two-room temperatures within the range from -25°C to +60°C and the relative humidity between 10% and 95%. In this experiment, CdTe PV glazing unit was installed in the 'Test Wall' as is shown in Figure 2 (b), resulting in measurement of various parameters, such as surface temperature, air temperature and heat flux of the glazing unit, can be obtained during operation.

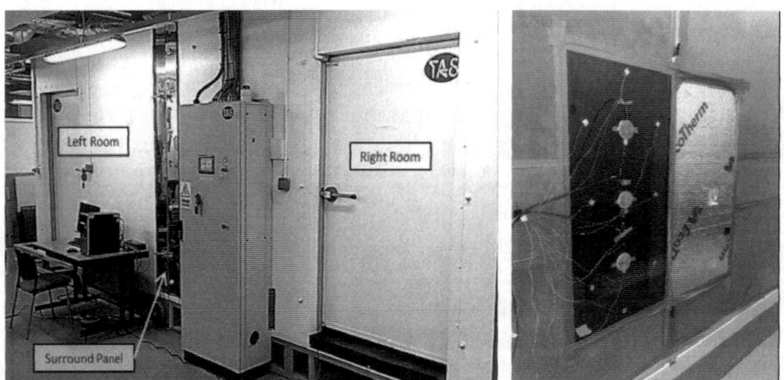

(a) (b)

Figure 2. (a) External view of the climatic chamber (b) CdTe PV glazing unit installed on 'Test Wall.'

Followed International Standard ISO 9869-1:2014 for In-situ thermal resistance measurements by using heat flow meters, The measurement method was designed [4]. Informed by International Standard ISO 12567-1:2012, which determines window and door thermal transmittance using the hot-box method, the apparatus was set up [5], as is shown in Figure 3. A 300mm-thick insulated wall (U-value has been measured as no more than 0.3 W/m^{2}K) was sandwiched between the two rooms, forming the initial 'Test wall'. According to ISO 12567-1:2012 [5], the internal surface of the glazing unit was mounted flush with the surface of the insulated wall. Silicone sealant covered with tape was used to seal all gaps between the window and the insulated wall and hold the window firmly in their position. Additionally, in order to diminish the convection effect caused by the fans on the air conditioning units in each room, two plywood baffles were installed in both the interior and exterior chambers, respectively.

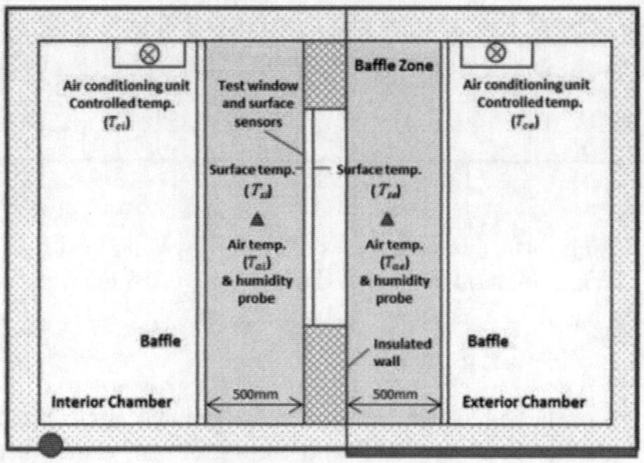

Figure 3. View of the system, the test specimen and sensors

1) External and internal surface temperature were measured by attaching 10 T type thermocouples (detector diameter < 0.3mm) on each side of the window unit [5], as is shown in Figure 2 (b), labelled as Tse (external surface temperature) and Tsi (internal surface temperature) shown in Figure 3, respectively.
2) Heat flux (q) measurement was measured by affixing heat flux meters (Hukesflux Type HFP01, thermal resistance <6.25×10-3 m-2K/W, measurement accuracy +/-5%) to the glass pane using a thin layer of thermal contact paste (Servisol, with thermal conductivity 0.77 W/mK) [4]. The thermal resistance of the heat flux meters is low enough so that the perturbation effect of the surface heat flow by positioning the heat flux meter can be negligible.
3) Air temperature (labelled as Tae and Tai showed in Figure 3) and humidity in the two baffles spaces were measured by positioning two temperature and humidity probes CS215 (accuracy +/- 0.4°C for temperature and +/-2% for humidity) closing to the window
4) Air velocities within the baffle zone were monitored by two hot wire air velocity sensors (testo 425, with measurement accuracy +/- 0.03m/s).

All of the thermocouples, heat flux meters and temperature and humidity probes were connected to a 24-channel data logger DT85, and the data were logged at 1-minute intervals.

Additionally, the Spectrometer USB2000+UV-VIS (Signal-to-noise ratio: 250:1 (at full signal), resolution: 0.1-10nm varies by configuration) was used to measure the spectral transmittance of each PV glazing window unit, and data were output by OceanView Spectrometer Operating Software.

2.3. Measurement procedure and data acquisition

Before the test, the instrumentation such as thermocouples was calibrated. At the beginning of the test, the relative humidity in the interior room was set at 30% to avoid condensation. The measured wind speed was less than 0.3m/s to represent that natural convection prevails [5]. During the test, a sufficient duration (i.e., over 72 hours) was assigned, in order to stabilise the environmental conditions in the test rooms and the heat flow through the window for each scenario, and then the measured data over a further period of 48 hours were used for analysis.

Since heat transfer between the two glass panes of the PV glazing unit is affected by both mean temperature and the temperature difference, two groups of tests containing 6 scenarios were undertaken as depicted in Table 1: group (a) has three scenarios with same temperature difference of 15 °C, while mean temperature was controlled as 5, 10 and 15 °C, respectively; group (b) has another three scenarios with the same mean temperature of 10 °C, and the temperature difference was set up to be 10, 15, and 30 °C, respectively.

Table 1. The arrangement of mean temperature and surface temperature difference for glazing systems.

Group	**(a) Same Temperature Difference**			**(b) Same Mean Temperature**		
Scenario	1	2	3	4	5	6
Mean temperature ($T_{si}+T_{se}$)/2 °C	5	10	15	10	10	10
Temperature difference ($T_{si}-T_{se}$) °C	15	15	15	10	15	20

2.4 Analysis of the data

Based on the measured glazing surface temperatures (Tsi and Tse) and heat flux (q) in the experiments, Data was analysed by average method [4], and the thermal resistance of the glazing system was calculated by using Equation (1):

$$R_T = \frac{\sum_{j=1}^{n}(T_{sij}-T_{sej})}{\sum_{j=1}^{n} q_j} \quad (1)$$

Where, the index j enumerates the individual measurement. The heat transmittance, U, can be obtained by using the calculated thermal resistance RT and the empirical values determined according to EN673[6].

3. Measurement results and discussion

3.1. Thermal resistance under varying temperature conditions

According to the surface temperatures and heat flux of each CdTe PV glazing unit (with the transparency of 10% and 50%) tested under the first three scenarios, the corresponding thermal resistance was calculated, and shown in Figure 4 (a). In these scenarios, the temperature difference between the interior and exterior surfaces of the PV glazing unit was controlled at constant 15 °C. As can be seen, with the increase of mean temperature difference from 5 °C to 15 °C, the calculated thermal resistance is slighting decreasing from approximately 0.68 to 0.65 (i.e., difference of 3%) for 50% transparency PV glazing. Although there is a deviation caused by experimental errors, the tendency of that for 10% transparency PV glazing is similar to the 50% one.

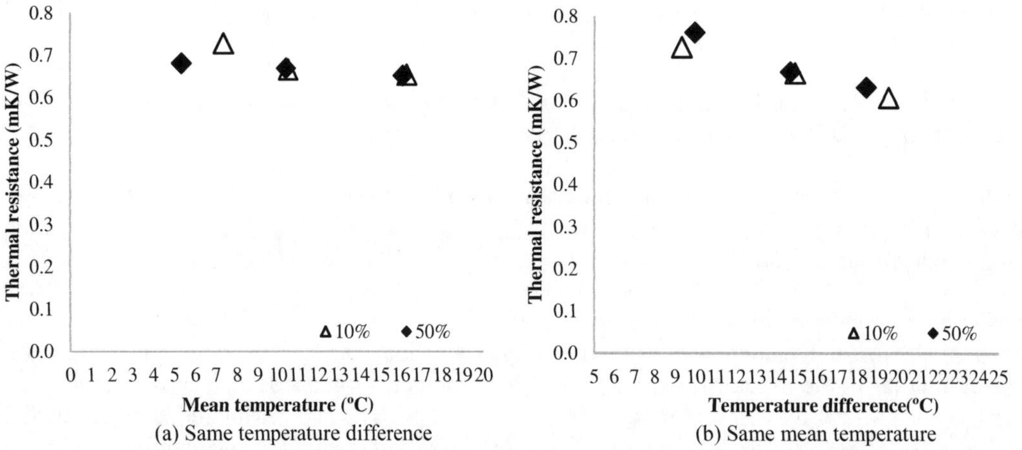

Figure 4. Calculated thermal resistance under the six scenarios of 2 groups

Figure 4(b) shows the relationship between thermal resistance and temperature difference, in which, a constant mean temperature of 10 °C across each glazing system was kept, within the three scenarios in the group (b). Increasing the mean temperature from 10°C to 20°C leads to a decrease of the thermal resistance by up to 13%. To sum up, compared with temperature difference, the mean temperature has a more significant influence on the thermal resistance of these PV glazing systems.

3.2. U-values of semi-transparent CdTe PV window units

Based on the standard boundary conditions from EN 673[6], where the temperature difference between two glazing panes is 15 °C, and the average glazing pane temperature is 10 °C, the overall heat transfer of the glazing system is calculated, and U-value is presented in Table 2. It can be seen that the experiment almost restored the standard conditions, and the corresponding U-value was calculated as 1.673 for CdTe PV glazing unit with 10% transparency, while 1.666 for that with 50% transparency. This means that increasing covered area with solar cells might improve heat transfer.

Table 2. U-value calculation under standard boundary conditions simulated in the experiment

CdTe Transparency	Target Temp. diff (°C)	Target mean temp. (°C)	Cold side surface temp. (°C)	Hot side surface temp. (°C)	Temp. diff in test (°C)	Mean temp. in test (°C)	U-value (W/Km)
10%	15	10	3.09	17.95	14.86	10.15	1.673
50%	15	10	3.10	17.72	14.62	10.41	1.666

3.3. Spectral transmittance of CdTe PV window units
The spectral transmittance within the visible spectrum (380-780nm) of CdTe PV window unit has been measured, and results are shown in Figure 5, which is labelled as CdTe_10% for 10% transparency, and CdTe_50% for the one with the transparency of 50%. It can be seen that the peak transmittance of CdTe_50% is 30.7% at the wavelength around 610 nm, while the transmittance of CdTe_50% is no more than 5% without apparent peak value.

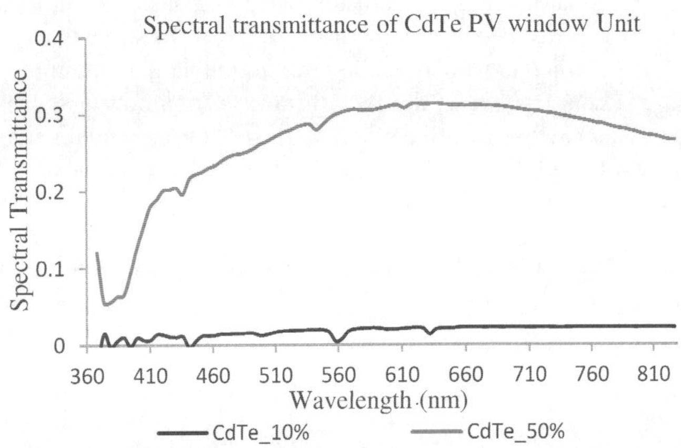

Figure 5. The spectral transmittance of CdTe PV window unit with 10% and 50% transparency

4. Economic analysis
Compared with normal double-glazing unit without low-e coated, the studied CdTe PV windows have similar U-value, as is shown in Table 3. The ones with low-e coated have lower U-value. For normal double glazing listed in the table, the price is ranging from £154 to £579 per m2 depending on the technologies they applied. CdTe technologies have the potential to achieve low cost through combing sufficient efficiency with lower module area cost. Additionally, compared with crystalline silicon modules, CdTe PV modules experience can increase their energy output by 5-9% annually[2]. The expected PV systems are between 25–30 years, CdTe PV modules can either be recycled or disposed of, when they reached their end-of-life. A CdTe PV recycling results in approximately 95% of semiconductor material recovery, and 90% of the glass can be reused[7].

Table 3. Introduction about current commercial window manufacture and products

Manufacture	**Pilkington**	
Product Name	Pilkington K Glass™ Range	Pilkington Optitherm™ S1 Plus double IGUs
Structure	Outer pane (4mm) + air space (16mm, Argon-filled 90%) + inner pane (4mm)	
Outer pane	Pilkington Optiwhite™ (low-iron extra clear float glass	Pilkington Optifloat™ Clear (float glass)

Inner pane	Pilkington K Glass^{TM} (low-emissivity coated thermal glass, $\varepsilon = 0.16$)	Pilkington Optitherm^{TM} S1 Plus (low-emissivity coated thermal glass, $\varepsilon = 0.013$)
U-value (W/m^{2}K)	1.2-1.5	1.0
Manufacture	**WALSHSGLASS**	
Product Name	Twin Seal IGUs clear	Pilkington Optitherm^{TM} S1 Plus double IGUs
Structure	Outer pane (6mm) + air space (12mm, air or argon-filled) + inner pane (6mm)	
Outer pane	Float glazing	Float glazing
Inner pane	Float glazing	Float glazing
U-value (W/m^{2}K)	2.7 with air filled 2.6 with argon filled	1.9 with air filled 1.6 with argon filled
Manufacture	**VENTA WINDOW**	
Product Name	Free combination outside opening windows	
Structure	Outer pane (4mm) + air space (16mm, Argon-filled) + inner pane (4mm)	
Outer pane	Clear glazing	
Inner pane	Low-e glazing	
U-value (W/m^{2}K)	1.3	

5. Conclusion

The CdTe window units with 10% and 50% transparency were investigated, respectively, by experimental measurement. Based on a series of standard, comprehensive assessment methods were developed in a controllable climate chamber. Both thermal (e.g., surface temperature, heat flux) and optical properties (e.g., spectral transmittance) were discussed. Results show that CdTe PV has a similar U-value of approximately 1.6, which is similar as the normal double glazing in the current market, and varying the temperature difference between the two surfaces of window unit has a more significant effect on thermal resistant than changing its mean temperature. Further studies about its application in the building will be conducted, including energy consumption, visual comfort and life cycle analysis.

Acknowledgement

This work was supported by the Faculty of Engineering, the University of Nottingham through a PhD studentship to Runqi Liang.

Reference

1. Skandalos, N. and D. Karamanis, *PV glazing technologies.* Renewable and Sustainable Energy Reviews, 2015. **49**: p. 306-322.
2. Fthenakis, V.M., *Life cycle impact analysis of cadmium in CdTe PV production.* Renewable and Sustainable Energy Reviews, 2004. **8**(4): p. 303-334.
3. Ordenes, M., et al., *The impact of building-integrated photovoltaics on the energy demand of multi-family dwellings in Brazil.* Energy and Buildings, 2007. **39**(6): p. 629-642.
4. Standardization, I., *ISO 9869-1:2014, Thermal insulation - Building Elements - In-situ Measurement of Thermal Resistance and Thermal Transmittance Part 1: Heat Flow Meter Method*. 2014.
5. Standardization, I., *ISO 12567-1:2012, Thermal performance of Windows and Doors - Determination of Thermal Transmittance by the Hot-box Method - Part 1: Complete Windows and Doors*. 2012.
6. EN673, *Glass in Buildinge Determination of Thermal Transmittance (U Value) - Calculation Method.* 2011: European Standard.
7. Sinha, P., *Life cycle materials and water management for CdTe photovoltaics.* Solar Energy Materials and Solar Cells, 2013. **119**: p. 271-275.

Experimental study on the thermal performance of ethylene-tetrafluoroethylene (ETFE) foil cushions

J-F Flor[1], Y Sun[1], P Beccarelli[1], C Rowell[2], J Chilton[1] and Y Wu[1]
[1]University of Nottingham, Faculty of Engineering, Department of Architecture and Built Environment, Nottingham, United Kingdom
[2] Architen Landrell, Station Road, Chepstow, United Kingdom

E-mail: jan.flor@nottingham.ac.uk, Yupeng.Wu@nottingham.ac.uk

Abstract. ETFE pneumatic foil constructions are used increasingly by designers and builders as an alternative to glass in state of the art building envelopes. Low weight, high transparency, mechanical resistance and self-cleaning properties of ETFE may contribute to the overall performance. However, reliable information on the thermal performance of ETFE cushions in building envelopes is scarce, limiting the performance prediction in energy simulations. The present study aims to investigate the thermal performance of air-inflated ETFE foil cushions and evaluates parameters of material properties and design which might affect the thermal transmittance. The paper reports on the experiments conducted to quantify the thermal performance using a climate chamber and full-size mock-ups. Based on available standards, tests were designed to compare the thermal performance of three different ETFE cushion designs, under changing climate conditions. The design variations, including frit prints and switchable shading mechanisms, were tested and compared at hot, temperate and cold weather scenarios. The test series provided detailed results on the thermal performance of ETFE cushions which may be of use for the comparison with the performance of other building components and materials, serve as input for energy simulations and provide a theoretical basis for future developments of novel building envelopes.

Keywords: solar radiation; energy, ETFE, thermal performance, building envelope

1. Introduction

Building light and flexible is a common approach in contemporary architecture. Benefits of saving energy and material while achieving attractive shapes and design features are made achievable with tensioned membrane structures. Pneumatic ETFE foil cushions are now considered state of the art for wide span canopy covers but are also increasingly applied in facades or conceived as a continuous enclosing building envelope. This development calls for new approaches of anticipating the thermal performance of the building envelope in order to meet the evermore demanding needs of reducing the energy consumption of buildings.

ETFE (ethylene-tetrafluoroethylene) is a material with outstanding characteristics of mechanical resistance, light transmittance and environmental performance. Buildings incorporating ETFE have achieved now, after four decades of successful implementation in permanent buildings, a high standard of building excellence and have allowed innovative design solutions to become reality. ETFE has been applied in a large number and variety of building types on a global scale and virtually matches any design brief with great flexibility. An example of a typical ETFE project mastering the complex site requirements for the modernization of a historical building can be seen in figure 1.

Figure 1. Nottingham High School for Boys, ETFE roof covering atrium of an enclosed courtyard in a historical building, 2009, Nottingham, UK © Architen Landrell

It is common practice in the building industry to indicate the thermal performance of components of the building envelope with the U-value or thermal transmittance with the aim to evaluate the energy performance of the whole building in advance of construction. U-values for ETFE building envelopes are frequently cited in the literature and compared to common clear double pane or reflective glazing [1]. Based on standard calculations ETFE cushions are reported to have a comparative or even better thermal performance than common glazing solutions [2]. However, due to the three-dimensional geometry and susceptivity to solar radiation of ETFE cushions, difficulties about the accuracy of the calculated values arise, as standard calculation methods represent the complexity of the heat exchange mechanisms in ETFE cushions only to a limited extent [3]. The experimental determination of the thermal performance appears to deliver a more realistic approach to provide data for a comprehensive energy simulation. In the past, various attempts to determine the thermal performance of ETFE foil cushions with experimental methods have been reported in the literature:

For the Eden Project (2000, St. Austell, UK) a thermal performance test of a ETFE double layer cushion with 200 μm foil, using the hot box method, was conducted and reported by Whalley [4]. Robinson-Gayle investigated later the environmental performance of transparent building envelopes and carried out a number of thermal performance tests on a triple layer ETFE cushion (150 μm - 50 μm - 100 μm) under laboratory conditions [5]. In an investigation into the soiling effects on ETFE transmittance, Mainini et al. [6] also used the hotbox method following ISO 9869 to determine thermal performance of a double layer ETFE cushion with 100 μm foil thickness. In another thermal transmittance assessment, a climatic chamber was used to test a double layer ETFE panel with parallel foils of different thickness (100 μm, and 250 μm) [7]. In a more recent study, which aimed for the thermal characterisation of lightweight building envelopes, again, the hot box method according to ISO 9869 was used to measure in situ and under laboratory conditions the thermal transmittance of a double layer ETFE cushion [8]. Overall, these studies provide evidence that this is the most frequently used method. Based on this conclusion this research campaign sets out to comprehensively investigate in a comparative way the thermal performance of different ETFE foil cushion designs under laboratory conditions. The aim is to determine the heat flux and surface temperatures of ETFE foil cushions under varying climate scenarios, which will allow designers to incorporate the data in future prediction studies for energy performance of buildings.

2. Methodology

To date various methods have been developed and introduced to measure the thermal transmittance of ETFE foil cushions. The methodology of this study was designed on the basis of ISO 9869-1-2014 [9]. Following this method, the heat flux through the samples is measured, which allows one to calculate in combination with surface and ambient air temperatures, the thermal resistance and transmittance for comparison with theoretical derived values. The details of the tested samples and experimental set-up are outlined in the following section.

2.1. Samples

Three ETFE cushions were tested in the climate chamber. All samples measured 1000 mm by 1000 mm with a rise of 100 mm of the main section curvature on each side of the centre plane, in a state of full inflation. The air filled volume is approximately 90 litres at a nominal pressure of 300 Pa. The cushion edges were clamped into an extruded aluminium profile and mounted on a structural frame to support the cushion and withstand tension forces when fully inflated. The whole unit measured 1200 x 1200 mm. All samples are composed of extruded ETFE foil with a thickness of 200 µm and a weight of 350 g/m^2. The detailed sample specifications are listed below:

- Sample 1 is composed of two layers of clear ETFE foil.
- Sample 2 is composed of two layers, one clear and one fritted, with a hexagon pattern of approximately 71% surface cover in a standard silver ink with a print density of 28%.
- Sample 3 is composed of three layers, one clear and two, with a square pattern of approximately 74% surface cover in a standard silver ink with a print density of 28%. The printed layers are shifted in a way that the printed areas of one layer overlap with the clear areas of the other. This makes it possible to modify the transmittance of the cushion, switching from open to closed state, by changing the position of the printed middle layer with air pressure.

2.2. Climate Chamber

A large environmental test chamber (fig. 2) was used to create different temperature scenarios. The climatic chamber is able to independently control the temperature and humidity of two walk-in-rooms, while the test samples are installed in-between. This enables one to simulate indoor and outdoor conditions with steady state or cyclic climate conditions.

2.3. Experimental Set-up

The samples were mounted in a separating insulated dry-wall between the two rooms of the climate chamber and shielded from convective effects by baffle walls on each side. The cushions were inflated and equipped with calibrated T-type thermocouples and heat-flux-meters. A total of eighteen temperature sensors were distributed in a square pattern, of three by three units, and attached to the inner and outer layer of the cushion. Three heat-flux meters were attached on a centred vertical axis over the outer layer of the cushion (fig. 3). An additional fourth heat-flux meter was attached at the centre of the inner layer for measurement control. The homogenous distribution grid of the sensors allowed any variations in heat-flux or surface temperature over the curved sample surface to be detected. The air temperature was measured with temperature sensors within the 500 mm baffle zone on both sides.

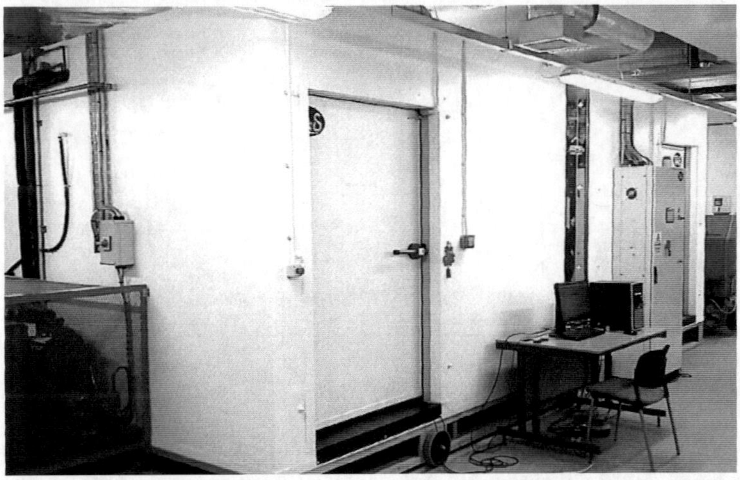

Figure 2. Climate Chamber

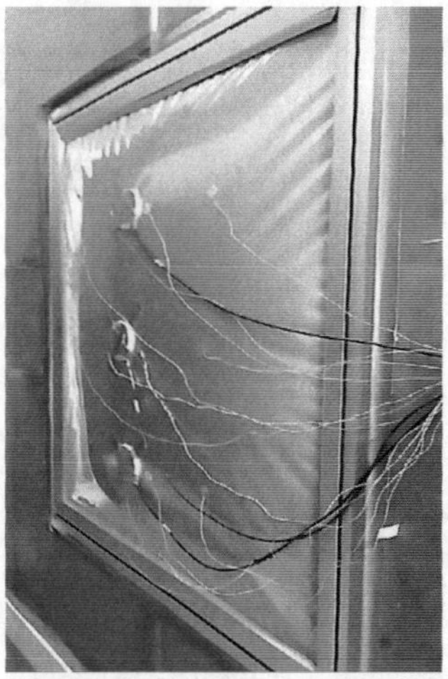

Figure 3. Instrumented ETFE Cushion (Sample 2, fritted)

IOP Conf. Series: Materials Science and Engineering **556** (2019) 012004 doi:10.1088/1757-899X/556/1/012004

2.4. Testing Scenarios

Thermal performance of lightweight structures is largely dependent on climate conditions. In order to comprehensively investigate the heat flux through ETFE cushions for varying conditions an array of testing scenarios was established. Departing from the standard boundary conditions for determining the thermal transmittance based on EN 673:2011 [10] the samples were tested under five scenarios with a varying range of mean surface temperatures and surface temperature differences (table 1). To achieve these temperature scenarios the two rooms of the climate were cooled and heated with HVAC units to pre-calculated temperatures. The surface temperatures were monitored continuously until steady state of the target values was achieved. Then the transversal heat-flux through the samples was measured. The results will be presented and discussed in the following section.

Table 1. Testing Scenarios

Surface Temperatures [°C]	Scenario 1	Scenario 2	Scenario 3	Scenario 4	Scenario 5
Outside	6.3	5.0	2.5	7.5	0.0
Inside	13.8	15.0	17.5	22.5	15.0
Difference	7.5	10.0	15.0	15.0	15.0
Mean	10.0	10.0	10.0	15.0	7.5

3. Results & Discussion

On continuation the preliminary results obtained from the heat flux measurements on the three cushion samples are outlined. The values obtained for the heat flux at different positions of the sample surface showed a variation according to the section height and width of the cushion at the measurement positions. For a more general analysis, these values were averaged and combined into a mean heat flux value for each sample and scenario. From the graphs in figures 4 and 5 it can be seen that all cushion samples showed a similar trend in the magnitude of heat flux for all temperature scenarios, ranging from the lowest heat flux of 38.4 W/m^2 to the maximum of 96.2 W/m^2, for sample 3 and 1 respectively. However, there was a significant difference between the three samples regarding the amount of energy transferred with a variation of approximately 5 to 10%, as can be seen in figure 4 and 5. Sample 1 presented the highest heat flux rate in all tested scenarios followed by sample 3 in closed mode, then sample 2 and with the lowest heat flux sample 3 in open mode. It seems possible that these results are due to the different emissivity of the printed and unprinted foil surfaces. Another important finding was that the mean temperature between the inner and outer surface had only a relatively small influence on the heat flux (fig. 4) the variation of the temperature difference had a larger impact with a significantly increasing heat flux at a higher temperature difference. While this result is not surprising from a physical

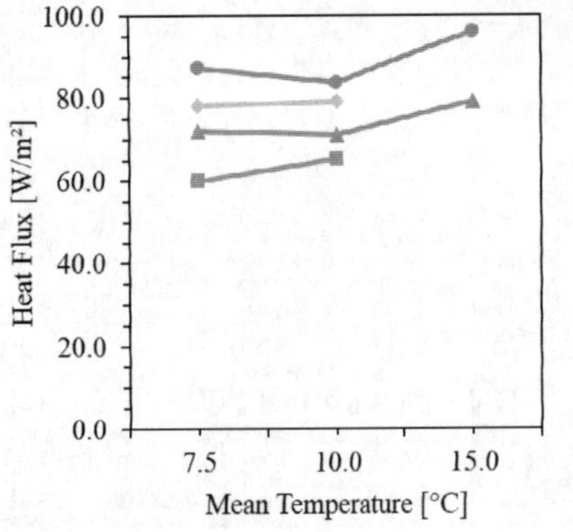

Figure 4. Transversal heat-flux through ETFE cushions at fixed surface temperature difference (15°C).

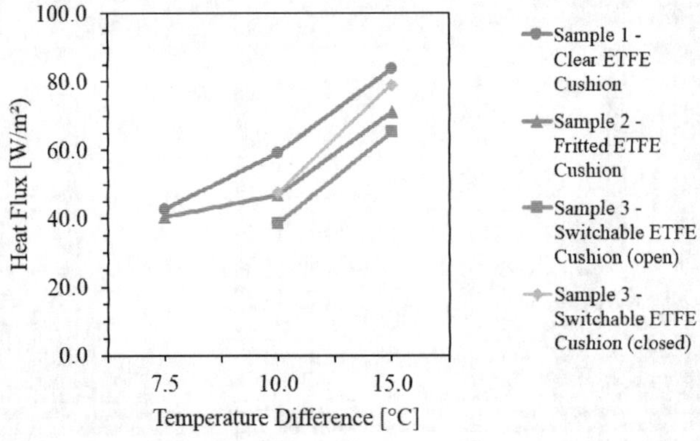

Figure 5. Transversal heat-flux through ETFE cushions at fixed mean surface temperature (10°C).

IOP Conf. Series: Materials Science and Engineering **556** (2019) 012004 doi:10.1088/1757-899X/556/1/012004

point of view it illustrates the high susceptibility of the thermal performance of pneumatic foil cushions to climatic conditions. The present findings seem to be consistent with other research which found a similar behaviour of other cushion designs. However, with the relatively small sample size, in comparison to usual building applications of ETFE, caution must be applied, as the findings might not be transferable without adjusting the measured values to the corresponding scale of the structure.

4. Conclusions

This study set out to investigate the thermal performance of printed and unprinted ETFE foil cushion designs under varying climate conditions. The most obvious finding to emerge from this study is the high susceptibility of the thermal performance of pneumatic foil constructions to changing conditions of in-/ and outdoor temperatures. This is of significance for the design of building envelopes using this type of construction and material. The comparison of the three distinctive designs also showed that the frit print reduces the heat flux of the whole cushion, and smart frit prints allow one to effectively switch between higher and lower heat flux states. The present study confirms previous findings and contributes additional evidence that suggests that reflective frit prints have a large impact on the overall thermal performance of ETFE foil cushions. Adaptive technologies like the switchable smart frit might offer even bigger benefits as they are capable of moderating effectively the heat flux under different climate conditions especially between high and low solar radiation. A further study with more focus on heat transfer through solar radiation is therefore suggested.

5. References

[1] Chilton J 2013 Lightweight envelopes: ethylene tetra-fluoro-ethylene foil in architecture. *P I CIVIL ENG - Construction Materials* **166** CM6 343–57

[2] Lippke R 2009 *Folien als Transparente Elemente in der Fassade mechanische und Bauphysik-lische Eigenschaften* (Berlin: Technishe Universität Berlin) p 272

[3] Afrin S 2016 *Thermal Performance Analysis of ETFE-foil Panels and Spaces Enclosed with ETFE-foil Cushion Envelope* (Nottingham: University of Nottingham) p 333

[4] Whalley A, Grimshaw & Partners N 2000 The Eden Project glass houses world environments In: Barnes M, Dickson M *Widespan Roof Structures* (London: Thomas Telford books) pp 75 - 84

[5] Robinson-Gayle S 2003 *Environmental Impact and Performance of Transparent Building Envelope Materials and Systems* (London: Brunel University) p 306

[6] Mainini AG, Poli T, Paolini R, Zinzi M, Vercesi L 2014 Transparent multilayer etfe panels for building envelope: Thermal Transmittance Evaluation and Assessment of Optical and Performance Decay due to Soiling *Energy Procedia* **48** 1302-10

[7] Martin BAJ, Lau B, Beccarelli P, Chilton J, Wu Y 2015 Modular transparent etfe panels for building envelopes thermal transmittance assessment and evaluation *Proceedings IASS 2015 Amsterdam Symposium: Future Visions* 1-9

[8] Dimitriadou E A 2015 *Experimental Assessment and Thermal Characterisation of Lightweight Co-Polymer Building Envelope Materials* (Bath: University of Bath) p 260

[9] ISO BS ISO 9869-1 2014 *Thermal insulation - Building Elements - In-situ Measurements of Thermal Resistance and Thermal transmittance - Part 1- Heat Flow Meter Method* p 48

[10] CEN BS EN 673 2011 *Glass in building - Determination of Thermal Transmittance (U value) - Calculation method* p 22

Acknowledgments

This work was supported by the Faculty of Engineering at the University of Nottingham through a PhD studentship awarded to Jan-Frederik Flor. The sample material and mock-up for the experiments was generously provided by Architen Landrell.

Photovoltaic Characteristics of Low Concentration CdTe Solar Cells

Shuai Wang[1], LiLi Wu[1]*, Xia Hao[1,2]*, Jingquan Zhang[1], lianghuan Feng[1], Tapas Mallick[3], Lingzhi Xie[2]
[1]College of Materials Science and Engineering, Sichuan University, Chengdu, China
[2]Institute of New Energy and Low-carbon Technology, Sichuan University, Chengdu, China
[3]Environment and Sustainability Institute, University of Exeter, Exeter EX4 4PY, U.K
E-mail: wulili@scu.edu.cn, hao.xia0808@scu.edu.cn

Abstract. It is a reasonable way to increase power generation and reduce the expensive solar cell materials by combining low-cost concentrator and solar cells. The characterization of low concentration CdTe solar cell under standard test condition (STC) and monochromatic light were carried out. The application of concentrator significantly increases the power output of CdTe solar cell. The monochromatic I-V results suggest that the FF of CdTe solar cell was critically affected in the wave length range of 525 to 780 nm. Also the effect of the anti-reflective film on the low concentration CdTe solar cells were revealed, and the results imply that the anti-reflective effect is magnified mainly at the wave length range of 525 to 780 nm.

Keywords: Low concentration CdTe solar cells, IV Characteristics

1. Introduction

Cadimium telluride (CdTe) solar cells are commercially available, low cost PV devices. CdTe has an ideal direct band gap of 1.45eV and a high absorption coefficient, which implies that 2 μm is enough to absorb most of the visible light. The highest efficiency of CdTe thin-film cell is 22.1% and average production module efficiency is up to 18.6%[1]. Improving the power output by further optimizing device performance will contribute to the reduction of production cost of CdTe solar cells. Alternatively, one may utilize a concentrator to improve the power output by increasing the incident light intensity. Currently the research on concentrator PV devices is mainly focused on GaAs, III–V multi-junction and silicon solar cells. The advantages of low temperature coefficient, high coefficient of absorption and good stability make CdTe solar cells quite competitive with the conventional Si concentrator solar cells[2]. Meanwhile, thin film solar cells can be deposited on flexible substrates[3], which will have a wide range of application at building integrated concentrating photovoltaic (BICPV) systems. However, the fundamental studies on thin-film concentrator solar cells are not carried out to date. In the present work, we studied the current density-voltage (J-V) characteristics of CdTe solar cell without and with concentrator under standard test condition (STC) and monochromatic light. The performance of CdTe solar cell at non-uniform radiation condition were also analysed.

2. Experimental

2.1. Device fabrications

The structure of CdTe thin film solar cells in this study was glass/SnO_2:F (FTO)/TO/CdS/CdTe/ZnTe:Cu/BC. The FTO with a thickness of ~400 nm substrates were purchased from Pilkington. FTO was mechanically polished for 15 minutes and successfully cleaned by acetone, ethanol and deionized water. After that, a ~40-nm-thick SnO_2 high resistivity transparent (HRT) layer was deposited followed by a CdS window layer deposition. The CdTe absorber films (~4μm) were deposited by vapour transport deposition and the ZnTe:Cu (~75 nm) back contact was deposited by thermal co-evaporation. Finally, the Au back electrodes were deposited by vacuum evaporation. For the devices utilized to clarify the effect of antireflective layer, an extra 80-nm-thick MgF_2 was deposited on the front side of FTO by thermal evaporation. The compound parabolic concentrator with an theoretical concentration ratio of 4 × was placed on top of the CdTe solar cells while demonstrating the concentration effects, as shown in Figure 1. The detailed parameters of 4× concentrator be found in the literature[4].

2.2. Characterization

The current density-voltage (*J-V*) characteristics were carried out under simulated AM1.5 illumination at 1 sun by using a class A solar simulator. The external quantum efficiency (EQE) of the device was determined by a QEX10 Solar Cell Spectral Response from PV Measurements, Inc. Monochromatic light collection efficiency (MCE) measurements were performed at 470, 525 and 780 nm. The intensity of monochromatic light was the same as the intensity of the corresponding wavelength in the solar spectrum of AM1.5.

Figure 1. The photo of the concentrator element

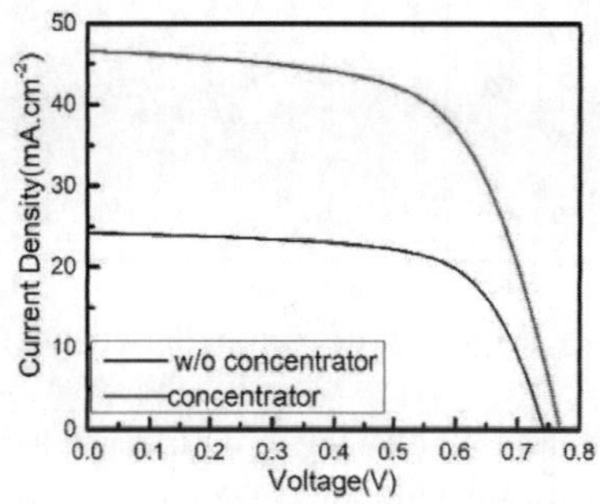

Figure 2. The current density-voltage characteristics of CdTe without (black line) and with (red line) concentrator

3. Results and discussion

3.1 J-V characteristics of CdTe solar cells without and with concentrators

Table 1. The photovoltaic parameters of CdTe solar cells without and with concentrator (Sample ID: X-1)

	Eff (%)	FF (%)	Voc (mV)	Jsc (mA·cm^{-2})	Pmax (mW)
W/o concentrator	11.93	64.65	721	25.57	11.93
With concentrator	4.37	63.58	746	36.80	17.48

In this study, the open-circuit voltage (Voc), short-circuit current (Jsc) and fill factor (FF) of the CdTe solar cell were 743 mV, 24.20 mA/cm^2, and 0.67, respectively. The photovoltaic parameters of CdTe solar cells without and with concentrator were summarized in Table 1. The device exhibits quite different performance after the application of concentrator, reflected by the varied photovoltaic parameters. This variation is mainly due to the increase of the incident light intensity. As shown in Figure 2, After concentrating, the Jsc and Voc increases significantly in turn to increase the power conversion efficiency. Rs and Rsh dropped obviously due the high-level injection at the CdS/CdTe junction. On the contrary, the electrical property of FTO keeps unchanged with the varied light intensities, which make the voltage drop on this layer[5]. Meanwhile, the actual concentration ratio (the $Jsc_{concentrated} / Jsc_{non\text{-}concentrated}$) was 1.44, which is seriously deviated from the theoretical value (4×). It

IOP Conf. Series: Materials Science and Engineering **556** (2019) 012005 doi:10.1088/1757-899X/556/1/012005

is suggested that when we simply placed the concentrator on top of the front side of solar cells, severe optical loss occurred at the glass/concentrator interface and this may be greatly improved if the encapsulation is introduced.

3.2 Wavelength dependent photovoltaic characteristics

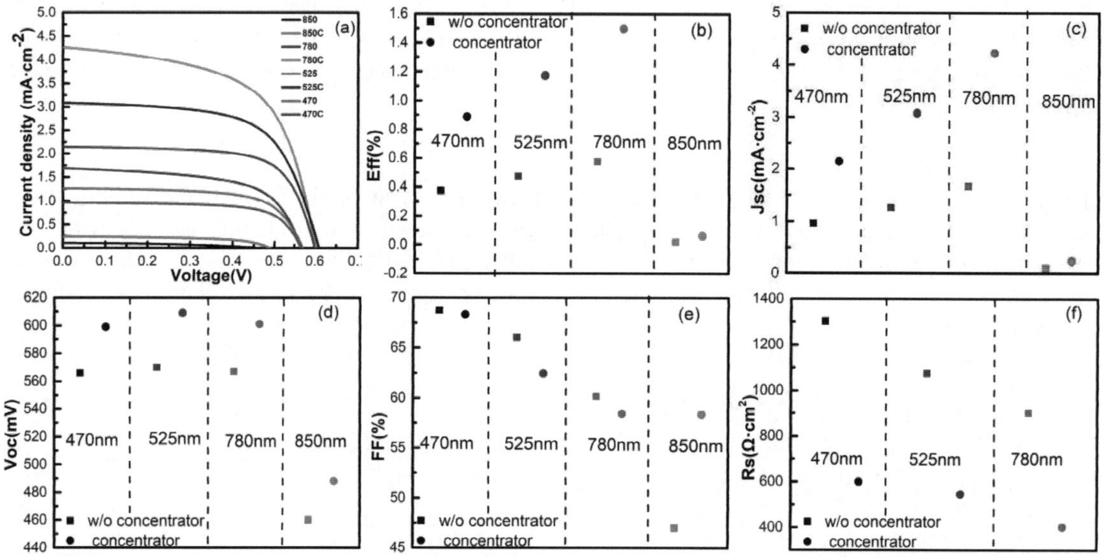

Figure 3. Wavelength dependent performance of CdTe solar cells: (a) J-V characteristics, (b) efficiency, (c) Jsc, (d) Voc, (e) FF and (f) series resistance (Rs)

Figure 3 shows the J-V characteristics of the devices under monochromatic light. In the case of CdTe solar cell without concentrator, The Jsc, Eff and Rs of non-concentrated devices increased with the increase of the wavelength below 780 nm and decreased while the wavelength keeps increasing to 850nm. Some of the light at the wavelength range of 470-525 nm was absorbed by CdS layer[6] and the generated photocurrent is small. As a result, low Eff and Jsc were obtained at 470 and 525 nm. High Eff and Jsc were achieved at the 780 due to the production and collection of generous photon-generated carriers, which also increases electrical conductivity and decreases series resistance, as shown in Fig. 3(f). However, in this case there is no obvious change in Voc. We also noticed that the performance of the solar cell at 850 nm is very poor. The possible reason could be that the photo-generated carriers were located at the back side of CdTe layer. These carriers are too far away from the built-in field. And as a result, the sweep-out speed of the carriers will be deceased. The carriers will then be trapped and recombined here rapidly. In case of the low concentrated case, the concentrator has no effect on the transmission of monochromatic light at 470-850 nm [7]. The wavelength dependence of the photovoltaic parameters is almost the same as the non-concentrating case. It is worth noted that the FF at the wavelength of 470 nm was no visible change while that at the wavelengths of 525-780 nm decreased obviously. One may speculate that in the wavelength range of 525-780 nm, the photo-generated carriers by the increased light intensity are suffered from severe charge trapping and recombination.

3.3 The effect of anti-reflection coatings on low concentration CdTe solar cell

IOP Conf. Series: Materials Science and Engineering **556** (2019) 012005 doi:10.1088/1757-899X/556/1/012005

Table 2. The photovoltaic parameters of CdTe soalr cells with anti-reflective layer (Sample ID: X-2) without and with concentrator.

	Eff (%)	FF (%)	Voc (mV)	Jsc ($mA \cdot cm^{-2}$)	Pmax (mW)
w/o concentrator	11.95	66.45	743	24.24	11.94
With concentrator	5.61	62.52	769	46.62	22.44

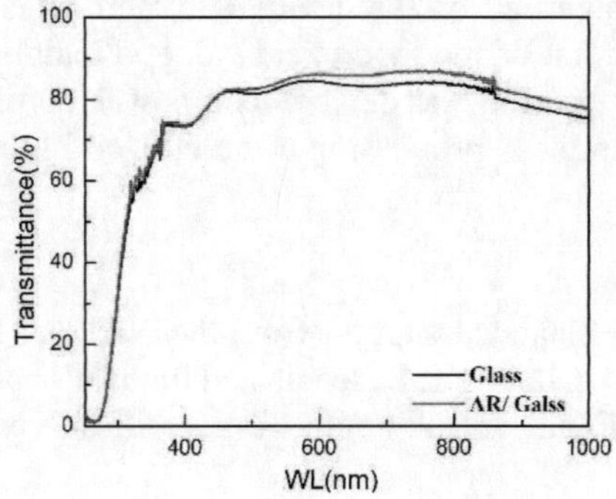

Figure 4. Transmittance spectrum of TEC-10 substrates with and without MgF_2 anti-reflective coating

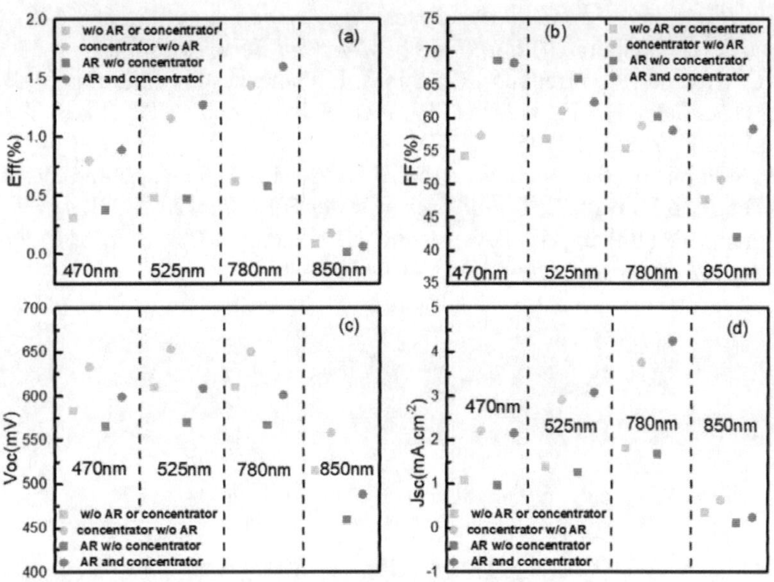

Figure 5. Wavelength dependent photovoltaic parameters of CdTe solar cells. (a) Eff, (b) FF, (c) Voc, (d) Jsc.

The effect of anti-reflective layer for concentrator devices was studied on CdTe solar cells with 80-nm-thick MgF_2 film deposited on the front glass of FTO. The transmittance spectra of glass substrate with and without MgF_2 film are dipected in Figure 4. One can find that the transmittance is mainly improved at the wavelength range of over 420 nm, which is in good agreement with the wavelength dependent J-V characteristics as shown in Figure 4. The Jsc (purple filled circles in Figure 4(d)) was sginificantly improved at the wavelength range of 525-780 nm. The anti-reflective coatings could reduce the optical losses at the wavelengths of 525-780 nm due to the increased transmittance of 500-800 nm. Similar to the device without anti-reflective layer, the optical loss also occurred at the interface of the concentrator and the front glass of FTO. Regardless of the optical loss induced by the uncompact contact between the concentrator and CdTe soalr cells, the actual concentration ratio is 1.92, which is better than the device without anti-reflective layer. The power output of devices are almost doubled by simply placing a low cost concentrator on a CdTe solar cell.

4. Conclusion
In this study, CdTe solar cell with concentrator under stand STC and monochromatic light were carried out. The power output was significantly increased after the application of concentrator and the improvement will be further magnified on the device with anti-reflective layer. The deviation of the actual concentration ratio from that of the theoretical value is mainly due to the optical losses at the interface between the concentrator and the device. The power output is finally doubled in the low concentration CdTe solar cell and it is promising to be utilized in the BIPV application after further structural oprimization.

5. Acknowledgement
The authors are grateful for the financial support from the National Key Research and Development Program of China: Newton Fund -China-UK Research and Innovations Bridges (No. 2016YFE0124500) and the Fundamental Research Funds for the Central Universities (No. YJ201722).

6. Reference
[1] M.A. Green, Y. Hishikawa, W. Warta, E.D. Dunlop, D.H. Levi, J. Hohl-Ebinger and A.W.H. Ho-Baillie 2017 *Progress in Photovoltaics: Research and Applications* **25** 668-676
[2] P. Singh and N.M. Ravindra 2012 *Sol Energy Mater Sol Cells* **101** 36-45
[3] L. Kranz, C. Gretener, J. Perrenoud, R. Schmitt, F. Pianezzi, F. La Mattina, P. Blosch, E. Cheah, A. Chirila, C.M. Fella, H. Hagendorfer, T. Jager, S. Nishiwaki, A.R. Uhl, S. Buecheler and A.N. Tiwari 2013 *Nat Commun* **4** 2306
[4] H. Baig, N. Sellami and T.K. Mallick 2015 *Sol Energy Mater Sol Cells* **134** 29-44
[5] W. Li, R. Yang and D. Wang 2014 *Sol Energy Mater Sol Cells* **123** 249-254
[6] K. Nakamura, T. Fujihara, T. Toyama and H. Okamoto 2002 *Japanese Journal of Applied Physics.pt Regular Papers & Short Notes* **41** 4474-4480
[7] Sarmah, Nabin, Bryce S. Richards, and Tapas K. Mallick. 2014 *Solar Energy* **103** 390-401

IOP Conf. Series: Materials Science and Engineering **556** (2019) 012006 doi:10.1088/1757-899X/556/1/012006

Annealing temperature dependence of properties of CdSe thin films by RF-sputtering

Ailing Wang[1], Chunxiu Li[1], Jingquan Zhang[1]*, Lili Wu[1], Wenwu Wang[1] and Lianghuan Feng[1]

[1]Institute of Solar Energy Materials and Devices, College of Materials Science and Engineering, Sichuan University, Chengdu, China
E-mail: zhangjq@scu.edu.cn

Abstract. Cadmium Selenide (CdSe) is a II-VI group compound semiconducting material with a direct band gap about 1.74 eV, which is widely applied in thin film transistors, light emitting diodes and photoelectrochemical (PEC) cells. In recent years, the application of CdSe thin films to CdTe thin film solar cells has attracted the attention of people. CdSe thin films were synthesized using radio frequency magnetron sputtering method on Corning glass substrates at room temperature. The as-deposited films were annealed at various temperatures (200°C, 300°C, 400°C and 500°C) in N_2 atmosphere for 1h and the effect of annealing temperature on structural, morphological, compositional, optical and electrical properties was studied. X-ray diffraction (XRD) indicates that the structure of CdSe thin films transforms from cubic to hexagonal structure when annealing temperature changes from room temperature to 500°C. Scanning electron microscope (SEM) images show that the crystallite size of CdSe thin films has no obvious increase with annealing temperature increasing from 200°C to 400°C, which is all about 33 nm. However, grains with size of 200 nm were precipitated on the surface of CdSe thin film annealed at 500°C, and the Se content of CdSe thin film decreased. The band gaps of CdSe thin films were found to firstly decrease, and then gradually increase with the increase of annealing temperature. The minimal band gap of CdSe thin film is 1.63eV when annealed at 300°C. The conductivity of CdSe thin film annealed at 500°C increased by two orders of magnitude compared with as-deposited CdSe thin film.

Keywords: CdSe thin films; annealing temperature; RF-sputtering

1. Introduction

CdSe thin films have been well applied in numerous fields such as thin film transistors[1, 2], light emitting diodes [3, 4] photoelectrochemical (PEC) cells [5] and solar cells [6-9]. In recent few years, CdSe thin films were used as window or absorption layers in CdTe solar cells, whose short circuit current density greatly improved [7-9]. Annealing has a great influence on the optical and electrical properties of the CdSe film [10-13]. So it is necessary to study the annealing of CdSe thin film. CdSe is a II-VI semiconductor material with a direct band gap (~ 1.74 eV), which is suitable for visible light spectrum in the solar spectrum and has a strong photoelectric effect when excited by visible light. CdSe thin films have been prepared using different methods such as chemical bath deposition (CBD) [11, 12], electrodeposition [13], thermal evaporation method [14], radio frequency (RF) magnetron sputtering [7] and pulsed laser deposition (PLD) [8]. CdSe thin film deposited by RF magnetron sputtering is compact and of greatly crystalline. The thickness of CdSe thin films is accurately controlled by changing sputtering power, sputtering pressure, and gas flow.

Annealing treatment provides great versatility for the preparation of CdSe thin films with tunable size, band gap, and surface appearance. Annealing in different atmosphere of CdSe thin films deposited by different methods have been widely reported. Nair et al have reported that an annealing process can enhance the photosensitivity of chemically deposited CdSe thin films [10]. The effect of annealing on structural, optical, and electrical properties of CdSe thin films deposited by CBD method was investigated by Kale et al [11]. They have reported that the structure of CdSe thin films changed from cubic to hexagonal phase with the decrease in value of band gap from 2.3 to 1.7 eV after annealing [15, 16]. The electrical resistivity of CdSe thin films was found in the order 10^5–10^6 Ω·cm which was observed decrease after annealing. Erat et al have reported the influence of annealing in nitrogen atmosphere and its dependency on structural, compositional, and electrical properties of chemical bath deposited CdSe thin films [12]. They found that the Se content in the film decreased gradually and activation energy became smaller with the increase of annealing temperature. A study on the effect of annealing on structural, optical, morphological and electrical properties of thermally evaporated CdSe thin films was reported by Purohit et al [14]. They have reported that the CdSe thin films have cubic phase with preferred orientation (111) and the surface morphology and roughness improved after annealing process. Somnath and Asit have reported the effect of annealing on structural, optical and photosensitive properties of electrodeposited CdSe thin films [13].They have showed that the sensitivity of CdSe thin films increases with annealing temperature increasing from room temperature to 350°C and then decreases.

An attempt has been made in this paper to enhance optical and electrical properties of CdSe thin films deposited by RF-sputtering. In this work, CdSe thin films were deposited by RF magnetron sputtering at room temperature, and the as-deposited films were annealed at various temperatures (200 °C, 300 °C, 400°C and 500 °C) in N_2 atmosphere for 1h and the effect of annealing on structural, morphological, compositional, optical and electrical properties was studied.

2. Experimental

2.1 Preparation

CdSe thin films have been deposited on corning glass substrates by RF magnetron sputtering. Substrates were cleaned in the deionized water for 10min and then cleaned in anhydrous ethyl alcohol for 5min assisted by ultra-sonic and, finally, they were dried in a drying oven before deposition. The CdSe target material (Chengdu Ultra-pure Application Materials Limited Liabilities Company) is more than 99.99% pure. The chamber was initially evacuated to a pressure of 8.0×10^{-4} Pa and the 99.999% pure argon was then introduced using a mass flow meters with the flow rate of 50 sccm to maintain the chamber pressure at 2.0Pa. CdSe thin films were sputtered for 2h with a 100W RF power at room temperature. The thicknesses of as-deposited CdSe thin films are about 440nm. The as-deposited films were annealed at 200°C, 300°C, 400°C and 500°C in nitrogen atmosphere for 1h in vacuum annealing furnace and followed by normal cooling to room temperature.

2.2 Characterization

The thickness of as-deposited and annealed CdSe thin films was measured with a step profiler (XP-2, Ambios Technology Inc. USA). The structure of CdSe thin films was characterized by X-ray diffraction (XRD, DX-2600 Dandong Fangyuan Instrument Co. Ltd., China) system with the Cu Kα radiation (λ=0.15406 nm) and the 2θ range between 10° and 70°. The Surface morphology and composition of CdSe thin films were studied using a scanning electron microscope (SEM, Hitachi S-4800) and dispersive X-ray spectroscopy (EDS, INCA, Inca Oxford) attached with the same SEM system respectively. The optical transmission spectra were measured in the wavelength range of 300–2500nm with a UV–Vis–NIR grating spectrophotometer (Perkin-Elmer Lambda 950). The electrical resistivity of CdSe thin films in dark was performed using two-probe method in the temperature range 300-573K.

3. Results and discussion

IOP Conf. Series: Materials Science and Engineering **556** (2019) 012006 doi:10.1088/1757-899X/556/1/012006

3.1 structure, morphology and composition

Figure 1(a) shows the XRD pattern of as-deposited CdSe thin film deposited by RF sputtering which is crystalline with cubic sphalerite structure. The intense peak at 25.65°corresponding to the plane (111) suggests a dominant orientation of nanocrystalline phase of CdSe thin film with cubic zinc blende structure (PDF standard card, JCPDS 88-2346). The CdSe thin films deposited by CBD [17, 18], thermal evaporation [14] and electrodeposition [19] respectively at room temperature have been reported to be cubic zinc blende structure.

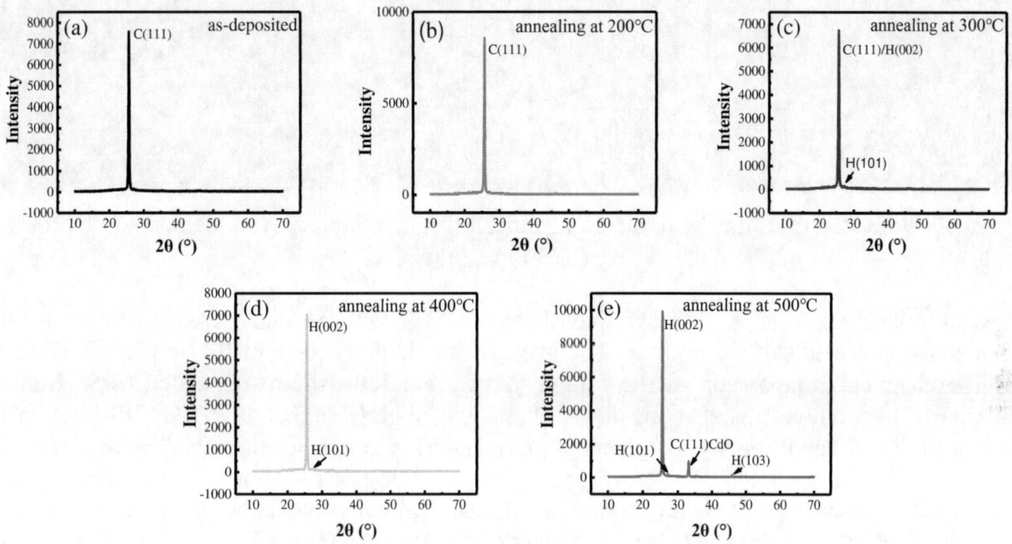

Figure 1. XRD patterns of CdSe thin films with different annealing temperature: (a) As-deposited, (b) 200°C, (c) 300°C, (d) 400°C, (e) 500°C.

Figure 1(b)-(e) show the XRD patterns of annealed CdSe thin films. The increase of peak intensity at (111) indicates that the film has been converted to a stable cubic phase with better crystallinity after annealed at 200°C [figure 1(b)]. In figure 1(c), the waning of the highest peak intensity and appearance of peak at (101) indicated the structure of CdSe thin film gradually changed from cubic to hexagonal phase. CdSe thin film annealed at 300°C was polycrystalline with a mixture cubic and hexagonal structure. Further, with the increase of annealing temperature (400°C-500°C), the highest peak intensity at (002) in figure 1(d)-(e) gradually strengthened and the peak at (101) and (103) are observed. The results indicated that CdSe thin film has a stable hexagonal phase with (002) preferential orientation after annealed at 500°C (PDF standard card, JCPDS 77-2307). The especial peak at 33.12°corresponding to (002) plane (PDF standard card, JCPDS 75-0594) indicates that the part of CdSe phase was converted to CdO phase resulting from the accidental incorporation of oxygen. The crystallite size of CdSe thin films was calculated from Scherrer's formula [13]. Calculation results show that average crystallite sizes of CdSe thin films except for film annealed at 500°C vary from 31.9 nm to 34.3 nm. The grain size of CdSe thin film annealed at 500°C was too large to be calculated by Scherrer's formula, which was estimated to be about 200nm by SEM [figure 2(e)].

The surface morphology of as-deposited and annealed CdSe thin films is showed in figure 2(a)-(e). In figure 2(a), it is observed that the as-deposited thin film was flat and dense with homogeneous distribution of sphere-like particles. With the increase of annealing temperature [figure 2(b)-(d)], the grain sizes of CdSe thin films have varied slightly, which were all about 33nm estimated by SEM. The result is well agreement with the results by XRD. But there are some obvious cracks appearing on the thin films. The appearance of cracks is attributed to the expansion of the original cracks with thermal annealing and the change in volume caused by transformation from cubic to hexagonal phase. From figure 2(a)-(d), it can be see that the surface of films is smoother with the increase of annealing temperature, and the connections between grains are tighter. This is the result of grain consolidation with thermal annealing. A drastic change in surface morphology is observed for 100°C increase in annealing temperature with respect to figure 2(e). The area of rectangle in figure 2(e) shows the ball-like particle precipitates out on the surface of the film. These ball-like particles gradually grew and agglomerated together, and formed the single crystal like pyramid (the area of circle in figure 2(e)). There were similar reports in the research of Shyju [17] and Somnath [13]. Excellent single crystalline structures of size as big as 200nm with clear crystallographic faces are noticed to evolve on the surface of film.

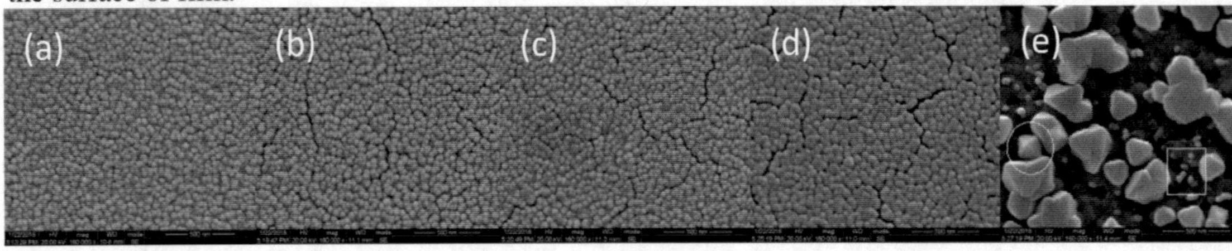

Figure 2. SEM of CdSe thin films with different annealing temperature: (a) As-deposited, (b) 200°C, (c) 300°C, (d) 400°C, (e) 500°C.

Energy-dispersive analysis by X-rays has been carried out to study the stoichiometry of as-deposited and annealed CdSe thin films. The average atomic percentage of Cd:Se was 50.61:49.39 showing the chemical composition of the sample is nearly 1:1 for the as-deposited CdSe film by RF-sputtering. The average atomic percentage of Se was slightly higher than Cd for the film of annealed at 300°C. A small amount of compositional variation is attributed to the rearrangement of atoms with thermal annealing [17]. The average atomic percentage of Cd:Se was 57.91:42.09 for film annealed at 500°C showing Se content decreased significantly, owing to evaporation of selenium from thin film with thermal annealing [12]. The deficiency of selenium content leads to cadmium-rich of CdSe thin film, making cadmium to dedicate electrons as a donor. In this way, the strengthening of n-type conduction of the film is beneficial to the formation of p-n junction in solar cells.

3.2 optical properties

The transmission spectra of as-deposited and annealed samples are illustrated in figure 3. Compared with CdSe thin films obtained by CBD [12] and electrodeposited [13], the CdSe thin films deposited by RF-sputtering have high transmittance, which can reach over 90%. Such CdSe thin film with high transmittance can be well applied to the window layer of the thin film solar cell, thereby increasing the short-circuit current density of solar cell. The interference fringe pattern in the near infrared (NIR) region owes to the constructive and destructive interference of the wave fronts from two interfaces air/CdSe film and CdSe film/corning substrate [20]. With thermal annealing, the sharper absorption edge varying between 700 nm and 800 nm indicates that the crystallization of films is getting better.

Absorption coefficient α can be obtained by a relation of $\alpha = -1/dlnT$ at the absorption edge where d is the film thickness and T is the transmittance[21]. The transmission data were analyzed to calculating the gap energies of CdSe thin films by using the following relation $\alpha = (\frac{k}{hv})(hv - E_g)^n$ where α is absorption co-efficient, hv is the photon energy, k is a constant, Eg is the optical band gap and n equals 1/2 and 2 for direct and indirect allowed transitions respectively[15]. The band gap energy of as-deposited and annealed CdSe thin film has been determined by the plot of $(ahv)^2$ versus hv based on the above formula as shown in figure 4.

Figure 4 showed band gaps with the increase of annealing temperature for CdSe thin films. The band gaps were found to firstly decrease slightly, and then gradually increase with thermal annealing. The minimal band gap of CdSe film is 1.641eV when CdSe film annealed at 300°C, which is favorable to apply CdSe thin film for CdTe solar cell as absorption layer. Due to the quantum size effect, the band gaps decreases gradually with the increase of grain sizes [20]. But in our study, there isn't large change in the Eg values for the thin films annealed at 200°C-300°C, owing to the small difference in the grain size obtained by SEM and XRD studies. It is interesting to note that the band gaps increased significantly for the thin films annealed at 400°C-500°C. There are two possible reasons for the increase of the band gaps: i) The increase of the band gaps may be attributes to a shift in stoichiometry which is encountered in the work of Mondal et al[22], ii) we can see the film annealed at 500 °C with the emergence of the single crystal from the SEM images above[figure 2(e)], and the value of band gap is found to be 1.7 eV for annealed films at 500 °C, which is quite closer to the value reported earlier for CdSe single crystal [23].

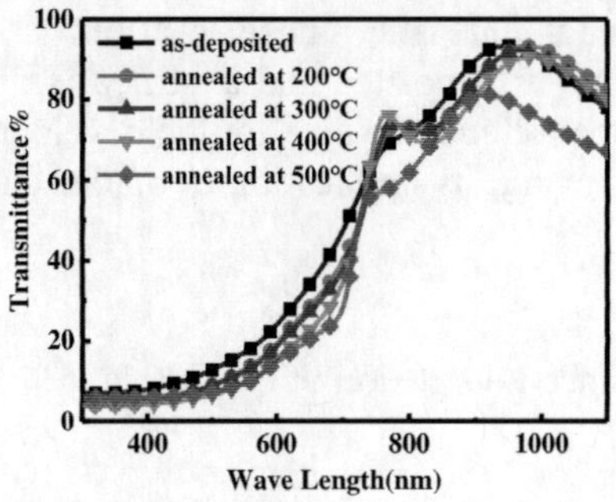

Figure 3. Optical transmittance spectra of as-deposited and annealed CdSe thin films.

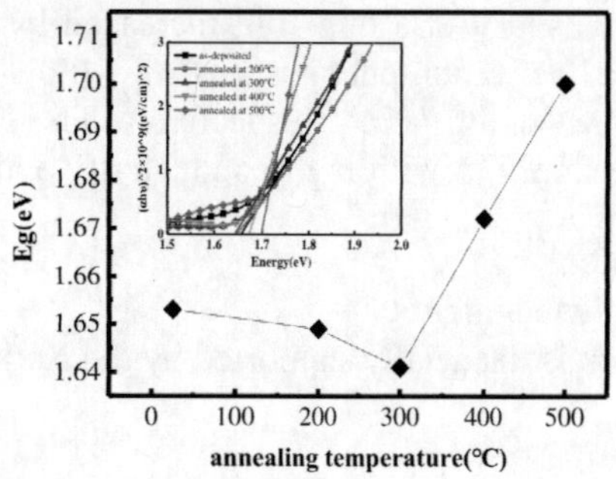

Figure 4. Tauc plots and band gaps with annealing temperature for as-deposited and annealed CdSe thin films.

3.3 electrical properties

The electrical resistivity of as-deposited CdSe thin films and CdSe thin films annealed at 500°C in dark was performed using two-probe method in the temperature range 300-573K. The resistivity of CdSe thin films decreases with the increase of annealing temperature, confirming semiconducting behavior of CdSe thin films [11]. The resistances of the films annealed at 200°C, 300°C and 400°C are too large to be measured. When the temperature rises to 433K, the resistivity of as-deposited film was measured to be 3.4×10^4 Ω·cm. The high value of resistivity of as-deposited film can be due to the dislocation, crystallite boundary discontinuities and presence of surface state. It is observed that the resistivity of the film after annealing decreased by two orders of magnitude compared with the as-deposited film in table 1. The decrease of resistivity may be attributed to the growth of grains, the decrease of defects and increase of carrier concentration (The deficiency of selenium content leads to cadmium-rich of CdSe thin film annealed at 500°C, and equivalents to increasing the donor impurity, thus the carrier concentration is increased.).

Table 1. Electrical resistivity and activation energy of dark conductivity of CdSe thin films as-deposited and annealed at 500°C

Samples	electrical resistivity ρ(Ω·cm)			activation energy Ea(eV)
	325K	433K	548K	
As-deposited	None	3.4×10^4	5.7×10^2	0.733
Annealed at 500°C	5.3×10^4	7.0×10^2	3.5	0.712

The activation energies of dark conductivity of thin films were calculated by using the following relation $\rho = \rho_0 exp(\frac{E_a}{kT})$ where ρ is the resistivity at temperature T, ρ_0 is a constant, k is Boltzman constant [11]. The calculated activation energies in 410K-480K of as-deposited CdSe thin films and CdSe thin films annealed at 500°C are 0.733eV and 0.712eV, respectively.

4. Conclusion

The effect of annealing temperature on the properties of CdSe thin films by RF-sputtering was studied.

The structure of CdSe thin films gradually transforms from cubic to hexagonal phase, and the crystallite sizes were maintained at around 33nm with the increase of annealing temperature (≤400°C). However, when the annealing temperature reached 500°C, the surface morphology of CdSe thin film changed significantly, and grains with size of 200 nm were precipitated on the surface of CdSe thin film. The Se/Cd ratio changed from 49.39:50.61 for as-deposited CdSe thin film to 42.90:57.91 for CdSe thin film annealed at 500°C. CdSe thin films maintained high transmittance after annealing. The band gaps of CdSe films firstly decrease, and then gradually increase with the increase of annealing temperature. The minimal band gap of CdSe thin film is 1.63eV when annealed at 300°C. The conductivity of CdSe thin film increased by two orders of magnitude after annealed at 500°C. The calculated activation energies in 410K-480K of CdSe thin films as-deposited and annealed at 500°C are 0.733eV and 0.712eV, respectively. These results are of great reference value for the development of CdSe thin film with high conductivity and high transparence and its application to optoelectronic devices.

5. Acknowledgments

This work is financially supported by the National Natural Science Foundation of China (No. 6170411 5).

6. Reference

[1] Gan F Y and Shih I 2002 *IEEE Trans. Electron Devices* **49** 15-8
[2] Van Calster A, Vervaet A, De Rycke I, De Baets J and Vanfleteren J 1988 *J. Cryst. Growth* **86** 924-8
[3] Schlamp M C, Peng X and Alivisatos A P 1997 *J. Appl. Phys.* **82** 5837-42
[4] Schreuder M A, Xiao K, Ivanov I N, Weiss S M and Rosenthal S J 2010 *Nano Lett.* **10** 573-6
[5] Shinde S K, Ghodake G S, Dubal D P, Lohar G M, Lee D S and Fulari V J 2014 *Ceram. Int.* **40** 11519-24
[6] Saha S K, Guchhait A and Pal A J 2012 *J. Appl. Phys.* **112** 1-7
[7] Paudel N R and Yan Y 2014 *Appl. Phys. Lett.* **105** 1-5
[8] Yang X, Liu B, Li B, Zhang J, Li W, Wu L and Feng L 2016 *Appl. Surf. Sci.* **367** 480-4
[9] Baines T, Zoppi G, Bowen L, Shalvey T P, Mariotti S, Durose K and Major J D 2018 *Sol. Energy Mater. Sol. Cells* **180** 196-204
[10] Nair M T S, Nair P K, Zingaro R A and Meyers E A 1993 *J. Appl. Phys.* **74** 1879-84
[11] Kale R B and Lokhande C D 2004 *Appl. Surf. Sci.* **223** 343-51
[12] Erat S, Metin H and Arı M 2008 *Mater. Chem. Phys.* **111** 114-20
[13] Mahato S and Kar A K 2017 *Journal of Science: Advanced Materials and Devices* **2** 165-71
[14] Purohit A, Chander S, Nehra S P and Dhaka M S 2015 *Physica E: Low-dimensional Systems and Nanostructures* **69** 342-8
[15] Kale R B and Lokhande C D 2005 *Semicond. Sci. Technol.* **20** 1-9
[16] Kale R B and Lu S-Y 2015 *J. Alloys Compd.* **640** 504-10
[17] Shyju T S, Anandhi S, Indirajith R and Gopalakrishnan R 2010 *J. Alloys Compd.* **506** 892-7
[18] Zapata-Torres M, Chale-Lara F, Caballero-Briones F and Calzadilla O 2005 *physica status solidi (c)* **2** 3742-5
[19] Athanassopoulou M D, Mergos J A, Palaiologopoulou M D, Argyropoulos T G and Dervos C T 2012 *Thin Solid Films* **520** 6515-20
[20] Mahato S, Shakti N and Kar A K 2015 *Mater. Sci. Semicond. Process.* **39** 742-7
[21] Manifacier J C, Gasiot J and Fillard J P 1976 *Journal of Physics E: Scientific Instruments* **9** 1002
[22] Mondal A, Dhar A, Chaudhuri S and Pal A K 1990 *Journal of Materials Science* **25** 2221-6
[23] Pal U, Zaldivar M H, Sathyamoorthy R, Manjuladevi V, Sudhagar P, Mohan S C and Senthilarasu S 2008 *Journal of Nanoscience and Nanotechnology* **8** 6474-80

IOP Conf. Series: Materials Science and Engineering **556** (2019) 012007 doi:10.1088/1757-899X/556/1/012007

LCOE analysis of tower CSP plants using different molten salts for TES in China

X R Zhuang[1], X H Xu[1,*], W R Liu[1], T L Liu[1] and W F Xu[1]
[1] School of Mechanical Engineering and Automation, Harbin Institute of Technology, Shenzhen 518055, China
E-mail: xuxinhai@hit.edu.cn

Abstract. In recent years, Chinese government is vigorously developing concentrating solar power (CSP) technology. It is one of the most promising renewable-based electricity generation technologies to deal with the increasing demand of electricity consumption and environmental sustainability. For the commercialization of CSP technology, low cost is a precondition. However, studies of electricity generation cost analysis for CSP systems, particularly for the tower systems, in China are quite limited. This paper conducts an economic analysis by applying the levelized cost of electricity (LCOE) model for 100 MW tower CSP plants in various locations in China with four different molten salts for thermal energy storage (TES). The results show that it's inappropriate to build a CSP plant nearby Shenzhen and Shanghai since their LCOEs are higher than 6 RMB kWh^{-1}. There is an optimal capacity of TES resulting in the lowest LCOE for a certain tower CSP plant. And the solar salt has lower LCOE than the other three new molten salts. In order to calculate the time when the grid parity will be reached, four scenarios for CSP development roadmap proposed by International Energy Agency (IEA) are considered in this study. It can be found that the grid parity of tower CSP electricity would be reached in the years of 2038-2041 in the case of no future penalties for the CO_2 emissions, and bring forward about 7-14 years when considering the penalties for the CO_2 emissions. This research can provide support for government to formulate incentive policies for the CSP industry.

Keywords: concentrating solar power (CSP); levelized cost of electricity (LCOE); China; molten salt; tower system

1. Introduction

Concentrating solar power (CSP) technology is one of the most promising renewable-based electricity generation technologies to deal with the increasing demand of electricity consumption and environmental sustainability. It can generate renewable electricity from direct sunlight and produce nearly none of greenhouse gas emissions. In addition, comparing to photovoltaic (PV) solar energy, the CSP systems can delay electricity generation in low or non-solar times by using thermal energy storage (TES). In recent years, Chinese government is vigorously developing CSP technology. The development of CSP in China has increased at a greater rate starting in 2013 [1] and the installed capacity is about 29.3 MWe by the end of 2017 with an increasing rate of 3.53% over 2016 which is higher than the global increasing rate (2.3%) [2]. Moreover, power tower technology is one of the most common technologies of CSP which can achieve higher operating temperature compared to parabolic trough and linear Fresnel technologies. It can yield greater efficiency of thermal-to-electric conversion in the power block and result in lower costs for storage [1].

For the commercialization of CSP technology, low cost of the electricity generation is a precondition. The electricity generation cost of CSP systems in China is approximately triple in comparison with the current grid [3-6]. As a key indicator to evaluate the economic feasibility of CSP systems, the analysis of the electricity generation cost is crucial and affects the development of the CSP generation industry

[7,8]. However, current researches [6,9,10] of electricity generation cost analysis for CSP systems, particularly for the tower systems, in China are quite limited. In addition, considering that no specific policies for CSP industry have been enacted yet in China [6], more data should be put forward for policy makers. Thus, this paper conducts an economic analysis by applying a levelized cost of electricity (LCOE) model for 100 MW tower CSP plants in China with four different molten salts for TES. In view of the geographic conditions, five locations are selected, including Huizhou near Shenzhen, Taicang near Shanghai, Yanqing in Beijing, Linxi in Inner Mongolia and Delingha in Qinghai. Four of them are situated at the eastern of China where the demand of electricity is relative high and near cities with good transportation and sufficient water resource. In order to analyse the economic feasibility of tower CSP plants built at the eastern of China, particularly near the first-tier cities of China, the site of Delingha is taken as reference in this paper, which has annual direct normal irradiance (DNI) near the bottom threshold (2000 kWh m^{-2} $year^{-1}$) for installing any form of CSP systems to achieve a reasonable economic performance [11]. Then, the calculations of both the present (2017) value of LCOE and its future evolution for the period (2018–2050) under four scenarios proposed by the International Energy Agency (IEA) (i.e. Blue Map [12], Global Outlook Advanced [13], Global Outlook Moderate [12], and Roadmap [14]) are presented. Moreover, this paper also estimates the time when CSP grid parities will be attained.

2. LCOE model for CSP systems

The LCOE is one of the most frequently used models to evaluate the electricity generation technologies which calculates the cost of solar electricity during the whole lifetime of the systems and takes into account the time value of money and the risks. It's equal to the sum of all the discounted costs incurring during the lifetime of the project divided by the units of discounted energy produced over the entire lifetime [15] and can be expressed as follows:

$$LCOE_t = \frac{C_t + L + \sum_{n=1}^{N}\left[(V+I)C_t(1+r)^{-n}\right]}{\sum_{n=1}^{N}\left[S \cdot TF \cdot \eta(1-DR)^n(1+r)^{-n}\right]} \tag{1}$$

where each variables in the equation are summarized on Table 1.

Table 1. Parameters and their values of the LCOE model.

Symbol	Description	Value	Units
$LCOE_t$	Levelized cost of energy of a tower CSP system installed in the year t between 2017 and 2050	Calculated by Eq. (1)	RMB kWh^{-1}
C_t	Total cost of the system installed in the year t between 2017 and 2050	Calculated by Eq. (2)	RMB W^{-1}
L	Land cost	On Table 2	RMB W^{-1}
V	Operation-maintenance cost	2 [5,16,17]	%
I	Insurance cost	0.5 [18]	%
S	Solar resource (DNI)	On Table 2	kWh m^{-2} $year^{-1}$
TF	Tracking factor	100 [16]	%
η	Performance factor	Calculated by SAM	m^2 W^{-1}
DR	Degradation rate	0.2 [19]	%
r	Discount rate	10 [17]	%
N	Lifetime of the system	30 [13,16,17]	years

The total cost of the system installed in a certain year, C_t, can be described as a function of the global cumulative installed capacity expressed as follows [5]:

$$C_t = C_0\left(\frac{Q_t}{Q_0}\right)^{\frac{\log(1-LR)}{\log 2}} \tag{2}$$

where Q_t is the cumulative installed capacity in a certain year; Q_0 is the cumulative installed capacity in the reference year; LR is the learning rate; C_0 is the total cost of the system installed in the reference year. In this paper, the year of 2017 has been taken as the reference year, and the system cost of the

LCOE present value is only determined by C_0 which is calculated by System Advisor Model (SAM) of NREL [20] in this paper. Besides, its future evolution also relates to the future cumulative installed capacity and the learning rate. The leaning rate is used a conservative estimation of 10% recommended by IEA [12,17]. According to CSPPLAZA (a major professional CSP industry website in China), the global cumulative installed capacity by the end of 2017 is reported to be 5.133 GW [2] which is Q_0 in this paper. In order to evaluate the future cumulative installed capacity during the years of 2018-2050, the functions proposed by Li et al. [10] that best fit the objectives of cumulative installed capacity for those four CSP development roadmap scenarios proposed by IEA are used in this paper.

3. Results and discussion

On the basis of the LCOE model described above, the economic analysis of 100 MW tower CSP plants in five locations in China with four different molten salts for TES are estimated in this paper. As a reference, a 100 MW project using solar salt (60% $NaNO_3$ + 40% KNO_3) with 6 h TES in Delingha is selected in this paper. When assessing the effect of one factor, the other factors are used the same as the reference case. Furthermore, unless otherwise stated, the values for the input parameters of Eq. (1) are used as specified on Table 1.

Table 2. Initial costs and LOCE present values for different locations in China.

Location	DNI (kWh m^{-2} year^{-1})	System cost (RMB W^{-1})	Land cost (RMB W^{-1})	LCOE (RMB kWh^{-1})	Capacity factor (%)
Huizhou	795.7	31.48	18.3	9.28	7.6
Taicang	762.9	31.45	13.97	8.49	7.7
Yanqing	1189.9	31.38	4.05	3.5	15.1
Linxi	1668.1	31.24	4.75	2.33	23
Delingha	2080.5	30.6	1.31	1.45	33.3

Table 2 shows the initial costs and LOCE present values for different locations under reference conditions. Since DNI is the most important energy input data for CSP systems, the specific DNIs for those locations are calculated by the method based on the air mass [21]. It can be seen that the initial costs of Huizhou and Taichang are much higher than that of Delingha which makes the LCOE higher. In addition, the influence of land cost on the LCOE is illustrated in Figure 1. The x-coordinate means the reduction percentage of the initial land cost for each location. It can be found that the land cost affects the LCOE significantly for Huizhou and Taichang as their land costs accounts for 38.8% and 30.8% of the total initial costs. Furthermore, even though their land costs are excluded from the capital cost, the LCOEs would be still higher than 6 RMB kWh^{-1}. It's mainly because their DNIs are less than 800 kWh m^{-2} year^{-1} which makes the capacity factor lower and system cost higher. Thus, it's inappropriate to build a tower CSP plant nearby Shenzhen and Shanghai. It can be also observed from Figure 1 that the land lost has slight influence on the LOCE when it is low such as in Yanqing, Linxi and Delingha. As Yanqing is the location of the pilot project for CSP plants in Beijing, the land cost is relatively low. The LCOE of Linxi has the lowest price of 2.33 RMB kWh^{-1} because of the highest DNI among the selected locations at the eastern of China. So the main factor resulting in low LCOE for different locations is the high DNI, due to more energy produced which increases the capacity factor of a plant and entails to an efficient TES.

TES is an important feature of CSP systems which makes for continuous electricity generation for solar energy. In this paper, four molten salts are selected to estimate the LCOE for tower CSP systems under reference conditions. These salts show the potential to be used in commercial CSP plants and have excellent thermodynamic properties which have been indicated in previous studies [22-26]. Their thermal properties and TES costs are displayed on Table 3. It can be seen that the chloride salt (8.1% NaCl + 31.3% KCl + 60.6% $ZnCl_2$) has the lowest TES cost due to the lowest salt price and largest density. And the TES costs of the fluoride salt (29.3% LiF + 11.7% NaF + 59% KF) and carbonate salt (32.1% Li_2CO_3 + 33.4% Na_2CO_3 + 34.5% K_2CO_3) are much higher since their salt prices are more than three times of the solar salt and chloride salt.

As the value of performance factor is relevant to the amount of TES, Figure 2 presents the effect of TES capacity on the LCOE present value for those four salts. It can be found that the chloride salt with the lowest performance factor has the highest LCOE when the CSP systems have no TES. Since a higher amount of TES improves the performance factor of CSP systems which has been confirmed by other studies [27,28], the LCOE of all molten salts decreases firstly as the capacity of TES increases. However, when the performance factor no longer improves with the increasing TES capacity, the LCOE goes up as the cost of the TES rises. Thus, there is an optimal capacity of TES resulting in the lowest LCOE for a certain tower CSP plant. The LCOE of the chloride salt reduces obviously at the lower range of TES capacity because of its lowest salt price. And the increasing of the LCOEs of the fluoride salt and carbonate salt become more distinct at the higher range of TES capacity due to their much higher costs of TES. The solar salt has the lowest LCOE among those four molten salts for different capacity of TES.

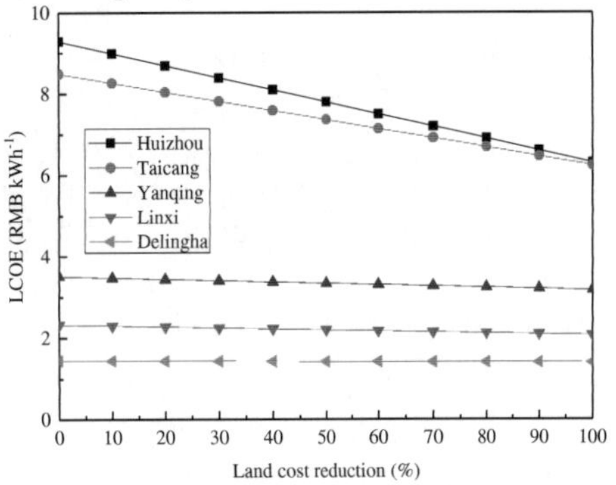

Figure 1. Effect of land cost on the LCOE for five different sites.

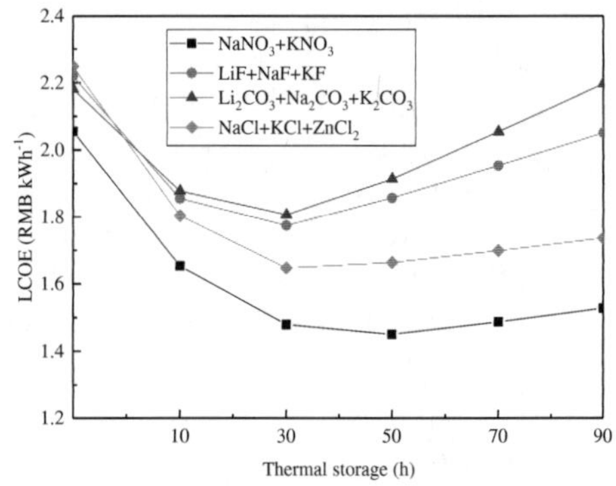

Figure 2. Influence of TES capacity on the LCOE for four different molten salts.

Table 3. Thermal properties and TES costs of selected molten salts.

Molten salt	$NaNO_3$, KNO_3	LiF, NaF, KF	Li_2CO_3, Na_2CO_3, K_2CO_3	NaCl, KCl, $ZnCl_2$
Composition by wt.%	60, 40	29.3, 11.7, 59.0	32.1, 33.4, 34.5	8.1, 31.3, 60.6
Melting point (°C)	220	454	400	229
Thermal stability (°C)	600	850	715	850
Density (kg m^{-3})	1708.4-1950.1	1851.6-2116.5	1959.7-2069.4	1946.2-2275.5
Heat capacity (kJ kg^{-1} K^{-1})	1.48-1.55	1.28-1.82	1.61	0.9-0.92
Viscosity (mPa s)	0.99-5.78	1.64-12.38	6.11-45.09	3.48-29.63
Thermal conductivity (W m^{-1} K^{-1})	0.33-0.40	0.05-0.27	0.45-0.49	0.30-0.38
Salt price (RMB kg^{-1})	5.80	16.66	15.02	4.68
TES cost (RMB kWh$_t^{-1}$)	156.00	316.79	490.15	111.69

Figure 3 displays the future evolution of the LCOE between 2018 and 2050 for four scenarios under reference conditions. It can be deduced that the future cost of tower CSP electricity decreases as the increase of cumulative installed capacity during the period of 2018-2050. Moreover, the future LCOE in the Blue map scenario presents slower reductions than other three scenarios, particularly at the initial years, owing to the lower objectives. Then, the differences of the LCOE in 2050 for all scenarios become not very large. The LCOE of tower CSP systems are 0.72, 0.64, 0.70 and 0.67 RMB kWh^{-1} in 2050 and represents 49.88%, 44.02%, 47.94% and 46.14% of its present value in 2017 for Blue map scenario, Global outlook advanced scenario, Global outlook moderate scenario and Roadmap scenario, respectively.

Figure 3 also illustrates the time when the electricity cost of tower CSP plants will reach the grid parity under reference conditions. Since more than three quarters of the electricity are generated in coal-fired power plants in China [17], the cost evolution of coal-fired electricity is used as the grid parity in this study. The cost of electricity produced by coal-fired power plants is 6.26 US cents kWh^{-1} [29] which is only considered the production costs and would increase linearly in the future with an annual conservative growth rate of 2.7% [10]. In addition, considering the CO_2 emissions by the coal-fired power plants, the carbon emission prices of 25 \$ ton^{-1} CO_2 [30] and 50 \$ ton^{-1} CO_2 [11] are added to estimate the grid parities with an emission factor of 0.9 kg CO_2 kWh^{-1} [17,31,32]. According to

Figure 3, the time when the grid parities of tower CSP systems would be achieved are summarized on Table 4. It can be observed that the cost of tower CSP electricity would reach the grid parity in the years of 2038-2041 in the case of no future penalties for the CO_2 emissions. And the grid parity would bring forward about 7-8 years for the carbon emission price of 25 \$ ton^{-1} CO_2 and about 13-14 years for the carbon emission price of 50 \$ ton^{-1} CO_2.

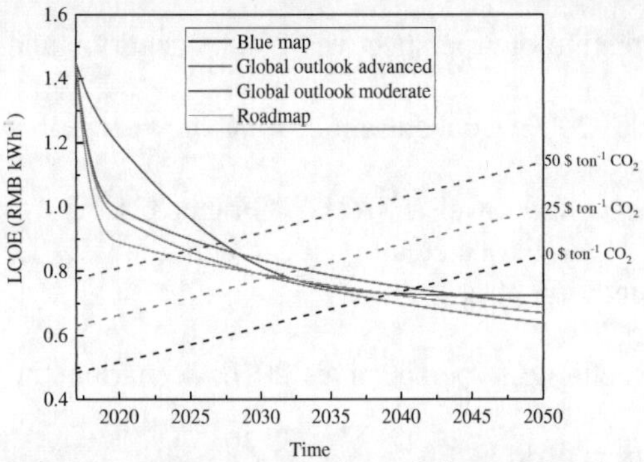

Figure 3. Future evolution of the LCOE for tower CSP systems for four scenarios and grid parities calculations.

Table 4. Time when the grid parities of tower CSP systems would be achieved.

Scenario	Time when LCOE equals to the grid parity		
	0 \$ ton^{-1} CO_2	25 \$ ton^{-1} CO_2	50 \$ ton^{-1} CO_2
Blue map	2040	2032	2028
Global outlook advanced	2038	2031	2025
Global outlook moderate	2041	2033	2027
Roadmap	2040	2031	2024

4. Conclusion

In order to provide support for Chinese government to formulate incentive policies for the CSP industry, we have presented the economic analysis for 100 MW tower CSP plants in five locations in China with four molten salts for TES in this paper based on a model of LCOE. From the results found in this paper, we can summarize the following conclusions:

(1) It's inappropriate to build a tower CSP plant nearby Shenzhen and Shanghai since even though their land costs are free, their LCOEs would be still higher than 6 RMB kWh^{-1} mainly owing to the DNIs lower than 800 kWh m^{-2} year^{-1}. When the land cost is low, it has slight influence on the LOCE. The main factor resulting in low LCOE for different locations is mainly the high DNI.

(2) There is an optimal capacity of TES resulting in the lowest LCOE for a certain tower CSP plant as the performance factor firstly increases with the increasing TES capacity, and then no longer rises. The solar salt has the lowest LCOE among those four molten salts for different capacity of TES.

(3) The cost of tower CSP electricity would reach the grid parity in the years of 2038-2041 in the case of no future penalties for the CO_2 emissions based on four scenarios for CSP development roadmap proposed by IEA. And the grid parity would bring forward about 7-8 years for the carbon emission price of 25 \$ ton^{-1} CO_2 and about 13-14 years for the carbon emission price of 50 \$ ton^{-1} CO_2.

Acknowledgments

Support from Shenzhen Science and Technology Research and development Funds (JCYJ20160318094917454) and Shenzhen Key Lab Fund of Mechanisms and Control in Aerospace (ZDSYS201703031002066) is gratefully acknowledged.

References

[1] Mehos M, Turchi C, Vidal J, Wagner M, Ma Z W, Ho C, Kolb W, Andraka C and Kruizenga A 2017 Concentrating solar power gen3 demonstration roadmap *NREL/TP-5500-67464* U. S. doi: 10.2172/1338899
[2] CSPPLAZA 2018 Global CSP installed capacity increased to 5133 MW by the end of 2017 China http://www.cspplaza.com/article-11387-1.html
[3] Du F L and Xie H 2013 *Adv. New Renew. Energy* **1(3)** 197–207 [in Chinese]
[4] Si J P 2015 *China Power Enterp. Manag.* **15** 54–7 [in Chinese]
[5] Hernandez-Moro J and Martínez-Duart J M 2013 *Renewable & Sustainable Energy Reviews* **20** 119–32
[6] Zhao Z Y, Chen Y L and Thomson J D 2017 *Energy* **120** 117–27

[7] Zhao Z Y, Li Z W and Xia B 2014 *Energy* **69** 179–85
[8] Frisari G and Stadelmann M 2015 *Energy Policy* **82** 12–22
[9] Zhu Z, Zhang D, Mischke P and Zhang X L 2015 *Energy* **89** 65–74
[10] Li Y Q, Liao S M, Rao Z H and Liu G *Renewable Energy* **69** 34–42
[11] Malagueta D, Szklo A, Borba B S M C, Soria R, Aragao R, Schaeffer R and Dutra R 2013 *Energy Policy* **59** 198–212
[12] International Energy Agency (IEA) 2008 Energy technology perspectives 2008: scenarios and strategies to 2050 *IEA/OECD* France p 1–650
[13] Greenpeace International, Solar Paces and Estela 2009 Concentrating solar power global outlook 09 *Power* 88
[14] Philibert C 2010 Technology roadmap: concentrating solar power *IEA/OECD* France p 1–52
[15] NREL 2014 Simple levelized cost of energy (LCOE) calculator documentation. U. S. https://www.nrel.gov/analysis/tech-lcoe-documentation.html
[16] Purohit I and Purohit P 2010 *Energy Policy* **38** 3015–29
[17] International Energy Agency (IEA) 2010 Energy technology perspectives 2010: scenarios and strategies to 2050 *IEA/OECD* France p 1–706
[18] Viebahn P, Lechon Y and Trieb F 2011 *Energy Policy* **39** 4420–30
[19] Navigant Consulting 2007 Levelized cost of generation model- renewable energy, clean coal and nuclear inputs *IEPR committee workshop on the cost of electricity generation* p 58
[20] NREL 2017 System Advisor Model (SAM) U. S. https://sam.nrel.gov/about
[21] PVCDRM 2017 http://www.pveducation.org/pvcdrom/properties-of-sunlight/air-mass
[22] Salanne M, Simon C, Turq P and Madden P A 2009 *J. Fluorine Chemistry* **130** 38–44
[23] An X H, Cheng J H, Yin H Q, Xie L D and Zhang P 2015 *Int. J. Heat & Mass Transfer* **90** 872–7
[24] Khokhlov V, Korzun I, Dokutovich V and Filatov E 2011 *J. Nuclear Materials* **410** 32–8
[25] Liu T L, Liu W R and Xu X H 2017 *IOP Conference Series: Earth and Environmental Science* **93** 012023
[26] Zavoico A B 2001 Solar power tower design basis document, revision 0 *Office of Scientific & Technical Information Technical Reports* U. S. doi:10.2172/786629
[27] Poullikkas A, Hadjipaschalis I and Kourtis G 2010 *Renewable and Sustainable Energy Reviews* **14** 1469–76
[28] Krishnamurthy P, Mishra S and Banerjee R 2012 *Energy Policy* **48** 407–19
[29] Staley B C, Goodward J, Rigdon C and MacBride A 2009 Juice from concentrate: Reducing emissions with concentrating solar thermal power *World Resources Institute in conjunction with Gold man Sachs* http://www.wri.org/sites/default/files/pdf/juice_from_concentrate.pdf
[30] Schellekens G, Battaglini A, Lilliestam J, McDonnell J and Patt A 2010 100% renewable electricity: A roadmap to 2050 for Europe and North Africa *PricewaterhouseCoopers* UK
[31] Pacca S, Sivaraman D and Keoleian G A 2007 *Energy Policy* **35** 3316–26
[32] Bhawan S and Puram R K 2011 CO2 baseline database for the Indian power sector *Central Electricity Authority, Ministry of Power, Government on India* Version 6.0

IOP Conf. Series: Materials Science and Engineering **556** (2019) 012008 doi:10.1088/1757-899X/556/1/012008

Study on Lighting -Heating-Electricity Coupled Energy Saving Potential for STPV Window in Southwest China

Hao Tian[1], Wei Zhang[1*], Lingzhi Xie[2], Yupeng Wu[3] ,Yanyi Sun[3], Wei Wang[1], Mo Chen[1]

1. College of Architecture and Environment, Sichuan University, Chengdu, China; 610065
2. Institute of New Energy and Low-carbon Technology, Sichuan University, Chengdu, China; 610065
3. Department of Architecture and Built Environment, Faculty of Engineering, The University of Nottingham,

University Park, Nottingham; NG72RD, UK

*Corresponding Author Email: Wei Zhang, xskin821@163.com

Abstract: Building integrated photovoltaic (BIPV) is an important application of solar cells. In this paper, the experiments were conducted to test the performance of photovoltaic panels installed in buildings. The experiments were carried out in Chengdu, Southwest China which located in climate zone of hot summer and cold winter. The real-time power generation, the simultaneous generation of heat and the impact on indoor lighting are important factors for building energy consumption. Two identical rooms were built up as experimental units, and one of which was fitted with amorphous silicon PV windows, and another was fitted with conventical window. According to the test results, it can be seen that, the PV film windows could save the 30% building load on typical sunny days, and provide additional 0.41 kWh/per day.

Key words: BIPV, PV film window, Power generation, Heating load, Lighting load

1.Introduction

PV used to meet the demand for electricity has been widely recognized. It is a key development direction in the energy field in China's 13th Five-Year Plan. The application of PV technology in the construction field is particularly important. The current building energy consumption accounts around 30% of the total energy consumption [1]-[3]. The research and development of BIPV application could improve living standards with the villages and towns in Southwest China.

In recent years, the attention of many researchers has been attracted by the semi-transparent PV (STPV) windows, and which can not only generate electricity, but also reduce cooling and heating loads. P. W. Wong et al [4] studied STPV as the roof of residential applications. The experimental results show that

PV roof can save indoor energy consumption. Wei Liao et al [5] compares amorphous silicon glass with traditional glass in different buildings. Considered the influence of building conditions, it was showed in the experiment that STPV windows are more advantageous than traditional glass windows.
For the energy consumption of STPV windows, Tin-Tai Chow et al [6] built an experimental system for a typical office environment in Hong Kong. It was showed in the test results that the innovative natural ventilation photovoltaic double glazing can significantly reduce indoor energy consumption by 28%. Wei He et al [7] compared the performance of single-layer PV windows and double-layer PV windows in Hefei, China. Based on the test result, double glaze PV has the better thermal comfort. Peng et al [8-9] set up the experimental units in Hong Kong to test the heating performance of STPV windows in winter. The simulation model set up by Peng [10] for an optimized double- PV window shows that the annual power output can be doubled even with ventilation.
Vartiainen et al. [11] found that the solar panel coverage of solar cells increased, and the annual average DA of buildings increased significantly. For dynamic lighting indicators (UDI), the improvement was not significant. Olivieri et al. [12] studied the relationship between photovoltaic cell transmittance and building interior lighting by establishing an ideal model. The results show that when the transmittance of STPV windows is reduced from 40% to 10%, the indoor lighting coefficient is gradually reduced, but the uniformity of illumination is improved.

According to the previous study, it was shown that STPV windows can reduce building load significantly. However, the comprehensive energy saving potential for STPV window in southwest china has not yet been evaluated. This article is aimed to evaluate the lighting heating-electricity coupled energy saving of film STPV windows. It would contribute to develop clean energy, reduce carbon emissions, promote green building development in
Southwest China.

2 Experimental methods

2.1 Parameters of PV module

The double-layer STPV window was designed and produced by our research team and the Hanergy Company. The film PV module is amorphous silicon. It has the advantages of low light absorption and better performance.

The double-skin STPV window consists of one layer of single STPV windows, and one layer of conventional window with a vacuum air layer. The key characteristics of the PV module are shown in Table 1.

Table 1. Parameters of PV module

PV type scribed semi-transparent module light transmittance	20%
Material	a-SiGe
Maximum power under STC (W)	50

IOP Conf. Series: Materials Science and Engineering **556** (2019) 012008 doi:10.1088/1757-899X/556/1/012008

Photoelectric conversion rate	6.7%

2.2Test rig

To test the performance of the double-skin STPV window, a test rig was built in Sichuan University is shown in the Fig.1. The test rig included two identical test units, which is 3m(depth)×3m(width)×3m(height), and the 75 mm thick sandwich rock wool board has been used as the materials of wall and roof to meet the requirement of the thermal insulation. The windows are installed on the south facing wall.

Measurement instruments of the test rig are shown in Fig. 1, the wireless daylighting illuminance sensors were used to measure the daylighting illuminance on the working surface. The experimental data except electrical parameters were collected by wireless Multi-channel data recorder with a sampling interval of 1 min. For the indoor energy consumption, STPV windows not only generate electric, but also have influence on the indoor light environment and thermal environment. To comprehensive analyze the indoor energy consumption, the power consumption of indoor air-conditioner and lighting have been collected by the electric quantity recorder.

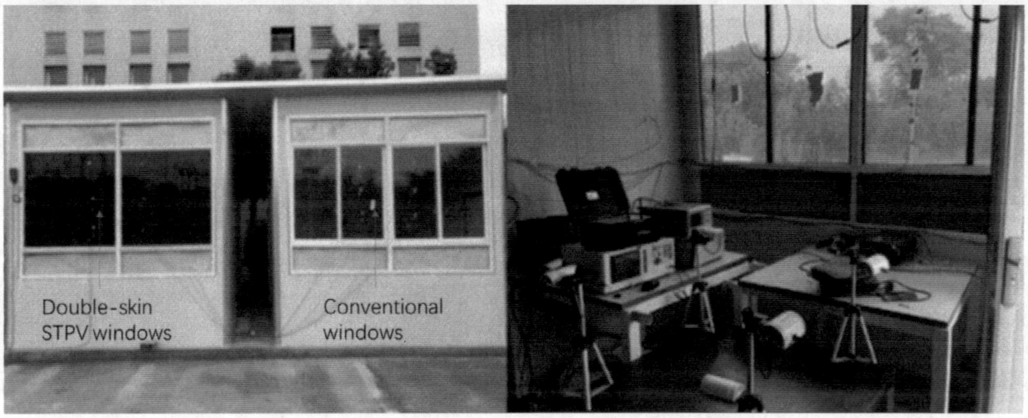

Fig 1. Test rig and measurement instruments

3. Experimental results and analysis

Experiments were conducted to test the heating energy consumption and lighting load of two units different working conditions in cloudy and sunny days from March to April.

3.1 Heating Load test

It is showed in the experiment that the power consumption of the test unit is more than that of the comparison unit when it is sunny. Because the shading of solar radiation by the STPV window increase the air conditioning load of the unit. The power consumption of air conditioners in two units is shown in the Fig. 2. On sunny days, the average power consumption of the test unit and contrast unit is 70W and 49W. On cloudy days, the average power consumption of the test unit and contrast unit is 206W and 219W, have almost the same amount of electricity.

IOP Conf. Series: Materials Science and Engineering **556** (2019) 012008 doi:10.1088/1757-899X/556/1/012008

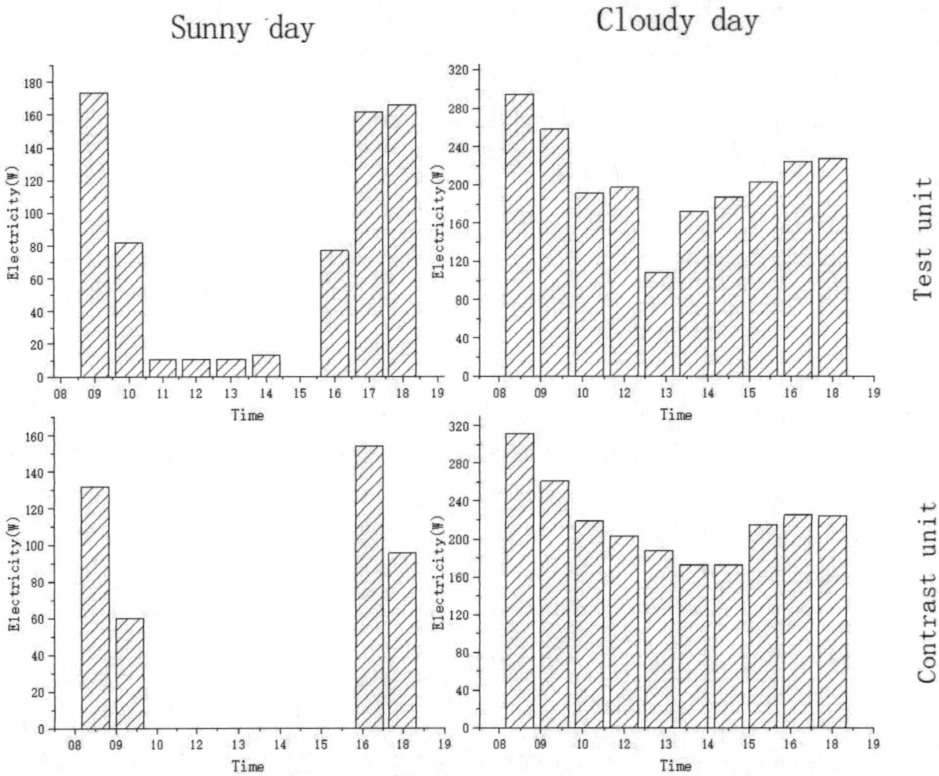

Fig 2 Electricity for air-conditioner

It shows that when the STPV windows were used in winter hot summer and cold winter, the air conditioning energy consumption is more in the experimental unit than that in the comparison unit.

3.2 lighting load test

Compared with the experiment tests of the daily lighting, indoor lighting would be reduced due to the shading of STPV windows. When the illuminance does not meet 300lx, the lighting is turned on automatically, and the electric power meter is used to record the electricity consumption of the lighting. As can be found in the experiment, in the sunny days, STPV windows would not affect indoor lighting, and the average illuminance reaches 1583 lx. In cloudy days, the light intensity is not high, and the average illuminance in the room is only 208 lx. Therefore, when the light intensity is insufficient, electric light lighting needs to be turned on. The Fig. 3 shows the consumption of lighting load in cloudy days.

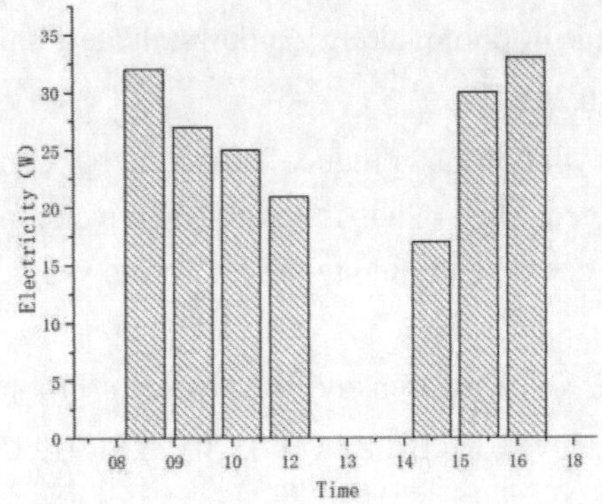

Fig 3 Electricity for lighting

3.3 Building load Analysis

In sunny days, the electricity generation of PV windows can reach 50W, and the power generation can have an income of around 0.41 kWh a day. The energy saving of experimental units will be better than cloudy days. The lighting in the unit will not be affected by PV windows. Natural lighting can fully meet the needs of indoor lighting. In cloudy days, there is little power generation, and the indoor lighting is affected by PV windows. Lighting is required to be turned on throughout the day.

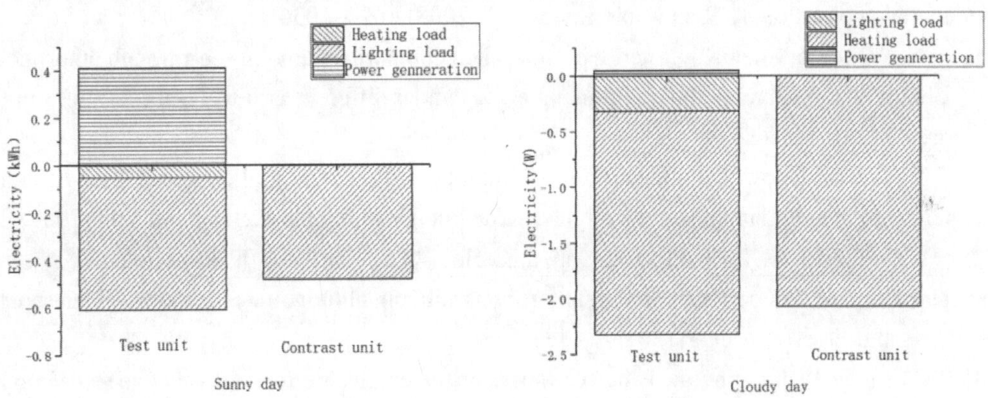

Fig 4 Overall power consumption of two units

4 Conclusion

The current work has studied the application potential of solar film windows in Southwest China, including the impact on daylighting and indoor heating loads. The results show that it has a high potential for application.

In sunny days, the power consumption of the experimental unit is more than that of the comparison unit, and the STPV windows can generate electricity, which can make up for the lighting electricity consumption in the unit. According to the test results, it can be seen that, the PV film windows could save the 30% building load on typical sunny days, and provide additional 0.41 kWh/per day.

In cloudy days, the test unit and the comparison unit have the same power consumption of airconditioner. Due to the sunshade of photovoltaic windows, the test unit needs to turn on the lighting, and the consumption is 0.24 kWh.

In breve, the use of photovoltaic windows can reduce indoor energy consumption by 30%. In cloudy days, indoor lighting will be affected. The a- SiGe is absent on conversion efficiency, and silicon which has higher conversion efficiency could be chosen in the PV windows in the future.

Acknowledgments:

The project is funded by National Key Research and Development Program of China: Newton Fund -China-UK Research and Innovations Bridges (No. 2016YFE0124500)

Reference:

[1] Tang Yanming. The integration of photovoltaic building ushered in the development of spring [J]. Urban Housing, 2009 (12) :56-59.

[2] Li Shunmei, Mei Jun, Yao Yong. Research Status and Development Trend of Solar Photovoltaic Building Integration[J]. Sichuan Architecture, 2013, 33(06): 38-41.

[3] Xiao Yu, Li Deying. Application status and development trend of solar photovoltaic building integration [J]. Energy Saving, 2010, 29 (02): 12-18+2.

[4] P.W. Wonga, Y. Shimodab,* , M. Nonakaa, M. Inouea, M. Mizunob .Semi-transparent PV: Thermal performance, power generation, daylight modelling and energy saving potential in a residential application[J]. Renewable Energy 33 (2008) 1024–1036

[5] Wei Liao, Shen Xu*.Energy performance comparison among see-through amorphoussilicon PV (photovoltaic) glazings and traditional glazings under different architectural conditions in China[J]. Energy 83 (2015) 267e275

[6] Tin-Tai Chow , Zhongzhu Qiu, Chunying Li. Potential application of "see-through" solar cells in ventilated glazingin Hong Kong[J]. Solar Energy Materials & Solar Cells 93 (2009) 230–238

[7] Wei He*, Y.X. Zhang, Wei Sun, J.X. Hou, Q.Y. Jiang, Jie Ji. Experimental and numerical investigation on the performance of amorphous silicon photovoltaics window in East China [J]. Building and Environment 46 (2011) 363e369

[8] Meng Wang, Jinqing Peng. Comparison of energy performance between PV double skin facades and PV insulating glass units. Applied Energy 194 (2017) 148–160

[9]Meng Wang,Jinqing Peng,Nianping Li,Lin Lu,Hongxing Yang. Experimental Study on Thermal Performance of Semi-transparent PV Window in Winter in Hong Kong[J]. Energy Procedia,2017,105.

[10] Jinqing Peng, Dragan C. Curcija. Numerical investigation of the energy saving potential of a semi-transparent photovoltaic double-skin facade in a cool-summer Mediterranean climate. Applied Energy 165 (2016) 345–356

[11] Eero Vartiainen,Kimmo Peippo,Peter Lund. Daylight optimization of multifunctional solar facades[J]. Solar Energy,2000,68(3).

[12] L. Olivieri,E. Caamaño-Martin,F .Olivieri,J. Neila. Integral energy performance characterization of semi-transparent photovoltaic elements for building integration under real operation conditions[J]. Energy & Buildings,2014,68.

Preliminary Simulation Investigation of Building Energy Consumption of Solar PV Windows in Hot Summer and Cold Winter climate zone in China

M Chen[1], W Zhang[1*], L Xie[2], Y Wu[3], Z Ni[4],QWei[5], W Wang[1], H Tian[1]

1. College of Architecture and Environment, Sichuan University, Chengdu, China; 610065

2. Key Laboratory Deep Underground Science and Engineering Foundation of Sichuan University, Sichuan University, Chengdu, China; 610065

3. Institute of New Energy and Low-carbon Technology, Sichuan University, Chengdu, China; 610065

4. Department of Architecture and Built Environment, Faculty of Engineering, The University of Nottingham, University Park, Nottingham; NG72RD, UK

5. Suzhou Talesun Solar Technologies Co., Ltd, Suzhou, China, 215542

*Corresponding Author Email: xskin821@163.com(Wei Zhang)

Abstract. Photovoltaic building integrated windows can provide electric energy, heat energy and lighting for buildings, which will change the way that buildings consume energy. Aiming at the influence of photovoltaic windows on building energy consumption in hot summer and cold winter conditions, a simplified energy consumption simulation model based on light-heat-electricity is studied. The building energy consumption model, daylighting model, and Sandia photovoltaic model are integrated. Based on the tested building thermal performance parameters, the window's power generation performance and its optical performance parameters, the impact of double-layer crystalline silicon PV windows on building energy consumption is simulated. The test results were used to validated the simulation model. According to simulation results, PV window could take advantages of the positive of blocking solar radiation, and compensate the negative effect. Considered the whole year energy saving, the PV window could save 470.03kWh, and reduce 466.5 kg of CO_2.

Keywords: hot summer and cold winter area; STPV; light-heat-electricity; annual energy consumption

1、Introduction

Building integrated photovoltaics, which uses photovoltaic power generation technology to convert solar energy into electrical energy to meet the building energy consumption, is the focus of research and application in recent years [1-3]. Photovoltaic windows, as a technical form of BIPV, can provide buildings with clean electric power while meeting the architectural aesthetic design, and reduce the energy consumption of building[4-5].

In the researches of light-heat-electricity performance of double-layer PV window, most research[6-11] mainly established a heat transfer model to simulate the thermal and electrical performance of double-layer STPV. Jong-Hwa Song et al. [6-7] established a numerical simulation model for double-layer STPV windows based on actual meteorological parameters in numerical simulation software TRNSYS. Simulation results showed that installation angle is an important factor affecting thermal and electrical performance, and 30° is the best installation angle. Tady Y.Y. Fung et al. [11] simulated thermal conduction of double-layer PV windows using one-dimensional transient thermal equilibrium equation. The results showed that when the coverage of PV modules is close to 80%, the heat gain can be reduced by 70%. It can be seen that the factor that has the greatest impact on indoor heat is the coverage area of PV modules.

Double-layer PV window as semi-transparent enclosure, it's light-electricity performance is another focus of research. Shen Xu et al. [12] simulated the effects of different crystalline silicon component coverage on the photovoltaic performance. The results show that the higher photovoltaic coverage rate is suitable for buildings with large window and wall ratio and large-and-deep buildings, which increases the total power generation but reduces the photoelectric conversion efficiency. K. Kapsis et al. [13] used Daysim to study the indoor lighting of different film transmittances. The results showed that the optical-electrical performance of the thin film photovoltaic window was optimal when the light transmittance was 30%.

According to previous research, most of the simulations of photovoltaic double-layer windows are light-electrical studies or thermal-electrical performance studies. There are fewer light- thermal-electricity simulations of STPV windows.

In the hot summer and cold winter climate zone, the presence of solar windows will greatly affect the solar energy through the windows, and affects the load conditions at different times, which is mainly shown by decrease of the cooling load in summer and the increase of heating load in winter. At the same time, photovoltaic windows will also reduce the natural lighting in the room when they block solar radiation, which will increase the lighting load. STPV windows can also have effects on indoor loads due to self-heating, which affects the power generation efficiency. Taking into account the influence of double crystalline silicon photovoltaic windows on indoor load, daylighting, and power generation, the energy savings compared with conventional double-glazed windows are studied.

Therefore, in light of the coupling mechanism of light-heat-electricity, the simulation of double-layer crystalline silicon PV windows and outdoor measurements were conducted in hot summer and cold winter regions. A three-factor model was established, and the simulation model was verified using the test results. The simulation results for the room with double-layered crystalline silicon photovoltaic windows were compared with the simulation results for the room with ordinary double-glazed windows to study the energy saving potential of the STPV. The development of this research

IOP Conf. Series: Materials Science and Engineering **556** (2019) 012009 doi:10.1088/1757-899X/556/1/012009

is very important in theoretical research and has practical application value.

2. Methodology

The translucent double-layer crystalline silicon photovoltaic window was used as the research object. Its structure is shown in Fig. 1. The common double hollow window structure is similar to it, except that the photovoltaic glass layer is replaced with a 4mm ordinary glass layer. In order to simulate the overall performance of the double-layer crystalline silicon window in hot summer and cold winter regions, a light-heat-electricity three-factor model was established using the lighting model, heat transfer model and Sandia model in EnergyPlus, as shown in Fig.2. The room model is shown in Fig.3. The temperature of the glass panel, the indoor illumination, and the power generation performance of the photovoltaic window were tested. The test results were used as verification of the simulation model.

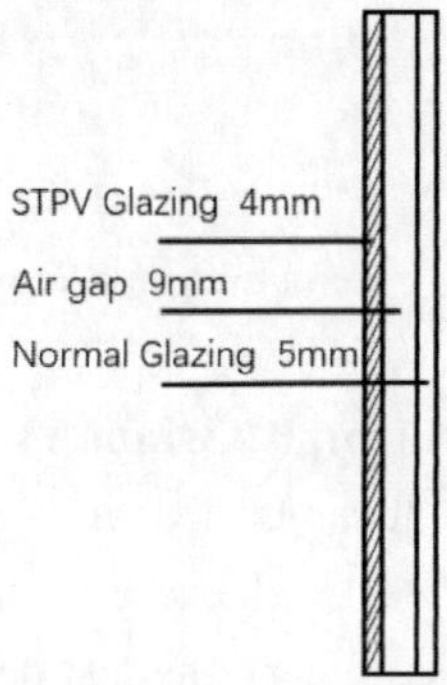

Fig.1 structure of STPV

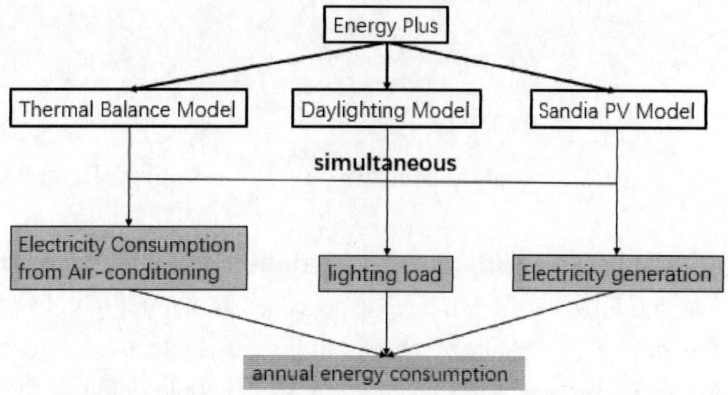

Fig.2 Light-Heat-Electricity model in EnergyPlus

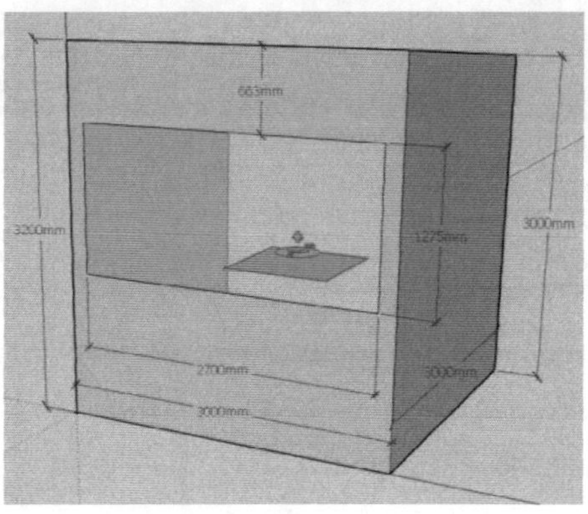

Fig.3 Room Model

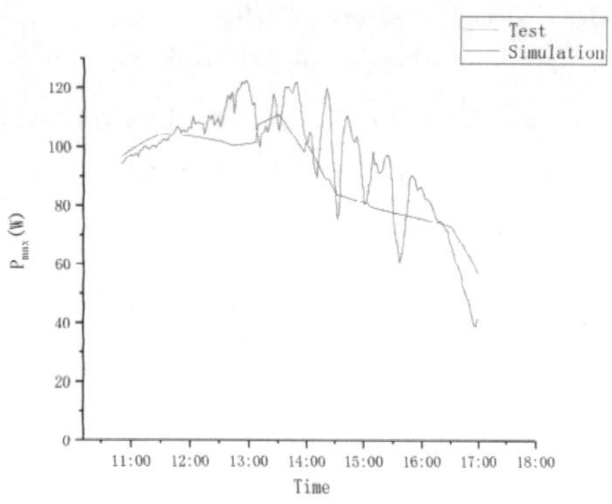

Fig.4 Electricity Generation Power Comparison

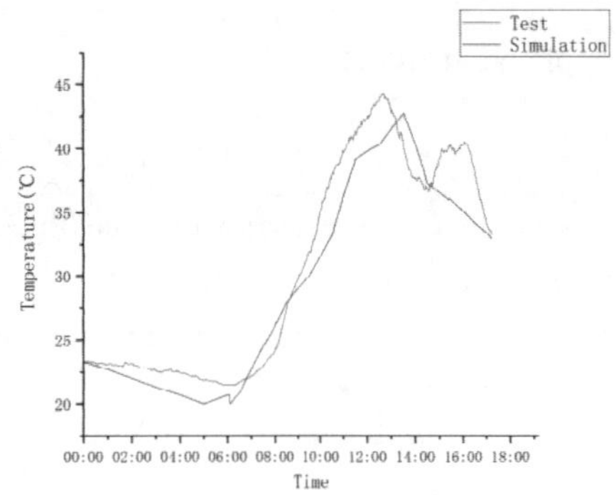

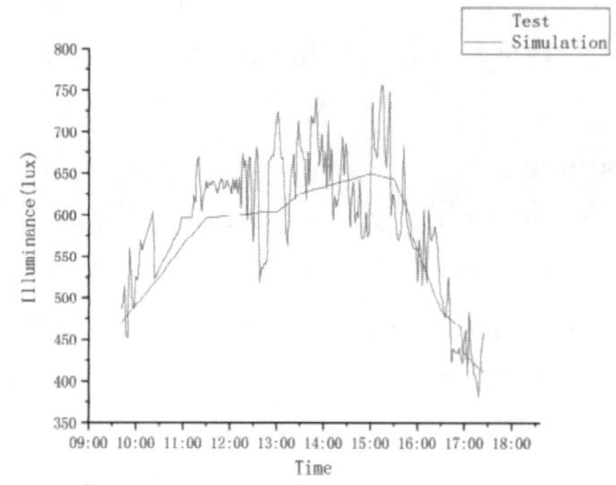

Fig.5 Temperature comparison

Fig.6 Reference Point Illuminance comparison

3. Simulation results and test validation of double-layer crystalline silicon PV windows

The model is verified in three aspects, as shown in Fig.4, Fig.5 and Fig.6. It can be seen in Fig.4 that the power of simulated photovoltaic panel electricity generation and test value are within the acceptable range. After calculation, the average value of the power generated by the test data and the average of the simulation results are within 15%. From Fig.5, the measured value of the photovoltaic window outer surface temperature is close to the simulated value. This shows that the window heat transfer conditions in the simulation match the test. From Fig.6, the simulation results are similar to the test results. The calculation results show that the difference between the simulation results and the test results are within 15%. Through the above experimental verification, it can be shown that the simulation model can simulate the real operation of the room. Therefore, the validated simulation model is used for annual energy analysis.

4. Annual Energy Consumption

The validated simulation model above was used to study the overall performance of the two-layer crystalline silicon window and the lab's annual energy consumption. The simulation result of room with STPV window and room with ordinary window is shown in Fig.7 and Fig.8.

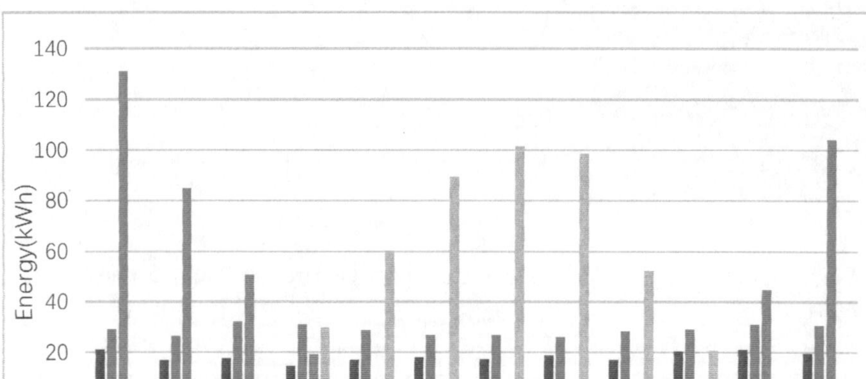

Fig.7 STPV Window: Energy consumption and generation

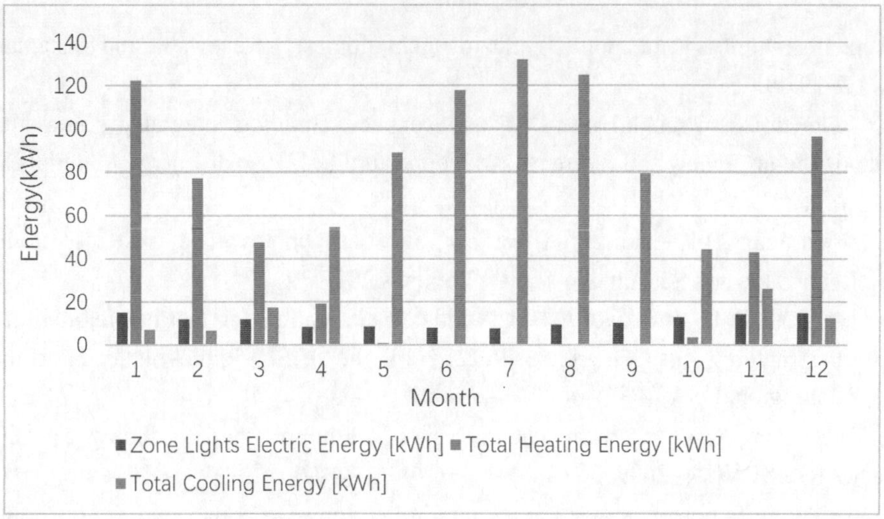

Fig.8 Common Window: Energy consumption

From the monthly energy consumption shown in Fig.7, the energy required for lighting in winter is generally greater than that in summer, but the difference is not significant. Mainly due to the relatively large window-to-wall ratio, natural lighting can meet indoor lighting requirements during the day with the photovoltaic window in 20% transmittance.

Comparing the annual energy consumption of a room with a STPV window and a room with an ordinary double-layer window, it can be seen that the difference between the thermal loads is 30.14 kWh, the difference between the cooling loads is -244.9 kWh, and the difference in lighting power consumption is 88.23 kWh. The total energy consumption of a room with a STPV window and a room with an ordinary double-glazed window is compared. The former is 1125.6 kWh and the latter is 1252.13 kWh. In this case, the room with STPV window has a decrease of 126.53 kWh. The reduction of the cooling load caused by blocking the solar radiation can completely offset the increase of the heat load and the increase of the lighting load brought about by the same time.

5. Conclusion

According to the simulation results, the effects of STPV window in light-heat-electricity aspect in the hot summer and cold winter regions should be taken into account, especially on the cooling load. According to the results of the annual energy consumption simulation, the decrease of cooling load can offset the increase of lighting energy consumption and heating energy consumption. Subtracting PV power generation, the annual comprehensive energy consumption is 779.11kWh. The annual energy consumption of a room with an ordinary window is 1252.13 kWh. The energy savings from photovoltaic windows is 476.03 kWh. The energy savings of the simulation results were converted into standard coal and carbon dioxide. The use of double-layer STPV windows can reduce 466.5 kg of carbon dioxide and save about 190.4 kg of standard coal, and the energy saving rate was 38%. Therefore, photovoltaic windows contribute a lot energy saving potential in hot summer and cold winter regions.

6. Reference

[1]Nikolaos Skandalos,Dimitris Karamanis. PV glazing technologies[J]. Renewable and Sustainable Energy Reviews,2015,49.

[2]Bjørn Petter Jelle,Christer Breivik,Hilde Drolsum Røkenes. Building integrated photovoltaic products: A state-of-the-art review and future research opportunities[J]. Solar Energy Materials and Solar Cells,2012,100.

[3]J. Siecker,K. Kusakana,B.P. Numbi. A review of solar photovoltaic systems cooling technologies[J]. Renewable and Sustainable Energy Reviews,2017,79.

[4] Yang H X, Marshall R H, Brinkworth B J. Validated simulation for thermal regulation of photovoltaic wall structures[A]. Conference Record of the IEEE Photovoltaic Specialists Conference[C], Washington, USA, 1996, 1453 – 1456.

[5] Pei Gang , Ji Jie , Jiang Aiguo . Performance analysis of PV double-glazed window [J]. ACTA ENERGIAE SOLARIS SINICA , 2009, 30 (4): 441 – 444.

[6]Jong-Hwa Song,Young-Sub An,Soek-Ge Kim,Sung-Jin Lee,Jong-Ho Yoon,Youn-Kyoo Choung. Power output analysis of transparent thin-film module in building integrated photovoltaic system (BIPV)[J]. Energy & Buildings, 2008,40(11).

[7]Jong-Ho Yoon,Se-Ra Shim,Young Sub An,Kwang Ho Lee. An experimental study on the annual surface temperature characteristics of amorphous silicon BIPV window[J]. Energy & Buildings,2013,62.

[8] Li Meia, David Infield. Thermal modelling of a building with an integrated ventilated PV façade. Energy and Buildings 35 (2003) 605–617.

[9] Re´mi Charron, Andreas K. Athienitis. Optimization of the performance of double-facades with integrated photovoltaic panels and motorized blinds. Solar Energy 80 (2006) 482–491.

[10] Mehdi Shahrestani. Experimental and numerical studies to assess the energy performance of naturally ventilated PV façade systems. Solar Energy 147 (2017) 37–51.

[11]Tady Y.Y. Fung,H. Yang. Study on thermal performance of semi-transparent building-integrated photovoltaic glazings[J]. Energy & Buildings,2007,40(3).

[12]Shen Xu,Wei Liao,Jing Huang,Jian Kang. Optimal PV cell coverage ratio for semi-transparent photovoltaics on office building façades in central China[J]. Energy & Buildings,2014,77.

[13]K. Kapsis,V. Dermardiros,A.K. Athienitis. Daylight Performance of Perimeter Office Façades utilizing Semi-transparent Photovoltaic Windows: A Simulation Study[J]. Energy Procedia,2015,78.

Acknowledgments

The project is funded by National Key Research and Development Program of China: Newton Fund -China-UK Research and Innovations Bridges (No. 2016YFE0124500), the National Natural Science Foundation of China (No. 51508352) and the Key Laboratory Deep Underground Science and Engineering Foundation of Sichuan University, China (No. DUSE201702).

IOP Conf. Series: Materials Science and Engineering **556** (2019) 012010 doi:10.1088/1757-899X/556/1/012010

Study on the flow and heat transfer in a thermal shielding radiator

Feiyang Sun, Yiqiang Wu, Fubo Xie, Peng Qian, Minghou Liu*

Department of Thermal Science and Energy Engineering, University of Science and Technology of China, Hefei, Anhui, China.

Email:mhliu@ustc.edu.cn

Abstract: A thermal shielding radiator, which has an extra cuboid cavity between channels and sealplate, is proposed to achieve a large temperature difference between the hot and cold surfaces and high temperature uniformity on the cold surface. To get uniform distribution of the flow inside each channel, the thermal shielding radiator with different inlet/outlet types (namely C-, Z-, Y- and I-type) are numerically simulated. It is found that I- type has the best performance in velocity uniformity. Due to the cavity, the temperature difference between two sides of the thermal shielding radiator improves significantly. Besides, the mini-channel with equidifferent fin thickness is also proposed, and more uniform flow and the temperature distribution on the cold surface are achieved. Moreover, effect of geometry, operating parameters on the flow and heat transfer performance for both the thermal shielding radiator and the heat radiator are studied and compared.

1. Introduction

Mini-channel and micro-channel heat radiators are always used to transfer heat from the rapid advanced electronics. Micro-channel which has an astonishing heat flux always come up with a high pressure drop[1] while mini-channel has intermediate level of pressure drop and heat flux. Usually, the topic micro-channels is ranged from 10 to 200 μm while the mini-channels is ranged from 200 μm to 3 mm[2]. On the other hand, Kew[3], Ong[4] used dimensionless parameters to definite the distinction of the channel. The confinement number of 0.5 and the Eötvös number around 0.2 are used to identify micro- or mini-channel flow.

Micro-channel was firstly raised up by Tuckerman and Pease. They reported the reduction of the hydraulic diameter of channel from macro-scale to micro-scale cause the increase of heat flux[5]. Gunnasegaran[6] and Xia[7] studied the effect of geometric structures on fluid flow and heat transfer performance. Three different shapes (rectangular, trapezoidal and triangular) of microchannel heat radiators, four kinds of header shapes (triangular, trapezoidal and rectangular) were numerically studied and found rectangular shaped microchannel heat radiators showed the highest heat transfer coefficient and Poiseuille number, while rectangular head shape provided the better velocity uniformity. Chen[8] performed various inlet/outlet arrangements to investigate the flow and temperature distribution. Zhou[9] installed an experiment on the transient heat transfer of mini-channel and found that the outlet temperature became higher with the increase of the cooling water velocity. Al-Neama[10] presented a study on the influence of chevron fin structures in the mini-channel. They found that, with the decreasing of the chevron fin oblique angle, the thermal resistance reduced furtherly. Ho[11] designed a divergent rectangular mini-channel and found it had a negligible impact on the heat transfer at low flow rate.

From the literature mentioned above, micro or mini-channel has done a great contribution in heat transfer. However, it is rarely used as a cold plate to achieve the thermal shielding function. In this paper, a thermal shielding radiator with a cavity in the middle of the sink is proposed to enlarge the temperature difference between the cold and hot side.

2. Numerical modeling of mini-channel thermal shielding radiator

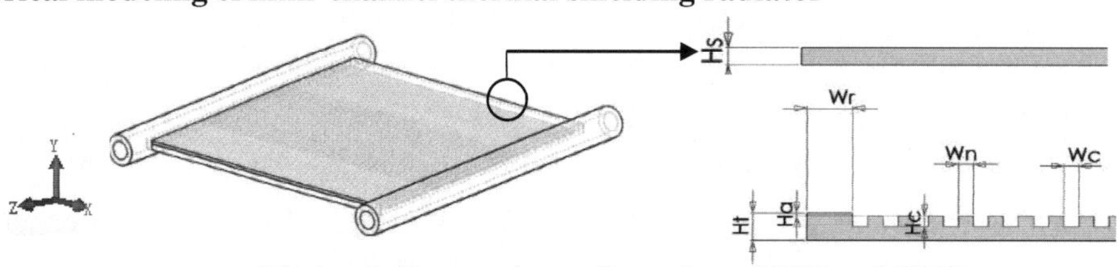

Figure 1. Geometric configuration of HIR and EHIR.

The heat insulation radiator(HIR), which has 48 channels with a cavity between the channel and the sealplate, is 100mm in length, 101mm in width and 3mm in thickness(figure 1). Cross section of traditional heat radiator and heat insulation radiator are displayed in figure 2. The equidifferent fins heat insulation radiator(EHIR) has the same size as HIR in the geometry. However, the fin width is growing as 1.05 times with the first fine width of 0.55mm. In present study, the heat insulation radiator is likely a mini-channel heat radiator which has enough height to arrange the inlet /outlet on the side wall. Four kinds of mini-channels(I、Z、Y、C) are proposed to study the influence of inlet/outlet position. The details are presented in figure 3. Geometrical dimensions are summarized in table 1.

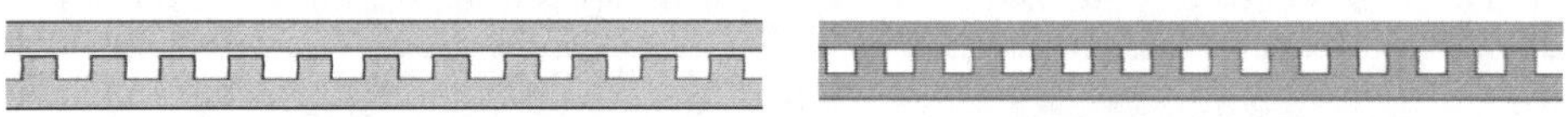

(a)HIR. (b)Mini-channel.

Figure 2. Four types of inlet/outlet arrangement.

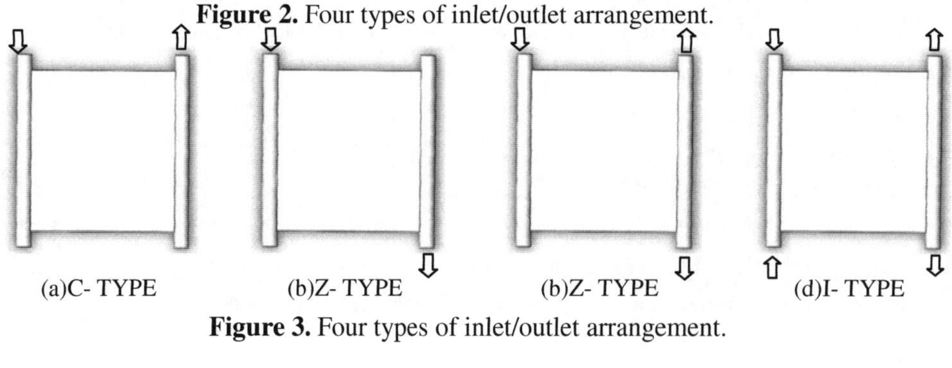

(a)C- TYPE (b)Z- TYPE (b)Z- TYPE (d)I- TYPE

Figure 3. Four types of inlet/outlet arrangement.

Table 1. Dimensions for HIR and EHIR thermal shielding radiator.

Characteristic	Size(mm)
Geometry width×length×height	100×101×3
Top plate height H_s	1
Bottom plate height H_t	2
Width between channel and side wall H_w	HIR:3; EHIR:1.7;
Fins width W_n	HIR:1; EHIR: W_n=$W_1 \cdot q^n$ [a]
Channel width W_c	1
Channel height H_c	1
Cavity height H_a	0.2

[a]W_1 =0.55, q^n1.05, n is positive integer

3. Governing equations

In present numerical study, the following assumptions are made: 1) Fluid flow and heat transfer are steady-state and three-dimensional. 2) The flow is laminar; 3) Properties of fluid and the radiator are temperature-independent. 4) Except the top plate where connected with the heat source, all the other surfaces exposed to the surroundings are heat insulated.

Based on above assumptions, the continuity, the momentum and the energy equation can be written as followings:

$$\frac{\partial u}{\partial x}+\frac{\partial v}{\partial y}+\frac{\partial w}{\partial z}=0 \quad (1)$$

$$\rho(\vec{V}\cdot\nabla\vec{V})=-\nabla p+\mu\nabla^2\vec{V} \quad (2)$$

$$\rho c_p(\vec{V}\cdot\nabla T)=k\nabla^2 T \quad (3)$$

We assemble the entrance of the insulation at z=0, and sign the channel from 1 to 48. The radiator is made by aluminum, with a rubber layer between the top plate and the bottom plate for thermal resistance. The bottom/top plate is connected with the heat source which is taken as 400 K. In the mini-channel section, the inlet velocity is 2 m/s and the inlet water temperature is 300 K. The outlet is set as pressure out.

The flow fluid is deionized water and stays 300K by thermostatic water bath. The properties of solid material are as follows: $k_s = 148\, W/(m\cdot K)$, $c_{ps} = 712\, J/(kg\cdot K)$, $\rho_s = 2329\, kg/m^3$ respectively. The properties of water in $k_f = 0.612\, W/(m\cdot K)$, $\rho = 996.5\, kg/m^3$,$c_p = 4177\, J/(kg\cdot K)$, $\mu = 943.25\times 10^6\, Pa\cdot s$.

The detail of parameter of the mini-channel is displayed in table 1. The hydraulic diameter D_h of the fluid flow channel is defined as $D_h = 4(W_c + H_c)/(2W_cH_c)$ and the Reynolds number Re is defined as $Re = \rho U D_h/\mu$, where U is the average velocity ranging from 0.68-1.37 m/s, μ is dynamic viscosity of fluid. The Re based on the hydraulic diameter and water velocity ranges from 718-1447.

To verify the independence of gird, three sets of grids (1.96× 10^6, 3.49× 10^6, 4.49× 10^6) are tested. The average and maximum temperature of top plate are316.14/314.28 K, 313.74/325.77 K, 320.63/321.98 K, respectively. It is clear that almost identical results are predicted for different grid systems. On account of the result from figure 4, the computation model with 1.96× $10^6$is used to obtain fluid flow characteristics accurately.

4. Results and discussion

4.1 flow uniformity

The thermal performance of the heat insulation is characterized based on the average top plate temperature while flow uniformity affects the temperature uniformity directly. Subsequently, Average flow flux of each channel is presented in figure 4 to show the flow uniformity. The cross section flow flux at the inlet of channel 1 to 48 revels that the inlet/outlet location of C-, Z-, Y-, I-type affects the fluid velocity. For I-type, the inlet/outlet location has higher average velocity as expected. Fundamentally, flow velocity is reassigned due to the opposite fluid flows that mix eventually. Figure 5 showed the pressure drop of four type mini-channels. It is found that the pressure drops of I-, Y-type are much smaller compared to C-, Z- type. When the fluid flows into the insulation by one side, recirculation zone can be obviously found near the center of the channel. It is because that the local pressure is smaller than surrounding pressure. Due to the uniformity of fluid flow, the I-type has been chosen for further study.

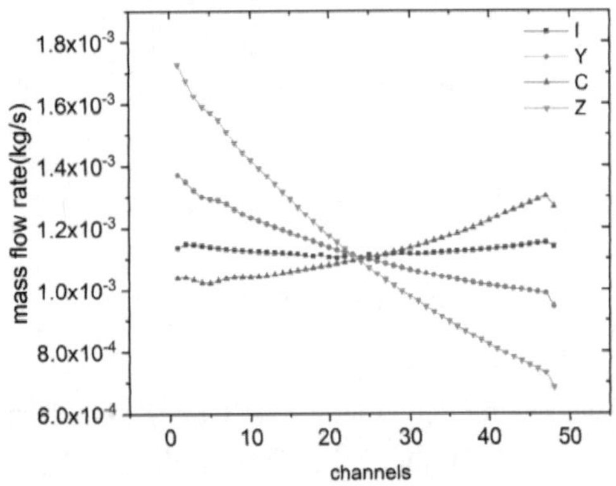

Figure 4. Comparison of the flow uniformity.

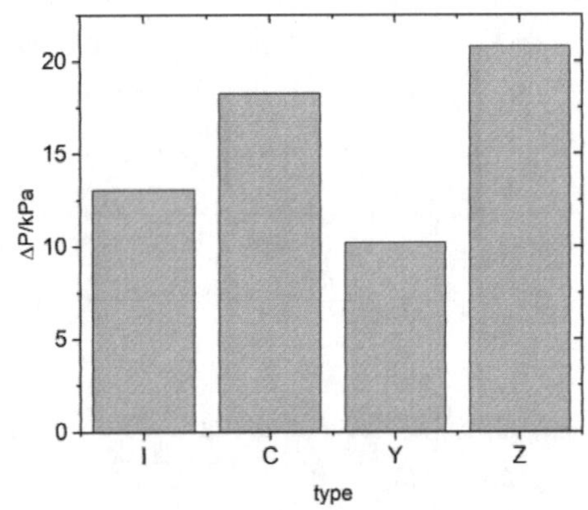

Figure 5. Comparison of pressure drop.

4.2 Temperature distribution

The top plate temperature contours of HIR, EHIR and traditional mini-channel heat radiator are shown in figure 6. The maximum, minimum, average temperature and the temperature difference between the top plate and the bottom plate are presented in table 2.

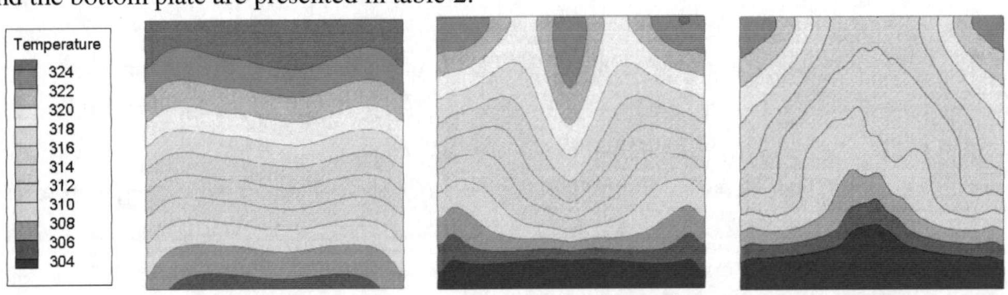

(a)mini-channel (b) HIR (c) EHIR

Figure 6. Temperature contours at the top plate.

According to figure 6, it is found the high temperature region occurs at the outlet for three type thermal shielding radiators due to temperature increasing of the fluid. The low temperature region occurs at the inlet due to the low inlet fluid temperature and the high local heat transfer coefficient. In figure 6(b), a high temperature located at the middle of the outlet region is found. This is due to the low mass flow rate at the middle channels. Figure 6(c) discloses that the equidifferent fins make the flow redistribution and eliminates the local high temperature. The table 2 showed that, compared with mini-channel, the HIR and EHIR get the lower average temperature because of the cavity used as a huge thermal resistance between the heat source and the surroundings. It can reduce the heat transfer to the surroundings and enlarge the temperature difference. In addition, the maximum temperature of HIR is higher than others due to the velocity reduces when the cross section of water area increase.

Table 2. Heat transfer performance of the threeheat radiators.

Type	T_{max}(K)	T_{min}(K)	T_{ave}(K)	ΔT(K)
Mini-channel	325.77	305.35	316.14	83.86
HIR	324.10	301.11	313.10	86.80
EHIR	323.48	301.05	312.32	87.68

4.3 Effect of heat source location

For HIR or EHIR, the heat source can be located either in seal plate side (top) or in channel side (bottom) which caused different location of the cavity. The heat dissipation, pressure drop and thermal shielding performance, characterized by the minimum, maximum or average temperature in the cold surface are shown table 3.

Table 3. The thermal shielding performance of HIR and EHIR.

Heat source	T_{max}(K)	T_{min}(K)	T_{ave}(K)	$Q(W/m^2)$	ΔP(Kpa)
HIR(top)	324.10	301.1	313.40	5.1×10^5	4.71
HIR(bottom)	335.70	301.24	322.47	7.8×10^5	4.71
EHIR(top)	323.48	301.05	312.32	4.3×10^5	4.69
EHIR(bottom)	329.26	302.29	317.62	6.3×10^5	4.69

For HIR or EHIR thermal shielding radiators, the heat dissipation is about 52.94% or 46.51% augment if heat source is located in channel side (bottom). However, the maximum temperature for HIT or EHIR increases about 11.60 K or 5.78 K when heat source is located in the seal plate side (top), respectively. For both thermal shielding radiators, the pressure drops are nearly the same. It is suggested that, for higher thermal shielding purpose, the heat source should be located in seal plate side, which can achieve larger temperature difference between the hot and the cold side. And it can get higher heat dissipation by changing the location of the heat source to the channel side.

5. Conclusions

To achieve a large temperature difference between the hot and cold surfaces and high temperature uniformity on the cold surface, a heat insulation radiator is proposed. To get uniform distribution of the flow inside each channel, the mini-channel thermal shielding radiator with different inlet/outlet types (namely C-,Z-,Y-, I types) are numerically simulated. The conclusions are summarized as:

(1) The temperature uniformity of the thermal shielding radiator cold face is affected by inlet/outlet

type due to changing of the flow distribution in the channel. Among C-, Z-, Y-, I-type of the heat radiator, the I-type shows uniform mass flow rate. For Y-, I-type, however, they can reduce the pressure drop.

(2) For higher thermal shielding purpose, the heat source should be located in seal plate side, which can achieve larger temperature difference between two sides for both HIR and EHIR. For heat dissipation purpose, however, the heat source should be located in the channel side.

(3) Due to the equidifferent fins, the flow and temperature are more uniform in EHIR.

Acknowledgments

The author sincerely acknowledge the help from the Supercomputing Center of University of Science and Technology of China.

References

[1] Qu, W. and I. Mudawar, Experimental and numerical study of pressure drop and heat transfer in a single-phase micro-channel heat sink. *International Journal of Heat and Mass Transfer,* 2002. **45**(12): p. 2549-2565.

[2]. Kandlikar, S.G.,Fundamental issues related to flow boiling in minichannels and microchannels.*Experimental Thermal and Fluid Science*, 2002. **26**(2): p. 389-407.

[3]. Kew, P.A. and K. Cornwell, Correlations for the prediction of boiling heat transfer in small-diameter channels.*Applied Thermal Engineering,* 1997. **17**(8): p. 705-715.

[4]. Ong, C.L. and J.R. Thome, Macro-to-microchannel transition in two-phase flow: Part 1-Two-phase flow patterns and film thickness measurements. *Experimental Thermal And Fluid Science,* 2011. **35**(1): p. 37-47.

[5]. Tuckerman, D.B. and R.F.W. Pease, High-performance heat sinking for vlsi. *Electron Device Letters,* 1981. **2**(5): p. 126-129.

[6]. Gunnasegaran, P., et al., The effect of geometrical parameters on heat transfer characteristics of microchannels heat sink with different shapes.*International Communications in Heat and Mass Transfer,* 2010. **37**(8): p. 1078-1086.

[7]. Xia, G.D., et al., Effects of different geometric structures on fluid flow and heat transfer performance in microchannel heat sinks.*International Journal of Heat and Mass Transfer,* 2015. **80**: p. 439-447.

[8]. Chein, R. and J. Chen, Numerical study of the inlet/outlet arrangement effect on microchannel heat sink performance. *International Journal of Thermal Sciences,* 2009. **48**(8): p. 1627-1638.

[9]. Zhou, Z., X. Xu, and X. Liang, Experiments on the transient heat transfer of minichannel heat sink under high heat flux density in an enclosed loop. *Experimental Thermal and Fluid Science*, 2010. **34**(8): p. 1409-1414.

[10]. Al-Neama, A.F., et al., An experimental and numerical investigation of chevron fin structures in serpentine minichannel heat sinks. *International Journal of Heat and Mass* Transfer, 2018. 120: p. 1213-1228.

[11]. Ho, C.J., et al., Thermal and hydrodynamic characteristics of divergent rectangular minichannel heat sinks.*International Journal of Heat and Mass Transfer,* 2018. **122**: p. 264-274.

The research of post-treatment on $Cd_{1-x}Mg_xTe$ thin films and CdTe solar cells

Ye Feng, Mingzhe Yu, Jian Huang, Xia Hao, Wei Li*, Jingquan Zhang and Lili Wu
College of Materials Science and Engineering, Sichuan University, Chengdu, China
E-mail: waylee2000@scu.edu.cn (W. Li)

Abstract. As for it higher band gap, $Cd_{1-x}Mg_xTe$ (CMT) material has been taken into consideration to improve the open-circuit voltage of CdTe solar cells as an electron reflector layer. An essential step in the development of this new interlayer is the passivation process. X-ray Diffraction (XRD) and scanning electron microscope (SEM) results give us the suitable annealing temperature range of CMT thin films from 350°C to 400°C, and suggest that the annealing process is beneficial to grain growth and recrystallization. The concentration of $CdCl_2$ solution has an influence on the crystallinity of the thin films and coating method would contribute to electrical performance of the devices. The CMT/CdTe cells which underwent the two-step annealing procedure at high purity nitrogen atmosphere showed the significant improvements in device performance. Furthermore, the CdTe solar cells with the CdTe cap layer and Te/Cu back contacts also demonstrated higher performance.

Keywords: $Cd_{1-x}Mg_xTe$, Passivation, Annealing process

1. Introduction

Cadmium telluride (CdTe) is a photovoltaic material with great development prospects over recent decades and is attracting the interest of many researchers in the area of thin film solar cells due to the incredible progress achieved by First Solar in pushing the efficiency reached 22.1%, but it still lags behind that of c-Si and GaAs solar cells [1]. Despite the records, laboratory CdS/CdTe solar cell efficiencies are still below the theoretical limits, primarily due to the poor current, fill factor and low open-circuit voltage [2]. K. Hsiao and T. Coutts [3, 4] have proposed the new ideas that develop a tandem structure or an electron reflector layer into the devices to improve the open-circuit voltage. So the $Cd_{1-x}Mg_xTe$ (CMT) material has been taken into consideration because its higher band gap and can be easily varied by changing the Mg/(Cd+Mg) fraction. CMT can be used as the absorber in higher band gap top cells in tandem junction devices and as a higher band gap layer to make the needed conduction band barrier in electron reflector (ER) structures [5]. In order to gain an in-depth understanding of CMT, researchers have carried out a series of studies. R. Dhere [6] showed the growth of CMT thin films by co-sublimation method, R. Palomera [7] did some research of the Mg incorporation and the related material properties. Furthermore, P. Kobyakov [8] discussed the passivation treatment effects on CMT thin films and found that $CdCl_2$ passivation is crucial to improve performance of CdTe solar cells which has improved device efficiency from ~1% to ~13% at Colorado State University.

We deposited CMT thin films by the co-evaporation technique and investigated the effects of annealing temperature and concentration of $CdCl_2$ solution on the properties of thin films in this study. Finally, we made some research on the post-treatment of the CdS/CdTe/CMT solar cells to find optimal passivation conditions and obtain better photovoltaic characteristics.

2. Experimental

2.1. Film fabrication
CMT thin films were prepared by co-evaporation method in this study and the Mg content is been controlled in 0.18 [9]. Substrate temperature was held at approximately 350°C and the base pressure in vacuum chamber was held below 10^{-3} Pa during deposition. The structure of CdTe/CMT thin film solar cells was glass/SnO_2: F(FTO)/CdS/CdTe/CMT/BC/Au, in which CdS window layer was deposited by chemical bath deposition method and CdTe absorber layer was deposited by vapour transport deposition method. A 80-nm-thick CMT was deposited on the back of CdTe thin films and then annealing five minutes in tubular annealing furnace after the $CdCl_2$ treatment [8]. Finally, a layer of back contact, ZnTe: Cu or Te/Cu, was deposited and Au was used as the back electrode.

2.2. Characterization
Structural characterization was performed using a DX-2600 X-ray diffractometer with Cu-Kα radiation scanning in the 10°~70° configuration and a Hitachi S-4800 field emission scanning electron microscopy (SEM) was used for imaging. The cells were analyzed by current density-voltage (J-V) characteristics used ABET technologies Sun 2000 solar light simulator under simulated AM1.5 illumination.

3. Results and discussion

3.1 The exploration of annealing temperature on CMT thin films
In order to study the effect of different annealing temperatures on the properties of the thin films, the samples were prepared under the same condition with the thickness of about 300 nm. The X-ray diffraction patterns (figure 1) indicate that the annealing temperature does not change the preferential orientation of the films and the (111) peak location [10]. However, the intensity of the (111) peak increases first and then decreases with increasing the temperature. The increase of the annealing temperature has made the grain crystallized and the preferential orientation tends to be obviously. But higher annealing temperature can damage grain growth, which degrades the film quality. As can be clearly seen from figure 2, the shape and size of the grain does not change significantly and the annealing process has no effect on the film structure when the film are annealed at 300°C~ 350°C. The grain boundary gradually becomes inconspicuous with the appearance of the pores until the temperature rises up to 400°C. As the temperature continue raising, it is obvious that the pores are dwindling and the grains become visible again due to the recrystallization of the grains during this temperature process. The experimental results show that the phenomenon of grain growth and recrystallization of the thin films is beneficial to the reduction of the grain boundary recombination. However, the film surface structure has been damaged under 500°C annealing temperature and the presence of grains was invisible at this time. Combine the analysis with figure 1, it is presumed that the optimum annealing temperature should be between 350°C and 450°C.

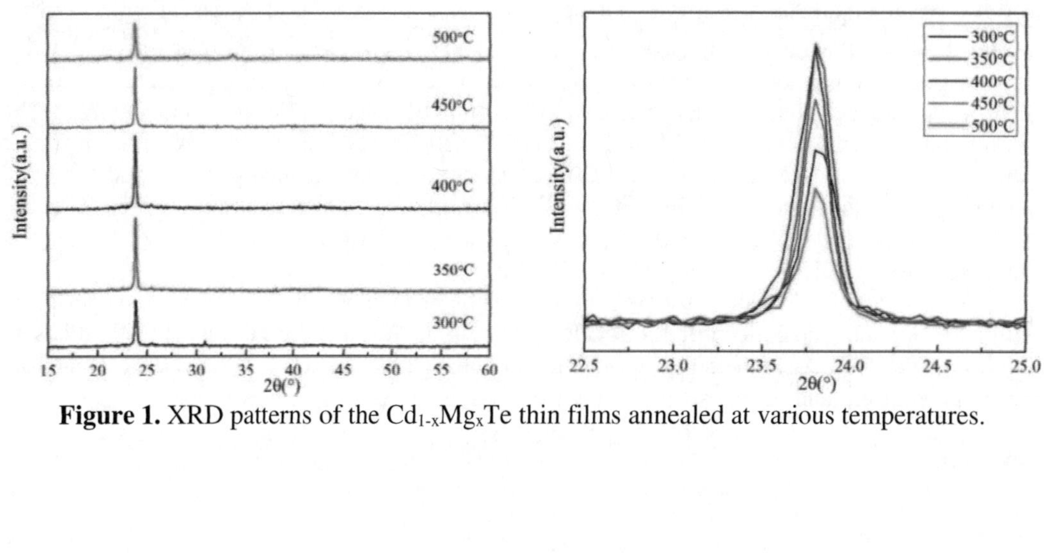

Figure 1. XRD patterns of the $Cd_{1-x}Mg_xTe$ thin films annealed at various temperatures.

IOP Conf. Series: Materials Science and Engineering **556** (2019) 012011 doi:10.1088/1757-899X/556/1/012011

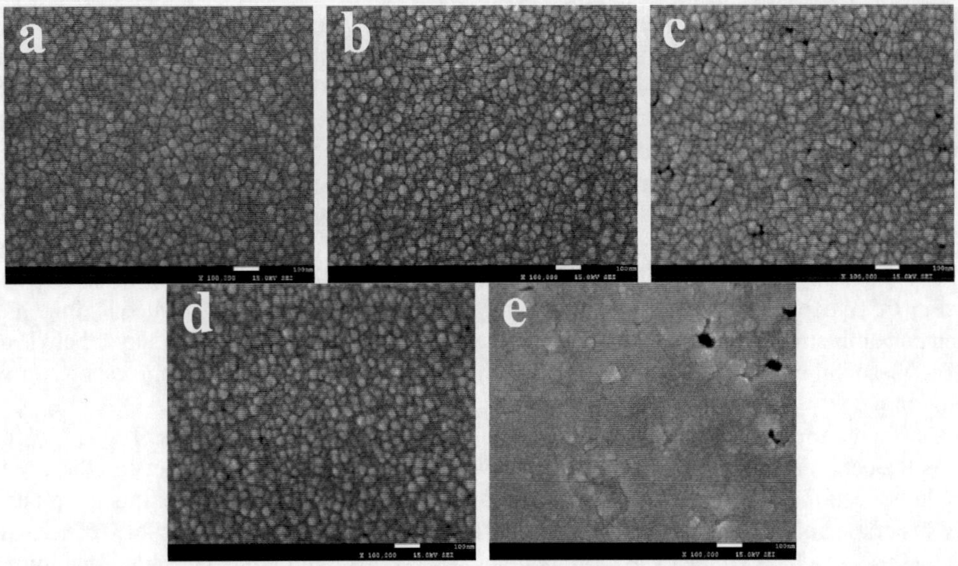

Figure 2. SEM images of $Cd_{1-x}Mg_xTe$ thin films annealed at various temperatures: (a) 300°C, (b) 350°C, (c) 400°C, (d) 450°C and (e) 500°C.

3.2 The $CdCl_2$ post-processing of CMT thin films

It can be seen from the figure 3 that the films only have a strong diffraction peak at 23.75^0 without the $CdCl_2$ treatment, which corresponds to CdTe (111) planes. On the contrast, the relatively weak diffraction peaks appeared at 27.6^0 and 38.4^0, which are also from CdTe thin films after annealing at 350°C. Then the 39.4^0 and 49.5^0 peaks are originated from Te (154) or Te (152) diffractions when the annealing temperature increased up to 400°C. However, the films without $CdCl_2$ treatment have no diffraction peak at these 2θ positions as shown in the JCPDS card (15-0770), indicated that $CdCl_2$ treatment is beneficial to grain recrystallization. In addition, it is also observed that the (111) peak intensity is not strong at the lower $CdCl_2$ solution (0.03 and 0.06 mol/L) treatment. But it becomes stronger with the increase of the $CdCl_2$ concentration and the diffraction peak intensity is approximately the same as that of the films without the $CdCl_2$ treatment, which indicates that the post-treatment has an effect on the crystallinity of the thin films. In a certain range, the greater the concentration of $CdCl_2$ solution, the higher the crystallinity of the thin films.

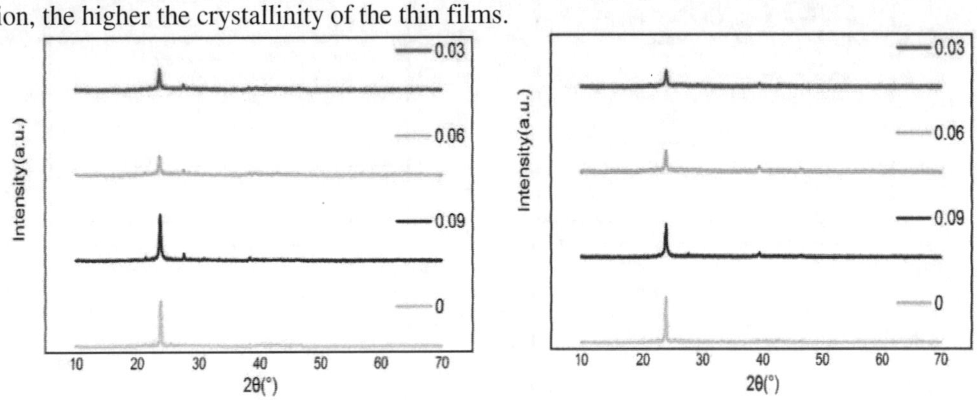

Figure 3. XRD patterns of the $Cd_{1-x}Mg_xTe$ thin films annealed at 350°C(left) and 400°C(right) with various $CdCl_2$ concentrations.

3.3 Current-voltage (J-V) characteristics of CdTe solar cells after post-treatments

In this part of the experiment, the same color of the box chart represents the nearly identical configuration cell. When CMT deposited on the back of CdTe absorber thin films, the $CdCl_2$ treatment need to be carried out before the cell annealing. Firstly, we made a comparison of two $CdCl_2$ processing methods, as the red box chart shown. The cell that has been coated with $CdCl_2$ has better J-V

IOP Conf. Series: Materials Science and Engineering **556** (2019) 012011 doi:10.1088/1757-899X/556/1/012011

performance while the spraying method exhibits breakdown characteristics of PN junction and its J-V curve is almost linear. The possible reason for this phenomenon is that the CMT is a kind of matierial which is easily oxidized, the film surface in the air during the spraying procedure while the solution coated on surface of the film make the isolation environment from the air to a certain extent. In order to reduce CMT oxidation, the addition of a thin CdTe cap layer was introduced into the back of the CMT thin films corresponding to the orange box chart as shown in figure 4 [11]. Another kind of protection method is to deposit a thicker CMT layer and then etching the oxidized part before the back contact deposition. It can be seen that two methods have their pros and cons. The latter technique exhibits the optional J-V performance in the efficiency and fill factor while the other prevails in the open-circuit voltage. For CdTe cap, the fill factor is reduced due to the increased series resistance and the presence of oxygen contamination in the film during the reheating which form an extra oxide layer between CMT and CdTe. Meanwhile, the addition of CdTe cap in the open circuit voltage improment is a worthy consideration [12].

ZnTe:Cu is a mature back contact material in CdTe solar cells. To some content, Te/Cu back contact also shows the better side. The green box chart show the effect of two back contacts on the CMT solar cells. It can be seen that the Te/Cu back contact show the better J-V performance, this is probable that the CMT material can form a larger valence band offset (VBO) blocking the holes and reducing the fill factor of the device, a back contact with a higher electron affinity and doping is needed to minimize the VBO formation and the tellurium is a potentially ideal contact for CMT cells. It is theorized that the holes may diffuse from the tellurium layer into the CMT layer moving the bands and minimizing the VBO [13, 14].

Then, we explored the effect of two annealing process: one-step procedure and two-step procedure. $CdCl_2$ annealing treatment before the deposition of CMT can promote the recrystallization and the diffusion processes of sulfur from the CdS [15]. Then $CdCl_2$ dealed with the cell again after the preparation of CMT aimed as passivating the grainboundary and facilating the carrier separation. For this count, two-step annealing procedure enhanced the higher cell performance which shown by the blue in figure 4. Pink box deliver the point that high purity nitrogen atmosphere would build a better environment than low vacuum condition during the annealing process. In addition, we can also choose the other inert gases, such as argon to act as protective gases to prevent the external environment from affecting the cell performance during this procedure.

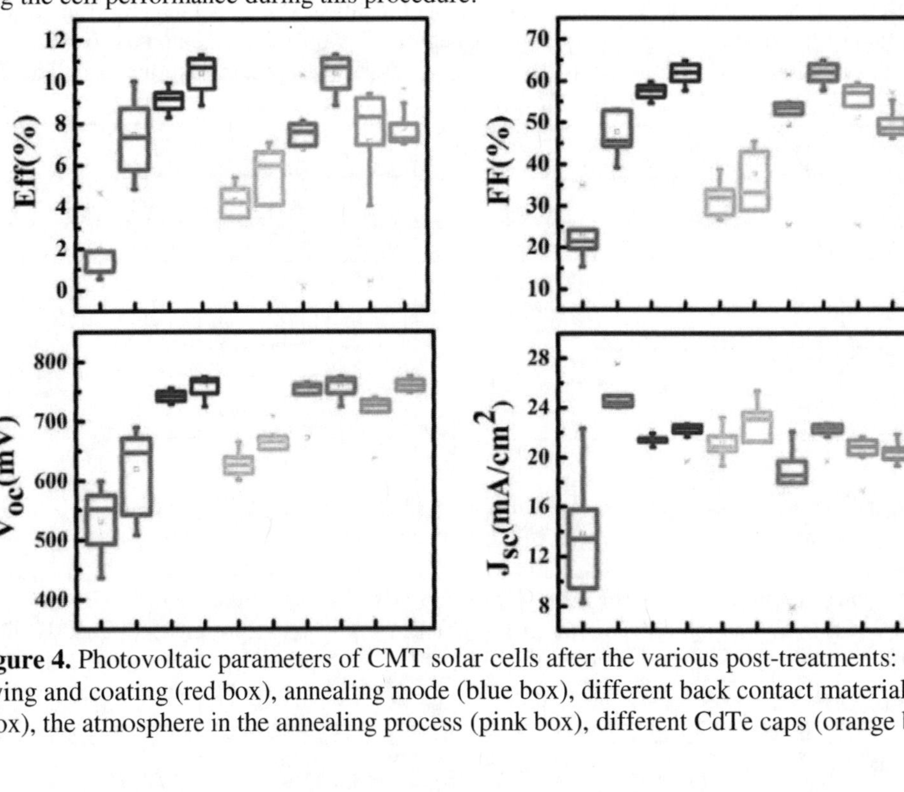

Figure 4. Photovoltaic parameters of CMT solar cells after the various post-treatments: $CdCl_2$ spraying and coating (red box), annealing mode (blue box), different back contact materials (green box), the atmosphere in the annealing process (pink box), different CdTe caps (orange box).

4. Conclusion

Based on the research of the CMT thin-film post-treatment process, the following conclusions have been drawn. The optimum annealing temperature of CMT films was obtain between 350°C and 450°C. The $CdCl_2$ treatment could promote the grain recrystallization during the annealing process and it had an appropriate concentration in order to obtain the higher crystallinity of the CMT thin films. A post-treatment using two-step annealing at high purity nitrogen atmosphere is an alternative passivation for CdTe based photovoltaic cells. It is also found that the $CdCl_2$ coating process can obtain better performance in the devices. Furthermore, the CMT/CdTe cells with a CdTe cap layer and Te/Cu back contact show significant improvement in device performance.

Acknowledgement

This work was supported by the Science and Technology Program of Sichuan Province, China (Grant No. 2017GZ0414).

References

[1] Green M A, Y Hishikawa, E D Dunlop, D H Levi, J Hohl-Ebinger 2018 *Progress in Photovoltaics: Research & Applications* **26** 3-12
[2] Morales-Acevedo 2006 *Sol. Energy* **80** 675-681
[3] Coutts T J, J S Ward, D L Young, K A Emery, T A Gessert 2003 *Progress in Photovoltaics Research & Applications* **11** 359-375
[4] Hsiao K J, J R Sites 2012 *Progress in Photovoltaics: Research & Applications* **20** 486-489
[5] Hsiao K J, J R Sites 2009 *IEEE 34th Phot. Spec. Conf. (Pvsc)* 1846-1850
[6] Dhere R, K Ramanathan, J Scharf, D Young 2007 *Mrs Proceedings* 1012
[7] Palomera R C, O S Martinez, J Pantoja-Enriquez, X Mathew 2017 *Appl. Therm. Eng.* **114** 1169-1175
[8] Kobyakov P S, D E Swanson, J Drayton, J M Raguse, K L Barth, W S Sampath *2013 IEEE 39th Phot. Spec. Conf. (Pvsc)* 386-391
[9] Kobyakov P S, A Moore, J M Raguse, D E Swanson, W S Sampath 2014 *J. Vac. Sci. Technol. A.* **32** 021511-021517
[10] Shimpi T M, J M Kephart, D E Swanson, A H Munshi, W S Sampath 2016 *J. Vac. Sci. Technol. A.* **34** 051202
[11] Reich C, D Swanson, T Shimpi, J Drayton, A Munshi 2016 *IEEE Phot. Spec. Conf. (Pvsc)* 487-491
[12] Sites J, A Munshi, J Kephart, D Swanson, W S Sampath 2016 *IEEE Phot. Spec. Conf. (Pvsc)* 3632-3635
[13] Swanson D E, A Abbas, A H Munshi, J A Drayton, J M Raguse 2015 *Mrs Proceedings* **1771** 133-138
[14] Zhao Y, M Boccard, S Liu, J Becker, X-H Zhao 2016 *Nature Energy* **1** 16067
[15] McCandless B E, L V Moulton, R W Birkmire 1997 *Prog. Photovoltaics* **5** 249-260

IOP Conf. Series: Materials Science and Engineering **556** (2019) 012012 doi:10.1088/1757-899X/556/1/012012

Feasibility study of $MgSO_4$ + zeolite based composite thermochemical energy stores integrated with vacuum flat plate solar thermal collectors for seasonal thermal energy storage

D Mahon[1], P Henshall[2], G Claudio[1] and P Eames[1]
[1]Loughborough University, Loughborough, UK
[2]Oxford Brookes University, Oxford, UK
E-mail: d.mahon@lboro.ac.uk

Abstract. A primary drawback of solar thermal technologies, especially in a domestic setting, is that collection of thermal energy occurs when solar irradiance is abundant and there is generally little requirement for heating. Thermochemical Energy Storage (TCES) offers a means of storing thermal energy interseasonally with very little heat loss. A combination of Solar Thermal Collectors (STC) and TCES systems will allow a variety of different heating applications, such as domestic space and hot water heating as well as low temperature industrial process heat applications to be met in a low carbon way. This paper describes and assesses the feasibility of two novel technologies currently under development at Loughborough University; i) an evacuated flat plate STC and ii) composite TCES materials, coupled together into a system designed to store and supply thermal energy on demand throughout the year. The predicted performance of an evacuated flat plate STC is described. The objective of this paper is to evaluate the economic, energy and carbon saving potential of conceptual STC + TCES systems suitable for domestic use. This research uses experimental results from Differential Scanning Calorimeter tests to evaluate the total enthalpy, dehydration enthalpy and sensible component enthalpy of composite TCES materials. The experimental results along with predicted performance of STC are used within a developed model to assess key metrics of conceptual STC + TCES systems feasibility, including; charging time, payback time, cost/ kWh, energy savings and CO_2 savings. Preliminary results suggest the combination of these two technologies has significant potential for domestic applications.

Keywords: solar; energy; storage; thermochemical; solar-collector; feasibility

1. Introduction

Nearly half of the UK's total energy consumption is used for heating purposes [1], with 26% of the UK's total energy consumption used specifically for Domestic Space Heating (DSH) and Domestic Hot Water (DHW)[1]. 88% of the energy for DSH and DHW comes directly from gas and oil with only 2% of the energy required for heating generated from renewable energy sources [1]. Thermochemical Energy Storage (TCES) stores thermal energy in reversible chemical reactions and can be used to help decarbonise the UK's DSH and DHW by storing summer generated thermal energy for use in the winter time to meet demand. $MgSO_4$ is an abundant, non-toxic, relatively cheap salt hydrate, with a high theoretical energy density ($2.8GJ/m^3$ / $778kWh/m^3$) that can be used as a thermochemical energy storage material [2]. $MgSO_4$ is interesting for domestic interseasonal TCES as it will dehydrate (charge) when exposed to a temperature of 150°C [2]. There has been some recent research into the potential of $MgSO_4$ [2,3] however, studies have shown problematic characteristics of such material. Zeolite is an absorbent material which has the potential to be used as a standalone

TCES material [4]. Due to the porous structure and typically large surface area zeolites tend to be very adsorbent [5]. Zeolites have been specifically used to enhance the characteristics of $MgSO_4$ [6].
STC are commonly used to heat water to be used for DSH and DHW supply. They usually come in two conventional non-concentrating varieties. One type, known as Flat Plate Collector (FPC) STC employs a thin metal sheet with a selective surface as a solar absorber, which fills a large proportion (>90%) of the gross collector area [7]. A FPC loses energy directly from the absorber via conduction, convection and radiation heat loss mechanisms. The other type of STC is an Evacuated Tube Collector (ETC) STC which has an absorber that is divided so that each section of the absorber can fit inside a glass tube. The glass tube is evacuated surrounding the absorber by a high vacuum. This results in the suppression of both convection and gas conduction heat loss mechanisms. ETC's are capable of achieving higher operational temperatures and higher efficiencies in terms of absorber area compared to FPC's [8]. However, ETC's have less absorber area per gross collector area in comparison to FPC's. A recent innovation in STC technology has been the development of Vacuum Flat Plate Collectors (VFPC) STC [9] which combine the benefits of FPC's and ETC's via the use of a flat and robust enclosure that surrounds a Flat Plate solar absorber. The FPC efficiency is improved thanks to the thermal insulation properties of the surrounding vacuum layer; similar to an ETC.
Composite materials of zeolite-Y containing various wt% of $MgSO_4$ have been created and experimentally characterized using a Differential Scanning Calorimeter (DSC). The generated DSC data was used to calculate the feasibility of a combined TCES and STC system for domestic inter-seasonal thermal energy storage and annual heat delivery. This study presents the potential of a TCES and VFPC thermal energy storage system for providing DSH and DHW energy. The proposed system has the advantage of offering a secure supply of heat and can potentially be used off-grid.

2. Materials and methods

To assess the energy density of the TCES materials a TA Instruments Discovery DSC was utilised. For the DSC tests the sample mass used was 6-10mg. The DSC is not configured to provide a humid air flow; therefore hydration of the samples took place in a custom built microcontroller regulated hydration chamber. The hydration conditions, were ~56% (+/- 3%) Relative Humidity (RH) at 20°C (pH_2O =~1.3kPa) for a minimum of 18 hours. The DSC samples were dehydrated to 150°C using a double DSC dehydration approach, (explained in more detail in a paper by Mahon et al. [10]), to establish i) the effective specific heat capacity, ii) the sensible enthalpy component and iii) the dehydration enthalpy of the TCES samples.

3. Solar collector information

The useful energy gain (Q) from a STC is given generally by equation (1) [11]:

$$Q = A[S - U_L(T_C - T_a)] \quad (1)$$

where A is the area of the absorber, S is the absorbed solar energy per unit area, U_L is the thermal loss coefficient of the collector, T_C is the collector temperature and T_a is the ambient temperature. The efficiency of a STC in terms of aperture area (η_a) can be evaluated by equation (2):

$$\eta_a = \frac{Q}{A_a G} \quad (2)$$

where G is the local solar irradiance and A_a is the area of the aperture. The efficiency in terms of gross collector area (η_c) can be evaluated by equation (3):

$$\eta_c = \frac{Q}{A_c G} \quad (3)$$

where A_c is the gross collector area. From equations (1-3), for a given solar irradiance, the efficiency of a STC decreases as the temperature of the collector increases or if A_c is used to calculate efficiency rather than A_a (as $A_c > A_a$). In order to model and compare the three types of domestic STC (FPC, ETC, VFPC) previously discussed, the characteristics of two commercial STC's were identified: a FPC [7] and a ETC [12]. The characteristics of a VFPC in this study was derived from the characteristics of both these commercial collectors in that the VFPC has a similar U_L to the ETC whilst having a similar A_a/A_c ratio as the FPC. The area available for the STC was limited to ~8m^2 with the values of A_a and A_c scaled appropriately. Efficiency curves for commercial collectors are characterized by equation (4):

$$\eta = \eta_0 - k_1 \frac{(T_c - T_a)}{G} - k_2 \frac{(T_c - T_a)^2}{G} \tag{4}$$

where the values of constants η_0, k_1 and k_2 are given by manufacturers in terms of aperture area. The characteristics of the STC's used in the model can be found in table 1.

Table 1. STC properties in terms of aperture area.

Collector	A (m²)	A_a (m²)	A_c (m²)	η_0	k_1 (W/m²K)	k_2 (W/m²K²)
FPC [7]	2.01	1.97	2.15	0.775	3.73	0.0152
ETC [12]	2.01	2.16	2.77	0.75	1.18	0.0095
VFPC	2.01	1.97	2.15	0.75	1.18	0.0095

4. Feasibility analysis and assumptions

The TCES material in the system was assumed to be charged over the 3 months of summer by the STC and to be discharged throughout the winter using pumped humid ambient air. The maximum amount of TCES material which could be charged was calculated from the summer gains produced by the STC. The remaining 9 month gains produced by the STC were assumed to be utilised when generated to provide DSH or DHW. The calculated average domestic energy consumption for space heating was 14,373 kWh [1]. To calculate the Winter Space Heating Demand (WSHD) average degree days for the UK were used and the WSHD was calculated to be 5,737 kWh. The value used for the average CO_2 production (kg/kWh) from current DSH energy sources was 0.23kg/kWh, calculated from taking averages of the CO_2 production from each DSH energy source and then calculating a weighted average depending on the % each energy source was used for DSH. To calculate the yearly useful gain from the STC when they were not charging the TCES (i.e. the remaining 9 months of the year) it was assumed that the STC's were outputting an increase in temperature of 50°C. First the efficiency of each STC was calculated using equation 4 assuming $T_c - T_a$ equals 50°C, the value used for G was an average hourly irradiance value for Loughborough, UK [13]. Once the efficiency was calculated the useful kWh gains from the STC were calculated. The effective specific heat capacity used for the TCES materials changed every 5°C with each value an average value within the 5°C band calculated from the DSC data. It was defined as the "effective" specific heat capacity as it was a combination of the specific heat capacity of the material and also the dehydration enthalpy of the material. The STC's efficiency for the summer months were calculated using equation (4) taking into account the irradiance for each specific hour for that month. The delta T when charging the store (i.e. $T_c - T_a$) was equal to (temperature of the store + 5°C) – T_a. This means that the TCES material store was always being charged with a temperature 5°C above the temperature of the store. The ambient temperature used was a 24 hour monthly average for each month [13]. The heat into the store was given by equation (5):

$$Q_{in} = 8E_e\eta \tag{5}$$

where E_e is the irradiance for that hour of the day and η is the efficiency of the collector calculated for the required delta T, using equation (4). The temperature of the store at a given time was described by equation (6):

$$T_i = T_{i-1} + \frac{3600(Q_{in} - Q_{loss})}{mC_{p(T)}} \tag{6}$$

where $C_{p(T)}$ is the effective specific heat capacity at a given temperature for the TCES material and Q_{loss} is the heat loss to the surroundings due to conduction and convection. The store temperature was calculated hourly over the summer period. The model calculated the maximum WSHD% which could be stored within the TCES material. The calculated average cost for space heating used was 5.01p/kWh. The likely increase in the average space heating energy cost (£/kWh) over the next 30 years was calculated as 7%/annum, considering the average change in fuel costs over the past 20 years, the inflation value used was 3.6%. The system life time was assumed to be 30 years. With the assumption that the system was paid in full, the system was fitted alongside the current heating system and only the savings in energy costs were used to assess the financial feasibility of the system. To assess the cost of each system configuration the cost for each component was based on bulk prices which were found from several sources and then averaged.

5. Results and discussion

To alleviate some of the $MgSO_4$ issues composite materials of $MgSO_4$ + zeolite-Y were synthesized in the laboratory and tested. Figure 1 shows the average dehydration enthalpy for each composite material measured from the DSC dehydration experiments from 20-150°C. The values on Figure 1 were found by conducting a sigmoidal integration of the DSC enthalpy plots. The dehydration enthalpy for each material was derived from the total enthalpy minus the sensible component. With increasing $MgSO_4$ wt% the dehydration enthalpy of the composite samples increased. As the sample will be dehydrated (charged) in the summer months when, typically, heat is not required all of the sensible heat which is stored within the TCES material has been assumed to be lost and not used. The sensible component for the 35wt% and 15wt% composite materials was approximately 17% and 20% of the total enthalpy, respectively.

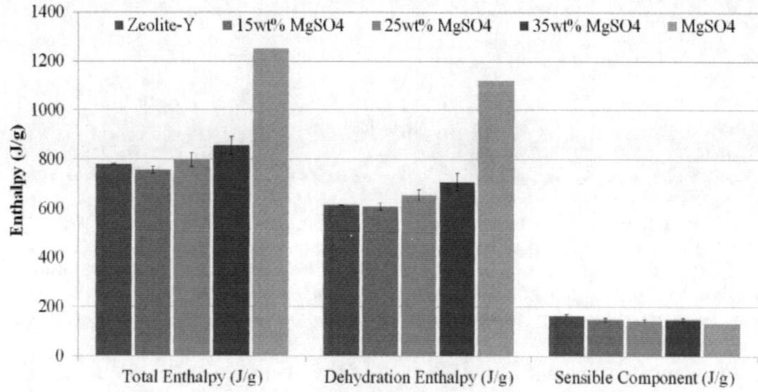

Figure 1. DSC measured average composite: total enthalpy, dehydration enthalpy and sensible component.

Table 2 shows the properties of each of the TCES materials used for the model.

Table 2. The properties of each of the TCES materials used for the model.

Material	$MgSO_4$-80°C	$MgSO_4$-150°C	35wt%-80°C	35wt%-150°C	Zeolite-150°C
Dehydration enthalpy (J/g)	484	1118	302	708	615
Density (kg/m^3)	2666	2666	1453	1453	800

Figure 2 shows the initial summer charging time of each of the TCES + STC configurations modelled. The key on figure 2 shows the volume and the WSHD% stored, within only the TCES material after the initial summer charge, for each system. The system which stored the highest WSHD% (23.4%) is the $MgSO_4$ system charged to 150°C using a VFPC ($MgSO_4$-150°C-VFPC). The store volume for this system was 1.62m^3. In comparison the 35wt%-80°C-VFPC system was able to store 19.3%, the second highest WSHD% with a required store size of 9.1m^3 making this system less likely to be used in a domestic environment. The FPC systems are not able to dehydrate any material to 150°C due to the low efficiency of the STC at high temperatures.

The most financially attractive system configuration was the $MgSO_4$-150°C-VFPC system, which also stores the most energy. The payback time for this system was 22 years. Over the lifetime of this system it should save the user over £4,300. Only five of the system configurations were financially viable. The financially viable systems were each of the VFPC systems, minus the 35wt%-80°C-VFPC

system, and the $MgSO_4$-150˚C-ETC system. The 35wt%-150˚C-VFPC system had a payback time of 26 years, meaning this system would also result in financial savings to the user.
The initial cost per kWh (£/kWh) is the amount of energy stored from each system over its 30 year lifetime divided by the cost of the system. In all cases the ETC had a lower £/kWh cost than the FPC systems. The VFPC systems had an energy cost of (0.05-0.09£/kWh) and the ETC and FPC systems an energy cost of (0.08-0.16£/kWh). The more competitive VFPC systems had an energy cost (£/kWh) which is competitive with the current average energy cost (0.05£/kWh) for current space heating energy sources in UK.

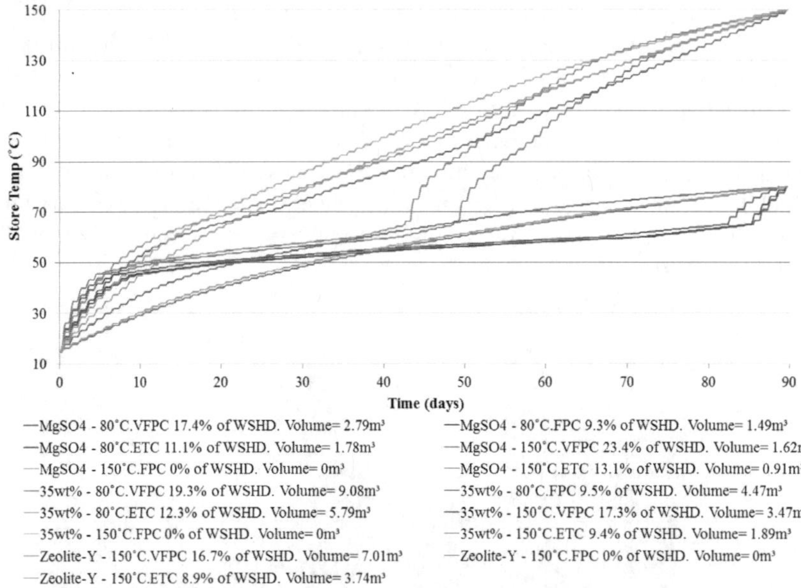

Figure 2. The store temperature with time for each of the proposed systems during initial charging.

For this investigation each of the systems were compared assuming it was beneficial to store the highest amount of WSHD% demand possible in the TCES material over the 3 summer months of charging. The WSHD% increased with the CO_2 saving however, the £/kWh did not necessarily decrease due to the material costs. For the 35wt%-150˚C-VFPC system the initial £/kWh of the system increased with increasing WSHD%. The 35wt%-150˚C-VFPC system was able to store a maximum of 17.3% WSHD%. The initial £/kWh of this system was 6.4p/kWh. If the WSHD% stored was decreased to 5% the initial £/kWh reduced to 6.0p/kWh. The initial payback time for this system was 26 years and with a store of 5% WSHD could potentially reduce to 25 years. Apart from the CO_2 savings of the system, storing the maximum WSHD% for this material does not represent a clear financial gain.
The average energy output of each system over its 30 year lifetime was calculated. The VFPC + TCES systems were able to meet around 30% of the total yearly Space Heating Demand (SHD) (4,312kWh/year) whilst each system using a FPC + TCES or an ETC + TCES were only able to meet less than 20% of the yearly SHD (2,875kWh/year). The choice of TCES material used in the VFPC systems did not have a significant impact on the amount of energy output from the systems, the best VFPC system ($MgSO_4$-150˚C-VFPC) and the worst VFPC system (Zeolite-Y-150˚C-VFPC) had an 8% annual difference in energy output. The selection of the TCES material should be a function of the available domestic storage volume and TCES material characteristics which in turn could favour the implementation of composite materials designed to reduce the agglomeration of the $MgSO_4$ while maintaining a high energy density.
The cumulative CO_2 savings were calculated for each system configuration assuming that a system was installed in 10% of all UK households. As the CO_2 savings were directly linked to the amount of energy output from each system the VFPC systems saved the most CO_2 over their lifetime (82.8-89.3MtCO_2e). The UK's CO_2 production from the residential sector for 2013 and 2014 was

74.4MtCO_2e and 62.3MtCO_2e , respectively [14]. This means a possible reduction of ~4.8% to the total residential CO_2 output if a TCES + VFPC system was installed in 10% of all UK households. The system with the most CO_2 savings was the VFPC-150˚C-$MgSO_4$ system which saved 89.3MtCO_2e over the operational life. The CO_2 savings from the ETC and the FPC systems was between 49.8-53.8MtCO_2e and 27.9-39.7MtCO_2e, respectively.

6. Conclusion

This study has shown that systems combining TCES and VFPC systems for domestic interseasonal heat storage can be financially viable and result in significant CO_2 savings. From a financial standpoint each one of the system configurations using a VFPC produced a saving to the user over its lifetime with the exception of the 35wt%-80˚C-VFPC system. Furthermore, the only system using either an ETC or a FPC which was financially viable was the $MgSO_4$-150˚C-ETC system. The best system choice was one which incorporated a VFPC due to its financial savings, energy output and CO_2 savings. The system which was the most financially feasible was the $MgSO_4$-150˚C-VFPC system which has the lowest system costs (£6,924), store volume (1.62m^3) and outputs the highest average amount of energy per year (4,622 kWh). The 35wt%-150˚C-VFPC system was a financially viable choice and saved the user around £2,250 with a payback time of around 26 years. Future work will be conducted to identify the characteristics of these TCES materials on a large scale and investigate the characteristics of potential large scale TCES reactor designs for domestic implementation.

7. References

[1] UK.GOVERNMENT. Estimates of heat use in the United Kingdom in 2012, 2013, p. 73–80.
[2] Van Essen VM, Zondag HA, Gores JC, Bleijendaal LPJ, Bakker M, Schuitema R, et al. Characterization of MgSO4 Hydrate for Thermochemical Seasonal Heat Storage. J Sol Energy Eng 2009;131. doi:10.1115/1.4000275.
[3] Van Essen M., Zondag HA, Schuitema R, van Helden WGJ, Rindt CCM. Materials for thermochemical storage : characterization of magnesium sulfate. 2009.
[4] De Boer R, Smeding SF, Zondag HA, Krol G. Development of a prototype system for seasonal solar heat storage using an open sorption process. Eurotherm Semin. #99, 2014, p. 1–9.
[5] Janchen J, Brosicke W, Ackermann D, Stach H. Studies of the water adsorption on Zeolites and modified mesoporous materials for seasonal storage of solar heat. Sol Energy 2004;76:339–44. doi:10.1016/j.solener.2003.07.036.
[6] Hongois S, Kuznik F, Stevens P, Roux JJ. Development and characterisation of a new MgSO4-zeolite composite for long-term thermal energy storage. Sol Energy Mater Sol Cells 2011;95:1831–7. doi:10.1016/j.solmat.2011.01.050.
[7] Kingspan. Kingspan Thermomax Flat Plate Collectors n.d.
[8] Henshall P, Moss R, Arya F, Eames P, Shire S, Hyde T. An evacuated enclosure design for solar thermal energy applications. Gd. Renew. Energy 2014, 2014.
[9] Shire FGS, Moss WR, Henshall P, Arya F, Eames CP, Hyde T. Renewable Energy in the Service of Mankind Vol II: Selected Topics from the World Renewable Energy Congress WREC 2014. In: Sayigh A, editor., Cham: Springer International Publishing; 2016, p. 859–66. doi:10.1007/978-3-319-18215-5_78.
[10] Mahon D, Claudio G, Eames PC. An experimental investigation to assess the potential of using MgSO4 impregnation and Mg2+ ion exchange to enhance the performance of 13X molecular sieves for interseasonal domestic thermochemical energy storage. Energy Convers Manag 2017. doi:http://dx.doi.org/10.1016/j.enconman.2017.03.080.
[11] Duffie JA, Beckman WA. Solar engineering of thermal processes. 3rd ed. New York etc.: Wiley; 1980.
[12] Kingspan_Solar. THERMOMAX HP400 EVACUATED TUBES n.d.
[13] Commission E. Photovoltaic Geographical Information System 2016. http://re.jrc.ec.europa.eu/pvgis/apps4/pvest.php.
[14] DECC. 2014 UK Greenhouse Gas Emissions, Provisional Figures 2015:1–33.

Numerical study of a thermal storage tank enclosed PCM capsules

Wei Lu[1], Shuqian Bi[1], Huakeer Wang[1], Jan-Frederik Flor[2], Yupeng Wu[2], Leyi Zhang[3]

[1]School of Energy and Power Engineering, University of Shanghai for Science and Technology, 516 Jungong road, Shanghai 200093, PR China
[2]Department of the Architecture and Built Environment, Faculty of Engineering, University of Nottingham, Nottingham NG7 2RD, UK
[3]Qinjun(shanghai) Industrial Co. Ltd, 152 Jiaoji Road, Shanghai, PR China
E-mail: wei.lu@usst.edu.cn

Abstract. Low-temperature thermal storage has a huge potential in residential use for regulating the energy supply. This paper proposes a new thermal storage system consisting of modular units. Users can combine different numbers of the units to achieve the desirable energy demand. In this study, a single unit was studied. The hydrated salt was used as the phase change material (PCM) and it was packed in spherical capsule. The phase change situation of each PCM balls were analysed with numerical simulation method. The results showed that the PCM balls in different position had different change rate. In the charging process, the thermal stratification happened after the middle part of the unit, the down layer PCM balls took 30% more time than upper layer to be melted completely. In discharging process, considered the user requirement, the low inlet velocity may fit the single unit. The thermal stratification also happened in the discharging process, and it is most obvious in the middle part.

Keywords: phase change material; low-temperature thermal storage; hydrated salt; numerical calculation

1. Introduction

To overcome time and space restrains of energy use, energy storage plays an increasingly important role in energy systems [1-3]. In some situations, energy storage systems can be proposed which allow for a high energy density and a small temperature drop during the heat storage processes [4]. The confirmed capacity of phase change materials (PCMs) to store high density energy, by making use of latent energy, allows to shrink the volume of storage containers and employ this energy storage technique at large scale [5].

Encapsulating PCM in spheres is a method which has been widely adopted to avoid any instability within the PCM as well as preventing corrosion [6]. F.L Tan [7, 8] conducted a study on the melting process of spherical PCM capsules in constrained and unconstrained situations, analysing with experimental and computational approaches the heat transfer characteristics of PCM. E. Assis [9] investigated experimentally and numerically the melting of a spherical PCM capsule formed of mixed salt materials.

With the development of computational fluid dynamics (CFD), more details of phase change materials and its behaviour have been studied [10]. This study proposed a new thermal storage system and developed a CFD model of the device, numerical analysis was used in this paper as a theoretical foundation for the future design and experiment.

2. The scheme of systems and simulation model

2.1. Scheme of the latent thermal storage system

A prototype model was designed in this work. Figure 1 (a) displays the three-dimensional (3-D) structure of the latent thermal storage component. The PCM is packaged within a spherical shell shown as Figure 1 (b). The diameter of each PCM ball is 42 mm and every four PCM balls are defined as one group. Each component have nine groups of PCM balls. In this case, it was assumed that all PCM balls are motionless and they are arranged orderly with a 2 mm distance between each other. The size of each storage component is 90 mm×90 mm with a height of 405 mm.

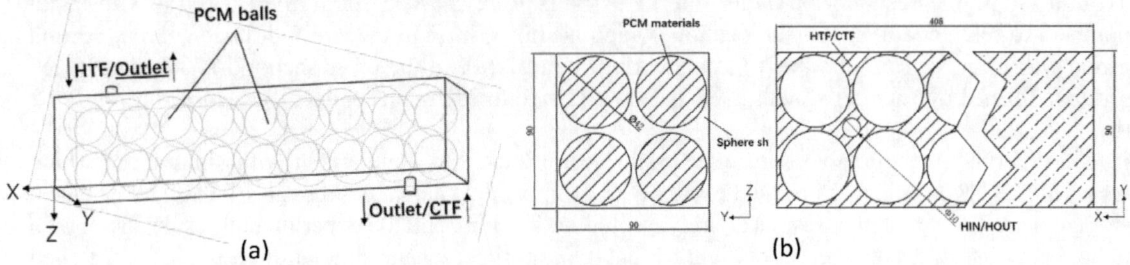

Figure 1. Scheme of the latent thermal storage system.
(Gravity acts along the positive Z direction, underline flow direction represents discharging process.)

The inlet tube, with a diameter of 10 mm, is placed between the first and second group. The outlet tube is placed in the symmetric place between the eighth and ninth group. During the charging process, the HTF (water in 353K) flows into the component from the top, which is shown as a solid line arrow in Figure 1 (a). While during the discharging process, the CTF (water in indoor temperature, 300K in this case) flows from the bottom, represented by a dotted line arrow. A complete cycle of the charge and discharge process is considered for the analysis in this study.

2.2. Physical situation and mathematical model

The following assumptions were made for the mathematical model: 1) Both solid and liquid PCM phases are homogeneous and isotropic. 2) The shell of the PCM balls is not built in the model, while its thermal conductivity is calculated in the simulation process and its thickness is 0.5 mm. 3) The liquid phase is a viscous Newtonian fluid. 4) The PCM balls are full of phase change material. 5) The changes of density and heat conductivity coefficient during the melting and solidification process are neglected. 6) The walls of the storage component are assumed as heat insulation.

The PCM balls were offered by Jinli New Energy Technology. The material they used is called SXC-CZ, which is a hydrated salt. Water is used as HTF and CTF, their properties are shown in Table 1. Moreover, phase change shell material is hard plastic, which properties are shown in Table 2.

Table1. Physical properties of liquid materials.

Material	Density (kg/m^3) (Boussinesq)	Cp ($J/kg \cdot K$)	Thermal conductivity ($W/m \cdot k$)	Viscosity ($kg/m \cdot s$)
Water	998	4198	0.6	1.03×10^{-3}
SXC-CZ	900	3000	0.95	5.1×10^{-3}
	Pure solvent melting heat (J/kg)	**Solidus Temperature(K)**	**Melting Temperature (K)**	**Thermal expansion coefficient (1/K)**
Water	/	/	/	0.00033
SXC-CZ	202000	331	331	2×10^{-3}

Table2. Physical properties of solid materials.

Material	Density (kg/m^3)	Cp ($J/kg \cdot K$)	Thermal conductivity ($W/m \cdot k$)
PCM shell	1200	1600	0.6

This case used ANSYS FLUENT as numerical calculation software. With the Boussinesq model [11], the natural convection is considered in this case. Also, the solidification & melting model with enthalpy-porosity [12-13] were used to calculate the variation of PCM.

3. Results and discussion

3.1. Comparison of experiment and simulation

To understand the results thoroughly, the PCM balls were marked for further analysis with serial numbers to reference their typical location, which is shown in Figure 2. In X direction, every second group was indexed from one to five as shown in Figure 2 (a). In the cross section in Y-Z plane, each group is formed by four balls, with a letter code assigned to each unit (a, b, c, d) as shown in Figure 2 (b).

In the experiment, the temperature of PCM balls are took into comparison with simulation, which represent the beginning, middle and the end of the component. There are three sets of phase change heat storage data. The simulation results are represented by full line, and the experimental results are shown in hollow points. It can be seen from Figure 3 that the experimental and simulation results were matched well. However, to study the phase change situation of each PCM balls, the simulation method was used in the study.

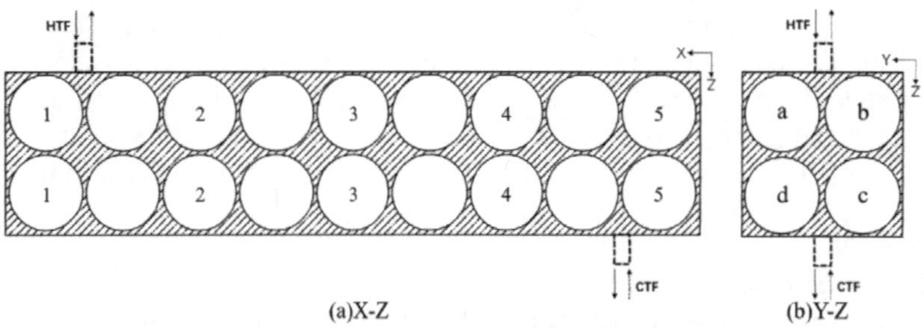

Figure 2. Serial number of each PCM balls.

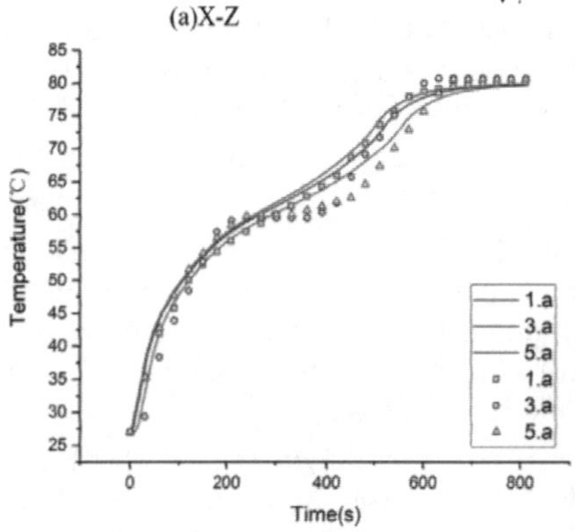

Figure 3. Comparison of experimental and simulation.

3.2. Charging process

The figure 4 shows the change of the liquid fraction for each PCM balls during the charging process which the inlet velocity is 1m/s. All curves on the figure shows a very similar tendency but with different melting rate. In general, along the flow direction the melting rate is decreased. It is because that the temperature difference and fluid velocity decreased along the flow direction.

Overall, the curves in figure4 (a) (b) is different like in (c) (d). The PCM balls in down layer (3, 4, 5 in group c, d) had slower melting rate than upper layer (3, 4, 5 in group a, b). It is because that the thermal stratification happened after the middle part of the unit. The warmer water tend to flow to the top and the cooler water stay at the bottom. This phenomenon will accelerate the melting rate of the upper layer PCM balls, but also extend the time for the unit to melt completely. In the middle part of the unit, the upper layer took 33% less time than down layer PCM balls to melt completely.

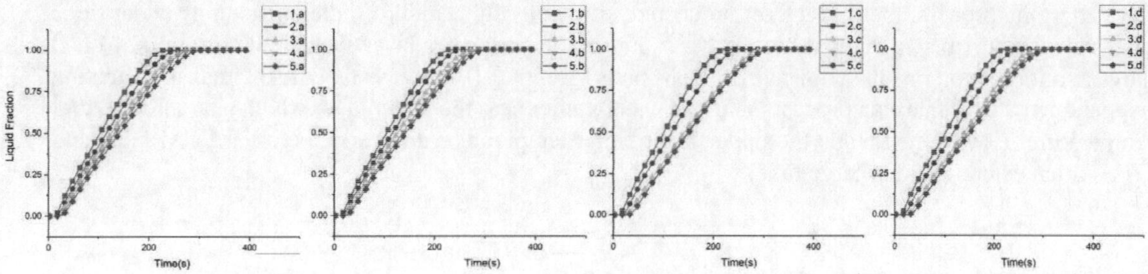

Figure 4. Liquid fraction of PCM balls ordered by X direction (v_{inlet} =1 m/s)

3.3. Discharging process

Different like in charging process, the discharging process is face to the users. So the outlet temperature need to be considered. In Figure 5 the outlet temperature variations are shown versus different inlet flow velocities. The inlet and outlet was reset and it was further assumed that the discharging process just happened after charging process. The initial temperature of the component is 353 K and the CTF is 300 K.

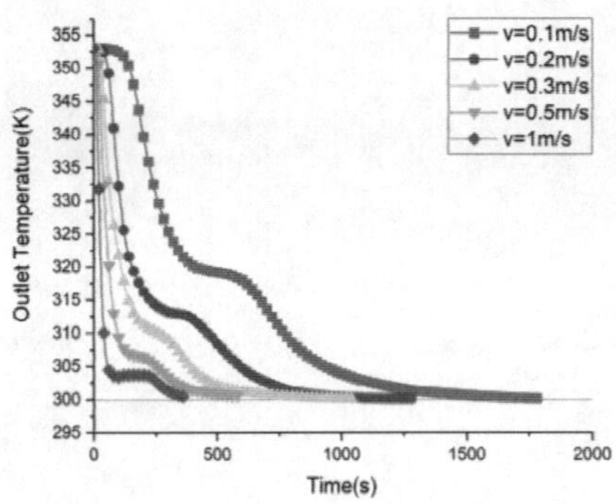

Figure 5. Temperature variations of outlet in different inlet velocities.

The curves plotted in Figure 5 can be divided into three stages. During the first stage, the CTF replaced remaining hot water in the tank, which temperature was dropped to a lower rate and sensible heat of the PCM balls started to release at this stage. In the second stage, the tank was filled with CTF and the outlet temperature dropped rapidly. Then the latent heat started to release and reached an equilibrium point which the temperature of the outlet stabilized at a relatively constant temperature. However, the heat was continuously transferred to the CTF, solidifying slowly the PCMs which would form a shell [14], which eventually would become thicker during the discharging process. Therefore, the internal heat exchange area became smaller and the process arrived at the final flat stage where the temperature dropped fast to meet the temperature of the component inlet.

The curves in Figure 6 shows when the CTF velocity is low the flat stage will show up at higher temperature. In Table 3 different standard inlet velocities have been listed in a comparative way and were set in relation to the total flow volume and time. Considered the user requirement, the 313K was set as lowest temperature for the outlet temperature. The scenarios with a low-velocity inlet had achieved a total flow with relatively minor difference, while as the flow velocity increased, the total hot flow started to decrease rapidly.

Table. 3 The comparison of total (lowest 313 K) heat flow in different inlet velocities.

Inlet velocity (m/s)	Flow time (s)	Total flow (L)
0.1	711	5.584
0.2	324	5.089
0.3	232	5.466
0.5	81	3.181
1	37	2.906

Like charging process, the liquid fraction change situation of group 3 (middle of the unit) under three inlet velocity (0.1m/s to 0.3m/s) are shown in Figure 6. The upper layer PCM balls (group a, b) had slower solidification rate than down layer PCM balls (group c, d). It shows that the thermal stratification happened in the discharging process. Warmer water gather near the upper layer which guaranteed outlet temperature. However, it made the upper layer small temperature difference between PCM balls and CTF which extend the discharge time.

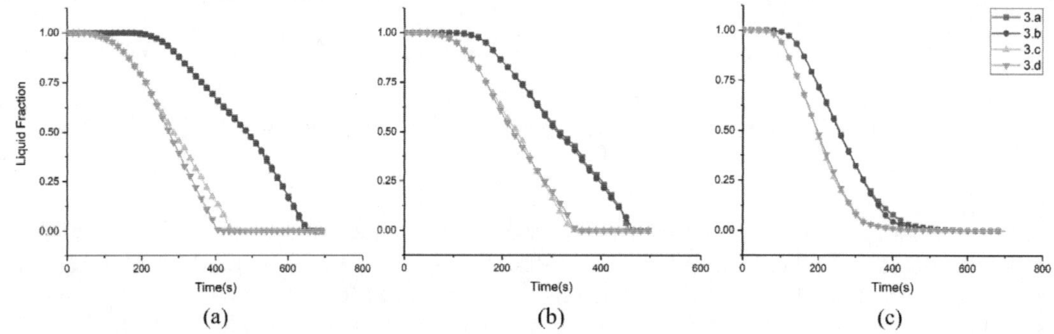

Figure 6. The variation of liquid fraction in (a) v_{inlet} =0.1 m/s, (b) v_{inlet} =0.2 m/s, (c) v_{inlet} =0.3 m/s

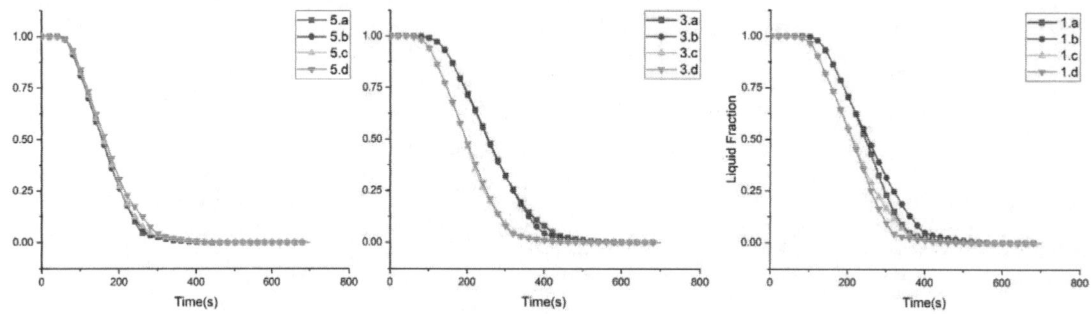

Figure 7. The variation of liquid fraction in different locations (v_{inlet} =0.3 m/s)

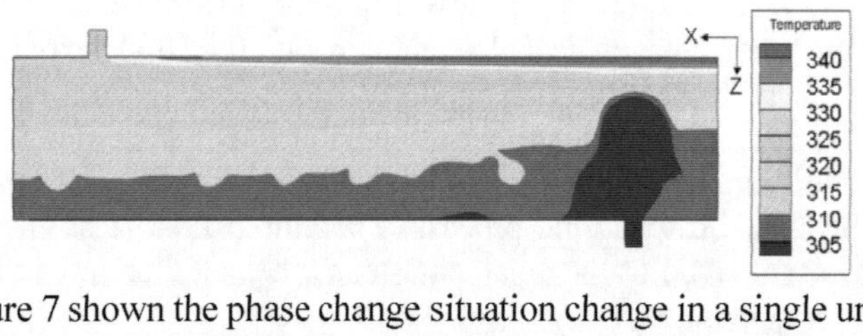

Figure 8. The distribution of temperature in the energy storage component at 300s

Figure 7 shown the phase change situation change in a single unit with inlet velocity of 0.3m/s. It picked up group 5, 3, 1 PCM balls which is the beginning, middle and end of the unit. From the Figure, the thermal stratification change along the flow direction. At the beginning of the unit, the CTF had the negative Z direction velocity near the inlet which it can reach the upper layer PCM balls as shown in Figure 8, so the solidification rate of these PCM balls almost same. At the middle part, after the CTF fully developed the thermal stratification appeared. At the end part, because of the outlet, the stability of CTF is broken. The solidification rate of each balls tend to be uniform.

4. Conclusions

This study uses numerical analysis to study one single unit of the detachable latent heat storage system, using hydrated salts as PCM filling in the sphere and water as HTF/CTF. The results show that during the charging process, the HTF heat the PCM balls rapidly and keep a relatively small thermal drop (about 4K in the steady situation) at the outlet. The melting rate of each PCM balls decrease along the flow direction. Moreover, the study showed that the thermal stratification phenomenon appeared after the middle section of the component and resulted in longer melting time for down layer PCM balls.

Different inlet velocities were compared during the discharging process. In comparison to high inlet velocities, lower inlet velocities resulted in prolonged periods of high outlet temperatures and a higher quality of total flow (higher than 313 K). However, thermal stratification took place throughout the low inlet velocity cases, resulting in a longer discharging time and partly heat waste at the end stage of the discharge. Combining these two factors, a flow velocity between 0.2m/s to 0.3m/s might result to be the better choice for one single unit.
Overall, this 3-D numerical model of the energy storage tank offered a high efficiency approach towards a thorough understanding of the phase change process and heat transfer in energy storage tank. To enhance the heat storage performance in this system, more works are expected to be carried out by optimizing the complex system, including choosing more appropriate geometric configurations and suitable sizes.

References

[1] Mohammed M. Farid, Amar M. Khudhair, et al. 2004, *A review on phase change energy storage:materials and applications*. Energy Conversion & Management 45 1597-1615

[2] Mohammed B., Kacem M., 2013, *Exact solution of thermal energy storage system using PCM flat slabs configuration*, Energy Convers Manage **76(5)** pp. 88-98

[3] AMARAL C, VICENTE R, MARQUES P P, et al, 2017, *Phase change materials and carbon nanostructures for thermal energy storage: A literature review*. Renewable & Sustainable Energy Reviews **79** 1212-28

[4] Kamal A.R. Ismail., Fátima A.M. Lino., 2014. *Experimentally validated two-dimensional numerical model for the solidification of PCM along a horizontal long tube*, International Journal of Thermal Sciences **75** pp. 184-193

[5] Cardenas B, Leon N, 2013. *High temperature latent heat thermal energy storage: phase change materials, design considerations and performance enhancement techniques*. Renew. Sust. Energy Rev **27** 724-737

[6] N.H.S. Tay., M. Belusko., 2015, *Investigation of the effect of dynamic melting in a tube-in-tank PCM system using a CFD model*, Applied Energy **137** pp. 738–47

[7] F.L. Tan., 2008, *Constrained and unconstrained melting inside a sphere*, International Communications in Heat and Mass Transfer **35** pp 466–75

[8] F.L. Tan., S.F. Hosseinizadeh., 2009, *Experimental and computational study of constrained melting of phase change materials (PCM) inside a spherical capsule*, International Journal of Heat and Mass Transfer **52** pp. 3464–72

[9] E. Assis., L. Katsman., 2007, *Numerical and experimental study of melting in a spherical shell*, International Journal of Heat and Mass Transfer **50** pp. 1790–1804

[10] Zhuo Li, Zhi-Gen Wu, 2015, *Analysis of HTFs, PCMs and fins effects on the thermal performance of shell-tube thermal energy storage units*, Solar Energy **122** 382-395

[11] Patankar SV, 1980, *Numerical heat transfer and fuid fow*. Hemisphere Publishing Corp, Washington

[12] Voller VR, Cross M, Markatos NC (1987) *An enthalpy method for convection/diffusion phase change*. Int J Numer Meth Eng **24** 271—84

[13] Brent AD, Voller VR, Reid KJ (1988) *Enthalpy-porosity technique for modeling convection diffusion phase change:application to the melting of a pure metal*. Numer Heat Trans **13** 297-318

[14] Liu Z Y, Yao Y P, et al, 2013, *Numerical modeling for solid-liquid phase change phenomena in porous media: shell-and-tube type latent heat thermal energy storage*, Applied Energy **112** 1222-32

Acknowledgement

This work is supported by National Natural Science Foundation of China (No.51406121 and No.51736007) and Eastern Scholar (No.QD2015017).

IOP Conf. Series: Materials Science and Engineering **556** (2019) 012014 doi:10.1088/1757-899X/556/1/012014

Numerical model of applying PCM boards for energy saving in lightweight buildings in Shanghai

Huakeer Wang[1], Wei Lu[1]*and Zhigen Wu[2]
[1] University of Shanghai for Science and Technology, 516 Jungong Road, Shanghai, 200093, PR China
[2] Tongji University, 1239 Siping Road, Shanghai, 200092, PR China
E-mail: wei.lu@usst.edu.cn

Abstract. The lightweight building industry develops fast, but it has competitively worse thermal comfort than traditional buildings because of less thermal mass. The use of phase change material (PCM) can help to solve the problem and boost the energy efficiency. In this paper, the effect of applying PCM-boards in air-conditioned lightweight buildings in Shanghai has been evaluated. A parametric study is performed considering different PCM-boards configurations, using an EnergyPlus single-zone model. The daily and annual situation have been simulated in this study, and the predictions show that the energy saving depends on the phase change temperature of PCM and seasonal temperature. The application of PCM can save up to 20.88% of energy for air conditioning system in buildings. The daily assessment reveals that the passive PCM technology inside the wall saves more energy under heating conditions than the cooling condition because of the overheating.

Keywords: PCM-boards, lightweight buildings, EnergyPlus, air conditioning energy saving

1. Introduction

Recent years, lightweight exterior wall panels are widely used in the construction industry [1]. However, they have low thermal mass, which increase the energy consumption of air conditioning. Meanwhile, these consumption generally takes up over one-third of energy consumption in buildings [2]. Use phase change material (PCM) to improve heat comfort is considered as a promising technology to solve the problem and reduce the power demand of air conditioning system [3]. The PCM can provide relatively high heat storage capability but with small mass and volume of material. Moreover, the temperature of the PCM remains almost constant during the phase change [4]. These features made it could be used to improve the thermal performance of building walls, reduce heating and cooling loads at the peak time, shift loads to low tariff hours and improve the indoor environment [5, 6]. In recent years, the PCM-enhanced building envelopes have been studied by increasingly researchers [7, 8]. The thermal performance of PCM wall is affected by many factors such as outdoor environment, the wall materials and the indoor environment [9, 10]. In this paper, a model with an innovative PCM enthalpy-temperature function has been built up to simulate a single PCM-boards enhanced room. The daily and annual energy saving have been predicted with the result of numerical simulation under weather situation in Shanghai.

2. Methodology

2.1. Numerical model

A standard 3-D building model was selected to simulate different conditions which is specified as ASHRAE-140 case-600. It is an air-conditioned single-zone model with the interior dimensions are 8m×6m×2.7m=129.6 m^3 as shown in Figure 1. All surfaces are external walls. This model assumed to have their windows on two sides to reflect current residential construction practices in Shanghai. On the south side, there are two windows while on the north side there is one window in the middle with the same size and shown in Figure 1. Moreover, the model is not obstructed by neighbouring buildings and the ground reflectance is equal to 0.2. The external walls properties and structure are shown in Table.1 and Figure.2. In the PCM-enhanced room, only walls and ceiling were installed with PCM-boards.

Table1. Properties of construction materials

	Conductivity (W/m · k)	Density (kg/m^3)	Specific Heat (J/kg · K)	Thickness (m)
EPS	0.039	20	1380	0.05
Gypsum Board	0.31	1100	1160	0.012
OSB	1.21	650	2000	0.02
Lightweight Concrete	0.52	1300	920	0.05
Roof Membrane	0.16	1121.29	1460	0.0095
Wood	0.15	608	1630	0.0254
PCM	0.6	1500	2000	

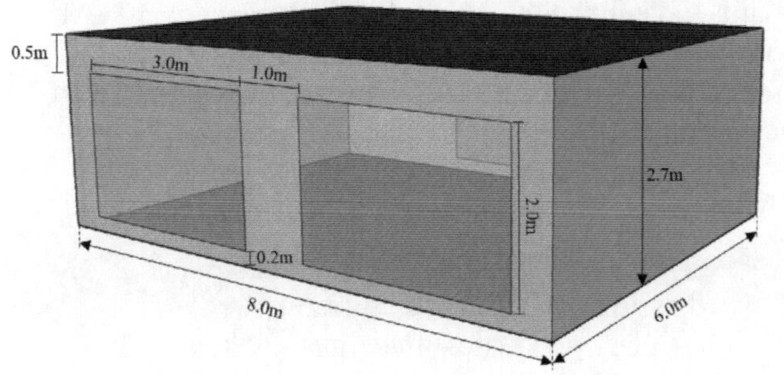

Figure 1. The building structure of the simulation model

Regarding internal heat gains, the building model was occupied by a maximum of 1 people per 1m^2 in sedentary activity with a constant metabolic rate of about 126W·person^{-1}. A maximum design lighting level of 10Wm^{-2} was considered for the equipment thermal gain. The air conditioner system in this study was *ideal loads air system* model in the EnergyPlus simulations. Regarding actual air temperature control, the thermostat was set with a cooling set point temperature of 26°C, and the heating set point temperature is 18°C. The air ventilation was set as one time per hour in this study.

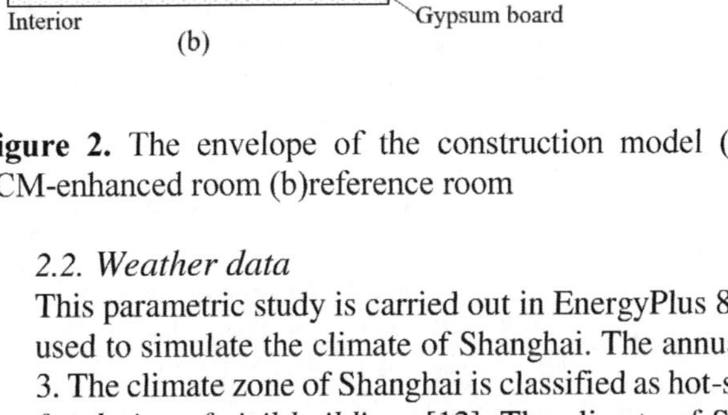

Figure 2. The envelope of the construction model (a) PCM-enhanced room (b)reference room

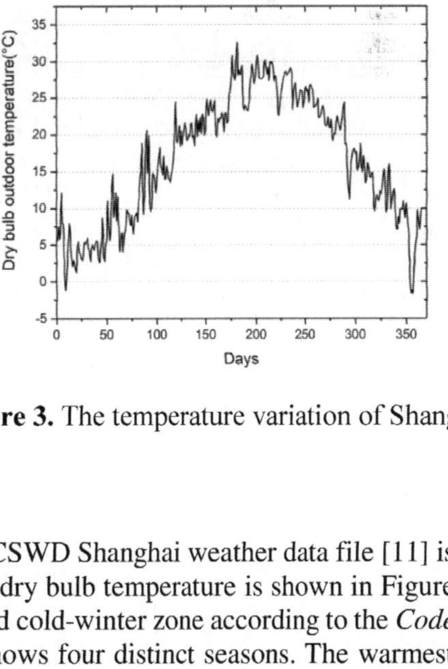

Figure 3. The temperature variation of Shanghai

2.2. *Weather data*

This parametric study is carried out in EnergyPlus 8.9.0. The CSWD Shanghai weather data file [11] is used to simulate the climate of Shanghai. The annual outdoor dry bulb temperature is shown in Figure 3. The climate zone of Shanghai is classified as hot-summer and cold-winter zone according to the *Code for design of civil buildings* [12]. The climate of Shanghai shows four distinct seasons. The warmest week, on average, is in the July with an average temperature of 27.49°C. The maximum temperature during the summer can reach 37°C. The most cooling mouth on average is January, with an average temperature of about 4.51°C, the minimum temperature can reach -4.3°C. More statistics information about the climate of Shanghai can be found in the weather data file [11].

2.3. Mathematical model

The annual cooling and heating energy demands are calculated in the lightweight reference model without PCM and compared with the PCM-enhanced lightweight model. The one-dimensional conduction finite-difference (ConFD) solution algorithm was adopted in this study which was validated against multiple test suites [13] in simulate the phase change process. In this study, according to Tabares et al [14], 1-minute time step was used in the simulation for assessing the annual and daily energy demand. Moreover, for the accuracy of hourly analysis, the space discretization was set to 0.4 for the CondFD model. This model read the inputs of the enthalpy-temperature information which was shown in the Figure 4.

The DuPont^{TM} PCM product was considered as the reference PCM. As proposed by Tabares and Soares [14, 15], based on the nonlinear enthalpy-temperature function of reference material, a new linear function can be plotted for a hypothetical PCM-boards to facilitate the parametric analysis. Seven PCM with the same latent heat but different melting peak temperature were considered in this study. The detailed enthalpy-temperature is showing in Figure 4, which is an ideal value referenced from Nelson et al [16] (pcm18 for the PCM with peak melting temperature of 18°C, and so on in a similar fashion).

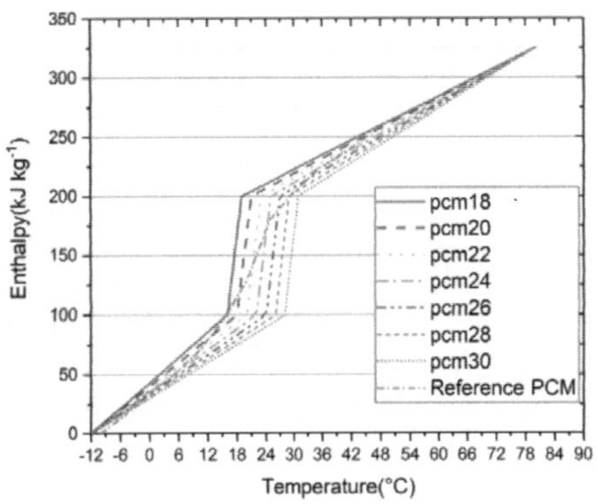

Figure 4. Enthalpy-temperature for the seven hypothetical materials of PCM-board *Reference data for DuPont^{TM} Energain® PCM product obtained from differential scanning calorimeter (DSC) measurements with a heating rate of 0.05°C min^{-1}[17]

3. Results and discussion

3.1. Annual assessment for PCM-enhanced room and reference room without PCM

Compared with the reference room in which the air-condition cooling load is $E_{cool,ref}$ =137.45 kWh m^{-2} year^{-1}, and heating load is $E_{heat,ref}$ =59.96 kWh m^{-2} year^{-1}, the PCM enhanced room reduced the cooling load(Figure 5 (a)) and heating load(Figure 5 (b)), which can be found in Figure 5 ordered with different thickness and melting temperature. The Figure 5 (a) reveals that the best PCM for the cooling season is the pcm26 and pcm28. Maximum 10.67% annual cooling energy saving can be achieved by using pcm28 and thickness is 50mm. Figure 5 (b) shows the prediction in the heating season, the pcm20 is the best. Up to 56.75% energy saving can achieve in the thickness of 50mm. Moreover, simulations show that

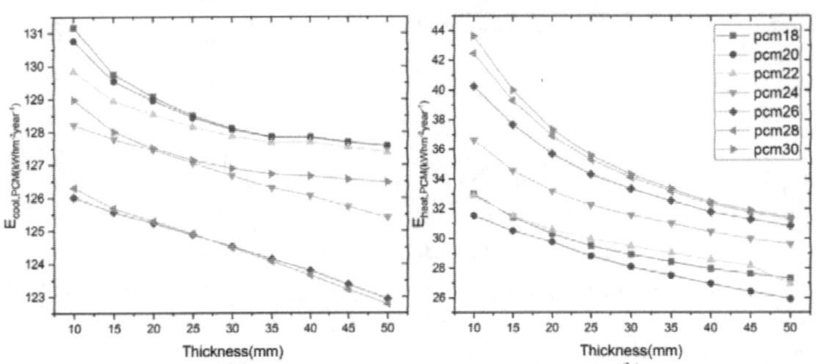

Figure 5. Annual (a) cooling and (b) heating energy demand for the PCM-enhanced room with different melting peak temperature and PCM-boards thickness

the energy load decreased with the increase of PCM thickness. It may be a good compromise to adopt 30mm instead of 50mm thick as the energy savings achieved with 30mm are close to those obtained with 50mm. This could be particularly important when taking into account the cost of the price of PCM-boards.

It is shown in Figure.6 that the annual air-condition energy consumption with different thickness of PCM. The pcm20 is the best for annual energy saving. The 20.88% energy saving is achieved in the thickness of 30mm. In addition, it is clear to see that the cooling energy saving has more influence on the annual energy saving, it will be discussed in the next section.

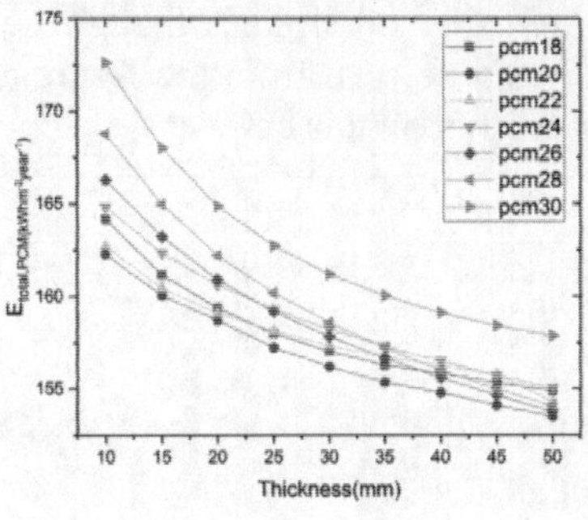

Figure 6. Annual air-condition energy demand for the PCM-enhanced room considering different melting peak temperature and PCM-boards thickness

3.2. Daily assessment basis for the effect of different melting temperature of PCM

To study the effect of different PCM on cooling and heating load of air conditioning system, three continuous days were picked up both in summer and winter. The pcm20 and pcm26 were employed as PCM. Figure 7 shows the case of summer from June 24th to 26th, and Figure 8 shows the prediction of winter from January 6th to 8th. The outdoor temperature have a big variation during these days. The area is used to indicate the cooling power consumption (Figure 7) in summer and heating power consumption (Figure 8) in winter time (read on left axis). The outdoor temperature, the temperature at the inside surface of PCM and the temperature of gypsum board at the same position in the reference room (read on right axis) are also plotted in lines.

In summer, due to the *ideal loads air system* model, the room temperature is maintained at the set point of 26°C. Figure 7(a) shows the case of the pcm20, because outdoor temperature and room temperature are always above 20°C so that PCM was in melting state. Therefore, PCM can be regarded as a layer of insulating material only. It can be seen that during the high outdoor temperature around noon, the energy consumption is reduced, but in the night, the cooling load of the PCM-enhanced room is higher than that of the reference room. However, the cooling load of PCM room during three days is similar to the reference room, but about 12.7% cooling load has been shifted from peak hour (8:00-21:00) to off-peak hour (22:00-6:00).

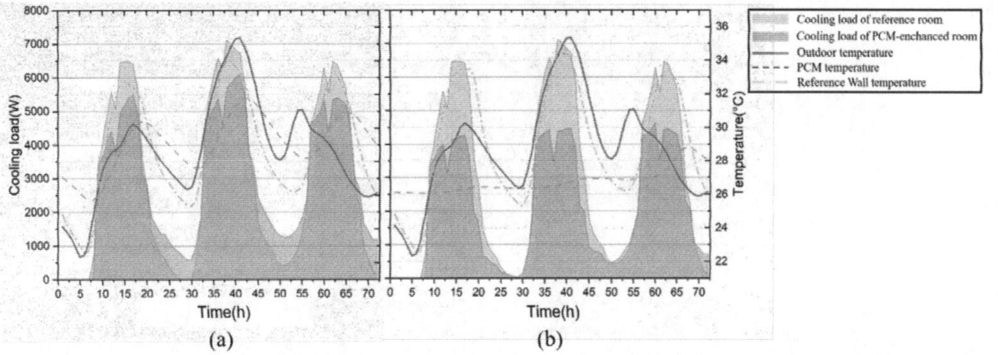

Figure 7. Hourly cooling load profiles and evolution of the estimated temperatures for the summer days (June 24th to 26th) for both the reference and the PCM-enhanced models (a) pcm20 (b) pcm26

In Figure 7(b), the temperature of the pcm26 is stabilized at melting temperature of 26°C in the first two days, when the PCM is absorbing heat. However, around the noon on the third day (65h), the temperature

IOP Conf. Series: Materials Science and Engineering **556** (2019) 012014 doi:10.1088/1757-899X/556/1/012014

of the PCM exceeded 26°C, indicating that the PCM began to overheat. In the next night (70h), it can be seen that the temperature of PCM is higher than melting temperature and cooling load in PCM-enhanced room is higher than that of the reference room. This negative effect happened because PCM boards release heat at night.

In winter, the room temperature is maintained at 18°C. Figure 8(a) shows the prediction for room with pcm20. The temperature of PCM is maintained at about 20°C. Unlike in summer, energy savings mainly happened at night. Moreover, the temperature of the wall keeps steady around 20°C, which is good for room comfort and energy saving. Figure 8(b) shows the effect of the pcm26 during the winter. At this case, the outdoor temperature is lower than 26°C, PCM remains solidified. Although the PCM increases the heat capacity and thermal resistance of the wall, it can be seen from the figure that the outdoor temperature fluctuation still affects the indoor wall temperature in this case.

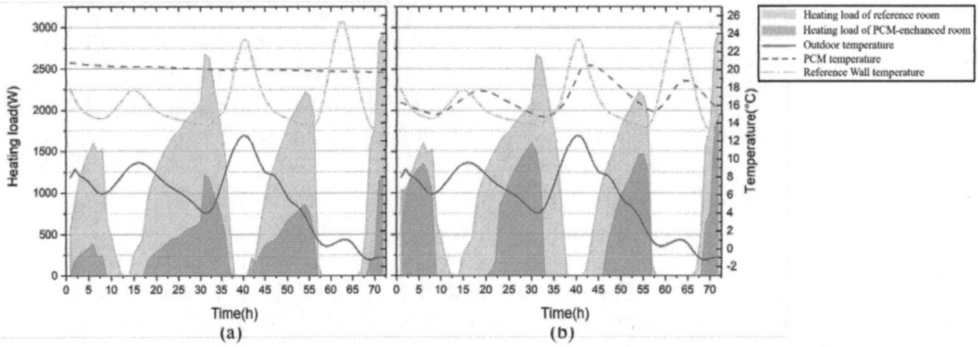

Figure 8. Hourly heating load profiles and evolution of the estimated temperatures for the winter days (January 6th to 8th) for both the reference and the PCM-enhanced models (a) pcm20 (b) pcm26

Combined with the summer and winter simulation, the following conclusions can be drawn: Firstly, when the outdoor temperature fluctuates around the melting temperature, the PCM has a weakening effect on the thermal fluctuation. Secondly, in summer, the most energy saving of the PCM happens around noon, and in the winter, it mainly happens at night. So the PCM can be chosen depending on the different service time of the room. Thirdly, Figure 9 shows the factors affecting the PCM melting and solidification in two seasons. The main difference is the solar radiation. In summer, the PCM is easily overheat under continuous high temperature because the sun keeps heating PCM, and it will have a negative impact on the power consumption for cooling when the PCM fails. Therefore, for summer, active PCM system may be a better choice. However, in winter PCM works stable because the sun can melt the PCM during daytime and the melted PCM is good for the night heat preservation.

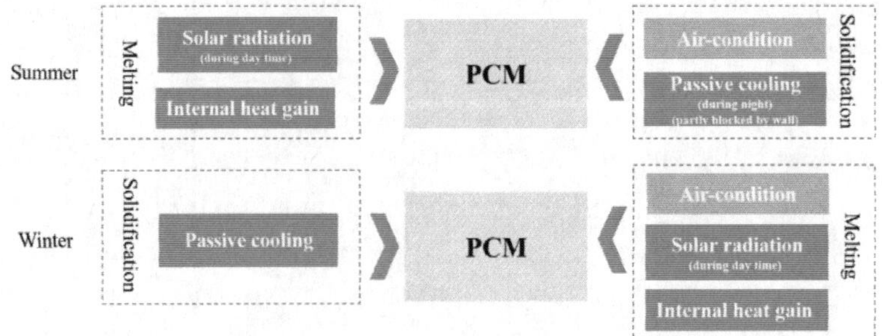

Figure 9. The different charging and discharging factors in summer and winter

4. Conclusion

This extensive EnergyPlus based parametric simulation study found that PCM-boards in lightweight construction in Shanghai helps to create annual savings of energy for cooling, heating up to 20%. The universal PCM-boards system with a thickness of 30mm and the melting temperature of the PCM at 20°C is recommended for the annual energy saving. It is also found that the passive PCM-boards does not appropriate for Shanghai in summer. It may have a negative impact when the PCM is fully melted. The active PCM system might be a better choice for summer energy saving. However, in winter the passive PCM have a good effect.

References

[1] J S Sage-lauck, D J Sailor 2014. Evaluation of phase materials for improving thermal comfort in a super-insulated residential building. Energy and Buildings 79 32-40

[2] International Energy Agency 2013. Technology roadmap: energy efficient building envelops, Oecd

[3] Anisur MR, Mahfuz MH, kibria MA, Saidur R, etc 2013. Curbing global warming with phase change materials for energy storage. Renew Sustain Energy Rev 18 23-30

[4] Mehing H, Cabeza LF 2008. Heat and cold storage with PCM: an up to date introduction into basics and applications. 1st ed. New York: Springer

[5] Genc Zk, Canbay CA, Acar SS, Sekerci M, Genc M 2015. Preparation and thermal properties of heterogeneous composite phase change materials based on camphene-palmitic acid. J Therm Anal Calorim 42 1759-72

[6] Castell A, Martorell I, Medrano M, et al 2010. Experiment study of using PCM in brick constructive solutions for passive cooling. Energy Build 42 531-40

[7] Saffari M, De Gracia A, Ushak S, et al 2017. Passive cooling of buildings with phase change materials using whole-building energy simulation tools: A review. Renewable & Sustainable Energy Reviews 80 1239-55

[8] Cabeza L F, Castell A, Barreneche C, et al 2011. Materials used as PCM in thermal energy storage in buildings. Renewable & Sustainable Energy Review 15(3) 1675-95

[9] Hasan M I, Basher H O, Shdhan A O 2018. Experimental investigation of phase change materials for insulation of residential buildings. Sustainable Cities and Society 36 42-58

[10] Kong X, Jie P, Yao C, et al 2017. Experimental study on thermal performance of phase change material passive and active combined using for building application in winter. Applied Energy 206 293-302

[11] Yi J. April 2005.China Standard Weather Data for Analyzing Building Thermal Conditions. Beijing. China Building Industry Publishing House

[12] Ministry of construction of the People's Republic of China. Code for design of civil buildings. GB50352-2005 7-8

[13] Tabares-Velasco P 2012. Energy impacts of nonlinear behavior of PCM when applied into building envelop. ASME 6TH International Conference on Energy Sustainability & 10th Fuel Cell Science, Engineering and Technology Conference. San Diego

[14] Tabares-Velasco P, Christensen C, Bianchi M 2012. Verification and validation of EnergyPlus phase change material model for opaque wall assemblies. Building and Environment 54 186-196

[15] Soares N, Gaspar AR, et al 2014. Multi-dimensional optimization of the incorporation of PCM-drywalls in lightweight steel-framed residential buildings in different climates. Energy and buildings 70 411-421

[16] Nelson S, Christoph F R, Ali H 2017. Simulation-based analysis of the use of PCM-wallboards to reduce cooling energy demand and peak-loads in low-rise residential heavyweight buildings in Kuwait. Build Simul 10 481-495

[17] Cao S, Gustavsen A, Barreneche C, et al 2010. The effect of wall-integrated phase change material panels on the indoor air and wall temperature-Hot box experiments. Renewable Energy Research Conference. Trondheim Norway.

Acknowledgement

This work is supported by National Natural Science Foundation of China (No.51406121 and No.51736007) and Eastern Scholar (No.QD2015017)

A study of overcast, partly cloudy and clear skies by global illuminance and its variation features

Siwei Lou[1,*], Danny H W Li[2] and Wenqiang Chen[2]
[1]School of Civil Engineering, Guangzhou University, 230 Guangzhou Higher Education Mega Centre West Outer Ring Road, Panyu District, Guangzhou, China
[2]Building Energy Research Group, Department of Architecture and Civil Engineering, City University of Hong Kong, Tat Chee Avenue, Kowloon, Hong Kong SAR, China
*E-mail: swlou2-c@my.cityu.edu.hk

Abstract. The diffuse radiance and luminance distribution in the skydome can determine the solar energy on the building envelope in various orientations and under complex obstructions. The diffuse sky is anisotropic for clear and partly cloudy skies, while isotropic for overcast skies. Thus, identifying the sky conditions are essential to the sky radiance and luminance studies. Previously developed approaches using statistic data suffered from data shortage and inaccuracy. Sunlight under cloudy skies may frequently be blocked, causing the high-frequency illuminance fluctuations. The illuminance should variate smoothly under clear skies without clouds. In this work, we studied the horizontal irradiance time series, its variation frequency, and its relationship to the sky conditions determined by the luminance scan. The work contributes to identifying the sky conditions and skylight distributions using the horizontal global daylight illuminance that is measured by the fixed sensors.

Keywords: Diffuse skylight, Building energy saving, anisotropic sky, Fourier transfer

1. Introduction

The building stocks use half of the total energy for subtropical cities [1]. The solar energy is essential to the building energy savings. The solar heat gain through building envelop leads to the cooling needs in summer [2] while compensates part of the heating demands in winter. The visible light saves the artificial lighting energy and prevents the resulted heat dissipation. The solar heating and photovoltaic techniques provide renewable energy on site [3]. The solar and daylight energy on building envelop in various directions are thus the cornerstone for active and passive building energy saving techniques.

The solar energy on building envelop are composed of the direct, diffuse and reflected parts. Studies showed that half of the solar radiation on horizontal and vertical planes was composed by the diffuse part [4]. However, the diffuse radiation and daylight data were usually estimated by models because of the high measurement costs, and the difficulties to take measurements for many different directions.

Different regions in the sky dome may have different sky radiance and brightness, and contribute to the solar energy and daylight evaluations differently. For clear skies, the forward scattering by air leads to an anisotropic sky radiance distribution that peak around the sun and weaken rapidly for the sky far away. For overcast skies, however, the cloud and mist may lead to azimuthal homogenous radiance and luminance distribution. For buildings in cities, the surrounding buildings may obstruct a part of the skydome, and weaken the diffuse skylight irregularly. In such condition, the sky radiance and luminance from unobstructed sky regions can be important for radiation and daylight on surfaces.

Previously developed skylight models differed from the early-stage uniform diffuse sky model to the later models [5, 6] that assumed anisotropic sky radiance variation discontinuously. In 1998, Kittler et al. [7] proposed a set of sky models that described continuous sky luminance variations in different sky regions for 15 different conditions from overcast to partly cloudy and clear. In 2003, the International Commission on Illumination (CIE) accepted the models as the CIE Standard Skies [8].

Many worldwide studies reported the Skies described the skylight distribution well [9], and the model was more accurate than other similar ones [10]. Since the daylight is the visible part of the solar energy, the model for skylight luminance has been used in sky radiance studies as well.
Though the models were well developed, there is no official criterion to judging the overcast, partly cloudy, and clear conditions, and selecting the CIE Sky models accordingly. Measuring sky luminance distribution [11] can be inclusive yet costly, and such data were accessible in a few places in the world. Alternative approaches identified the conditions by vertical diffuse radiation and skylight illuminance [12] suffered from data measurement problems as well. An approach [13] using horizontal sunlight measurements and its indices mixed a part of the partly cloudy skies with either clear or overcast skies. For overcast and partly cloudy skies, the randomly distributed cloud may cause the horizontal global daylight data to fluctuate in high frequency. For clear skies, in contrast, the daylight and radiation would vary smoothly by solar altitude angle. Fourier Transfer [14] is fundamental to study the data series variation frequencies. It is thus possible to identify the sky conditions via the Fourier Transfer analysis of the global illuminance on the horizontal surfaces. The approach was studied in the current work using the CIE Standard Sky by the sky luminance scan in Hong Kong in 2004. The daily time series of horizontal global illuminance were analyzed using the Fourier Transfer for the variations in high and low frequencies. Features of the daylight clearness index series in the frequency domain for overcast, partly cloudy and clear skies were reported and discussed.

2. Data measurements and pre-processing

The diffuse sky luminance of 145 sky directions was measured by sky scanner EKO MS-301LR. The scanner scrolls its luminance and radiance sensors in altitude and azimuth directions. The sensor view angle is 11 degrees, and the 145 measurements covered 68% of the skydome. Direct sunlight greater than 35 kcd/m^2 was prevented to protecting the sensor. Both the sky luminance and the horizontal global illuminance were recorded every 10 minutes at City University of Hong Kong (CityUHK) from Jan. to Dec. 2004. Our previous work described the measurements in detail.
The data quality control excluded datasets with negative luminance recordings, and the horizontal illuminance less than zero or recorded at solar altitude angle less than 4 degrees. Low illuminance data were conserved to keep its time series complete, and preventing interpolations for the missing or rejected data. In total, 21,809 sets of luminance data and 23,083 sets of horizontal global illuminance data were properly recorded. Both measurements were in a 10-minute interval.

3. Methodology

The work evaluates horizontal global daylight illuminance and its clearness index K_t of different CIE Standard Skies. K_t is the ratio of horizontal global daylight illuminance on the ground to horizontal extraterritorial illuminance. The horizontal global daylight illuminance daily series in 10 minutes interval was analyzed by the Fast Fourier Transfer (FFT) for its components that variated in high and low frequencies. The CIE models describe the luminance distributions of the five overcast, five partly cloudy and five clear skies by its well-defined equations. Comparing the sky luminance scan against the 15 CIE Standard Sky models determines the overcast, partly cloudy and clear sky conditions.

3.1. The CIE Standard Skies

The 15 CIE Sky models depict the relative luminance ($L_{\alpha\phi}$, cd/m^2) of a sky point at sky altitude α and azimuth ϕ to that at the zenith point (L_Z, cd/m^2), as given by Eq. 1. Z is the zenith angle of the sky element at α, $Z = \pi/2 - \alpha$. $\varphi(Z)/\varphi(0)$ is the relative gradation function, and $f(\chi)/f(Z_S)$ is the relative indicatrix function, given by Eqs. 2 and 3, respectively. χ is the angular distance between the sky element at (α, ϕ) and the sun, $\cos\chi = \cos Z_S \cos Z + \sin Z_S \sin Z \cos|\phi - \phi_S|$, Z_S is the solar zenith angle, and ϕ_S is the solar azimuth angle. The gradation function depicts the sky luminance variation from sky zenith to horizon. The indicatrix function marks the luminance attenuation from the sun to the sky parts that are far away. Coefficients a to e can be adjusted for different variation rates in different parts of the sky dome, which represents different sky conditions from overcast to clear.

$$\frac{L_{\alpha\phi}}{L_Z} = \frac{f(\chi)\varphi(Z)}{f(Z_S)\varphi(0)} \tag{1}$$

$$\frac{\varphi(Z)}{\varphi(0)} = \frac{1 + a\exp(b/\cos Z)}{1 + a\exp b} \tag{2}$$

$$\frac{f(\chi)}{f(Z_S)} = \frac{1 + c\left[\exp(d\chi) - \exp(d\pi/2)\right] + e\cos^2\chi}{1 + c\left[\exp(dZ_S) - \exp(d\pi/2)\right] + e\cos^2 Z_S} \tag{3}$$

3.2. Identifying the CIE Standard Skies by luminance scan

The scanner measures the sky luminance (L_i, cd/m^2, i =1 to 145) in 145 sky directions for 68% of the sky. The most straightforward approach to identifying the CIE skies is comparing the measured luminance distribution against the CIE models. However, for sites with a low latitude of 22.3 degrees, the sun would approach the sky zenith at summer solar noon, which imposes sunlight instead of the skylight to the zenith measurements. Thus, the measured luminance was normalized by the horizontal illuminance instead of the zenith luminance L_Z using the Tregenza approach [15]. The CIE sky model that gave the least root mean square error against the normalized measurements represented the luminance distribution, and points out the overcast, partly cloudy or clear sky condition.

3.3. Frequency analysis by Fast Fourier Transfer

Fast Fourier Transfer (FFT) in MATLAB toolbox was used for discrete data series [14]. The absolute value of the FFT results denotes the frequency amplitude in the time-domain. For time series with N illuminance I_j samples, the kth term of the Fourier transfer is denoted as S_k (j and k = 1 to N). Given sampling frequency F_s (144/day for 10-min interval), the frequency for k was $F_s(k-1)/N$.

It can be difficult to evaluate the Sky type and the variation feature correlations by profiles. Thus, we weighted the daylight illuminance S_k at different frequencies by w_k defined as Equation 4, and defined the frequency factor F given as Equation 5. The ratio of (k−1) to (N/2) gives the relative significance of frequency k to the highest frequency under study. Higher weights were given to S_k at high frequencies, leading to a significant F for daylight profile that variated in a high frequency. a and b are constant coefficients. b defines the high and low-frequency threshold while a gives the weight variation speed around the threshold. We set a as 20 for the current study, which depicts a fast increase of the weight from 0 to 1 at the threshold b and a "hard" cut off for the low-frequency terms.

$$\omega_k = \frac{1}{1 + \exp\left(-a\left[(k-1)/(N/2) - b\right]\right)} \tag{4}$$

$$F = \left[\sum(w_k S_k)\right] / \sum S_k \tag{5}$$

Apart from the cloud coverage, however, the high frequency could result from the noises and data series discontinuities even though the data series is smooth. A solution is removing the low amplitude terms in the frequency domain. However, it can be difficult to set a hard threshold for removing the low amplitude terms without harming the data integrity. A systematic approach will be discussed later.

4. Results and discussions

4.1. CIE Standard Skies frequency of occurrence

Figure 1 gives the occurrence frequency of the CIE Standard Skies according to the sky luminance scan in the 10-minute interval. The figure shows that 37, 40 and 23% of the data sets were overcast, partly cloudy and clear skies, respectively. It can be surprising that the partly cloudy skies took such a high frequency. The most frequent skies of overcast, partly cloudy and clear skies were Skies 1, 8 and 13. The results were similar to another local [11], and an overseas work in similar climate [9].

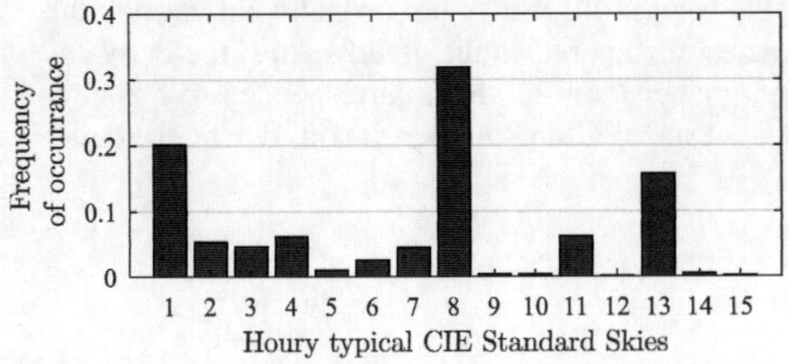

Figure 1. The occurrence frequency of CIE Standard Skies in the 10-minute interval

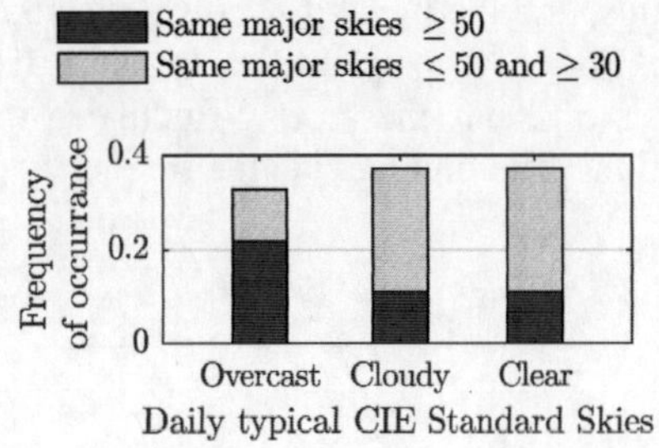

Figure 2. Occurrence frequency of daily overcast, partly cloudy and clear skies

Major sky type of the day plotted in Figure 2 was the most frequent overcast, partly cloudy or clear skies throughout the day. For the 73 measurements from 6:00 to 18:10, the major daily sky determined by more than 50, and 30 to 50 measurements in 10-minute interval were plotted separately. In total, 318 days with more than 30 consistent sky recordings in 10-miniute interval were accepted. The remaining 48 days were excluded since their 10-minute sky types were inconsistent during the day.

4.2. *Daily clearness index profiles in time and frequency domains*

Figure 3 (a) shows the daily horizontal global illuminance (I_{HG}) and its clearness index (K_t) profiles in the time domain for cloudy skies. The profile of I_{HG} in the frequency domain was similar to K_t. For the day dominated by the partly cloudy skies, both I_{HG} and K_t variated in high frequency throughout the measurement period. The randomly distributed clouds would block the direct sunlight recurrently, which probably lead to the irregular illuminance drops in its time series and the illuminance fluctuations in high frequency. Figure 3 (b) gives the I_{HG} components in various variation frequencies. We analyzed I_{HG} instead of K_t because I_{HG} was close to a normal distribution especially for clear skies, and there would be less high frequency terms due to the data discontinuity. Considering the sampling interval of 10 minutes and a total sample size of 36 during the daytime, the frequency of 10, for example, indicates an illuminance drop every 72 minutes. The figure quantifies I_{HG} at different variation frequencies and reveals noticeable I_{HG} components at frequencies greater than 10 times per day. The highest I_{HG} component at low frequency was due to daily solar altitude variation.

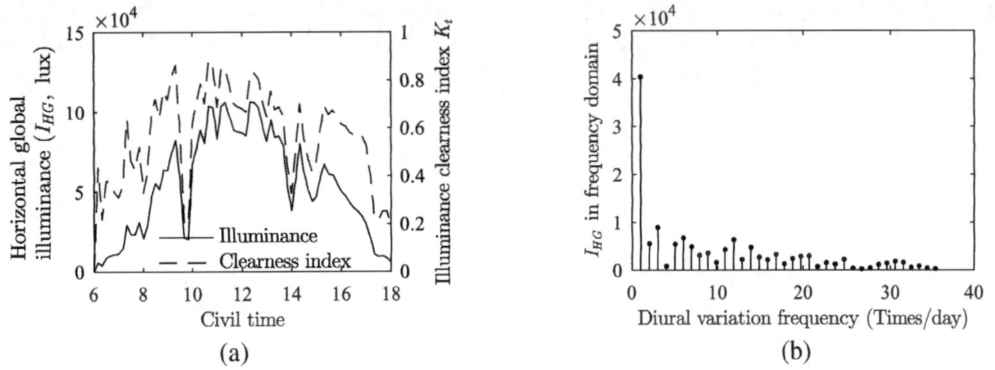

Figure 3. (a) Daily horizontal global illuminance and its clearness index in the time domain for cloudy skies; (b) Horizontal global illuminance in the frequency domain for cloudy skies

Figure 4 (a) gives the typical I_{HG} and K_t variations in time and frequency domains under clear skies. For the day that is dominated by the clear skies, the illuminance variated smoothly throughout the day, and its daily profile was in a bell shape. Compared to the partly cloudy skies in Figure 3, less I_{HG} and K_t variations in time domain were found for the day of clear skies. For clear skies, the sky was free of clouds, and the I_{HG} and K_t were affected mainly by the solar altitude that variated smoothly during the day. Figure 4 (b) quantifies I_{HG} in different frequencies. Compared to the cloudy skies, the daily I_{HG} series of clear skies rarely fluctuated in frequencies greater than 2, which corresponded to a 6-hour interval. The findings via the frequency profile indicated stable illuminance variations during the day.

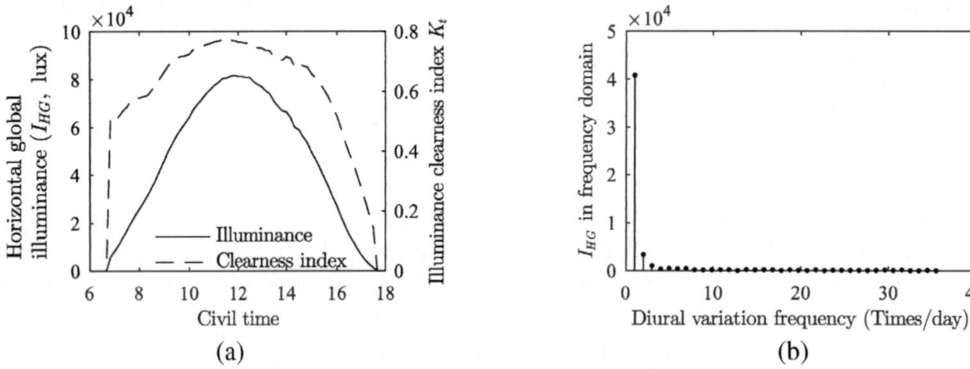

Figure 4. (a) Daily horizontal global illuminance and its clearness index in the time domain for clear skies; (b) Horizontal global illuminance in the frequency domain for clear skies

4.3. Daily clearness index and variation features for different skies

Figure 5 plots the daylight clearness index K_t and fluctuation factor F CIE overcast, partly cloudy and clear skies for the whole day. The cutting threshold b for high and low frequencies in Equation 7 was set as 0.6 by empirical trial and error tests. We removed the frequency terms that were less than two times of the 4th lowest frequency term to avoid noise impacts. The figure shows that K_t = 0.4 was a good criterion for separating the overcast and daily partly cloudy skies. K_t = 0.6 can be a potential threshold to separating the daily partly cloudy and clear skies. However, there were a few partly cloudy days whose K_t was greater than 0.6, though their daily sky conditions were supported by less than 50 sky scans per day. It is noticeable that the fluctuation feature of K_t, denoted by F, can be a supplementary criterion to identifying the clear and partly cloudy skies. Most of the clear skies were featured by an F less than 0.07 due to the stable daylight and its index K_t during the days clear of clouds. Thus, F would improve separating the partly cloudy and clear skies at high daily K_t. The overlapped partly cloudy and clear skies in the figure was probably because of the sky condition changing during the day. Selecting different thresholds b for high and low-frequency terms can lead to different F, yet its relative significance for overcast, partly cloudy and clear skies would not change.

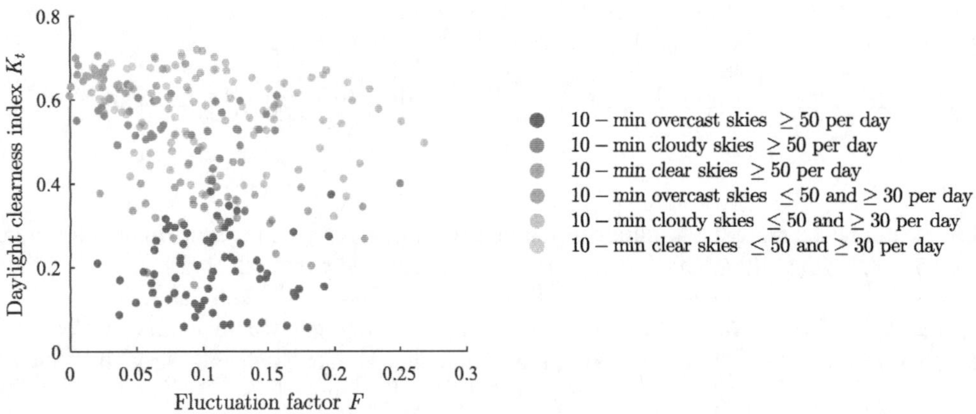

Figure 5. F and K_t of CIE overcast, partly cloudy and clear skies in the daily interval

5. Conclusions and future works

The current work attempted to identifying the daily CIE Standard overcast, partly cloudy and clear skies using the horizontal global illuminance data only. We analyzed the daily series of the horizontal global illuminance by transferring it from the time domain to the frequency domain. The daily global horizontal illuminance I_{HG} series of clear skies was featured by the low amplitude terms in high frequency, owing to the stable sunlight from the sky clear of clouds. The overcast and partly cloudy skies, in comparison, were featured by high-frequency terms greater than the clear skies, probably due to the frequent sunlight coverage by clouds. The frequency domain profiles were summarized by assigning greater weights to the high-frequency terms for a weighted total, which was defined as the fluctuation factor. For I_{HG} in the frequency domain, we assigned weight close to 1 for those whose frequency was greater than 60% of the maximum, and setting 0 for the rest. Finally, we found greater fluctuation factors for overcast and partly cloudy skies compared to the relative lower fluctuation factor for clear skies. The fluctuation factors for most clear skies were less than 0.075 by the current weight settings, which can improve identifying the partly cloudy and clear skies whose daily K_t was greater than 0.6.

A limitation of the current work is that the data series were in the daily interval. However, the data series for an hour or half day may not begin and end at zero. Such discontinuity at the series head and tails can lead to many low-amplitude, high-frequency terms even though the data series is smooth in the time domain. A systematic approach to solve this problem will be presented in future work.

Acknowledgments

The work was under the project *A Fundamental Study of Anisotropic Sky Radiance and its Effect on Building Envelopes in City Environments* supported by National Natural Science Foundation of China (ID: 51808139).

References

[1] EMSD 2017 *Hong Kong Energy End-use Data* (Hong Kong: Electrical & Mechanical Services Dept.)
[2] Lam J C 2000 Energy analysis of commercial buildings in subtropical climates *Building and Environment* vol 35 pp 19-26
[3] Li D H W, Chow S K H and Lee E W M 2013 An analysis of a medium size grid-connected building integrated photovoltaic (BIPV) system using measured data *Energy and Buildings* vol 60 pp 383-387
[4] Liang H, Li S K, Ma J Y and Luo Y 2011 Variation trend of direct and diffuse radiation in China over recent 50 years *Acta Physica Sinica* vol 60(6) pp 069601
[5] Klucher T M 1979 Evaluation of models to predict insolation on tilted surfaces *Solar Energy* vol 23 pp 111-114
[6] Perez R, Ineichen P, Seals R, Michalsky J and Stewart R 1990 Modeling daylight availability and irradiance components from direct and global irradiance *Solar Energy* vol 44 pp 271-289
[7] Kittler R, Darula S and Pereze R 1998 *A set of standard skies* (Bratislava: Polygrafia SAC)
[8] CIE 2003 *Spatial Distribution of Daylight - CIE Standard General Sky* (Vienna: CIE Central Bureau)
[9] Wittkopf S K and Soon L K 2007 Analysing sky luminance scans and predicting frequent sky patterns in Singapore *Lighting Research & Technology* vol 39 pp 31-51
[10] Yun S I and Kim K S 2018 Sky Luminance Measurements Using CCD Camera and Comparisons with Calculation Models for Predicting Indoor Illuminance *Sustainability* vol 10(5) pp 1556
[11] Ng E, Cheng V, Gadi A, Mu J, Lee M and Gadi A 2007 Defining standard skies for Hong Kong *Building and Environment* vol 42 pp 866-876
[12] Li D H W, Chau T C and Wan K K W 2014 A review of the CIE general sky classification approaches *Renewable and Sustainable Energy Reviews* vol 31 pp 563-74
[13] Lou S, Li D H W and Lam J C 2017 CIE Standard Sky classification by accessible climatic indices. *Renewable Energy* vol 113 pp 347-56

[14] Frigo M and Johnson S G 1998 FFTW: an adaptive software architecture for the FFT *Proceedings of the International Conference on Acoustics, Speech and Signal Processing* vol 3 pp 1381-1384
[15] Tregenza P R 2004 Analysing sky luminance scans to obtain frequency distributions of CIE Standard General Skies *Lighting Research and Technology* vol 36 pp 271-9

IOP Conf. Series: Materials Science and Engineering **556** (2019) 012016 doi:10.1088/1757-899X/556/1/012016

Holistic Optimisation of Solar-assisted Multiple Effect Evaporators (MEE)

Soundaram Ramanathan[1] and Dibakar Rakshit[2]
[1]Centre for Science and Environment, New Delhi, India
[2]Indian Institute of Technology - Delhi, New Delhi, India
E-mail: dibakar@ces.iitd.ac.in

Abstract. The effects of heat transfer performance of systems can be studied in combination with thermodynamic performance. The design solution of maximizing heat transfer always leads to increased entropy generation else if minimizing entropy was aimed then heat transfer got minimized. Hence there is need for holistic optimization. This paper attempts to merge the two competing criteria's - (i) heat transfer and (ii) the thermodynamic performance of a solar assisted multiple effect evaporator system and optimize study parameters for its energy efficient operation. Solar assisted multiple effect evaporators use steam as source of energy to evaporate wastewater.

The idea of amalgamating these two factors is to observe how much reduction in the performance of one factor will improve the performance of the other. This study couples the two criteria, which has been dealt with separately, to arrive at the parameters of interest which must be fine-tuned for ensuring system performance.

Mass flow rate of steam and wastewater appeared to chiefly influence heat transfer and entropy generation of the system, negligible influence was observed on changing the concentration of the waste water or optical efficiency of the solar concentrator in single legged analysis. However through this holistic optimization study, it was found that variations in Prandtl number, i.e., the concentration of the waste water and optical efficiency of solar concentrator drastically shift the optimization balance of entropy generation and heat transfer rate.

Keywords: Multiple effect evaporators, Holistic optimisation, Solar, Exergy, Entropy, Wastewater

1. Introduction

Multiple effect evaporation process is a thin film evaporation process, in which the vapor formed in one chamber, or effect, condenses in the next one, providing a heat source for further evaporation. (See Figure 1. Multiple effect evaporator-Concept (MC Georgiou et al)) [1] Advantage of the process is steam economy. It is estimated that for a single effect evaporators nearly 0.8 kg of water can be evaporated with 1 kg steam. [2] To treat the reverse osmosis rejects multiple effect evaporators were found the best option. This device is indispensable in wastewater treatment systems because to do any end-of the life functions with concentrated salt rejects – separation of salt and water is necessary. Vaporisation and separation is the best possible technique to achieve this. [3] Hence it is agreed upon by researchers that further advancements must be made in multiple effect evaporators. Unlike traditional boilers where heat is supplied to vaporize products, in single-effect evaporator systems steam provides energy for vaporization instead of direct heat and the vapor product is condensed and removed from the system, while in double-effect evaporator, the vapor product off the first effect is used to provide energy for a second vaporization unit. [4] This cascading of effects continues for many stages. Vapor from Effect I will be used to heat Effect II, which consequently will operate at lower pressure. These continue through the train: pressure drops through the sequence so that the hot vapor will travel from one effect to the next. Effects are numbered beginning with the one heated by steam which will have the highest pressure.

Normally, all effects in an evaporator will be physically the same in terms of size, construction, and heat transfer area. [5] Unless thermal losses are significant, they will all have the same capacity as well.

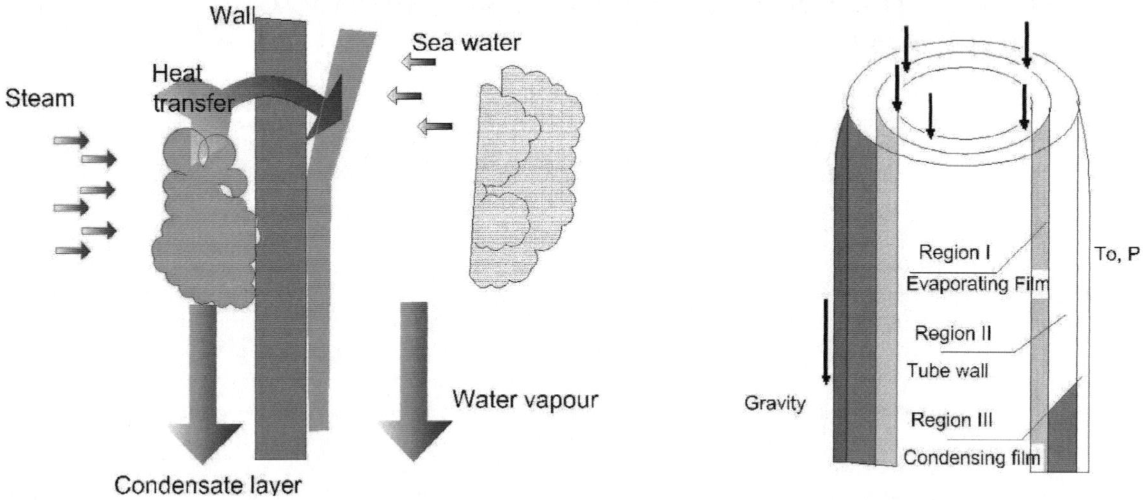

Figure 1. Multiple effect evaporator-Concept (MC Georgiou et al)

2. Multiple effect evaporators: System description

Theoretically evaporators may be built with any arbitrary number of stages, however the stages are limited in practical cases by constraints such as space, cost, etc. Several configuration models have been developed in the past depending on the configuration of feed flow. [6]
Equation 1 gives the heat transfer up-to ‘n’ effects/chambers/evaporators-

$$Q_n = m_s\lambda_s = U_{Dn}A_n(T_{sn} - T_{Bn}) = U_{Dn}A_n\Delta T_n \tag{1}$$

Specific entropy of the steam was calculated using simplified equation given by Affandi et al (2012) depending on the temperature of the steam [7],

$$\ln S_g = a + b\,(0.35 * \ln(\frac{1}{T_s})) + \frac{c}{{T_s}^2} + \frac{d}{{T_s}^4} + \frac{e}{{T_s}^5} \tag{2}$$

where a = 1.47735, b=0.53242, c=-0.01923, d=0.02974, e=-0.00802

And exergy using the entropy calculated above,

$$X = h - h_0 - T(S - S_0) \tag{3}$$

3. Parameters influencing the performance

A general formula for heat transfer coefficient during evaporation processes in the first effect of the multiple effect evaporator attached with a solar concentrator can be put in the form:

$$Nu = a(Re)^b\,(Pr)^c(\eta_0)^d\left(\frac{A_a}{A_r}\right)^e) \tag{4}$$

where a, b, c, d and e are constants.

Similarly the function for entropy was obtained as

IOP Conf. Series: Materials Science and Engineering **556** (2019) 012016 doi:10.1088/1757-899X/556/1/012016

$$S_{gen} = f(Re)^g (Pr)^h (\eta_0)^i \left(\frac{A_a}{A_r}\right)^j) \quad (5)$$

In the model equation (5), the below parameters were altered and the associated dimensionless number were recorded (see. **Table 1.** Constraints Range Sampled). Based on the below described constraints which were obtained from industry, equation (4) and (5) were solved for correlation.

Table 1. Constraints Range Sampled

S. No.	Constraint	Range
1.	Optical efficiency of the solar collector	40-80%
2.	Concentration Ratio	2 - 600
3.	Mass flow rate	20 – 335 kg/h
4.	Concentration	1-52 g/kg
5.	Temperature of the steam in 1st effect	150 – 300 ^{0}C
6.	Number of effects	3 – 15
7.	Overall heat transfer coefficient of the material in the 1st effect of the evaporator	2,270 – 6,000 W/m-K

The below correlation was obtained through the multiple regression analysis,

$$Nu = a\,Re^b\,Pr^c\,\eta_0{}^d\,\text{concentration ratio}^e$$
$$r^2 = 0.97 \quad (6)$$
where a = 0.183, b = 1.692, c = 0.513, d = 0.283, e = 0.18

The correlation had a good parity with the model data with standard deviation error of 0.1-8% and R-square of 0.97. Similarly correlation was obtained associating entropy of the steam generated with Reynolds number, Prandlt number, optical efficiency and concentration ratio for evaluating the performance of the system by second law of Thermodynamics.

$$S_{gen} = 0.015\,Re^{0.604}\,Pr^{0.345}\,\eta_0{}^{0.015}\,\text{concentration ratio}^{0.117}$$
$$r^2 = 0.98 \quad (7)$$

4. Holistic Optimisation

The effects of heat transfer performance of systems is to be studied in combination with thermodynamic performance. The design solution of maximizing heat transfer always leads to increased entropy generation else if minimizing entropy was aimed then heat transfer got minimized. Hence need for holistic optimization was recognized. The idea of amalgamating these two factors is to observe how much reduction in the performance of one factor will improve the performance of the other [8]. As far as the heat transfer of a system is concerned the performance parameter of choice is the heat dissipated per unit mass of the system and with regard to irreversibility, the entropy generation rate is taken as the parameter for comparison [9].

To obtain the heat transfer in the first effect of the MEE, the Nusselt number [10] is used to obtain heat content of the steam as,

$$h = 0.183kRe^{1.692}Pr^{0.513}\eta_\circ^{0.283}CR^{0.18} \quad (8)$$

It is to be noted that this heat content of the steam is basically the heat transferred to the wastewater for steam generation neglecting the sensible heat requirements as per the model hence instead of the conventional ΔT method to obtain heat transfer, the indirect heat content measurement has been used here, to obtain the heat transfer neglecting sensible heat requirements of the phase change. Similarly analysis was extended to estimate the entropy generated.

For the range of parameters discussed in Table 1, the corresponding energy and entropy generation were computed. This was followed by obtaining the ratio of heat transfer to maximum possible heat transfer (q/qmax) and ratio of the minimum possible entropy generation to the minimum entropy generated (Smin/S). The ratios presented the penalty percentage, if 50% is the penalty range considered, i.e the system needs to operate for say 50% of best heat transfer rate possible, with 50% of minimized entropy generation rates for the given parameter range then the corresponding was obtained by plotting graphs of q/qmax, Smin/S against the parameter sampled. This can give a preliminary idea on the solution range, however this would be insufficient. For holistic optimization penalty ranges must be varied and weightages considered for the ratios must be altered as in Rakshit et al. [11],

$$\Phi = \alpha\left(\frac{q}{q_{max}}\right)^{a} + (1-\alpha)\left(\frac{S_{min}}{S}\right)^{b} \tag{9}$$

where, $0 \leq \alpha \leq 1$

In order to obtain a holistic optimum, Φ needs to be minimized, a and b are constants and can have the same or different values. For the present case, a = b = 1 is considered which signifies equal weightage for heat transfer per unit mass maximization and entropy generation minimization. To determine the optimized parameters the value of α needs to be identified such that Φ is minimum and this determines the minimum total penalty.

5. Results and Discussion

5.1. Single-legged analysis

Exergy efficiency of the device depends on the amount of solar insolation. Also the exergy efficiency and heat transfer of the system is chiefly related to Reynolds number or waste water flow in the system. Change in Prandtl number did not significantly affect the energy need of the system as change in the properties of the fluid only marginally shifted the boiling temperature of the fluid by about 50°C which did not alter the energy needs as change in mass stream velocity in Reynolds number. This was also verified by parametric study where for a range of Prandtl and Reynolds number the Nusselt number showed significant concave linear variation for change in Reynolds number rather than Prandtl number which signifies change in concentration of the waste water flow would not affect significantly the Nusselt number.

Similarly the entropy had a convex linear variation with change in Reynolds number. Upto a value of Reynolds number 400, change in optical efficiency does not alter the Nusselt number, however post 400 there are variations in the order of 1/10000 for different optical efficiency for the same Reynolds number. (See. **Figure 2.** Variation of Nusselt number with Reynolds number and Optical efficiency). Similar range of variation was observed when the concentration of the property of waste water was kept constant with varying optical efficiency. No variation in Entropy however was observed with varying optical efficiency and constant Reynolds number and Prandtl number respectively. Also from the study one can understand that the variation in concentration ratio of solar collector or optical efficiency of the solar thermal collector system did not significantly influence the Nusselt or exergy efficiency of the system which clarifies hourly variations in solar radiations does not significantly affect the performance of the solar assisted MEE system as a whole.

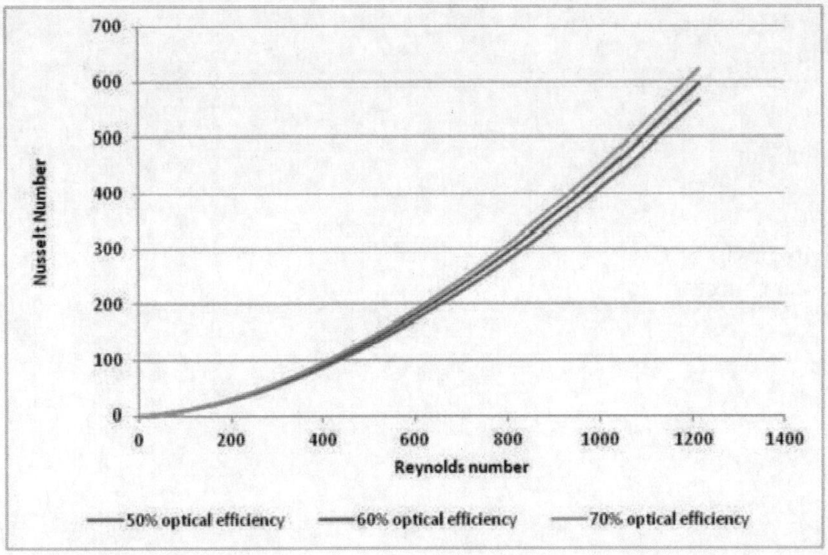

Figure 2. Variation of Nusselt number with Reynolds number and Optical efficiency

5.2. Holistic Optimisation

In the present case where equal weightage was given for both heat transfer per unit mass maximization and entropy generation minimization, the minimum total penalty in heat transfer (approx.) was 20% and reduction in entropy generation (approx.) was 80%. The least value of **Φ** was obtained in the case of Reynolds number (see. **Table 2.** Parameter range and values of **Φ**). This signifies that the parameter holds in the case of arriving at the holistic optimization.

Table 2. Parameter range and values of **Φ** (a=1, b=1, **α**=0.2)

Parameter	Range	**Φ**
Reynolds number	3,270-16,370	0.03
Concentration ratio	2-10	0.5
Prandtl number	6.8-7	0.9
Optical efficiency	60-65%	0.9

6. Conclusion

The above results thus signifies that changes in Prandtl number, i.e the concentration of the waste water and optical efficiency of solar concentrator can drastically shift the optimization balance of entropy generation and heat transfer rate. While when these parameters were studied separately it appeared their changes had negligible influence on change in heat transfer or entropy generation. Also it was observed that no significant changes appeared on altering weightages, which implies keeping the mass flow rate low or making sure the heat transfer rates are bare minimum is the best way to destroy entropy generation.

7. Nomenclature

m	flow rate in kg/s
H	enthalpy in kJ/kg-^{0}C
λ	latent heat in kJ/kg
U_D	heat transfer coefficient of the evaporator in kJ/ sq.m-^{0}C-s
T	temperature in ^{0}C
Q	heat transfer in each effect in kJ

A	heat transfer area in the evaporator in sq.m
CR	Concentration Ratio
S_g	Entropy in kJ/kg
X	Exergy in kJ/kg
μ	dynamic viscosity in N/m
l	length in m
C_p	Specific heat in kJ/kg-°C
k	Thermal conductivity in kJ/kg-°C-s
η_c	Efficiency of the solar collector
Re	Reynolds number
Pr	Prandtl number
Nu	Nusselt number

Subscript

f	feed flow rate
s	steam
B	brine flow rate
1,2,..n	effect/chamber/evaporator number.

8. References

[1] MC Georgiou, The Cyprus Institute, 2013 DOI: http://www.poemconference.org/poem2013/uploads/presentations/08-10-2013/Special_Session-Solar_thermal_cogeneration_of_electricity_and_water_The_STEP-EW_project/Thermal_desalination_performance_analysis.pdf

[2] Golder AK, Goud VV. Chemical Engineering Design-II-Web course.

[3] Vishnu G, Palanisamy S, Joseph K. Assessment of fieldscale zero liquid discharge treatment systems for recovery of water and salt from textile effluents. Journal of Cleaner Production. 2008 Jul 31;16(10):1081-9.

[4] El-Dessouky HT, Ettouney HM. Fundamentals of salt water desalination. Elsevier; 2002 Mar 20.

[5] Sarma G, Barma SD. Energy Management in Multiple–Effect Evaporator System: A Heat Balance Analysis Approach. Gen. 2010 Dec;1(2):84-8.

[6] Khanam S, Mohanty B. Energy reduction schemes for multiple effect evaporator systems. Applied Energy. 2010 Apr 30;87(4):1102-11.

[7] Marwan Affandi, Normahira Mamata, Siti Nurul Aqmariah Mohd Kanafiaha , Nurul Syahirah Khalida. Simplified Equations for Saturated Steam Properties for Simulation Purpose, Malaysian Technical Universities Conference on Engineering & Technology 2012, MUCET 2012 Part 6 - Science

[8] Sasikumar M, Balaji C. A holistic optimization of convecting-radiating fin systems. Journal of heat transfer. 2002 Dec 1; 124(6):1110-6.

[9] Sasikumar M, Balaji C. Optimization of convective fin systems: a holistic approach. Heat and Mass transfer. 2002 Nov 1; 39(1):57-68.

[10] Jyoti G, Khanam S. Simulation of heat integrated multiple effect evaporator system. International Journal of Thermal Sciences. 2014 Feb 28; 76:110-7.

[11] Rakshit D, Balaji C. Thermodynamic optimization of conjugate convection from a finned channel using genetic algorithms. Heat and Mass transfer. 2005 Apr 1; 41(6):535-44.

Calibrating Perez Model Coefficients Using Subset Simulation

M Meshkinkiya[1], **R.C.G.M. Loonen**[2], **R Paolini**[3] and **J.L.M Hensen**[2]
[1]Politecnico di Milano, Milan, Italy
[2]Eindhoven University of Technology, Eindhoven, Netherlands
[3]University of New South Wales, Sydney, Australia

E-mail: maryam.meshkinkiya@polimi.it

Abstract. The Perez sky diffuse model has been validated in many locations, however, studies have shown that the precision of estimation can be improved through localization of the model coefficients. This paper studies the effect of tuning Perez irradiance coefficients based on local information for estimating the incident solar radiation on tilted surfaces. The added value of a calibrated Perez model is highlighted by evaluating the heating and cooling energy needs of an office building. Calibration is performed by using Subset Simulation, i.e. a sampling technique based on Markov Chain Monte Carlo. The versatility of the Subset Simulation technique allows for handling multivariant calibration problems and is therefore suitable for finetuning Perez irradiance coefficients. Measurements of global horizontal, direct normal and diffuse horizontal solar irradiation, as well as the global solar irradiation on 90°, 30° and 15° tilted surfaces form the basis of the calibration. It is found that in the studied location, the default Perez model overestimates the incident solar radiation on vertical surfaces facing south. The calibrated coefficients are then embedded in the source code of EnergyPlus. Simulations with a reference office show that the effect of calibrating Perez coefficients can be significant, because it leads to approximately 12% difference in predicted annual cooling energy use and 9% difference in peak cooling loads, respectively.

Keywords: Perez model; Calibration; Subset Simulation; Building energy

1. Introduction

Solar radiation has a significant influence on a building's energy balance, as it affects both the potential for on-site energy generation (e.g. PV or solar thermal) and the indoor contribution of solar heat gains and daylighting. While solar irradiation on horizontal surfaces is measured in numerous locations around the world, the incident solar irradiation on tilted surfaces is mostly calculated through white-box models [1]. Specifically, the diffuse component of global solar radiation on tilted surfaces is calculated by models that transpose the diffuse horizontal solar radiation to diffuse solar irradiation on tilted surfaces, generally referred to as diffuse solar irradiation models. Solar models are mainly developed based on the data available in a few selected locations, which are then extrapolated to the entire globe [2]. The increasing availability of incident solar irradiation measurements on tilted surfaces, especially at PV farms, creates the opportunity to assess the accuracy and performance of solar models in different locations [3–7]. Comparing solar models in different studies reveals that for obtaining more accurate estimations of solar irradiation on inclined surfaces, each location may perform better with a specific (or even a hybrid) solar model [4]. Therefore, finding the best model for a site is not easy since the mismatch between measurements and calculations can come from either the model or the measurements [8].

IOP Conf. Series: Materials Science and Engineering **556** (2019) 012017 doi:10.1088/1757-899X/556/1/012017

Among solar models, the Perez Model is frequently used in building energy simulation [9]. The model calculates diffuse irradiation on tilted surfaces based on estimations of the brightness and clearness of the sky [10]. Different approaches have been proposed to calibrate the Perez model to a specific location. Sun et al. proposed an uncertainty quantification of Perez parameters through fitting regression on a selection of parameters [11], resulting in an improved version of the Perez model for the intended location. As an alternative approach, Yang et al. used a "Least Square Method" to optimize the Perez irradiance coefficients for tropical sky conditions, stressing that the intended location was not among the original experimental data used to develop the Perez model [12].
In this study we propose a new method to tune the Perez model for a specific location. This process involves Subset Simulation, which is based on Markov Chain Monte Carlo and Importance Sampling. The effect of using a calibrated Perez model is evaluated on the incident solar irradiance on vertical surfaces, as well as cooling and heating loads of a building. The method proposed in this study can successfully adapt Perez coefficients to local measurements and is easy to implement in building energy simulation models.

2. Calibration method

2.1 Subset Simulation

In this study, we opt for Subset Simulation (SuS) to calibrate Perez solar coefficients. The application of SuS for calibration of different models has previously been adopted by Gong et al. [13]. This technique is specifically useful for complex models with high dimensions of uncertain variables and can reduce the computational time of the process. The method shifts rare probabilities to more frequent events, by resorting to Markov Chain Monte Carlo sequence, i.e. rejecting unwanted samples that have higher errors [14]. The SuS framework adopted in this study is presented in Figure 1.

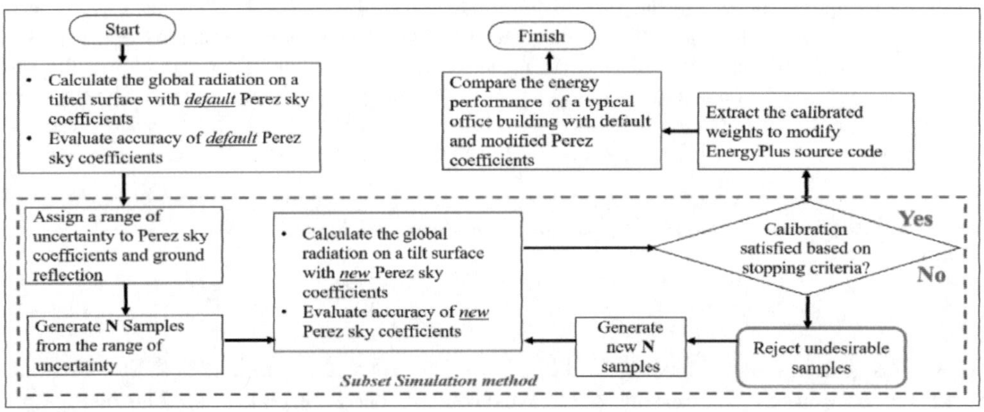

Figure 1. Framework of calibration method using Subset Simulation

2.2 Uncertain parameters

Global irradiation on tilted surfaces is calculated from direct, sky diffuse and ground reflected irradiation. Among the three components, sky diffuse irradiation has been reportedly associated with the largest amount of uncertainty [11]. In the Perez model, sky diffuse radiation is calculated through a series of predefined irradiance coefficients (See Table 1 in [10]). Irradiance coefficients in Perez model are presented for eight sky categories (e-Bins), which vary from overcast to clear sky conditions. To decrease the number of uncertain parameters and avoid multi-modality, second order curves are fitted to e-Bins values. Converting the bins to curve equations reduces the number of uncertain parameters from 48 to 18. Therefore, instead of using the actual values of Perez coefficients, the coefficients of polynomial equations are used in the SuS calibration. The total radiation incident on a tilted surface is also affected by the reflected solar radiation from the ground. Since the ground reflected radiation is composed of both direct and diffuse radiation and considering that the diffuse component is the subject of the study, the albedo of ground is also associated with uncertainty and subject to calibration.

2.3 Gaussian Mixture Model

Ground reflection and Perez irradiance coefficients are the parameters that are associated with uncertainty for calibration. Therefore, the study deals with a multivariate distribution of uncertain variables, composed of 19 different parameters in total. To represent the covariation of all uncertain parameters within a single matrix, we opted for a Gaussian Mixture Model (GMM). The GMM creates a k multivariant distribution, where k is the number of random (or uncertain) parameters [15]. In this study, MATLAB scripting is used to create a pipeline of functions for implementing the SuS.

3. Case study

3.1 Data description

This study is developed based on sub-hourly measurements of solar irradiation, which are provided by SEAC/SolarBEAT. The measurement site is located at Eindhoven University of Technology with 51.4° N, 5.5° E latitude and longitude, respectively. All measurements are performed with secondary class pyranometers from EKO Instruments B.V., which are installed perfectly due south. Table 1 provides an overview of datasets utilized in this study. It should be noted that for performing building energy simulations, we used measured horizontal diffuse irradiation rather than the traditional diffuse irradiation in typical weather data files. Therefore, the global, direct and diffuse horizontal solar irradiation in the TMY file are replaced with that of measured data. This is important as we want to assess the specific effects of calibrating Perez with real measurements.

Table 1. Availability of measured data

Quantity	Tilt	Time steps	Period of data
Global, diffuse and direct irradiation	0°	15 minutes	2015-2017 / 24 months
	0°	1 minute	2017-2018 /10 months
Global irradiation	90°	15 minutes	2016-2017 /15 months
	30°	15 minutes	2015-2016 /12 months
	15°	1 minute	2017-2018 /10 months

3.2 Model implementation

Prior to performing the calibration, measured solar irradiation datasets are divided in two portions; a training set and a test set. The training data consists of 70% of the measurements which are randomly selected from the entire dataset. The remaining 30% are used solely for testing the performance of the model after calibration. To increase the probability of sampling from the "area of interest" with smaller errors, it is necessary to define a threshold, based on which unwanted samples are rejected. In this study the performance of the MCMC is evaluated based on the Euclidean distance between absolute zero and a custom measure dubbed the "Threshold", i.e. equation (Eq.1). The Threshold is calculated from the CV_RMSE (%) and MAE of all datasets with various tilts.

$$Threshold = \begin{bmatrix} (CV\_RMSE_{Tilt\ 90})^2 + (MAE_{Tilt\ 90})^2 + (CV\_RMSE_{Tilt\ 30})^2 + \\ (MAE_{Tilt\ 30})^2 + (CV\_RMSE_{Tilt\ 15})^2 + (MAE_{Tilt\ 15})^2 \end{bmatrix}^{1/2} \quad \text{(Eq.1)}$$

Where:

$$RMSE = \left[\frac{1}{n}\sum_1^n (GI_{Measured} - GI_{Simulated})^2\right]^{1/2} \quad \text{(Eq.2)}$$

$$CV\_RMSE = RMSE\left(\frac{1}{n}\sum_1^n GI_{Measured}\right)^{-1} \quad \text{(Eq.3)}$$

$$MAE = \frac{1}{n}\sum_1^n |GI_{Meas} - GI_{Simulated}| \quad \text{(Eq.4)}$$

The output of SuS are the calibrated parameters i.e. ground albedo and Perez sky coefficients. The calibrated ground reflectance returns a value of 0.1, which is notably smaller than the default value (0.2) but can be explained by the fact that a black rubber mat is placed below the pyranometers in the experiment. The Perez sky clearness coefficients fitted to Eindhoven sky conditions are presented in Table 2. The performance of the calibration process is reported in Table 3. The improvement of estimation using a calibrated model, is evaluated on the test dataset. The highest improvement was observed for 30° of tilt, reducing the RMSE from 57.6 W/m^2 to 36.6 W/m^2. This improvement was slightly smaller for the 90° estimations with the calibrated model.

Table 2. The numeric value of Perez irradiance coefficients before and after calibration for Eindhoven sky. Default coefficients (left) and calibrated coefficients (right).

e-Bin	F11	F12	F13	F21	F22	F23
1	-0.008	0.588	-0.062	-0.06	0.072	-0.022
2	0.13	0.683	-0.151	-0.019	0.066	-0.029
3	0.33	0.487	-0.221	0.055	-0.064	-0.026
4	0.568	0.187	-0.295	0.109	-0.152	-0.014
5	0.873	-0.392	-0.362	0.226	-0.462	0.001
6	1.132	-1.237	-0.412	0.288	-0.823	0.056
7	1.06	-1.6	-0.359	0.264	-1.127	0.131
8	0.678	-0.327	-0.25	0.156	-1.377	0.251

e-Bin	F11	F12	F13	F21	F22	F23
1	0.2437	0.7074	-0.1881	0.0966	0.4489	-0.0015
2	0.2895	0.6334	-0.2001	0.1138	0.4253	0.0046
3	0.3733	0.4996	-0.2221	0.1453	0.3824	0.016
4	0.5033	0.2962	-0.2567	0.1944	0.3166	0.0345
5	0.7094	-0.0111	-0.3127	0.2737	0.2153	0.0665
6	1.0058	-0.3888	-0.3988	0.3932	0.0816	0.124
7	1.2305	-0.5363	-0.4759	0.4957	0.0063	0.1924
8	0.9992	0.4676	-0.469	0.4632	0.2442	0.2741

Table 3. Statistical analysis

				Training set			**Test set**		
	Original Perez			**Calibrated Perez**			**Calibrated Perez**		
	South 90°	South 30°	South 15°	South 90°	South 30°	South 15°	South 90°	South 30°	South 15°
RMSE	37.56	57.67	55.46	32.15	36.62	40.17	32.93	36.17	40.49
CV-RMSE (%)	13.95	13.91	21	11.93	8.83	15.21	12.38	8.87	15.67
MAE (W/m^2)	27.17	40.07	40.37	22.33	25.67	28.07	22.3	25.42	27.74

4. Result and discussion

The effects of opting for a calibrated Perez model and how it impacts building's estimated energy performance is discussed in this section. Therefore, the original sky clearness coefficients from Perez are replaced with the calibrated coefficients in EnergyPlus software, and both the original and calibrated sky models are tested on a typical building energy model. In this study, we perform the comparisons on an energy model that is characterized based on ASHRAE's large office reference building. To have a more robust observation of the calibration's impact, a single thermal zone oriented due south is selected. The area of the studied zone is 340.72 m^2 and the studied zone is preserved from the adjacent indoor spaces by considering internal boundary conditions as adiabatic. In this study, the term "Base-EP" refers to the default EnergyPlus code with original Perez coefficients, and a ground reflection of 0.2. The term "Modified-EP" refers to the modified EnergyPlus code with calibrated Perez parameters and a ground reflectance of 0.1.

Figure 2 (Left) indicates that the calibrated implementation of the Perez model returns different estimations of incident solar irradiation on vertical surfaces. It is understood that the Base-EP overestimates incident solar irradiation on south facade. Also, it can be clearly observed how the calibrated model Perez shifts the estimations of solar irradiation with Modified-EP to real measurements. In free-floating conditions, indoor comfort is strongly correlated by the building's boundary conditions, and therefore, any changes in the surrounding environment can affect the indoor comfort conditions.

This said, the estimations of solar irradiation are among the parameters that affect the solar heat gains, and therefore the indoor temperature as displayed in Table 4. The Modified-EP reported a smaller frequency of temperatures in the comfort bound while the frequency of heating hours (T < 20 °C) is increased.

Table 4. Indoor temperature frequency in free-floating conditions

Indoor air temperature (°C)	Base-EP (hours)	Modified-EP (hours)
T < 20	5805	6067
20 ≤ T ≤ 26	2715	2522
T > 26	240	171

We also compared the building's heating and cooling loads using both default and modified EnergyPlus models. The results of annual heating and cooling loads reveal that the overestimation of solar irradiation by EnergyPlus mostly affects the cooling loads (Table 5). Consequently, the default Perez model overestimates the annual cooling load by approximately 12%, in the considered climate context. Moreover, the difference in the peak cooling loads obtained by the two models is significant (9%). On the other hand, it can be argued that the variation of annual heating energy use and peak heating load are negligible.

Table 5. Effect of calibrated Perez model on the estimations of heating and cooling loads

Case	Annual energy (MWh)		Peak load (kW)	
	Heating	Cooling	Heating	Cooling
Base-EP	21.72	15.52	37.82	11.83
Modified-EP	22.48	13.75	37.87	10.85

Since the frequency of peak loads can affect the system sizing process, a cumulative distribution of all cooling loads is provided in Figure 2 (Right). It is observed that when opting for a 90% reliability, the Base-EP returns a cooling load of 5569 W, which corresponds to a cooling system with at least 5500 W of capacity. Meanwhile, the Modified-EP returns a cooling of 4933 W for the same level of reliability (90%), which corresponds to a smaller system with capacity of 5000 W.

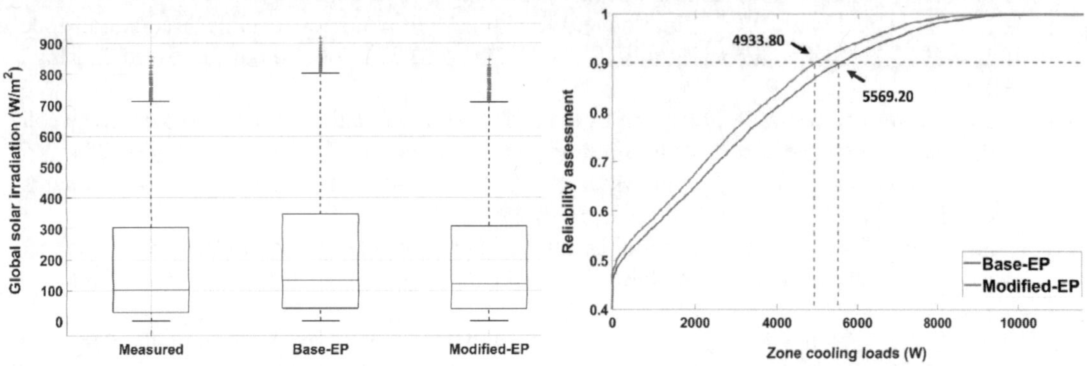

Figure 2. Comparison of incident solar irradiation on vertical surface (Left), Reliability assessment of the proposed cooling system (Right).

5. Conclusion

In this study, we proposed a method for calibrating the Perez sky diffuse model using Subset Simulation. The calibration method focused on ground reflection and Perez irradiance coefficients, by using measured global solar irradiance obtained from three different tilts and a horizontal surface. Also implementing the calibrated Perez model into building energy simulation tool EnergyPlus, revealed an

overestimation of incident radiation on vertical walls where annual mean absolute error is reduced from 14 to 4 W/m^2. This resulted in overestimating the cooling loads by approximately 12 % for Eindhoven climate. More accurate calculation of solar irradiations in the context of building energy performance can also be important in estimating the performance of BIPVs, and therefore expanding this study to the envelope's solar potential can further underline the importance of using calibrated solar models.
Since in this study the measured data were only available for the south orientation, concluding about the numerical effect of calibration on building's energy performance is not finalized. Therefore, the proposed framework can be adapted to other case studies with measured data in all orientations and is a potential for future studies.

Acknowledgment
Maryam Meshkinkiya would like to thank Fondazione Fratelli Confalonieri for their support of her PhD studies through the scholarship "Borsa per dottorandi di ricerca delle Universita Milanesi"

References
[1] Boland J, Ridley B, Brown B. Models of diffuse solar radiation. *Renew Energy* 2008; 33: 575–584.
[2] Dervishi S, Mahdavi A. Computing diffuse fraction of global horizontal solar radiation: A model comparison. *Sol Energy* 2012; 86: 1796–1802.
[3] Noorian AM, Moradi I, Kamali GA. *Evaluation of 12 models to estimate hourly diffuse irradiation on inclined surfaces*. Epub ahead of print 2008. DOI: 10.1016/j.renene.2007.06.027.
[4] Demain C, Journée M, Bertrand C. Evaluation of different models to estimate the global solar radiation on inclined surfaces. *Renew Energy* 2013; 50: 710–721.
[5] de Simón-Martín M, Alonso-Tristán C, Díez-Mediavilla M. Diffuse solar irradiance estimation on building's façades: Review, classification and benchmarking of 30 models under all sky conditions. *Renew Sustain Energy Rev* 2017; 77: 783–802.
[6] Wattan R, Janjai S. An investigation of the performance of 14 models for estimating hourly diffuse irradiation on inclined surfaces at tropical sites. Epub ahead of print 2016. DOI: 10.1016/j.renene.2016.02.076.
[7] Li Z, Xing H, Zeng S, et al. Comparison of Anisotropic Diffuse Sky Radiance Models for Irradiance Estimation on Building Facades. *Procedia Eng* 2017; 205: 779–786.
[8] Gueymard CA, Myers DR. Validation and Ranking Methodologies for Solar Radiation Models, https://link.springer.com/content/pdf/10.1007/978-3-540-77455-6_20.pdf (accessed 22 May 2018).
[9] Loutzenhiser PG, Manz H, Felsmann C, et al. Empirical validation of models to compute solar irradiance on inclined surfaces for building energy simulation. *Sol Energy* 2007; 81: 254–267.
[10] Perez R, Ineichen P, Seals R, et al. Modeling daylight availability and irradiance components from direct and global irradiance. *Sol Energy* 1990; 44: 271–289.
[11] Sun Y, Su H, Wu CFJ, et al. Uncertainty quantification of solar diffuse irradiation on inclined surfaces for building energy simulation. *Proc BS 2013 13th Conf Int Build Perform Simul Assoc* 2013; 1493: 71–78.
[12] Yang D, Ye Z, Nobre AM, et al. Bidirectional irradiance transposition based on the Perez model. Epub ahead of print 2014. DOI: 10.1016/j.solener.2014.10.006.
[13] Gong ZT., DiazDelaO FA., Beer M. Bayesian model calibration using subset simulation. *Risk, Reliab Saf Innov Theory Pract - Proc 26th Eur Saf Reliab Conf ESREL 2016* 2017; 50.
[14] Papadopoulos V, Giovanis DG, Lagaros ND, et al. Accelerated subset simulation with neural networks for reliability analysis. Epub ahead of print 2012. DOI: 10.1016/j.cma.2012.02.013.
[15] Srivastav A, Tewari A, Dong B. Baseline building energy modeling and localized uncertainty quantification using Gaussian mixture models. *Energy Build* 2013; 65: 438–447.

A climate zone approach to global solar radiation modelling using artificial neural networks

Liu Yang [1], Xujie Huo [1], Danny H W Li [2], Joseph C Lam [2]
[1] School of Architecture, Xi'an University of Architecture and Technology, Shaanxi 710055, China
[2] Building Energy Research Group, Department of Architecture and Civil Engineering, City University of Hong Kong, Tat Chee Avenue, Kowloon, Hong Kong SAR, China

E-mail: hxujie2002@gmail.com

Abstract. Information on solar availability is crucial in the study of both passive building designs and active solar energy systems. There are still many locations within different climate zones worldwide that do not have solar radiation measurements. Correlation between solar radiation and the more commonly measured meteorological variables such as temperature and sunshine hours is useful for locations with no measured solar radiation data. This is also useful for locations with measured solar radiation data, in that any missing data due to equipment breakdown or malfunction can be modelled. Global solar radiation (GSR) was modelled for 96 cities in different climate zones across China using artificial neural networks (ANNs). The novelty of this study is the climate zone approach, by which locations with similar climates were modelled together. Climate classification was based on both the traditional thermal climates and the solar climates. Two sets of models were developed based on measured diurnal temperatures and sunshine hours. Model performance in terms of the predictive power of ANN solar radiation models was evaluated through the Nash-Sutcliffe efficiency coefficient (NSEC). Error analysis of the predicted solar radiation as compared with the measured data was also conducted for each of the 96 cities.

Keywords: Energy use; Solar radiation; Solar and thermal climate zones; Artificial neural networks; temperatures and sunshine hours

1. Introduction

Recently there is a growing awareness of the likely rise in energy consumption in the built environment due to climate change [1] and an increase in the desire for better thermal comfort through space conditioning [2]. These have resulted in more attention being paid to the development and application of renewable technologies when formulating energy and environmental policies. For instance, the Energy Performance of Buildings Directive requires all new buildings to be "nearly zero energy buildings" by the end of 2020 in EU countries, in which energy-efficient building designs and solar energy applications (in terms of photovoltaic (PV) and solar thermal) tend to play an important role in the development of zero energy buildings. Information on solar availability is essential in the assessment of the potential of any solar applications [3]; whether it is related to passive solar building designs, PV/building integrated photovoltaic (BIPV), etc. Because of the costs and regular maintenance (including calibration) of the equipment, long and sustained records of solar radiation data are less readily available compared with other meteorological variables such as temperatures, cloud cover and sunshine hours. As a result, a number of empirical models have been developed to estimate the global solar radiation using the more readily available measured weather data.

One of the most widely used climatic parameters to estimate global solar radiation (GSR) is the possible bright sunshine. Compared with solar radiation measurements, sunshine data tends to be simpler and more readily available from measuring stations in many locations worldwide. A number of researchers had employed regression techniques to correlate global solar radiation with the corresponding possible bright sunshine [4-7] These studies were largely based on linear and/or non-linear regression techniques (e.g. the two-parameter Angstrom-Prescott regression model [8, 9]). Subsequent to these earlier works, there had been two new developments. First, an alternative modelling approach based on artificial neural networks (ANNs) was used in estimating solar radiation for locations with different latitudes and climates [10, 11]. In general, the results indicated an improvement in solar radiation modelling accuracy compared with the conventional regression techniques. Second, temperature data (especially daily minimum and daily maximum) in general are much more readily available than both solar radiation and sunshine hours, particularly in developing countries/regions. Recently, it had been shown that daily global solar radiation correlated reasonable well with daily diurnal temperature for some locations [12]. This work was based on the much earlier work on solar radiation modelling for agricultural and forest meteorology applications. The general approach was to express the daily global solar radiation on a horizontal surface as a function of the corresponding diurnal temperature and several empirical coefficients. These empirical coefficients were then calibrated or determined for a particular location from measured global solar radiation data using regression technique and/or the ordinary least square method by comparing the observed and estimated data.

In the present work, we have developed solar radiation models for different climate zones across China using ANN technique. Both the much more readily available meteorological data (i.e. daily minimum and maximum temperatures) and sunshine duration were considered. Most of the previous modelling works were individual-location-based (i.e. one model for each location using solar radiation data measured at that location). Our work is climate-zone-based. In other words, measured solar radiation data from all the locations with the same climate zones were considered in the ANN modelling. The aim was to develop one solar radiation estimation model for each of the major climate zones, so that any location with no solar radiation measurement but lies within the same climate zone could use the same model to assess its solar potential. This is an extension of our earlier work on regression and ANN solar radiation models for major climate zones in China [10]. Model performances of the temperature-based and sunshine-based ANN models were compared to get a better understanding of the differences in modelling accuracy between these two types of solar radiation models. Comparative studies were conducted for each major climate zone as well as individual cities.

2. Climate classifications and major climate zones

In China, for the thermal design of buildings, the national design guidelines stipulate five major climate types - severe cold (SC), cold(C), hot summer and cold winter (HSCW), mild(M), and hot summer and warm winter (HSWW) [13]. Figure 1 shows an overall layout of the nine thermal climate zones and sub-zones. For assessing solar potentials among different locations, a clearness index (K_t) was used to describe the prevailing solar climates in China (details can be found in Ref. [14]). Five major solar climates were identified with annual mean monthly-average-daily clearness index ranging from 0.3 in the Sichuan Basin to 0.65 in the west and north-west. Both Zone II and Zone III have two regions – Basin/Plateau and Plain/Plateau, respectively. Figure 2 shows the seven solar climate zones identified.

3. Data gathering and quality control

In this study, daily mean, minimum and maximum temperatures, relative humidity, GSR and sunshine duration measured at 96 stations in different climate zones across China were gathered and analysed. The period of records ranged from 5 to 55 years covering the period between 1958 and 2012. The measured GSR and sunshine duration data were checked for errors and inconsistencies. This quality control was to eliminate spurious data and inaccurate measurements [15].

IOP Conf. Series: Materials Science and Engineering **556** (2019) 012018 doi:10.1088/1757-899X/556/1/012018

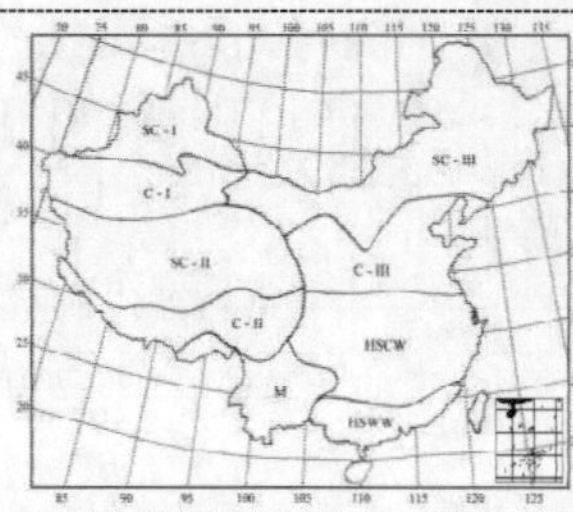

Figure 1. Geographical layout of the nine thermal climate zones and sub-zones (Refs. [10, 13]).

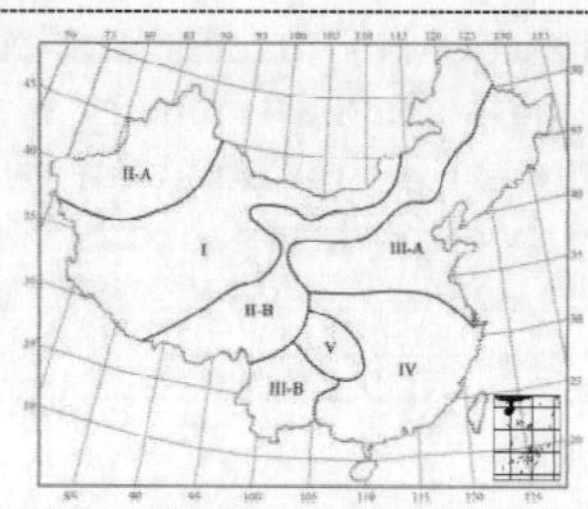

Figure 2. Geographical layout of the seven sola climate zones and sub-zones (Refs. [10, 14]).

4. Artificial neural networks

A total of six geographical and climatic variables were used as the input parameters for the input nodes of the input layer. Both the diurnal temperature model and the sunshine duration model have four geographical variables (i.e. the day-number, latitude, longitude and altitude). The climatic variables for diurnal temperature model were relative daily mean humidity and daily diurnal temperature. Daily mean temperature and normalized daily sunshine duration (i.e. measured sunshine duration ÷ daily maximum possible sunshine duration) were adopted for the sunshine duration model. There was one single node at the output layer with the estimated daily GSR as the output. The transfer function adopted for the neurons was radial basis function.

In this study, 70% and 30% of the data were used for training and testing, respectively. ANNs were trained and two corresponding GSR prediction models (one using diurnal temperature and the other sunshine duration) were developed for each of the nine thermal climate zones and sub-zones. Likewise, two ANN GSR prediction models (one using diurnal temperature and the other sunshine hours) were developed for each of the seven solar zones and sub-zones.

To get a general idea about the predictive power of the ANN models developed, the Nash-Sutcliffe efficiency coefficient (NSEC) was determined. A summary for the thermal and solar model climate zones is shown in Tables 1, respectively. It can be seen, in Table 1 (thermal zone models), that NSEC for the ANN models based on temperature varies from 0.7 in SC-II climates to 0.81 in SC-I with a zone average of 0.76. NSEC based on sunshine duration tends to be larger, ranging from 0.83 also in SC-II to 0.94 in SC-I with a zone average of 0.89. Similarly, in the solar climate zones, NSEC for the ANN models based on diurnal temperature varies from 0.7 in solar zone IIIB to 0.82 in solar zone V with a zone average of 0.76. Again, NSEC based on sunshine duration tends to be larger, ranging from 0.83 in Zone IIIB to 0.92 in Zone IIA with a zone average of 0.89. It suggests that sunshine-based ANN models in general would have better performance than temperature-based. It is interesting to note that the overall mean NSEC (i.e. average NSEC of the climate zones) is the same for the thermal and solar climates, 0.76 for diurnal temperature and 0.89 for sunshine duration. This indicates that there is no significant difference in model performance, in terms of predictive power, whether the ANN solar radiation models are developed using thermal or solar climate classification.

5. Error analysis of the ANN models for individual cities

To have a better understanding of how well the ANN models could predict GSR, performance of the thermal climate zone ANN models based on diurnal temperature was evaluated using measured temperature and solar radiation for the three-year period 2010-2012, which were excluded in the ANN training and testing. To quantify the performance of the ANN models and ascertain whether there was any underlying trend in the different climates, mean bias error (MBE) and root-mean-square error (RMSE) were determined for the 96 cities.

A summary of the MBE and RMSE of the temperature-based ANN models for the 96 cities grouped into the nine thermal climate zones and sub-zones is shown in Table 2. In terms of the actual GSR, Sanya (HSWW) has the largest underestimation of -5.03 MJ/m^2 and Hami (SC-I) the largest overestimation of 1.94 MJ/m^2. Most of the MBEs (about two-third) are within the ±5% range and most of the RMSEs (about three-quarter) are less than 25%. In general, cities with warmer climates (i.e. HSCW and HSWW) in the south tended to have larger MBEs and RMSEs. Similarly, an error analysis of the sunshine-based ANN models was conducted for the 96 cities within the nine thermal zones using sunshine duration, and a summary is shown in Table 3. Similar to the temperature-based ANN models, most of the MBEs (again about two-third) are within the ±5% range. RMSEs, however, are much smaller; nearly all of them (about 97%) are less than 25%.

Table 1. Nash-Sutcliffe efficiency coefficient (NSEC) for the ANN thermal and solar climate zone models.

Thermal zone	NSEC		Solar zone	NSEC	
	Diurnal temperature	Sunshine hour		Diurnal temperature	Sunshine hour
SC-I	0.81	0.94	I	0.76	0.89
SC-II	0.70	0.83	IIA	0.79	0.92
SC-III	0.78	0.91	IIB	0.76	0.90
C-I	0.80	0.93	IIIA	0.75	0.90
C-II	0.71	0.84	IIIB	0.70	0.83
C-III	0.75	0.91	IV	0.76	0.89
HSCW	0.79	0.90	V	0.82	0.88
Mild	0.72	0.87			
HSWW	0.75	0.86			

Likewise, error analysis results grouped into the seven solar zones for temperature-based and sunshine-based ANN models are shown in Tables 4 and 5, respectively. It can be seen from Table 4 that MBE varies from -30.3% underestimation in Sanya (solar zone IV) to 20.1% overestimation in Gushi (solar zone IIIA). RMSE varies from 12.9% in Lhasa (solar zone I) to 42.9% in Gushi (solar zone IIIA). Table 5 shows that MBE for sunshine-based ANN models ranges from -20.9% in Changning (solar zone IV) to 17.8% in Hami (solar zone I), and RMSE ranges from 7.9% in both Golmud (solar zone I) and Yinchun (solar zone IIB) to 28.7% in Gushi (solar zoneIIIA). Again, similar to thermal zone results, about two-third of the MBEs are within the ±5% range for both the temperature-based and sunshine-based models, and about three-quarter of the RMSEs are less than 25% for the temperature-based models and nearly all (97%) for the sunshine-based models.

6. Comparisons of results and discussion

6.1. Individual cities

From the error analysis results shown in Tables 4 and 5 for temperature-based models, there is very little difference in both the MBEs and the RMSEs for all the 96 cities. Similar finding has been observed for sunshine-based models shown in Tables 6. It suggests that grouping of cities into thermal or solar climate zones makes very little difference in terms of model performance for individual cities. There are, however, differences in the error analysis results between sunshine-based and temperature-based models. As for the MBE, there is no specific trend indicating whether temperature-based or sunshine-based model would perform better. Different cities perform differently. For example, in Altay (SC-I) temperature-based has slightly less underestimation (-6.6%, see Table 2) than sunshine-based (-6.9%, see Table 3), whereas in Tacheng (SC-I) temperature-based has a much larger overestimation (2.7%) compared with sunshine-based (1.5%). In terms of long-term performance, therefore, sunshine-based models do not tend to out-perform temperature-based. This finding does not tend to support the generally held view that sunshine hours is a better meteorological variable than temperature in solar radiation modelling. As for the RMSE, the situation is different. Sunshine-based models tend to have a much smaller RMSEs (mostly 30-50% smaller) than temperature-based. In terms of short-term performance, sunshine-based models would be more appropriate than temperature-based. For both the temperature-based and sunshine-based models, MBEs in most cities (92%) are within ±10% (Tables 4 and 5). This suggests that the ANN models developed should be able to predict long-term (e.g. annually) solar availability to within ±10%. For short-term prediction, however, the variations would be larger. For temperature-based (Table 2), MBE varies from 12.9% in Lhasa (C-II) to 42.5% in Gushi (HSCW). This implies that predicted daily GSR could deviate from the measured data by up to 12.9% in Lhasa and 42.5% in Gushi. Only 16 cities have RMSEs less than 20%. For sunshine-based models, the expected deviations are much smaller. Table 3 shows that RMSE ranges from 7.8% in Yinchuan (C-III) to 28.7% in Gushi (HSCW). Most of the cities (92 out of 96) have RMSEs smaller than 20%.

6.2. Zone averages

Performances of the temperature-based and sunshine-based ANN models were compared into terms of the zone averages. Table 6 shows a summary of the comparison for the nine thermal climate zones and

Table 2. Error analysis of individual cities within the 9 thermal climate zones (temperature-based models).

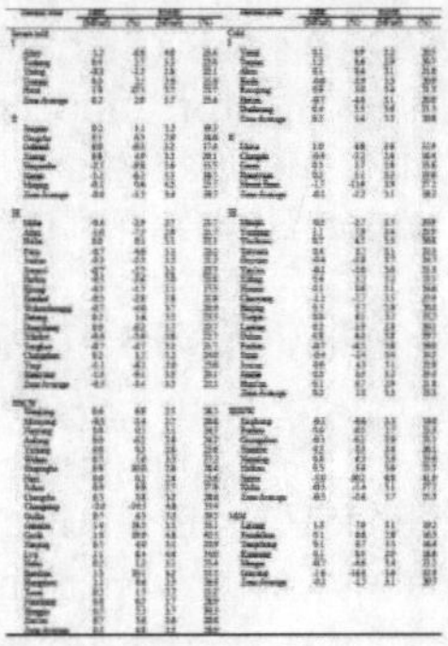

Table 3. Error analysis of individual cities within the 9 thermal climate zones (sunshine-based models).

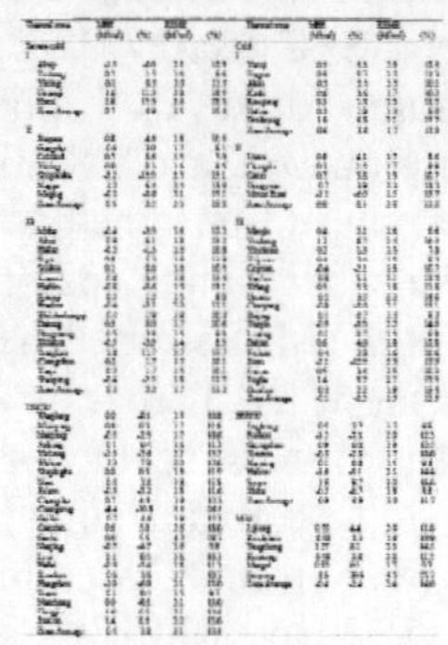

Table 4. Error analysis of individual cities within the 7 solar climate zones (temperature-based models).

Table 5. Error analysis of individual cities within the 7 solar climate zones (sunshine-based models).

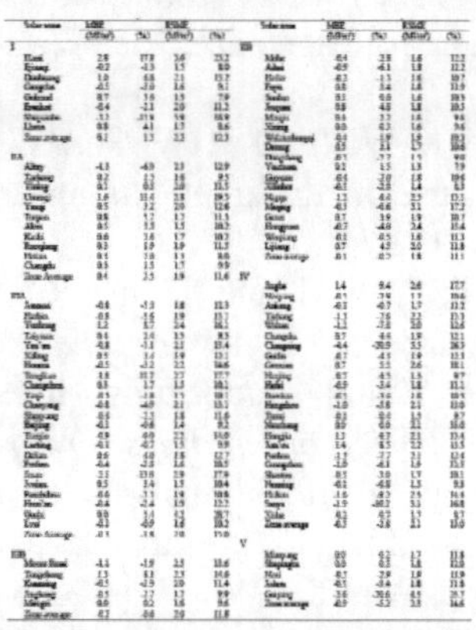

Table 6. Comparison of error analysis between temperature-based and sunshine-based models (thermal and solar climate zone averages).

Thermal zone	MBE (%)		RSME (%)		Solar zone	MBE (%)		RSME (%)	
	Temperature	Sunshine	Temperature	Sunshine		Temperature	Sunshine	Temperature	Sunshine
SC-I	2.0	4.9	23.4	15.4	I	0.8	1.5	18.3	12.3
SC-II	-1.3	-2.2	18.7	12.0	IIA	0.1	2.5	21.3	11.6
SC-III	-3.4	-2.2	22.3	11.2	IIB	-0.4	-0.2	21.4	11.1
C-I	1.4	3.8	20.8	11.0	IIIA	0.9	-1.8	24.5	13.0
C-II	-1.2	-0.1	18.2	11.2	IIIB	-3.3	-0.6	19.5	11.8
C-III	1.3	-0.3	23.3	12.3	IV	1.6	-2.8	27.0	13.0
HSCW	4.8	-1.8	28.0	13.4	V	0.2	-5.2	27.7	14.6
HSWW	-2.6	-4.9	25.3	11.7					
Mild	-1.5	-2.4	20.7	14.0					

sub-zones. As for the MBEs, there is no distinct pattern or trend indicating whether temperature-based on sunshine-based models would perform better. Similar to individual cities, sunshine-based models perform better than temperature-based. For temperature-based models, MBE ranges from 18.2% in thermal zone C-II to 28% in HSCW, whereas the range for sunshine-based models is from 11% in C-I to 15.4% in SC-I. Similar comparison was conducted for the solar climate zone results, and a summary of the zone-averaged MBEs and RMSEs is shown in Table 6. As for the RMSEs, sunshine-based models have smaller errors (between 30-50% smaller) than temperature-based. Again, it is interesting to note that there is very little difference between the overall average RMSEs for the two climate classifications. This is in with the observation based on the overall mean NSEC reported earlier.

7. Conclusions

- In terms of predictive power and error analysis MBEs and RMSEs, performance of the ANN solar models developed differ very little whether the ANNs are grouped under the more traditional thermal climate classification or the more solar-related solar climates.
- The ranges of NSEC are 0.7-0.81 for temperature-based ANN models (thermal climates), 0.83-0.94 for sunshine-based ANN models (thermal climates), 0.7-0.82 for temperature-based ANN models (solar climates), and 0.83-0.92 for sunshine-based ANN models (solar climates). In terms of predictive power, sunshine-based ANN models have larger NSECs and hence, in general, would perform better than temperature-based models.
- Comparisons of the MBEs for each of the 96 cities between temperature-based and sunshine-based ANN models show that there is no significant difference in the long-term performance. Both the temperature-based and sunshine-based models for most cities (92%) could predict annual GSR to within ±10% of the corresponding measured GSR data. The RMSE results, however, show a different picture. Sunshine-based models tend to indicate better short-term performance. Only 16 cities (about 15%) have RMSEs less than 20%. For sunshine-based models, the expected deviations are much smaller. Most of the cities (92 out of 96) have RMSEs smaller than 20%.

Acknowledgements

The research described in this paper was also supported by National Key Research and Development Program of China during the 13th Five-year Plan Period (NO. 2018YFC070004). The authors would like to thank their colleague Dr. Eric W.M. Lee for his advice on the artificial neural networks.

References

[1] Wan K K W, Li D H W, Pan W and Lam J C 2012 Impact of climate change on building energy use in different climate zones and mitigation and adaptation implications *Applied Energy* **97** 274-82

[2] Yang L, Yan H and Lam J C 2014 Thermal comfort and building energy consumption implications – A review *Applied Energy* **115** 164-73

[3] Muneer T and Kambezidis H 1997 *Solar radiation and daylight models for the energy efficient design of buildings* (Oxford: Achitectural Press)

[4] Gueymard C 1993 Analysis of monthly average solar radiation and bright sunshine for different thresholds at Cape Canaveral, Florida *Solar Energy* **51** 139-45

[5] Lam J C, Wan K K W, Lau C C S and Yang L 2008 Climatic influences on solar modelling in China *renewable Energy* **33** 1591-604

[6] Lam J C and Li D H W 1996 Regression Analysis of Solar Radiation and Sunshine Duration *Architectural Science Review* **39** 15-23

[7] Şen Z 2007 Simple nonlinear solar irradiation estimation model *Renewable Energy* **32** 342-50

[8] Prescott J A 1940 Evaporation from a water surface in relation to solar radiation *Trans Royal Soc South Australia* **61**

[9] Angstrom A 1924 Solar and terrestrial radiation *Quart J Royal Meteorol Soc* **24**

[10] Wan K K W, Tang H L, Yang L and Lam J C 2008 An analysis of thermal and solar zone radiation models using an Angstrom–Prescott equation and artificial neural networks *Energy* **33** 1115-27

[11] Lam J C, Wan K K W and Yang l 2008 Solar radiation modelling using ANNs for different climates in China *Energy Conversion and Management* **49** 1080-90

[12] Almorox J, Hontoria C and Benito M 2011 Models for obtaining daily global solar radiation with measured air temperature data in Madrid (Spain) *Applied Energy* **88** 1703-9

[13] *Thermal Design Code for Civil Building (GB 50176-93)* (Beijing: China Architecture and Building Press)

[14] Lau C C S, Lam J C and Yang L 2007 Climate classification and passive solar design implications in China *Energy Conversion and Management* **48** 2006-15

[15] Muneer T and Fairooz F 2016 Quality control of solar radiation and sunshine measurements – lessons learnt from processing worldwide databases *Building Services Engineering Research and Technology* **23** 151-66

Optical design and experiment evaluation of a novel asymmetric compound parabolic concentrator (ACPC) integration with PV for building south wall application

Q Xuan[1], G Li[1, 2*], G Pei[1*], Y Lu[1], Y Su[3], J Ji[1]
[1] Department of Thermal Science and Energy Engineering, University of Science and Technology of China, 96 Jinzhai Road, Hefei City 230026, China
[2] School of Engineering, University of Hull, Hull HU6 7RX, UK
[3] Institute of Sustainable Energy Technology, University of Nottingham, University Park, Nottingham NG7 2RD, UK
E-mail: ligq@mail.ustc.edu.cn; peigang@ustc.edu.cn

Abstract. Compound parabolic concentrator (CPC) as a typical low-concentration-ratio concentrator is an interesting topic for it can work as the static concentrator without any additional tracking systems or seasonal adjustments, which shows a promising concept of introducing the concentrating PV technology for building application. Therefore, based on the compound parabolic curves, a novel asymmetric compound parabolic concentrator (ACPC) is proposed in this paper for building south wall integration. The optical model of the ACPC is built by the software Lighttools to study its optical performance, and the structure parameters of it are optimized through the optical simulation. Then the prototype of the ACPC integrated with the PV cell as the ACPC-PV module is manufactured and assembled. The indoor experiments by the solar simulator (Oriel Sol3A Model 90943A) from the Newport Corporation at the dark environment with the temperature of 25 °C are conducted to study the electrical performance of the ACPC-PV module. The simulation and experiment results show that the acceptance range of the ACPC with the geometric concentration ratio of 2.4X can be 65° with high optical efficiency. A good agreement is observed between the simulation and experiment results despite the deviation of around 13% which is inevitable due to all kinds of errors, such as manufacturing errors, mismatch losses, series resistance losses, etc.

Keywords: Asymmetric; Compound parabolic concentrator; Concentrating photovoltaic; Optical performance; Electrical performance

1. Introduction

With the rapid development of the society and the vast consumption of the fossil fuels, the environmental problems are becoming more and more serious. It has been stated that the world energy consumption will soar into 750 million kilowatts in 2020 which would be 50–80% higher than 1990 levels and the energy consumption in buildings has been steadily increasing and contributing up to more than 40% of the total energy use in developed countries. For developing countries, the share of building energy consumption is less, but, as the growth of population, urbanization, and demands of building services and comfort levels, the sharp rise of building energy use is probably inevitable [1]. A useful measure to reduce building energy use is integrating solar energy technologies with building envelopes such as Building integrated photovoltaic (BIPV) or building attached photovoltaic (BAPV) systems [2].

It was predicted by the International Energy Agency (IEA) that building integrated solar PV technology which is considered during the design and construction of all types of buildings would be a future potential [3]. As it is revealed by a report conducted by IEA about the prospect of the building integrated PV systems (BIPV) in fourteen selected countries, a total potential BIPV area is around 23 billion m^2 which is able to generate about the electricity of 3 pWh annually [4]. It was concluded by

Oliver and Jackson that compared with the centralized PV plant, BIPV systems possess three main benefits [5]:

(1) The need of the land, fence, access road and other important support components can be avoided, for the PV panels are attached or replacing the building structure. The most buildings are close to the electricity grid which means that some cabling cost can also be avoided.

(2) When the electricity is generated by the BIPV systems, it will be consumed by buildings themselves, therefore the losses that caused by the transmission and distribution of the electricity can be minimized. As for the commercial buildings, their electricity demands coincide with the peak electricity generation from BIPV systems.

(3) The overall cost of the BIPV system can be further reduced due to the substitution parts of the building roof, window and façade.

In the recent years, researchers have realized the advantages of introducing the concentrating PV (CPV) into the BIPV systems as BICPV (Building Integrated Concentrating Photovoltaic), so the use of the concentrators have significantly increased in the last decades, such as sky lights, double glazing windows and solar blinds [6, 7]. The concentrator is a device usually makes use of geometrical optics in the design of reflective and/ or refractive types of concentrating devices to focus the solar flux onto a receiver module where the PV cell is attached [3]. In this area, the compound parabolic concentrator (CPC) is proved to be a more utility and economic concentrator [8].

Since the concept of the CPV proposed for the BIPV systems, there are a lot of researchers have committed themselves in designing and integrating solar concentrators with buildings. Carlo Renno et al designed a concentration PV thermal system that was able to provide the electricity for the domestic use and recover the solar cell thermal energy to both supply heat for domestic application and enhance the performance of the solar cell [9]. Mallick et al designed a novel asymmetric CPC which consists of two different parabolas, and the simulation and experiment results showed that it is a feasible way to integrate it with building façade at Northern Ireland (54°36'N, 5°37'W) and the experiments revealed that the asymmetric mirror CPC (with geometric concentration ratio of 2.0X) increased the PV power by 62% and a maximum power ratio of 2.01 was observed for a dielectric asymmetric CPC (with geometric concentration ratio of 2.45X) [10, 11].

It has been proved by Xuan et al. [12] and Li et al. [13] that asymmetric compound parabolic concentrator shows the promising concept of introducing the BICPV systems for building south wall integration. Based on the lens-walled structure, a novel asymmetric compound concentrator (ACPC) by adopting the rotation angle for the reflection lens-walled structure to increase the overall acceptance range is proposed in the paper. An optical model is built for the ACPC by the software Lighttools. The indoor experiments are conducted to study the optical and electrical performance of the ACPC-PV.

2. The geometric structure of the asymmetric compound parabolic concentrator (ACPC)

The lens structure of the asymmetric compound parabolic concentrator is shown in Fig.1, which consists of the asymmetric compound parabolic curves *AB* and *CD*. The equations of *AB* and *CD* are expressed in Eqn. (1) and (2). In order to increase the energy collection for the asymmetric concentrator by enlarging the front aperture, the parabolic curve *AB* is rotated by the lower end point *B* for a certain degree of 15°, then the curves of the $A'B$ and CD will be the outer contour of the ACPC. Finally, $A'B$ and CD are rotated by the up end points A' and C with the specific angle (usually 3-5°) to form the lens-walled structure for the ACPC. Considering the machining precision, the upper part of the ACPC are truncated at *EF* and *GH* to make it easier to manufacture. The geometric concentration ratio of the ACPC is 2.4X, according to the equation: EH / BD, the distance between $B'D'$ and BD is the base height of the ACPC.

$$-0.4x+0.917y+5.007=\frac{1}{28}(0.917x+0.4y+4.585)^2 \quad 5\le x\le 6 \tag{1}$$

$$-0.830x+0.550y+4.980=\frac{1}{36.67}(0.55x+0.83y+2.756)^2 \quad 5\le x\le 12.5 \tag{2}$$

According to Snell's law, the lens-walled structure of the ACPC can make use of the total internal reflection to collect sun rays. For this reason, the solar energy usage ratio can be increased obviously. Thus the lens-walled structure can achieve the same function as the dielectric compound parabolic concentrator but less material needed. However, due to the restriction of the incidence angle, not every sun ray can be collected through the total internal reflection, which will escape the concentrator. In order to collect the escape sun rays, an asymmetric mirror CPC is also integrated with the lens-walled structure to form as the ACPC, the geometric figure of which is shown in Fig. 2. In this way, the sun rays at various incidence angles can be collected through either the total internal reflection or the specular reflection.

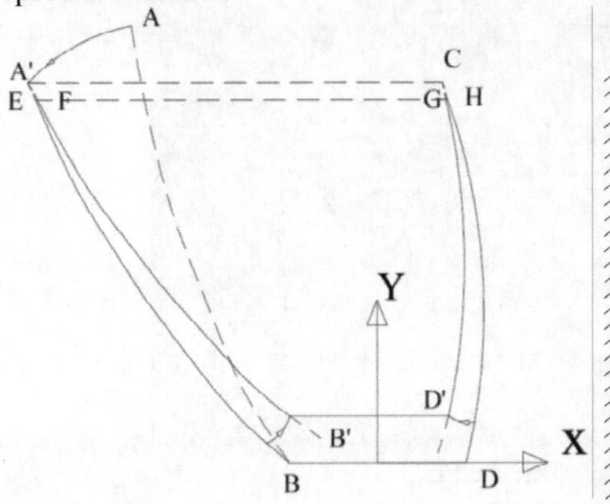

Figure 1 Formation process diagram of the lens-walled structure.

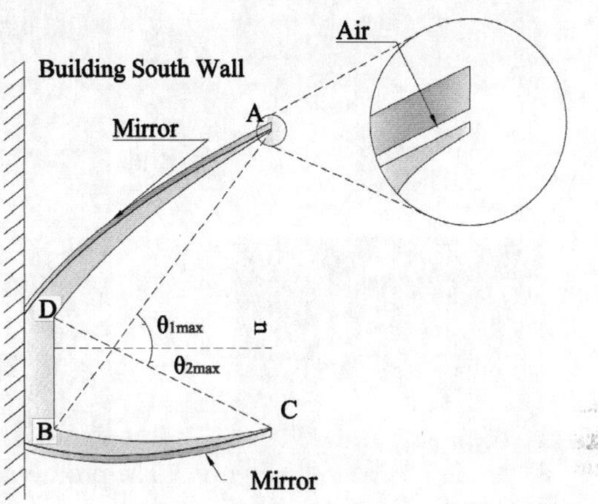

Figure 2 The structure of the ACPC.

3. Ray-tracing simulation and experimental investigation

The model of the ACPC is first designed in the SolidWorks and then transferred into Lighttools for the optical simulation. The optical efficiencies of the ACPC at different incidence angles are then determined. Lighttools is a fast and accurate ray-tracing photometric analysis program which provides the optical system modeling and performance evaluation for non-imaging optical design [12].

For the optical simulation, the PMMA is chosen as the material of the concentrator and the specular reflectivity is set as 90%. The light source generates a solar intensity of $1000Wm^{-2}$ with 10000 solar rays which are assumed to be parallel.

The prototype of the ACPC is manufactured by the CNC wire-cutting machine. The mirror reflectors are gotten from the evaporated aluminum coating, and a reflectivity of around 85% is achieved for the reflection mirrors. Photographs of the concentrator are presented in Fig. 3. Considering the machining precision, finally the ACPC prototype with the geometrical concentration of 2.3X is got, the total height, aperture width, length and absorber width of which are 22.8 mm, 34.5 mm, 70 mm and 15 mm respectively.

In the indoor experiment, a solar simulator (Oriel Sol3A Model 90943A) from Newport Corporation is adopted to generate a ray intensity of 1000 W/m^2 through a 450W Xenon lamp. The electrical characteristics of solar cells are derived with a Keithley 2420 digital source meter (Keithley, USA). The room temperature is kept at a constant value of 25℃ during the entire test period. The ray intensity of the solar simulator is calibrated by the PV reference system before the experiment test and in order to avoid the unpredictable influence of other light rays, the experiment is conducted in a dark environment.

4. Results and discussion

4.1 simulation results

The base height of the ACPC plays a vital role in the optical performance for it can influence the ray path near the base area. Based on the optical model built through the SolidWorks and Lighttools, the effect of the base height on the optical performance of the ACPC can be investigated. Detailed simulation results are presented in Fig. 4. From the results, it can be seen clearly that with the base height increases form 1.35 mm to 5 mm, the optical efficiency of the ACPC always shows a decreasing trend. With the base heights of 1.35-5 mm, the optical efficiency remains at high level while when the base height is larger than 2.35 mm, the optical efficiency decreases significantly. Considering the cost, weight and machine precise, the base height of 2.35 mm is a better choice. Thus, the ACPC with the base height of 2.35 mm will be detailed studied in the paper.

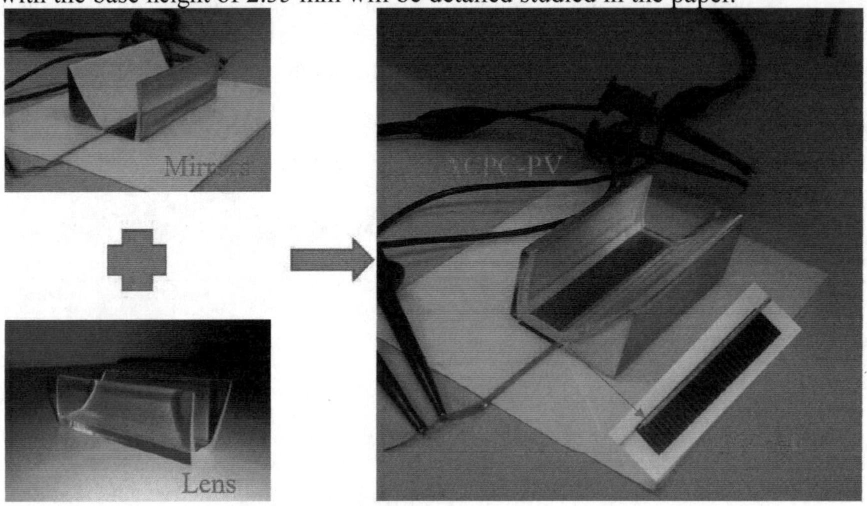

Figure 3 The prototype of the ACPC.

4.2 Indoor experiment results

For the actual engineering, the angular response of the ACPC-PV is very important because the solar incidence angles on the building south wall varies a lot throughout the year. In the following section, the electrical and optical performance of the ACPC-PV at the various incidence angles as compared with the non-concentrating PV are detailed analyzed.

The short-circuit current of the ACPC-PV and the bare cell at the incidence angles of 0°-65° are presented in Fig. 5. From the results, it can be seen clearly that the short-circuit current of the ACPC-PV and the bare cell all shows a decreasing trend as the incidence angle increases, which is mainly caused by the cosine effect [14]. It should be noted that during the experiments, the ACPC-PV and PV are titled from 0°-65° to create different incidence angles instead of titling the solar simulator. Thus as the title angle increases, the actual solar intensity on the front aperture will decrease which makes the short-circuit current decrease. But it's obvious that the short-circuit current produced by the ACPC-PV are always much larger than that produced by the bare cell within the incidence angles of 0°-65°.

The opto-electronic gain and the optical efficiency of the concentrator can both be used to identify the optical performance of a concentrator. The opto-electronic gain of a concentrator is defined as the ratio of the short-circuit current of the CPV module to that of the bare cell [3, 15]. The optical efficiency of the concentrator is the ratio of the solar radiation captured by the absorber of the concentrator to the total incoming solar radiation through the aperture of the concentrator and it can also be got by dividing the opto-electric gain by the geometric concentration ratio of the concentrator [16]. The maximum acceptance angle of a concentrator is defined as the angle when the gain reaches 90% of its peak value [17]. It has been proved that the short-circuit current is proportional to the solar radiation that falls on the PV, thus the actual optical efficiency of the ACPC can be expressed by:

$$\eta_{\text{opt,ac}} = \frac{1}{C}\frac{I_{sc}^{with}}{I_{sc}^{without}} \tag{3}$$

Where $\eta_{\text{opt,ac}}$ is the actual optical efficiency of the concentrator; C is the geometric concentration ratio of the concentrator; I_{sc}^{with} is the short circuit current of the concentrating PV; $I_{sc}^{without}$ is the short circuit of the non-concentrating PV. The opto-electronic gain and the optical efficiency of the ACPC are presented in Fig. 6 and Fig. 7 respectively.

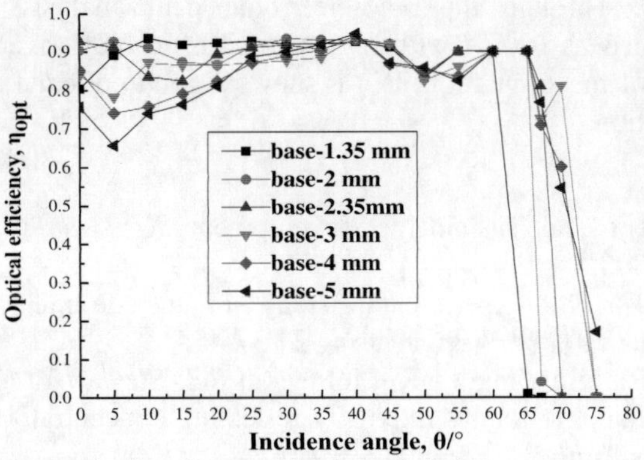

Figure 4 The effect of the base height.

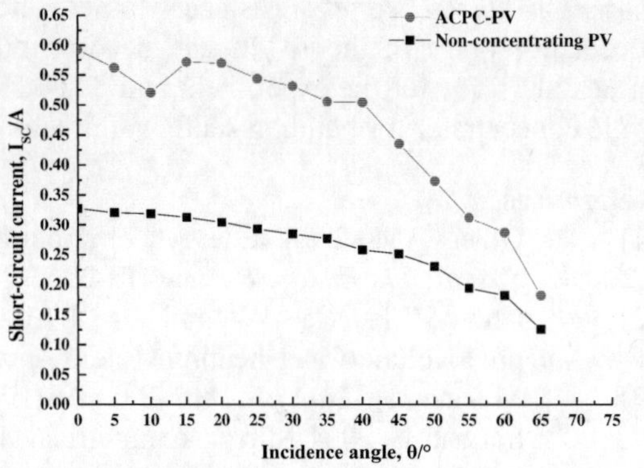

Figure 5 The short circuit current variation of the ACPC-PV and the bare cell at different incident angles.

It can be seen clearly from the results in Fig. 6, the opto-electronic gain of the ACPC-PV are all larger than 90% of the peak value within the incidence angles of 0°-65°. So it can be concluded that the acceptance range of ACPC is 0°-65°, which shows a good building south wall integration potential. And the average value of the optp-electronic gain of the ACPC within the acceptance range is around 1.74. The simulation results are also presented in Fig. 6. A similar trend of the experiment results is observed as the simulation results and they show a good agreement.

The simulation and actual optical efficiency of the ACPC at various incidence angles are shown in Fig. 7. From the results, it can be seen clearly that the experiment and simulation results show a good agreement with an average deviation of around 13%. The main reasons that cause the deviation can be concluded as [17]: 1) Manufacturing errors which makes the surfaces of the ACPC flawed; 2) the deviation of the ACPC and the PV cell. When soldering the ACPC on the PV cell base, assembly errors existed. 3) Test errors.

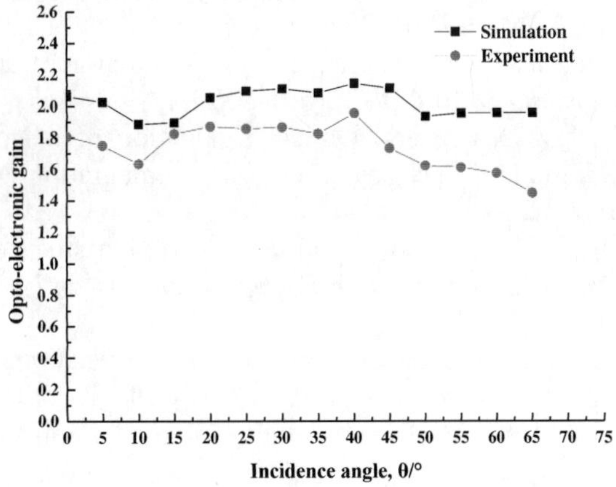

Figure 6 The opto-electronic gain of the ACPC at different incidence angles.

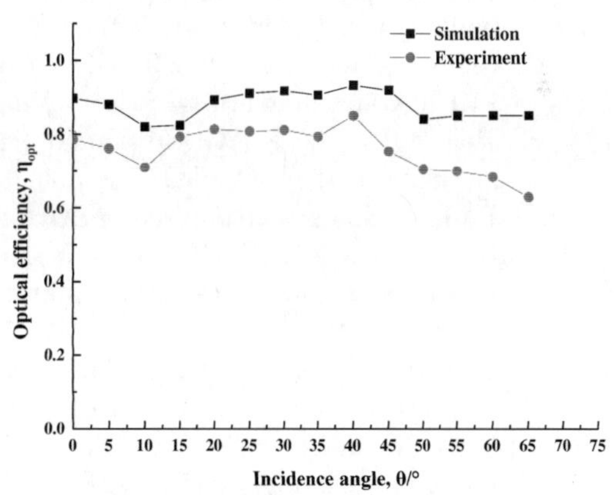

Figure 7 The optical efficiency of the ACPC at different incidence angles.

5 Conclusion

A novel asymmetric compound parabolic concentrator (ACPC) is proposed in this paper for building south wall integration. The optical model is built in the software Lighttools to study the optical performance of the ACPC and optimize the geometric parameters of it. The indoor experiment under a solar simulator is also conducted to analyze the actual electrical performance of it. A good agreement is observed between the experiment and simulation results. The ratio of the maximum power gotten from the ACPC-PV to that produced by the non-concentrating PV cell delivers an average concentration ratio of 1.74X at various incidence angles which is around 75.7% of the maximum theoretical geometric concentration ratio of the designed ACPC. The average deviation of the experiment results from the ray tracing results is around 13% due to manufacturing errors, mismatch losses, series resistance losses, etc. Through the opto-electronic gain analysis, the acceptance range of the ACPC can be determined as 0°-65° with high optical efficiency (average values of 75.7% for the experiment and 87.8% for the simulation) that it shows a good potential as a static concentrator for building south wall integration.

References:

[1] Omer AM 2008 Energy, environment and sustainable development *Renewable and Sustainable Energy Reviews* **12** 2265-2300.
[2] Chow TT, He W and Ji J 2007 An experimental study of façade-integrated photovoltaic/water-heating system *Applied Thermal Engineering* **27** 37-45.
[3] Muhammad-Sukki F, Abu-Bakar SH, Ramirez-Iniguez R, McMeekin SG, Stewart BG and Sarmah N 2014 Mirror symmetrical dielectric totally internally reflecting concentrator for building integrated photovoltaic systems *Applied Energy* **113** 32-40.
[4] International Energy Agency (IEA), 2002. Potential for building integrated photovoltaics. Report IEA-PVPS T7-4:2002 (Summary), IEA.
[5] Oliver M and Jackson T 2001 Energy and economic evaluation of building-integrated photovoltaics *Energy* **26** 431-439.
[6] Sellami N, Mallick TK and McNeil DA 2012 Optical characterisation of 3-D static solar concentrator *Energy Conversion and Management* **64** 579-586.
[7] Norton B, Eames PC, Mallick TK, Huang MJ, McCormack SJ and Mondol JD 2011 Enhancing the performance of building integrated photovoltaics *Solar Energy* **85** 1629-1664.
[8] Qiushi Wang, Ziye Zhu, Gang Wu, Xiang Zhang and Hongfei Zheng 2018 Energy analysis and experimental verification of a solar freshwater self-produced ecological film floating on the sea *Applied Energy* 224: 510-526.
[9] Renno C and Petito F 2013 Design and modeling of a concentrating photovoltaic thermal (CPV/T) system for a domestic application *Energy and Buildings* **62** 392-402.
[10] Mallick TK, Eames PC, Hyde TJ and Norton B 2004 The design and experimental characterisation of an asymmetric compound parabolic photovoltaic concentrator for building fa?ade integration in the UK *Solar Energy* **77** 319-327.
[11] Mallick T and Eames P 2007 Design and fabrication of low concentrating second generation PRIDE concentrator *Solar Energy Materials and Solar Cells* **91** 597-608.
[12] Xuan Q, Li G, Pei G, Su Y and Ji J 2017 Design and Optical Evaluation of a Novel Asymmetric Lens-Walled Compound Parabolic Concentrator (ALCPC) Integration with Building South Wall *Journal of Daylighting* **4** 26-36.
[13] Li G, Xuan Q, Lu Y, Pei G, Su Y and Ji J 2018 Numerical and lab experiment study of a novel concentrating PV with uniform flux distribution *Solar Energy Materials and Solar Cells* **179** 1-9.
[14] Sarmah N, Richards BS and Mallick TK 2014 Design, development and indoor performance analysis of a low concentrating dielectric photovoltaic module *Solar Energy* **103** 390-401.
[15] Ning X, Winston R and O'Gallagher J 1987 Dielectric totally internally reflecting concentrators. *Applied Optics* **26** 300-305.
[16] Buonomano A, Calise F, Dentice D'Accadia M and Vanoli L 2013 A novel solar trigeneration system based on concentrating photovoltaic/thermal collectors. Part 1: Design and simulation model *Energy* 2013 **61** 59-71.
[17] Mammo ED, Sellami N and Mallick TK 2013 Performance analysis of a reflective 3D crossed compound parabolic concentrating photovoltaic system for building façade integration *Progress in Photovoltaics: Research and Applications* **21** 1905-1103.

Acknowledgment
The study was sponsored by the National Natural Science Foundation of China (NSFC 51476159, 51776193, and 5171101721), International Technology Cooperation Program of the Anhui Province of China (BJ2090130038) and the Fundamental Research Funds for the Central Universities (WK6030000099). The authors would like to thank Prof. Zheng Hongfei (School of Mechanical Engineering, Beijing Institute of Technology, China) for his assistance in the software simulation.

Experimental study on thermal performance improvement of solar air heater using piezoelectric fan

Hu Jiayu[1], Jin Chengjun[1]

[1]College of architecture and environment, Sichuan University, Chengdu ,No.24 South Section 1 Yihuan Road. Zip:610065

Abstract. This paper presents experiment research about piezoelectric fan array in solar air heater (SAH) at the frequencies 45-60Hz by using lamp array as solar radiation simulator. The effectively heat transfer coefficient (HTC) of the traditional method such as strip rib on the solar collector and enhancement by piezoelectric fan are compared in detail with different mass flow rate of the air. A prototype solar collector with dimension of 300(L)×150(W)mm is fixed in a wind tunnel in the experiment. Three configurations of piezoelectric fan array have been tested, the results show that the piezoelectric fan can increase the HTC effectively, the HTC is increased 3-3.5 times compared with smooth plate and the additional power consumption of the piezoelectric fan as low as 0.6W for 2×2 piezoelectric fan array, whereas for the pressure drop of the SAH increased no more than 5%. The resonance frequencies of the two types of piezoelectric fans in experiment are 55Hz and 53Hz respectively, the maximum amplitude and the HTC achieved at the resonance frequency. It's an effective way for SAH application compares to other methods of heat transfer enhancement in SAH and rarely be reported in the area of SAH.

Keywords: Single&double flow; Solar air heater; Piezoelectric fan; Heat transfer enhancement; Thermal performance; Pressure drop

1. Introduction

Solar air heaters(SAH)is the popular device to utilize the solar energy. There are two types of SAH which include single and double pass mode[1], according to the airflow path around the solar absorber plate. The SAH may be further classified into four basic configurations are: a front-pass collector, a back-pass collector, a front and back pass collector and a through-pass collector. How to increase area or heat transfer rate of the absorber is the principal problem of the SAH design and optimization. Numerous methods for heat transfer enhancement and area increasing have been researched. The most popular extend surface is rib in SAH. The geometrical parameters of rib roughness have a significant influence on heat transfer and fluid friction, Anil Kumar Patil [2] reviewed a variety of artificially roughened solar air heater and suggested that transverse rib roughness produces maximum heat transfer enhancement, and lightweight can be achieved by multi V-rib roughness geometry. Alam[3] et al. researched effect of circularity of perforation holes in V-shaped blockages on heat transfer and friction characteristics of rectangular solar air heater duct. Other shape of the obstacles are also have been researched ,Bekele[4] et al. have proposed a research about delta shaped obstacles in SAH and concluded the thermo-hydraulic performance of the presented SAH is higher than those in previous studies. Some researchers are also investigate the optimization problem of the SAH. Yang, Ming et al.[5], presented numerically method to optimize the offset strip fins of the SAH. Bouadila et al [6]et al. gave the energy and exergy by using first law and second law to analysis solar air heater with latent heat storage and obtained the optimization energy and exergy efficiencies. Some researcher performed the experiments by using solar simulator, numerous methods to simulate the heat flux from solar were applied in literature. P Velmurugan et al [7] investigated different absorber surface geometries by using halogen lamp as solar simulator and got average radiation 645 W/m2, whereas M. Zukowski et al [8] used infrared lamp to simulate the solar and investigated thermal and flow characteristics of microjet air solar heater.

As mentioned above, most of the methodologies and researches of the heat transfer enhancement in solar air heater are passive method, namely, by increasing the heat transfer area or insert some extend surface to generate turbulence. Unfortunately, all of these methods will increase the pressure drop of SAH dramatically and lead to a lower overall energy efficiency. A piezoelectric fan or PE fan is new methodology in heat transfer enhancement, especially with the low power consumption. A PE fan is a flexible cantilever beam, and the vibration of the fan is actuated by piezoelectric material. There are plenty of studies about PE fans in literatures, especially in optimization of the location of the PE fans. Mark Kimber et al [9]gave the maximum heat transfer coefficient of the single PE fan and conducted that the heat transfer enhancement is only relate to amplitude and frequency, whereas the more effect to HTC as for the frequency; Ten times more efficient in converting the input power to useful energy compared to axial fan. Abdullah, M. K. et al.[10],[11],[12]investigated PE fans using CFD software and particle image velocimetry (PIV) system, the research conducted that the ratio of the height of PE fans from heat source to length of PE fans, tip gap, amplitude of vibration etc. are all critical factors and the related optimized values of these parameters was given by their research.

Although there are so many works in PE fans and SAH but the combination study about two subjects is rarely, this work will focus on the application of the PE fans in solar air heater, the capability of heat transfer enhancement by using PE fans in SAH has been demonstrated by using wind tunnel and lamp array. The experimental study has been presented with three different PE fan configurations.

2. Experiment apparatus and procedures

The experiment apparatus is composted by four sections: wind channel, absorb plate, heating element and measurement unit. The wind channel is made by plywood, acrylic and glass, the length of channel is 710mm,the interior cross section of the channel is 150mm × 150mm,it include a glass window on the top of the channel for simulate the solar radiation with the dimension of 300mm × 150mm and same as the absorber plate, the absorber plate is fixed in the middle of the channel, it can be replaced by different type absorber plate. Five different absorber plate are used as described in Figure 1.Three types of absorb plate used in experiment which the PE fans was fixed on board as heat transfer enhancement element. All of board are aluminium with black anodized, the finished absorber plate shown in Figure 2.The inlet temperature and outlet temperature is measured by two groups T type thermocouple. The air volume flow rate and system resistance is measure by wind tunnel system (Longwin 9266SR)which accordance to AMAC 210-07 standard[13].

The heating element includes eight Philips R63 reflector infrared bulbs with the power of 40W each for solar heat simulation and powered by 220V AC. The energy intensity was measured by infrared power meter (LS122) and the intensity is 3337W/m^2. Although the energy intensity is higher than some literatures but it is not affection to compare the heat transfer performance of the difference set-up, the more effective absorber plate will lead to higher heat transfer efficiency and higher temperature rise of the air under the same experimental configuration.

The air was sucked from ambient with the temperature of indoor by variable exhaust system, variation of the flow rate was performed by using variable frequency drive controlled by Longwin 9266SR, the air pass through the absorber plate and the temperature raised. Measurements of air volume flow rate were done using nozzles in 9266SR and static pressure differences was also recorded for preparing flow characteristics.

IOP Conf. Series: Materials Science and Engineering **556** (2019) 012020 doi:10.1088/1757-899X/556/1/012020

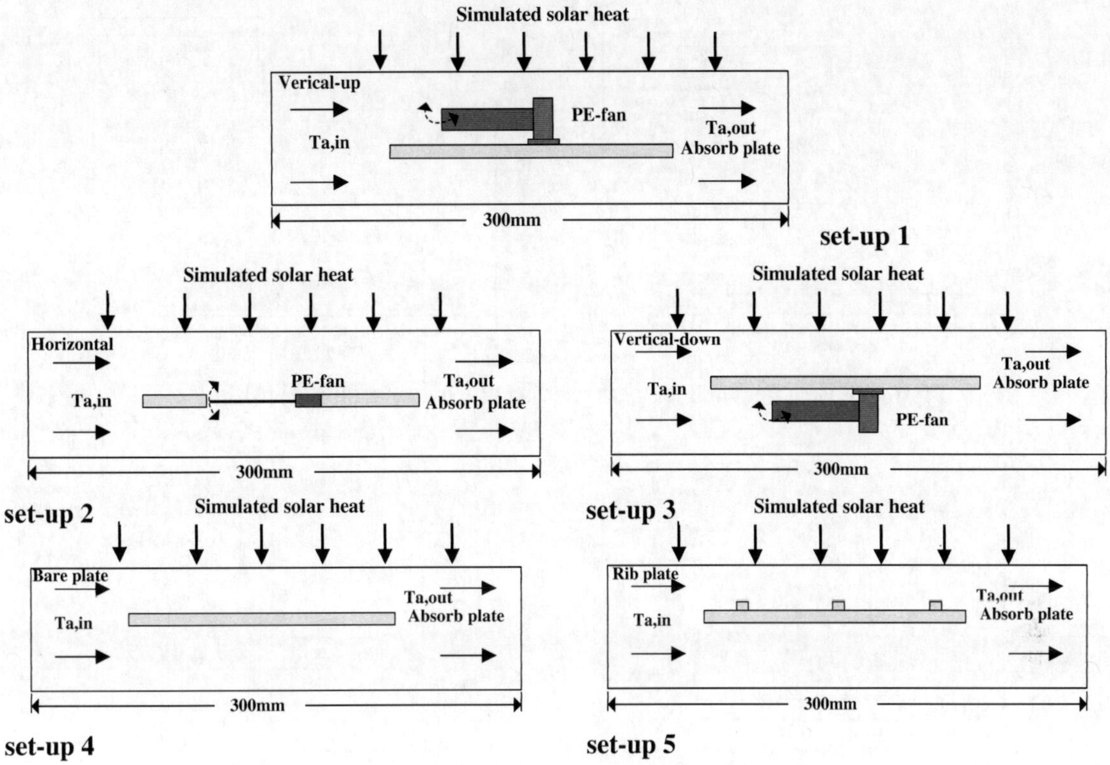

Figure 1. Five types of absorb in wind channel

Figure 2. The photo of four absorber plates using in experiment

The different pressure drop acquired by different absorber plate, different mass flow rate and different frequency of the PE fan. The experimental tests include a series of measurements, system pressure drop vs. volume flow rate was performed automatically with the Longwin 9266SR and the temperature difference was recorded by Agilent 34970A datalogger. The volume flow rate Q_a range from 40m^3 /h to 162m^3 /h(40.5 m^3/h,81 m^3/h,121 m^3/h,162m^3/h corresponding the air speed 0.5m/s,1m/s,1.5m/s), Q_f is the total heat gained by air.

3. Experiment results and discussion

Three set-ups of PE fans and two benchmark set-ups were used in experiment as previously described, the temperature rise and the heat dissipation with the different configuration was prepared as shown in Fig 3 ~Fig5. The benchmark group are also presented in these charts. The highest air temperature rise is 7℃ of set-up1 and higher than set-up 2 , set-up 3 and benchmark set-ups.

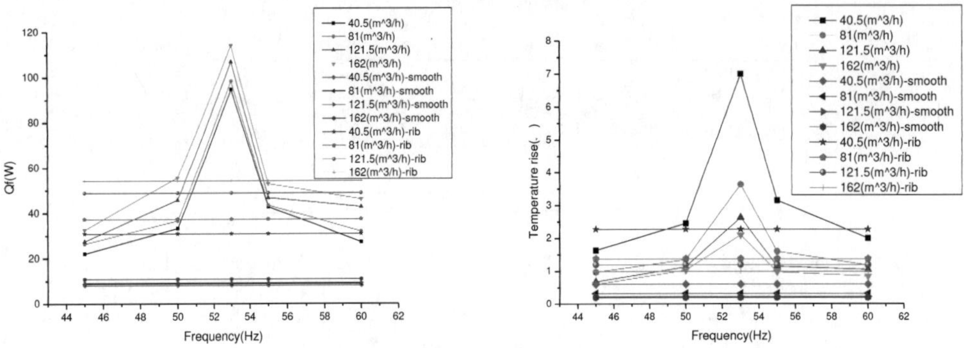

Figure 3. Heat gain and temperature rise of experimental set-up 1

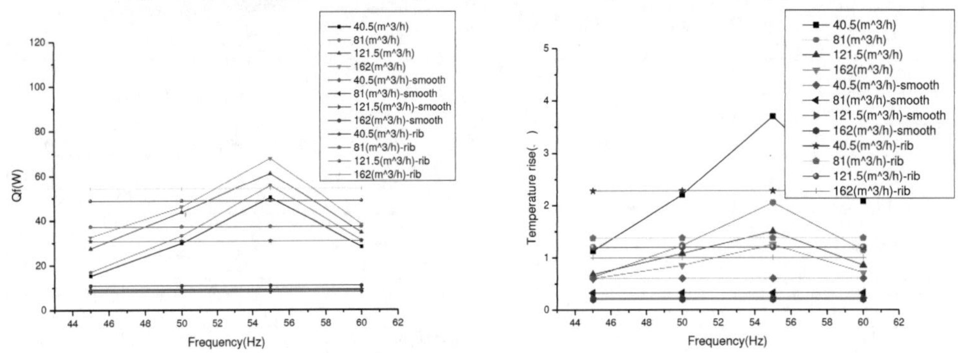

Figure 4. Heat gain and temperature rise of experimental set-up 2

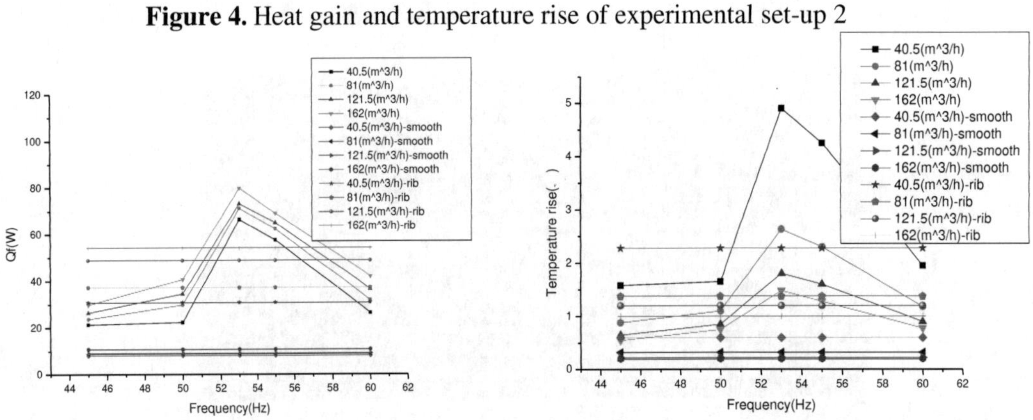

Figure 5. Heat gain and temperature rise of experimental set-up 3

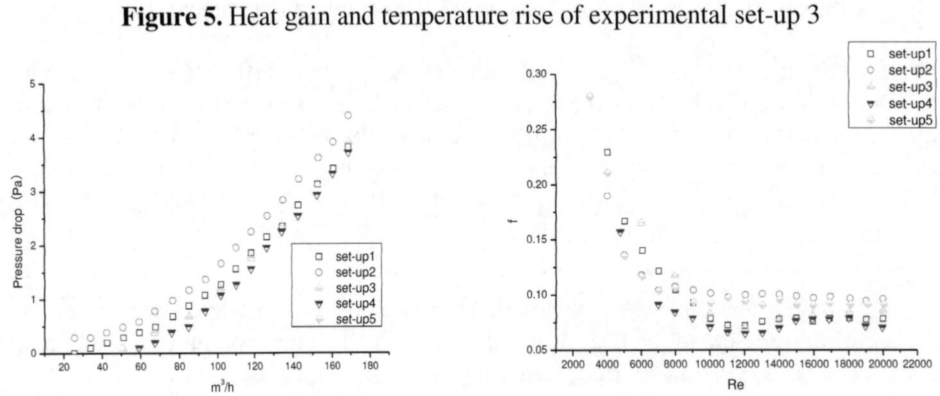

Figure 6. Pressure drop and friction factor of the different set-ups

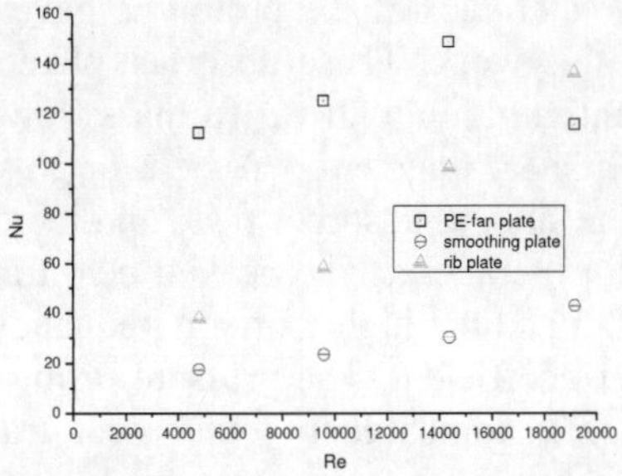

Figure 7. Comparison the Nu of the smooth absorber plate and PE fan enhancement absorber plate(set-up3 at the resonance frequency)

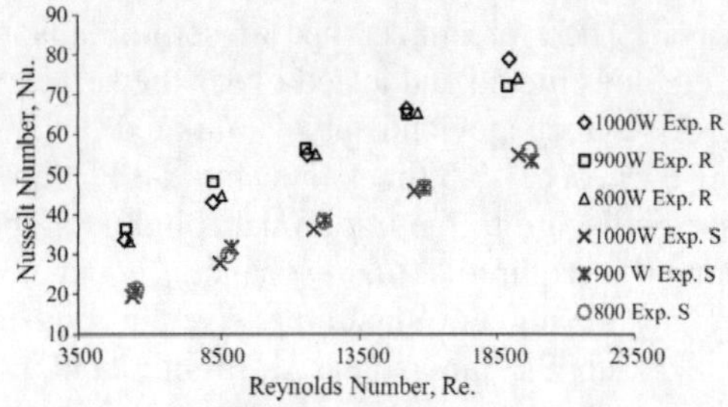

Figure 8. Comparison between the Nu versus Re in Rahmati Aidinlou's[14] experiment

The highest heat transfer quantity or temperature rise was acquired at the resonance frequencies of each set-up. In the same PE-fan set-up, higher frequency always brings higher heat transfer rate. The reason may be that the boundary was disturbed by the vortex of the PE fans and the higher vibration may lead to higher turbulence than the lower frequency. The set-up1 is superior to any other set-ups in our works. The set-up1 and set-up3 have different heat transfer rate, although they are all vertical configuration of PE fan. The cause the difference is because that the top glass cover with the high temperature for most of the long wave spectrum were absorbed hence that the vortex generated by PE fan enhanced the heat transfer not only for absorber plate cover but also for the top glass cover for the set-up 1, whereas absorber plate heat transfer was enhanced only for the set-up 3.The heat transfer coefficient was calculated using the vertical down (set-up 3) for the reason that PE fans enhanced heat transfer of absorber plate only in the set-up 3.The result is more conservative and reasonable for set-up 3 considering the main aim of our work is to investigate the heat transfer enhancement of the absorber plate.

The pressure drop of the five set-up was prepared as shown in Fig 6 and the measurement were taken at the resonance frequency (excepted the benchmark group). The set-up 2 with the horizontal configuration has the highest pressure drop and set-up 4 has the lowest pressure drop without any heat transfer enhancement device. The pressures drop of set-up 1 and set-up 3 higher slightly than smooth plate (set-up 4) whereas highest heat transfers performance. The set-up 1 and set-up 3 have almost the same pressure drop for the symmetry configuration. As shown in Fig 6, the friction factor also has the same trend. Fig 7 and Fig 8 presents the PE-fan significantly increased the heat transfer coefficient comparison the smooth absorber plate and traditional method such as rib in SAH. Nussuelt number is increased approximately 2-2.5 times corresponded the Re range from 15000~20000 compared with the Rahmati Aidinlou's[14] experiment.

Rahmati[14], Velmurugan[7], Zukowski[8] are all used the solar simulator to measure the efficiencies of solar air heater. We used the infrared lamp for the cost consideration but the distance to the glass cover is short and with small thickness of the top glass cover. These measures in our experiment are meant to increase the absorbed energy of the middle absorber plate and get more reasonable results about the heat transfer performances. The efficiency of SAH varied with the solar intensity according to [7] , the higher solar intensity are always the higher temperature rise. But a little affection for the heat transfer coefficient calculate of the absorber plate, because of the higher heat flux lead to higher surface temperature.

The independent parameters in this work are inlet and outlet temperatures, absorber plate temperature, pressure drop of the channel, pressure drop of the flow meter nozzle and simulated solar radiation, T-type thermocouple with accuracy 0.1℃, pressure gauge with 0.05% accuracy, pressure difference gauge with 0.06% accuracy and infrared power meter 4% accuracy were used. The heat transfer coefficient is ±4.29%,voulem flow rate is ±1.45%.

4. Conclusion

An experimental investigate was conducted to demonstrate the heat transfer enhancement of steady state SAH by using PE fan, thermal and hydraulic characteristics of the prototype have been developed. There are three type of set-ups absorber plate in this work. These absorber plate were designed, constructed and tested under the same experiment configuration with varied mass flow rate, varied drive frequency and same simulated solar intensity. The heat transfer coefficient or Nussuelt number increased 3-3.5 times by using the PE fan compare to the smooth absorber plate and increased 2.5 times compare to the traditional roughness absorber plate in literatures. The peaked heat transfer coefficient is acquired at the resonance frequencies of the PE-fan and the higher driver frequency can accompany the higher temperature rise and higher heat transfer coefficient. The pressure drop of the vertical set-up PE fan increased slightly and no more than 5% compared to other heat transfer enhancement measures. Therefore, the horizontal set-up PE fan has the highest pressure drop within our work. It can be conclude that PE fan is an effective way for SAH application compares to other methods of heat transfer enhancement.

Acknowledgments

The author would like to thank special work by Mr Li Yuxian for his experiment fixture design.

References

[1] M.O. Pavel Charvat, Tomas Mauder, Lubomir Klimes, A solar air collector with integrated latent heat thermal storage, in: EPJ Web of Conferences, EDP Sciences, 2012.

[2] A.K. Patil, Heat transfer mechanism and energy efficiency of artificially roughened solar air heaters—A review, Renewable and Sustainable Energy Reviews, 42 (2015) 681-689.

[3] T. Alam, R.P. Saini, J.S. Saini, Effect of circularity of perforation holes in V-shaped blockages on heat transfer and friction characteristics of rectangular solar air heater duct, Energy Conversion and Management, 86 (2014) 952-963.

[4] A. Bekele, M. Mishra, S. Dutta, Performance characteristics of solar air heater with surface mounted obstacles, Energy Conversion and Management, 85 (2014) 603-611.

[5] M. Yang, X. Yang, X. Li, Z. Wang, P. Wang, Design and optimization of a solar air heater with offset strip fin absorber plate, Applied Energy, 113 (2014) 1349-1362.

[6] S. Bouadila, M. Lazaar, S. Skouri, S. Kooli, A. Farhat, Energy and exergy analysis of a new solar air heater with latent storage energy, International Journal of Hydrogen Energy, 39 (2014) 15266-15274.

[7] R.K. P Velmurugan, Energy and exergy analysis in double-pass solar air heater, Sadhana, 41 (2016) 8.

[8] M. Zukowski, Experimental investigations of thermal and flow characteristics of a novel microjet air solar heater, Applied Energy, 142 (2015) 10.

[9] M. Kimber, K. Suzuki, N. Kitsunai, K. Seki, S.V. Garimella, Pressure and Flow Rate Performance of Piezoelectric Fans, Ieee Transactions on Components and Packaging Technologies, 32 (2009) 766-775.

[10] M.K. Abdullah, N.C. Ismail, M.Z. Abdullah, M.A. Mujeebu, K.A. Ahmad, Z.M. Ripin, Effects of tip gap and amplitude of piezoelectric fans on the performance of heat sinks in microelectronic cooling, Heat and Mass Transfer, 48 (2012) 893-901.

[11] M.K. Abdullah, B.H. Murni, M.Z. Abdullah, M.A. Mujeebu, F. Hussin, H. Yusoff, N.C. Ismail, K.A. Ahmad, Z.M. Ripin, HEAT TRANSFER ENHANCEMENT USING PIEZOELECTRIC FAN IN ELECTRONIC COOLING - EXPERIMENTAL AND NUMERICAL OBSERVATIONS, Isi Bilimi Ve Teknigi Dergisi-Journal of Thermal Science and Technology, 32 (2012) 41-50.

[12] M.K. Abdullah, M.Z. Abdullah, M.V. Ramana, C.Y. Khor, K.A. Ahmad, M.A. Mujeebu, Y. Ooi, Z.M. Ripin, Numerical and experimental investigations on effect of fan height on the performance of piezoelectric fan in microelectronic cooling, International Communications in Heat and Mass Transfer, 36 (2009) 51-58.

[13] ASHRAE, ANSI_ASHRAE 51-07 (ANSIAMCA 210-07) Laboratory Methods of Testing Fans for Certified Aerodynamic Performance Rating, (1997).

[14] A.M.N. H. Rahmati Aidinlou Heat flux: thermohydraulic investigation of solar air heaters used in agroindustrial applications, Heat Mass Transfer, 53 (2017) 12.

Solar-air hybrid source heat pump water heater: Performance characterization and comparison analysis

Jingyong Cai and Jie Ji
University of Science and Technology of China, Hefei, China
E-mail:cjy0806@mail.ustc.edu.cn, jijie@ustc.edu.cn

Abstract. Heat pump water heater(HPWH) can yield great energy saving for commercial and residential appliance. To enable the HPWH maintain efficient operation under diverse circumstances, the solar-air hybrid source heat pump water heater has been proposed. The configurations of solar-air series type, air-solar series type and solar-air parallel type have been discussed and compared in this paper. The dynamic model has been developed to identify the impacts of solar irradiation and environmental temperature. The results show that the solar-air hybrid source heat pump water heater can guarantee efficient operation under the ambient temperature of 10°C and solar irradiation of 100W/m^2, with the average COPs of 2.68 for solar-air series type HPWH, 2.71 for air-solar series type HPWH and 2.69 for solar-air parallel type HPWH. The rise of ambient temperature can improve the evaporating capacity and condensing capacity leading to the increase of power consumption and COP for each type of solar-air hybrid source heat pump water heater. With the rise of solar irradiation, the heat gain at evaporation side increases leading to the rise of condensing capacity and COP. The air-solar series type HPWH can achieve the highest COP under a range of working conditions and solar-air parallel type HPWH can provide the highest condensing capacity under lower solar irradiation.

Keywords: solar energy, hybrid source, air source, heat pump

1. Introduction

The power consumed by water heating takes over 30% of the overall energy consumption in building [1]. With the rising requirement for building energy saving, the efficiency improvement of domestic water heating has become increasingly significant. Heat pump water heater (HPWH) brings great energy conservation potential by recovering energy from low grade heat source [2].

To raise the stability and efficiency of the HPWH system, hybrid source heat pump system with multi-source acting as the complementary has been developed. Chargui et al. developed a dual source heat pump applied in building with the air evaporator and water evaporator working alternately [3, 4]. Qu et al. analysed the behaviour of a dual source heat pump integrated with PV/T experimentally, and the results suggested that the air-water mode should be adopted under lower solar irradiation [5]. Wang et al. proposed a dual source heat pump system utilizing a three-medium composite evaporator modified from tube-fin heat exchanger, which can recover energy from the water and environment [6]. Han et al. reported a multi-source coupled heat pump system which was capable to operate efficiently and continuously in cold region [7]. On this basis, the seasonal thermal storage was applied in the above system to balance the output characteristics of system and the variation of building load [8]. Ahn et al. compared the heating and dehumidification performance of heat pump, air conditioner and dual evaporator heat pump for cabin, and the dual source configuration exhibited the optimal performance [9, 10]. Zou et al. designed a dual-parallel-evaporator heat pump with two evaporators connected in parallel for the thermal management of electric vehicle [11]. Besides, the similar configuration was used as for water heating with air source and solar energy acting as heat source [12, 13]. Liu et al.

tested the operation performance of a multi-functional heat pump, which has four different heat source combinations: water source, air source, water source and air source in series, and water source and air source in parallel [14]. It is proven that applying hybrid heat source in vapor compression heat pump system is an effective way to realize stable and economical operation. However, it may lead to complicated structure or complex control strategy, which brings limitation in practical application.
In this paper, the performance of air-solar series type, solar-air series type and solar air parallel type hybrid source heat pump water heater is discussed and compared to identify the effect of ambient temperature and solar irradiation.

2. System description

The configurations of solar-air hybrid source heat pump water heater discussed in this paper include air-solar series type, solar-air series type and solar-air parallel type, as shown in Fig. 1. The main components of solar-air hybrid source heat pump include collector evaporator, air source evaporator, compressor, capillary and condenser. In series type, the refrigerant gets vaporized in air source/collector evaporator and collector/air source evaporator in sequence. The electric control valve(ECV) is utilized to adjust the flow distribution in the evaporators of the parallel type. The solar collector (with total area of 4.20 m^2) without any glazing or back insulation was used as evaporator for the refrigerant. The water tank with immersed condensing coil acts as the condenser. The air source evaporator is a finned tube heat exchanger.

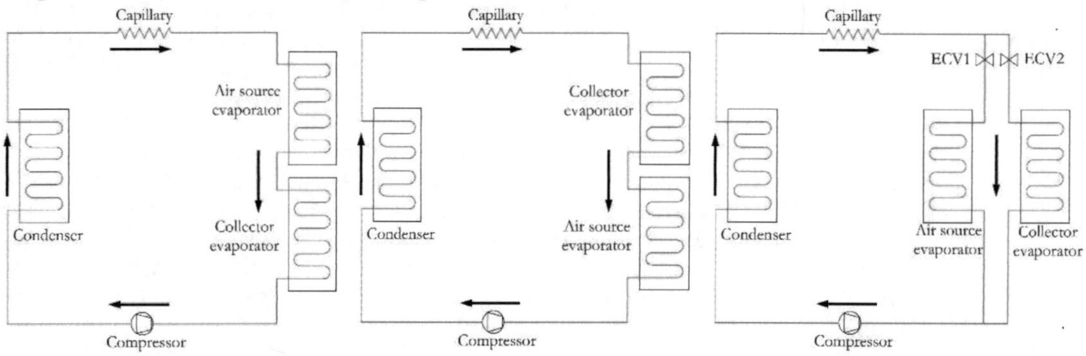

(a) Air-solar series HPWH (b) Solar-air series HPWH (c) Solar-air parallel HPWH

Fig 1. The schematic of solar-air hybrid source heat pump water heater

3. Modelling and evaluation

3.1 Mathematical modelling

For collector evaporator

The energy balance equation for the refrigerant pipeline is:

$$\rho_p c_p \frac{\partial T_p}{\partial t} = k_p \frac{\partial^2 T_p}{\partial z^2} + \frac{1}{A_p}\left[\pi D_{p,i}\alpha_{ref,p}\left(T_{ref} - T_p\right) + W\alpha_{a,p}\left(T_a - T_p\right) + \beta G W\right] \quad (1)$$

ρ_p, c_p, k_p, T_p, A_p and $D_{p,i}$ are the density, specific heat, heat conductivity, temperature, cross section area and inner diameter of the refrigerant pipe respectively, W is the tube pitch of collector evaporator, G is the solar irradiation, $\alpha_{ref,p}$ is the convective heat transfer coefficient between the refrigerant and pipeline, $\alpha_{a,p}$ is the convective heat transfer coefficient between the air and pipeline, T_a and T_{ref} are the temperatures of the ambient and refrigerant respectively, β is the absorptivity of collector evaporator.

For air source evaporator

The energy balance equation for the refrigerant pipeline is:

$$\rho_p c_p \frac{\partial T_p}{\partial t} = k_p \frac{\partial^2 T_p}{\partial z^2} + \frac{1}{A_p}\left[\pi D_{p,i}\alpha_{ref,p}\left(T_{ref} - T_p\right) + \pi D_{p,o}\alpha_{a,p}\left(T_a - T_p\right)\right] \quad (2)$$

$D_{p,o}$ is the outer diameter of the refrigerant pipe.

For compressor

The mass flow rate of refrigerant can be concluded as:

$$\dot{m}_{ref,com} = \eta_v \frac{nV_{com}}{v_{suc}} \quad (3)$$

V_{com}, n and η_v are the displacement, rotating speed and the volumetric efficiency of compressor respectively, v_{suc} is the specific volume of refrigerant at the suction of compressor.
The power consumed by compressor is given as [33]:

$$P_{com} = n\frac{p_{suc}V_{com}}{\eta_{com}} \frac{\gamma}{\gamma-1}\left[\left(\frac{p_{dis}}{p_{suc}}\right)^{\frac{\gamma-1}{\gamma}} -1\right] \quad (4)$$

η_{com} is the total operation efficiency of compressor, γ is the polytropic index, p_{suc} and p_{dis} are the suction and discharge pressures respectively.
In compressor, the refrigerant is compressed into the vapor with high temperature and pressure, which causes the temperature rise of the compressor cylinder and the heat loss to environment. Therefore, the energy balance equation of compressor is:

$$M_{com}c_{com}\frac{\partial T_{com}}{\partial t} = P_{com} - \dot{m}_{ref,com}\left(h_{com,out} - h_{com,in}\right) - U_{com}A_{com}\left(T_{com} - T_a\right) \quad (5)$$

M_{com}, c_{com}, T_{com} and A_{com} are the mass, specific heat, temperature, and external surface area of compressor respectively, U_{com} is the total heat transfer coefficient between the compressor and the environment.
For capillary
In the capillary, the throttling process is taken as isenthalpic, and the mass flow rate of refrigerant is determined by the length (L_{cap}) and inner diameter ($D_{cap,i}$) of the capillary, the subcooled temperature ($\Delta T_{cap,in}$) and the condensing temperature (T_{con}), given as [18]:

$$\dot{m}_r = C_1 D_{cap,i}^{\ C_2} L_{cap}^{\ C_3} T_{con}^{\ C_4} 10^{C_5\times\Delta T_{cap,in}} \quad (6)$$

C_1~C_5 are the empirical constant.
In condenser, the energy balance equation of condensing coil is:

$$\rho_p c_p \frac{\partial T_p}{\partial t} = k_p \frac{\partial^2 T_p}{\partial z^2} + \frac{1}{A_p}\left[\pi D_{p,i}\alpha_{ref,p}(T_{ref} - T_p) + \pi D_{p,o}\alpha_{w,p}(T_w - T_p)\right] \quad (7)$$

T_w is the temperature of the condensing water, $\alpha_{w,p}$ is the convectively heat transfer coefficient between the condensing coil and condensing water.
For the condensing water, the energy balance equation is:

$$M_w c_w \frac{\partial T_w}{\partial t} = A_{con}U_{con}(T_a - T_w) + A_{p,o}\alpha_{w,p}(T_p - T_w) \quad (8)$$

M_w and c_w are the mass and specific heat of condensing water respectively, A_{con} is the external surface area of the water tank, U_{con} is the total heat transfer coefficient between the water tank and ambient.
3.2 Performance evaluation
The evaporating capacity in collector evaporator is:

$$Q_{eva,coll} = \dot{m}_{ref,coll}\left(h_{coll,out} - h_{coll,in}\right) \quad (9)$$

$h_{coll,out}$ and $h_{coll,in}$ are the specific enthalpies at the outlet and inlet of the collector evaporator respectively.
The evaporating capacity in air source evaporator is:

$$Q_{eva,air} = \dot{m}_{ref,air}\left(h_{air,out} - h_{air,in}\right) \quad (10)$$

$h_{air,out}$ and $h_{air,in}$ are the specific enthalpies at the outlet and inlet of the air source evaporator respectively.
The condensing capacity can be calculated by:

$$Q_{con} = \dot{m}_{ref,con}\left(h_{con,in} - h_{con,out}\right) \quad (11)$$

IOP Conf. Series: Materials Science and Engineering **556** (2019) 012021 doi:10.1088/1757-899X/556/1/012021

$h_{con,out}$ and $h_{con,in}$ are the specific enthalpies at the outlet and inlet of the condenser respectively.
For the SA-HPWH system, the performance of coefficients (COP) is:

$$COP = \frac{Q_{con}}{P_{com}} \tag{12}$$

4. Results and discussions

4.1. Impact of ambient temperature

The performance of solar-air series type, air-solar series type and solar-air parallel type HPWH has been compared at the ambient temperature of 10°C, 20°C and 30°C with the solar irradiation of 100W/m^2, as shown in Fig. 2~3. It can be observed that the evaporating capacities in air source evaporator and collector evaporator increase with the rise of ambient temperature. The evaporating capacity of collector evaporator is the highest in solar-air series type HPWH, and the evaporating capacity of air source evaporator is the highest in air-solar series HPWH. Similarly, the condensing capacity increases with the rise of ambient temperature. With the ambient temperature rising from 10°C to 30, the average condensing capacity increases from 1819.97W to 2568.14W for solar-air series type HPWH, rises from 1825.42W to 2579.33W for air-solar series type HPWH, and increases from 1836.65W to 2613.45W for solar-air parallel type HPWH. It can be concluded that the solar-air parallel type HPWH can obtain the highest condensing capacity at the solar irradiation of 100W/m^2.

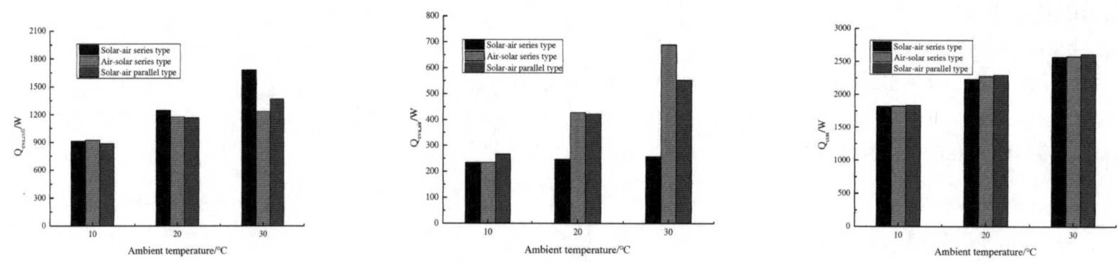

(a) $Q_{eva,coll}$ (b) $Q_{eva,air}$ (c) Q_{con}

Fig 2. The comparison of evaporating capacity and condensing capacity at specified ambient temperature

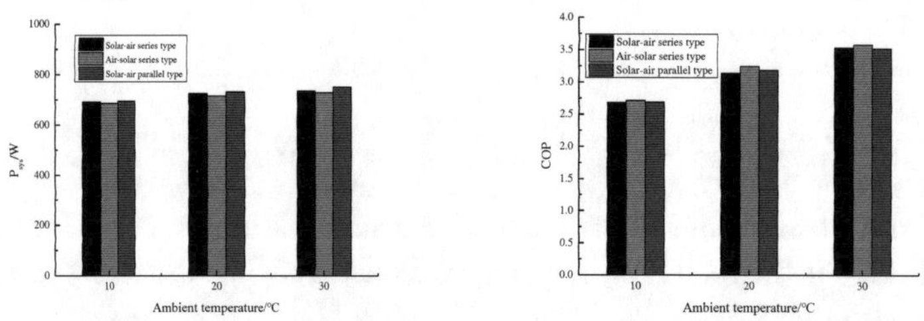

(a) P_{sys} (b) COP

Fig 3. The comparison of power consumption and COP at specified ambient temperature

The power consumption in each type of HPWH is approximative and increases with the rise of ambient temperature. With the ambient temperature rising from 10°C to 30°C, the average power consumption increases from 691.92W to 735.50W for solar-air series type HPWH, increases from 685.91W to 728.36W for air-solar series type HPWH, and increases from 695.62W to 750.88W for solar-air parallel type HPWH. Due to the rise of pressure ratio caused by the increase of ambient temperature, the COP in each type of HPWH increases with the rise of ambient temperature. With the ambient temperature rising from 10°C to 30, the average COP increases from 2.68 to 3.53 for solar-air series type HPWH, increases from 2.71 to 3.58 for air-solar series type HPWH, and increases from 2.69 to 3.52 for solar-air parallel type HPWH. It can be concluded that the air-solar series type HPWH can obtain the highest COP at the solar irradiation of 100W/m^2.

IOP Conf. Series: Materials Science and Engineering **556** (2019) 012021 doi:10.1088/1757-899X/556/1/012021

4.2. Impact of solar irradiation

The performance of solar-air series type, air-solar series type and solar-air parallel type HPWH has been compared at the solar irradiation of 100W/m^2, 200W/m^2 and 300W/m^2 with the ambient temperature of 10°C, as shown in Fig. 4~5.

It can be observed that the evaporating capacity in collector evaporator and the condensing capacity increase with the rise of solar irradiation. In air-solar series type HPWH, higher solar irradiation can improve the evaporating capacity in air source evaporator, while it plays the opposite role for solar-air series type HPWH and solar-air parallel type HPWH. With the solar irradiation rising from 100W/m^2 to 300W/m^2, the average condensing capacity increases from 1819.97W to 2071.46W for solar-air series type HPWH, the average condensing capacity increases from 1825.42W to 2215.45W for air-solar series type HPWH, and the average condensing capacity increases from 1836.65W to 2147.98W for solar-air parallel type HPWH. It can be concluded that the air-soalr series type HPWH can obtain the highest condensing capacity at the ambient temperature of 10°C.

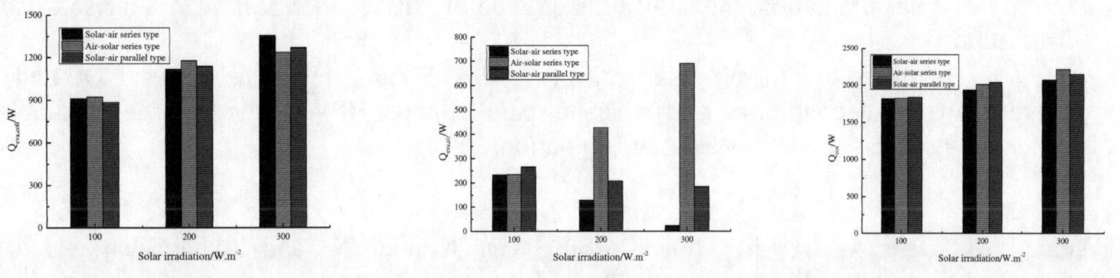

(a) $Q_{eva,coll}$ (b) $Q_{eva,air}$ (c) Q_{con}

Fig 4. The comparison of evaporating capacity and condensing capacity at specified solar irradiation

The power consumption increases with the rise of solar irradiation, and air-solar series type HPWH consumes the least power. With the solar irradiation rising from 100W/m^2 to 300W/m^2, the average power consumption increases from 691.92W to 722.12W for solar-air series type HPWH, increases from 685.91W to 700.62W for air-solar series type HPWH, and increases from 695.62W to 716.50W for solar-air parallel type HPWH. The COP in each type of HPWH increases with the rise of solar irradiation. With the solar irradiation rising from 100W/m^2 to 300W/m^2, the average COP increases from 2.68 to 2.92 for solar-air series type HPWH, increases from 2.71 to 3.32 for air-solar series type HPWH, and increases from 2.69 to 3.04 for solar-air parallel type HPWH. It can be concluded that the air-solar series type HPWH can obtain the highest COP at the ambient temperature of 10°C.

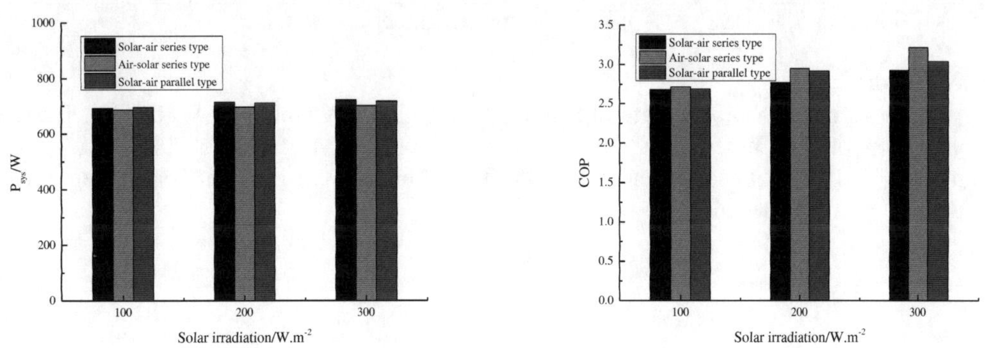

(a) P_{sys} (b) COP

Fig 5. The comparison of power consumption and COP at specified solar irradiation

5. Conclusion

In this paper, the solar-air hybrid source heat pump water heater with solar-air series, air-solar series and solar-air parallel configurations has been discussed and compared. The dynamic model has been developed to identify the impacts of solar irradiation and environmental temperature. The conclusion has been summarised as follows:

- Solar-air hybrid source heat pump water heater can guarantee efficient operation under the ambient temperature of 10°C and solar irradiation of 100W/m^2, with the average COPs of 2.68 for solar-air series type HPWH, 2.71 for air-solar series type HPWH and 2.69 for solar-air parallel type HPWH.
- The ambient temperature can improve the evaporating capacities of air source evaporator and collector evaporator leading to the rise of condensing capacity. The pressure ratio increases with the rise of ambient temperature leading to the increase of power consumption and COP for each type of solar-air hybrid source heat pump water heater.
- In air-solar series type HPWH, higher solar irradiation can improve the evaporating capacity in air source evaporator, while it plays the opposite role for solar-air series type HPWH and solar-air parallel type HPWH. As the overall heat gain at the evaporating side increases with the rise of solar irradiation, the COP of each type of HPWH increases with the rise of solar irradiation.
- It can be concluded that the air-solar series type HPWH can achieve the highest COP under a range of working conditions and solar-air parallel type HPWH can provide the highest condensing capacity under lower solar irradiation.

Reference

[1] Remeli M, Date A, Orr B, Ding L, Singh B, Affandi N. and Akbarzadeh A. 2016, Energy Conversion and Management, **111** 147.
[2] Chua K.J., Chou S and Yang W 2010, Applied Energy, **87** 3611.
[3] Chargui R and Awani S 2017, Energy Conversion and Management, **147** 40.
[4] Chargui R and Sammouda H 2014, Energy Conversion and Management, **81** 384.
[5] Qu M, Chen J, Nie L, Li F, Yu Q and Wang T 2016, Applied Thermal Engineering, **94** 819.
[6] Wang G, Quan Z, Zhao Y, Sun C, Deng Y and Tong J 2015, Energy and Buildings, **108** 175.
[7] Han Z, Ju X, Qu L, Liu J, Ma X and Zhang S 2017, Solar Energy, **155** 1208.
[8] Han Z, Qu L, Ma X, Song X and Ma C 2017, Applied Thermal Engineering, **116** 292.
[9] Ahn J, Kang H, Lee H, Jung H, Baek C and Kim Y 2014, Applied Energy, **119** 1.
[10] Ahn J, Kang H, Lee H and Kim Y 2015, Applied Energy, **146** 29.
[11] Zhang G, Qin F, Zou H and Tian C 2017, Energy Procedia, **105** 2390.
[12] Deng W and Yu J 2016, Energy Conversion and Management, **120** 378.
[13] Cai J, Ji J, Wang Y, Zhou F, Yu B 2017, Energy Conversion and Management, **148** 635.
[14] Liu X, Ni L, Lau S, Li H 2013, Applied Thermal Engineering, **51** 698.

Acknowledgments

Financial support from the National Natural Science Foundation of China (No. 51806092), Bureau of International Cooperation, Chinese Academy of Sciences (No. 211134KYSB20160005), National Key R&D Plan (No. 2016YFE0124800) and the DongGuan Innovative Research Team Program (No. 2014607101008) is acknowledged.

Synthesis of a carbazole-substituted diphenylethylene hole transporting material and application in perovskite solar cells

Juncong Li[1,2], Hongwei Zhu[1,2], Xianggao Li*[1,2], Shirong Wang[1,2] and Yin Xiao[1,2]

[1] Fine Chemicals Department, School of Chemical Engineering and Technology, Tianjin University, 92 Weijin Rd., Nankai Dist., Tianjin 300072, China

[2] Collaborative Innovation Center of Chemical Science and Engineering (Tianjin), 92 Weijin Rd., Nankai Dist., Tianjin 300072, China

Abstract. In this work, a carbazole-substituted stilbene derivatives hole transporting material, 9,9'-(ethene-1,1-diylbis(4,1-phenylene))bis(3,6-dimethoxy-9H-carbazole) (EPC) was designed and synthesized. The structure was characterized by FT-IR spectroscopy, ^{1}H NMR, ^{13}C NMR and MS. The glass transition temperature is 95.09°C and the 5% weight loss temperature is 423.37°C by DSC/TG analysis. Through the UV-Vis absorption spectrum combined with cyclic voltammetry, the energy level of the highest occupied molecular orbital (HOMO) of EPC and the lowest unoccupied orbital (LUMO) energy level were obtained (-5.38 eV, -2.19 eV). The energy level is well matched with methylamine lead iodine crystals. A mesoporous structure perovskite solar cell was fabricated based on the undoped EPC, and photoelectric conversion efficiency of 13.11% was obtained. After 600 hours of storage in a dark environment at 25°C and 60% relative humidity, 82% of the initial efficiency was obtained. This work provides new ideas for the development of undoped hole transporting materials.

Keywords: carbazole, hole transporting material, undoped, perovskite solar cell, stability

1. Introduce

Since 2009, organic-inorganic perovskite solar cells (PSCs) have been received attention due to their advantages in photovoltaic performance, preparation process, and cost. The highest efficiency has exceeded 22%. [1-5]

The hole transporting layer is an important structure in the perovskite solar cell and is a physical / energy barrier between the cathode and the perovskite layer, blocking the transfer of electrons to the cathode while promoting hole transporting. [6, 7] At present, the hole transporting materials for perovskite solar cells include p-type inorganic semiconductors (CuI [8], CuSCN [9], NiO [10]), polymers (P3HT [11], PTAA [12], PEDOT: PSS [13]), and organic small molecule compounds. [14-17] Because of easy purification and good film properties, triarylamine compounds are the most abundant hole transporting materials, and the highest photoelectric conversion efficiency (PCE) is obtained in the cells using *spiro*-OMeTAD. [18] However, the cost of *spiro*-OMeTAD is high, and the use of hydrophilic dopants will affect the stability of the cells. [19] It is necessity to develop lower cost and undoped hole transporting materials.

The carbazole derivatives are excellent in charge transporting ability, have multiple reaction sites, permit fine tuning of electrical and optical properties, and are low in the price of synthetic raw materials. [20-24] Therefore, carbazole derivatives have attracted much attention as hole transporting materials. In this work, a novel carbazole-based small molecule hole transporting material is reported. Its photophysical properties are measured, a mesoporous perovskite solar cell is fabricated with an efficiency of 13.11%, in which the hole transporting material is the synthesized compound EPC.

2. Results and discussion

EPC is synthesized by nucleophilic substitution and Wittig reaction (**Scheme** S1). The molecular structure is shown in **Fig.** 1a. The structure was fully characterized by ^{1}H NMR spectroscopy, ^{13}C NMR spectroscopy and MALDI-TOF mass spectrum (**Fig.** S1-4).

Fig. 1b shows the normalized ultraviolte-visible (UV-Vis) absorption and photoluminescence (PL) spectra of EPC in tetrahydrofuran (THF) solution (c=1×10^{-5} mol/L). The maximum absorption is at 313 nm. It is caused by the π-π* transition of electrons in the molecule. [25] In addition, the EPC fluorescence emission maximum intensity is at 391 nm.

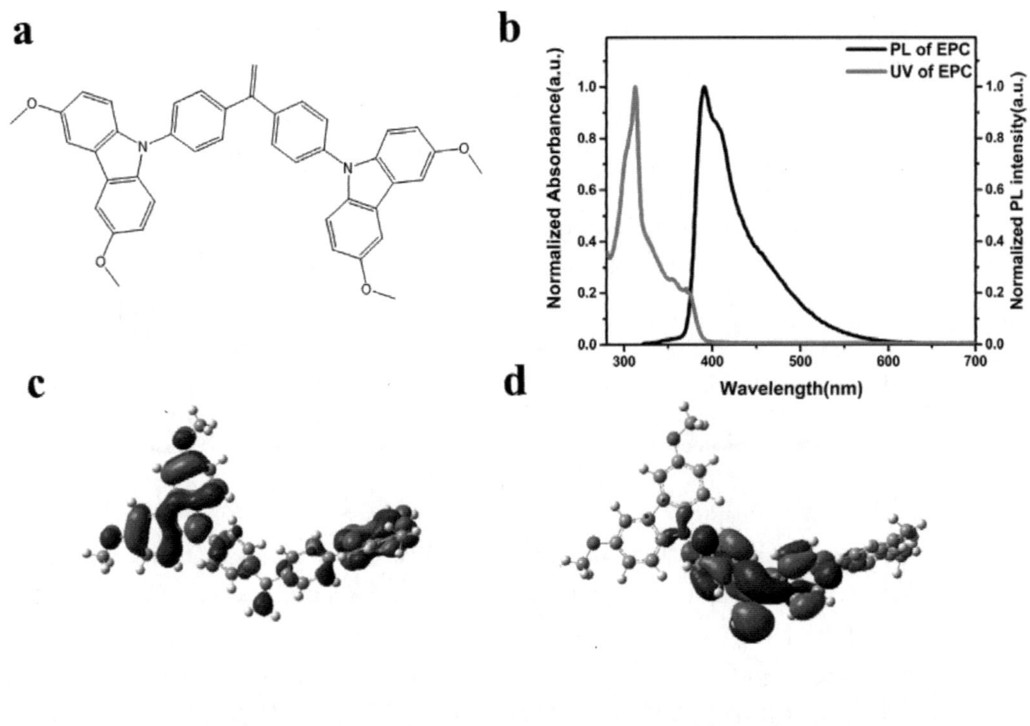

Fig. 1 (a) The molecular structure of EPC. (b) Normalized UV-Vis absorption and PL spectra of EPC in tetrahydrofuran (THF) solution. (c) HOMO distribution of EPC. (d) LUMO distribution of EPC.

In order to study the relationship between molecular structure and material properties and verify theoretically that the compound is suitable for perovskite solar cells, we used Gaussian09 calculation software, density functional theory (DFT), B3LYP as the calculation method, and 6-31G(d) as the basis set and obtained the optimal configuration of EPC. The results are shown in **Fig.** 1c-d. Due to the presence of the rigid group carbazole, the benzene ring attached to the C=C is not coplanar, showing a rotation of 67.02°, resulting in a large steric hindrance, not easy to form crystals, and conducive to the formation of an amorphous solid film. The highest occupied molecular orbital (HOMO) level of EPC is -4.82 eV, that mainly distributed on the benzene ring and carbazole at the periphery of the molecule. The lowest unoccupied molecular orbital (LUMO) level is -1.15 eV, that mainly distributed on the C=C in the center of the molecule and on the adjacent benzene ring, which facilitates carrier transmission.

Table 1. Photophysical, electrochemical and thermal properties of EPC

	λ_{abs} [nm]	E_g [eV]	HOMO [eV]	LUMO [eV]	T_g [°C]	T_d [°C]	μ [$cm^2 V^{-1} s^{-1}$]
EPC	313	2.05[a]	-4.73[a]	-1.68[a]	95.09	423.37	7.23×10^{-5}
		3.19[b]	-5.38[b]	-2.19[b]			

(a) calculation values (b) experiment values (HOMO level was measured by CV).

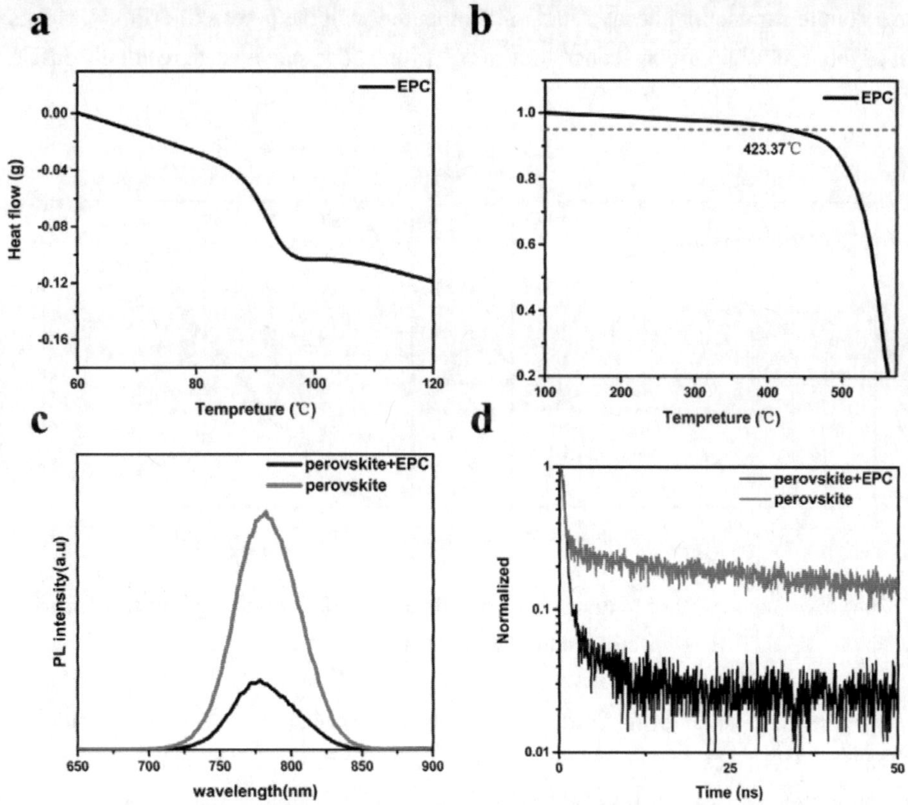

Fig. 2 (a) DSC curve. (b) TGA curve. (c) Steady-state PL spectra, excitation at 450 nm. (d) Time-resolved PL spectra, decay at 775 nm, excitation at 460 nm.

The electrical property of EPC was obtained by cyclic voltammetry (CV), and the energy levels were determined by CV and UV-Vis absorption (**Fig.** S5). The data are summarized in **Table** 1. The EPC has a HOMO energy level of -5.38 eV and a LUMO energy level of -2.19 eV. The HOMO level matches the $MAPbI_3$ HOMO level (-5.40 eV), which facilitates hole injection.

The glass transition temperature and the 5% weight loss temperature were measured by thermal gravimetric analysis (TGA) and differential scanning calorimeter (DSC), and the results are shown in **Fig.** 2a-b. The 5% weight loss temperature of EPC is 423.37 °C, and the glass transition temperature is 95.09 °C, which satisfies the application requirements of hole transporting materials for perovskite solar cells. The hole mobility of EPC was measured by space charge limited current method (SCLC). [26] The hole only device with the structure of ITO/MoO_3 (10 nm)/EPC (60 nm)/MoO_3 (10 nm)/Al (100 nm) was fabricated and the hole mobility of EPC is 7.23×10^{-5} cm^2 V^{-1} s^{-1} (**Fig.** S6).

The charge transfer process at the perovskite/EPC interface was investigated by the steady-state and time-resolved PL spectra, and the results are shown in **Fig.** 2c-d. When the EPC is deposited on top of the perovskite, the PL intensity is significantly reduced, reaching 28.41% of the perovskite film alone, indicating that between the perovskite and the EPC there is an efficient hole extraction process. [27, 28] The time-resolved photoluminescence spectra measured excitation wavelength of 460 nm and fitted the data through a double exponential decay function. Compared with the perovskite film (13.08 ns) alone, EPC can effectively reduce the fluorescence lifetime (1.17 ns). The steady state result is consistent with the PL spectrum.

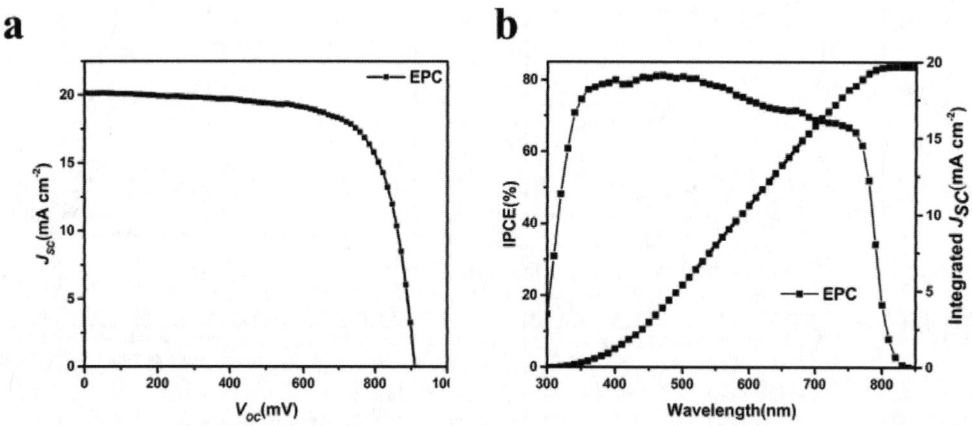

Fig. 3 (a) Current-voltage curve of perovskite solar cell based on EPC with the highest PCE. (b) IPCE spectrum and integrated current curve

Mesoporous perovskite solar cells were fabricated based on undoped EPC. The specific structure is FTO/TiO_2 compact/TiO_2 mesoporous/$MAPbI_3$/EPC/Au. The perovskite precursor solution concentration was 1.56 mol/L, the solvent was DMF/DMSO (v: v=9:1), chlorobenzene was used as the anti-solvent, and the EPC was 20 mg/mL chlorobenzene solution. The current-voltage curve (*J-V*) of cell with the highest PCE under AM 1.5 G (100 mW cm^{-2}) illumination is shown in **Fig.** 3a. The open-circuit voltage (V_{OC}) of the optimal cell is 909.43 mV and the short-circuit current (J_{SC}) is 20.17 mA cm^{-2}, the fill factor (*FF*) is 71.45%, and the photoelectric conversion efficiency (PCE) is 13.11%. In addition, **Fig.**

3b shows the incident photon current-to-electron efficiency (IPCE) spectrum of the cell based on EPC. The photocurrent density obtained from the IPCE integration is 19.68 mA cm^{-2}, which is consistent with the *J-V* curve.

Stability of EPC-based perovskite solar cells were obtained by storing cells without encapsulation in a dark state at 25 °C and 60% relative humidity and testing *J-V* curves at regular intervals. The change in PCE over time is shown in **Fig.** 4a. The test results showed that after 600 hours (25 days) of storage, the cell efficiency remained at the initial 82%. According to the previous reports, dopants can cause pinholes in the hole transporting layer and can induce decomposition of perovskite crystals. [29-31] In addition, the hydrophobic hole transporting layer can act as a barrier to decompose the perovskite of H_2O. The contact angle of the perovskite/EPC interface with water is 91.23°, which is much larger than the contact angle of the perovskite interface with water alone. This is advantageous to block the contact between H_2O and perovskite in the air and improve the stability of the cells.

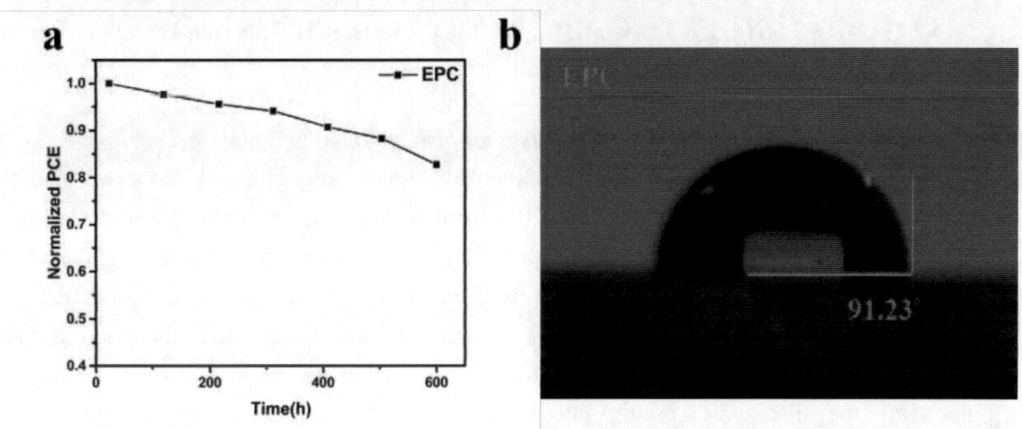

Fig. 4 (a) The stability of the cell based on EPC. (b) Contact angle between perovskite/EPC film with water in air.

3. Conclusions

In this work, a novel type of carbazole-substituted hole transporting material EPC was designed and synthesized. EPC has suitable hole mobility, good thermal stability, and energy levels that match $MAPbI_3$. The cells were fabricated based on the undoped EPC, the PCE of 13.11% was obtained, and the cell maintained its initial efficiency of 82% after 600 hours in a dark state at 25 °C and 60% relative humidity, showing a good stability.

4. Experiment Section

4.1. Synthesis of 3,6-dimethoxy-9H-carbazole

3,6-dibromo-9H-carbazole (6.0 g, 18.45 mmol), CuI (14.8 g, 77.50 mmol), and 70 mL DMF were charged into a three-neck flask. Methanol solution of sodium methoxide (prepared from 8.5 g, 369.5 mmol sodium and 100 mL methanol) was added under stirring and heated to reflux for 12 h. After the reaction system was cooled to room temperature, the solution was poured into 500 mL ice water to obtain a brown solid. The filter cake was filtered and washed with deionized water (3×100 mL). The solid was separated by column chromatography (petroleum ether: ethyl acetate=10:1) to give 3.11 g white solid with a yield of 81.12%. ^{1}H NMR (400 MHz, DMSO) δ10.80 (s, 1H), 7.79 (dd, J = 102.4, 2.4 Hz, 2H), 7.33 (d, J = 8.7 Hz, 2H), 7.08–6.85 (m, 2H), 3.86 (d, J = 16.5 Hz, 6H); MS(MALDI-TOF) =226.738.

4.2. Synthesis of bis(4-(3,6-dimethoxy-9H-carbazol-9-yl)phenyl)methanone

3,6-dimethoxy-9H-carbazole (2.08 g, 9.16 mmol), t-BuNa (4.00 g, 41.60 mmol) and 40 mL DMF were charged into a two-necked flask. After argon protection, the mixture was stirred at 60°C for 20 min, and 4,4-difluorophenone (0.95 g, 4.36 mmol) was added, then the temperature was raised to 110 °C, and the reaction was stirred for 12 h. After the reaction system was cooled to room temperature, the solution was poured into 400 mL ice water to obtain a yellow solid. The filter cake was filtered and washed with deionized water (3×100 mL). The crude product was recrystallized from methanol and dichloromethane to give 2.01 g yellow solid with a yield of 72.57%. ^{1}H NMR (400 MHz, $CDCl_3$) δ8.24–8.00 (m, 4H), 7.76 (d, J = 8.5 Hz, 4H), 7.56 (d, J = 2.5 Hz, 4H), 7.48 (t, J = 7.7 Hz, 4H), 7.08 (dd, J = 8.9, 2.5 Hz, 4H), 3.97 (s, 12H); MS MALDI-TOF) =632.126.

4.3. Synthesis of 9,9'-(ethene-1,1-diylbis(4,1-phenylene))bis(3,6-dimethoxy-9H-carbazole)

$CH_3P^+Ph_3Br^-$ (0.84 g, 2.37 mmol) and t-BuOK (0.36 g, 3.16 mmol) were charged into a two-neck flask which was placed in ice water, and the mixture was purged with argon. The stirring was started and 50 mL dry THF was slowly added dropwise through a constant pressure dropping funnel, and the reaction was performed for 2 hours. The reaction apparatus was transferred to an oil bath, and bis(4-(3,6-dimethoxy-9H-carbazol-9-yl)phenyl)methanone (1.00 g, 1.58 mmol) was added to the reaction system, and the mixture was heated to 50 °C and reacted for 10 hours. After the reaction system was cooled to room temperature, the solution was poured into 400 mL ice water to obtain a yellow solid. The filter cake was filtered and washed with deionized water (3×100 mL). The solid was then separated by column chromatography (petroleum ether: ethyl acetate = 10:1), 0.65 g pale yellow solid with a yield of 65.00% was obtained. ^{1}H NMR (400 MHz, DMSO) δ 7.86 (d, J = 1.9 Hz, 4H), 7.76–7.61 (m, 8H), 7.43 (d, J = 8.9 Hz, 4H), 7.08 (dd, J = 8.9, 1.9 Hz, 4H), 5.75 (s, 2H), 3.91 (s, 12H); ^{13}C NMR (100 MHz, DMSO) δ = 56.17, 103.86, 111.07, 115.79, 116.16, 123.92, 126.40, 129,99, 135.65, 137.69, 139.33, 148.17, 154.32 ppm; MS(MALDI-TOF) =630.155.

4.4. Characterization

EPC molecular structure was identified by NMR spectroscopy and MS. We used the INOVA 400 MHz spectrometer (Varian, USA) and tetramethylsilane (TMS) as internal standards to obtain NMR spectra. MS was performed by using microTOF-QII (Bruker Daltonic, USA). The UV-Vis spectrum of EPC in THF solution ($c=1\times10^{-5}$ mol/L) was obtained by using a Thermo Evolution 300 UV-Vis spectrometer (Thermo Electron, USA). DSC and TGA were recorded under a N_2 atmosphere by using a TA Q20 (TA Instruments, USA) thermal analyzer.

The device was measured for *J-V* characteristics using 450W Xenon lamp (Oriel) at 100 mW cm^{-2}. All devices were measured by masking the active area with a metal mask with an area of 0.09 cm^2. The IPCE of the device was measured by focusing light from a 300W Xenon lamp (ILC Technology, USA) through a Gemini-180 double monochromator (Jobin Yvon Ltd., UK). TRPL and PL spectra were measured by PL spectrometer, Edinburgh Instruments, Fluoroloy3, excited with a picosecond pulsed diode laser (EPL-460).

The CV measurements were performed on fresh dichloromethane solutions with 0.1 mol/L 1-Bu_4NPF_6 as the electrolyte by using a three-electrode system. Each solution was bubbled with Ar for 1 min before measurement. In the three-electrode system, the working electrode and the counter electrodes are Pt wires, and an Ag/0.01M $AgNO_3$ electrode (with acetonitrile as the solvent) is the reference electrode. The measurements were calibrated using ferrocene as a standard.

4.5. Device fabrication

The FTO glass was ultrasonically washed with deionized water, acetone and isopropanol successively. A titanium diisopropoxide bis(acetylacetonate) in n-butanol solution having a volume ratio of 1:16 was prepared and spin-coated at 2000 rpm for 30 s. The FTO glass was sintered at 500 °C for 30 min to obtain a TiO_2 compact layer. Then, a 30 nm TiO_2 ethanol suspension having a mass ratio of 1:6 was spin-coated at 5000 rpm for 30 s, and the glass was sintered at 500 °C for 30 min to obtain a TiO_2 mesoporous layer. The perovskite precursor solution concentration was 1.56mol/L, wherein the solvent was a DMF/DMSO mixed solution (v: v=9:1), and the perovskite spin coating process was 2000 rpm for 20 s and then 5000 rpm for 30 s. In the final 15 s of the second spin coating, 100 μL chlorobenzene solution was added dropwise, and the obtained perovskite film was annealed at 100 °C for 1 hour. Then, 45 μL chlorobenzene solution of EPC having a concentration of 20 mg/mL was dropwise added to the perovskite crystal layer at 4000 rpm and spin-coated for 15 s. Finally, an 80 nm gold electrode was deposited on the hole transporting layer by vacuum evaporation.

Acknowledgements

The authors gratefully acknowledge the financial support from the National Natural Science Foundation of China (21676188) and Key Projects in Natural Science Foundation of Tianjin (JCZDJC37100). The calculation in this work was supported by the high performance computing center of Tianjin University, China.

[1] Lee M M, Teuscher J, Miyasaka T, Murakami T N, Snaith H J 2012 *Science* **338** 643-7
[2] Green M A, Ho-Baillie A and Snaith H J 2014 *Nature Photonics* **8** 506-14
[3] Service R F 2011 *Science* **332** 293
[4] Williams S T, Rajagopal A, Chueh C C and Jen A K 2016 *J Phys. Chem. Lett.* **7** 811-9
[5] http://www.nrel.gov/ncpv/images/efficiency_chart.jpg
[6] Bakr Z H, Wali Q, Fakharuddin A, Schmidt-Mende L, Brown T M and Jose R 2017 *Nano Energy* **34** 271-305
[7] Zhu L, Xiao J, Shi J, Wang J, Lv S, Xu Y, Luo Y, Xiao Y, Wang S, Meng Q, Li X and Li D 2014 *Nano Research* **8** 1116-27
[8] Ye S, Rao H, Zhao Z, Zhang L, Bao H, Sun W, Li Y, Gu F, Wang J, Liu Z, Bian Z and Huang C 2017 *J Am Chem Soc* **139** 7504-12
[9] Arora N, Ibrahim D M, Hinderhofer A, Pellet N, Schreiber F, Mohammed Z S and Grätzel M 2017 *Science* **358** 768-71

[10] Xie Y, Lu K, Duan J, Jiang Y, Hu L, Liu T, Zhou Y and Hu B 2018 *ACS applied materials & interfaces* **10** 14153-9
[11] Narges Yaghoobi N, Fabio M, Lucio C, and Aldo D C 2017 *ChemSusChem* **10** 3854-60
[12] Ko Y, Kim Y, Lee C, Kim Y and Jun Y 2018 *ACS applied materials & interfaces* **10** 11633-41
[13] Xue Q, Liu M, Li Z, Yan L, Hu Z, Zhou J, Li W, Jiang X-F, Xu B, Huang F, Li Y, Yip H-L and Cao Y 2018 *Adv. Funct. Mater.* **28** 1707444
[14] Bi D, Xu B, Gao P, Sun L, Grätzel M and Hagfeldt A 2016 *Nano Energy* **23** 138-44
[15] Xu B, Bi D, Hua Y, Liu P, Cheng M, Grätzel M, Kloo L, Hagfeldt A and Sun L 2016 *Energy Environ. Sci.* **9** 873-7
[16] Gratia P, Magomedov A, Malinauskas T, Daskeviciene M, Abate A, Ahmad S, Gratzel M, Getautis V and Nazeeruddin M K 2015 *Angew. Chem. Int. Ed. Engl.* **54** 11409-13
[17] Xu B, Zhang J, Hua Y, Liu P, Wang L, Ruan C, Li Y, Boschloo G, Johansson E M J, Kloo L, Hagfeldt A, Jen A K Y and Sun L 2017 *Chem* **2** 676-87
[18] Yu Z and Sun L 2015 *Adv. Energy Mater.* **5** 1500213
[19] Rakstys K, Saliba M, Gao P, Gratia P, Kamarauskas E, Paek S, Jankauskas V and Nazeeruddin M K 2016 *Angew. Chem. Int. Ed. Engl.* **55** 7464-8
[20] Xu B, Sheibani E, Liu P, Zhang J, Tian H, Vlachopoulos N, Boschloo G, Kloo L, Hagfeldt A and Sun L 2014 *Adv. Mater.* **26** 6629-34
[21] Magomedov A, Paek S, Gratia P, Kasparavicius E, Daskeviciene M, Kamarauskas E, Gruodis A, Jankauskas V, Kantminiene K, Cho K T, Rakstys K, Malinauskas T, Getautis V and Nazeeruddin M K 2018 *Adv. Funct. Mater.* **28** 1704351
[22] Li D, Shao J Y, Li Y, Li Y, Deng L Y, Zhong Y W and Meng Q 2018 *Chem. Commun. (Camb.)* **54** 1651-4
[23] Benhattab S, Cho A-N, Nakar R, Berton N, Tran-Van F, Park N-G and Schmaltz B 2018 *Org. Electron.* **56** 27-30
[24] Daskeviciene M, Paek S, Wang Z, Malinauskas T, Jokubauskaite G, Rakstys K, Cho K T, Magomedov A, Jankauskas V, Ahmad S, Snaith H J, Getautis V and Nazeeruddin M K 2017 *Nano Energy* **32** 551-7
[25] Qi P, Zhang F, Zhao X, Liu X, Bi X, Wei P, Xiao Y, Li X and Wang S 2017 *Energy Technology* **5** 1173-8
[26] Kwon O K, Uddin M A, Park J H, Park S K, Nguyen T L, Woo H Y and Park S Y 2016 *Adv. Mater.* **28** 910-6
[27] Roldán-Carmona C, Gratia P, Zimmermann I, Grancini G, Gao P, Graetzel M and Nazeeruddin M K 2015 *Energy Environ. Sci.* **8** 3550-6
[28] Mosconi E, Grancini G, Roldán-Carmona C, Gratia P, Zimmermann I, Nazeeruddin M K and De Angelis F 2016 *Chem. Mater.* **28** 3612-5
[29] Li W, Dong H, Wang L, Li N, Guo X, Li J and Qiu Y 2014 *J. of Mater. Chem. A* **2** 13587-92
[30] Hawash Z, Ono L K, Raga S R, Lee M V and Qi Y 2015 *Chem. Mater.* **27** 562-9
[31] Berhe T A, Su W-N, Chen C-H, Pan C-J, Cheng J-H, Chen H-M, Tsai M-C, Chen L-Y, Dubale A A and Hwang B-J 2016 *Energy Environ. Sci.* **9** 323-56

IOP Conf. Series: Materials Science and Engineering **556** (2019) 012023 doi:10.1088/1757-899X/556/1/012023

Thermo-economic analysis of integrated linear Fresnel reflector gas turbine trigeneration power plants

Yousef N. Dabwan[1, 2] and Pei Gang[1]
[1]Department of Thermal Science and Energy Engineering, University of Science and Technology of China, 96 Jinzhai Road, Hefei, China.
[2]Department of Mechanical Engineering, Sana'a University, Sana'a, Yemen.

E-mail Yousefdabwan@gmail.com

Abstract. The present work provides an economic feasibility and annual performance of integrated linear Fresnel reflector gas turbine trigeneration power plant (LFR-GTPP) that is progressively being installed to produce electricity, steam and chilled water. In order to pinpoint the best integration mode, different sizes of gas turbine and solar collector's area have been examined and presented. Thermoflow software was used for evaluating the performance of each integrated design under consideration. The optimal solar integration sizes have been determined. Moreover, reduction in CO_2 emissions due to integrating the LFR system is estimated with respect to that of the conventional trigeneration plant. For the considered trigeneration plant (that is required to produce 120.5 MW of steam, and 2500kg/s of chilled water), the study revealed that LFR-GTPP with gas turbine sizes less than 190 MWe capacities have more economic feasibility and sufficient ability for utilizing solar energy. The levelized electricity cost (LEC) for the (LFR-GTPP) varied between 4.28 US¢ and 5.6 US¢/ kWh. Furthermore, the study revealed that integrating LFR system with a conventional gas turbine trigeneration power plant in Sun Belt regions leads to a considerable avoidance in CO_2 emissions compared to the conventional trigeneration plant.

Keywords: Gas turbine trigeneration power plant; solar thermal power plants; integrating linear Fresnel reflector; thermo-economic analysis.

1. Introduction

A Linear Fresnel Reflector (LFR) based concentrating solar power technology power plant has been considered as one of the most promising technologies for utilizing solar energy. A 100MW solar thermal power plant powered by a LFR system with thermal energy storage was evaluated by Bishoyi and Sudhakar [1]. Mokhtar et al.[2] preformed the thermal performance analysis of a solar water heating system based on LFR technology. Bachelier and Stieglitz [3] presented optimisation analyses in terms of LEC of several designs and configurations for direct molten salt technology plants based on the LFR system. The performance of an integrated gas turbine cogeneration power plant based on LFR technology was investigated using Thermoflow simulation software by Dabwan and Mokheimer [4].

Trigeneration for electricity, heating and cooling production systems integrated with solar energy resources have been investigated in some previous studies. For example, Wang and Yang [5] presented a hybrid-thermal power strategy of a trigeneration system driven by natural gas and solar energy. The system was used to produce electricity, space cooling/heating and domestic hot water. Tora and El-Halwagi [6] developed a systematic procedure for energy saving through the integration design of a trigeneration system powered by natural gas and solar energy. Al-Sulaiman, et al.[7, 8] performed a thermodynamic analysis of trigeneration power plant powered by parabolic trough collector (PTC) system with thermal energy storage. In another work, Zhai, Dai [9] evaluated annual energetic and exergtic efficiencies of a hybrid solar trigeneration system integrated with PTC system under Beijing weather data (China). Zhang, Li [10] analyzed a solar-biomass hybrid trigeneration power plant for producing chilled water, hot water and electricity. In another work, Baghernejad, et al.[11] compared three trigeneration systems in terms of multi objective optimization parameters. Meng, Yang [12] proposed a trigeneratgion system driven by industrial waste heat and solar energy. The performance of the system was theoretically evaluated. Yao, Wang [13] proposed a new trigeneration cycle based on compressed air energy storage system. The system was evaluated in terms of energy, exergy and exergoeconomic performance. It is obvious from the literature that the application of LFR system in the trigeneration power plant for steam, chilled water and electricity generation has been rarely investigated. In addition, most of the previous research works deal with the application of PTC system for co/trigeneration power plants. Thus, this paper presents the investigation of possible design modifications required for integrating LFR to a conventional trigeneration plant for the optimal operation under Al-Hodeidah weather conditions, Yemen.

2. Problem statement and system description

In order to perform thermo-economic analyses for an integrated LFR gas turbine trigeneration power plant (LFR-GTPP), a conventional gas turbine trigeneration power plants (GTPP) was considered as a reference cycle in this study. The reference plant used to produce 340MWe of electricity (250 MWe at gas turbine + 90MWe at steam turbine), 120.5MW of industrial process steam, and 2500 kg/s of chilled water (24440 tons). In this regard, several simulations have been conducted to assess the performance of the plant with different gas turbine sizes and to determine the optimal size of the LFR solar fields that can be integrated with each gas turbine size. The schematic diagram of an integrated solar gas turbine trigeneration plant considered (LFR-GTPP) is presented in Figure 1. The schematic diagram of the conventional gas turbine trigeneration plant (GTPP) is similar to the plant in the Figure 1, but without solar integration. Thermoflow software [14] has been used to simulate both configurations (GTPP and LFR-GTPP). The objective of the current study is to investigate the possible different designs of the conventional and integrated solar gas turbine trigeneration plants for optimal operation for a constant plant steam side load (electrical, thermal and cooling load). In this regard, different gas turbine sizes for the trigeneration plants described above have been investigated.

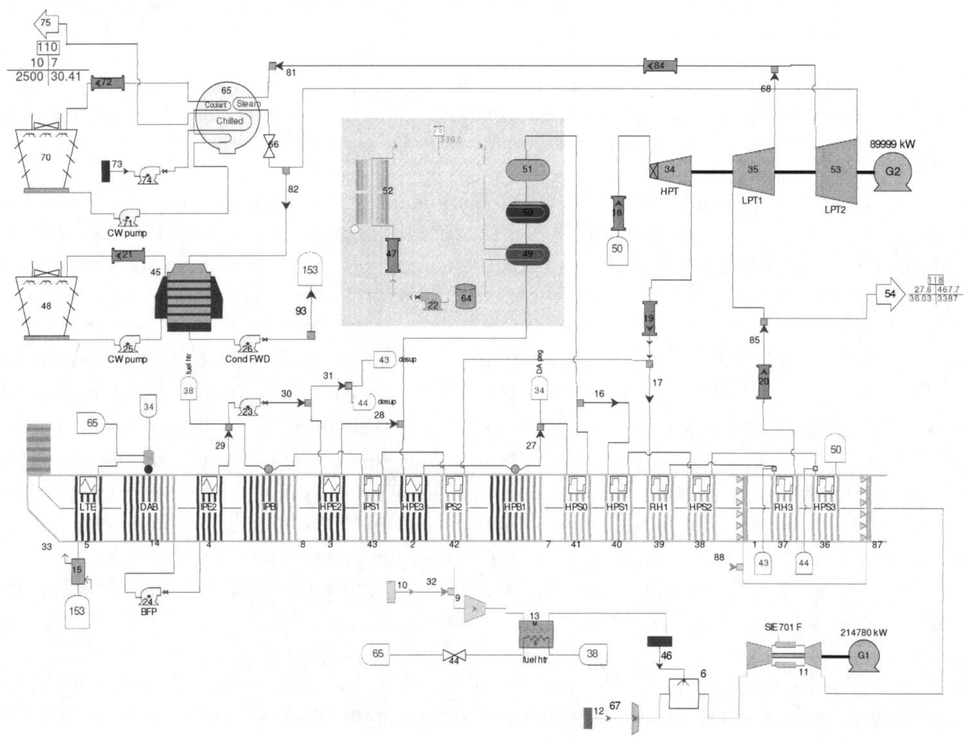

Figure 1: Schematic diagram of an integrated solar gas turbine trigeneration power plant based on LFR as simulated in Thermoflow.

The plants are originally designed to give the required electricity (at stream turbines), industrial process steam and chilled water at the required conditions for the large gas turbine size of 250 MWe. Therefore, for gas turbine sizes smaller than 250 MWe, the required plant steam side load was provided by the LFR system and/or a duct burners. The duct burners are installed in the HRSG to keep the same amount of steam at the same conditions produced when the solar generator is unavailable.

3. Performance parameters

Several thermo-economic parameters were used for evaluating the considered power plants.

Solar multiple is the ratio of the produced thermal power by the LFR system to the total thermal power produced by HRSG (equal to 391.9 MW) [15-17].

$$SM = \frac{P_{th,solar}}{P_{th,total}} \tag{1}$$

Instantaneous solar share can be estimated by the ratio between the power generated by the LFR system and the total power generated by both of fuel and solar energy input at the design hour, [16-19].

$$X_{net,eng,solar} = \frac{P_{gen,solar}}{P_{gen,LFR-GTPP}} = \frac{P_{gen,LFR-GTPP} - \eta_{GTPP} * P_{fuel,LFR-GTPP}}{P_{gen,LFR-GTPP}} \tag{2}$$

Annual solar share can be estimated by the ratio between the energy generated annually by the solar energy input and the total energy generated annually by the solar energy and fuel input [16-19].

$$SS = 1 - \frac{(Annual\ \ fuel\ \ consumption / \ kWh)_{LFR-GTPP}}{(Annual\ \ fuel\ \ consumption / \ kWh)_{GTPP,reference}} \tag{3}$$

Incremental CO_2 avoidance is the annual reduction of CO_2 emissions due to integrating LFR technology [16, 17, 19], and it is defined as:

$$\Delta CO_2 = \left(\frac{E_{gen,LFR-GTPP}}{\eta_{ref}} - E_{fuel,LFR-GTPP} \right) \times f_{CO_2} \text{ , or } \Delta CO_2(\%) = \frac{CO_{2\,GTPP,ref} - CO_{2\,LFR-GTPP}}{CO_{2\,GTPP,ref}} \times 100 \tag{4}$$

Where, $CO_{2\,LFR\text{-}GTPP}$ is annual CO_2 emissions from the LFR-GTPP, $CO_{2\,GTPP,ref}$ is annual CO_2 emissions from the GTPP; and f_{CO2} is the amount of CO_2 emissions per fuel's heating rate [20].

Levelized energy cost (LEC) can be estimated by the ratio between the total annual cost in USD$ and the total annual energy produced by the power plant. In a trigeneration power plant, the LEC of electricity can be estimated by subtracting the total annual price of steam and chilled as per their common local market price, as follows:

$$LEC = \frac{I_{tot} * fcr + OM_{ann} + F_{ann} - (thermal + cooling)_{cost}}{E_{el,\,ann}} \quad ; \quad fcr = \frac{K_d * (1+K_d)^n}{(1+K_d)^n - 1} + K_{insurance} \tag{5}$$

Where, $E_{el,ann}$ is annual total electric power (KWh), I_{tot} is total investment cost; F_{ann} is annual cost of fuel; OM_{ann} is annual cost of maintenance and operation; and f_{cr} is annuity factor; where K_d is the interest rate, $K_{insurance}$ is the annual insurance rate, and n is the depreciation period in years.

In current study, the steam and chilled water price have been assumed to be of 20% and 33% higher than that of the respective fuel price, respectively.

Solar levelized energy cost is used to assess the feasibility of integrating LFR technology to the power plant.

$$SLEC = \frac{LEC_{\text{LFR-GTPP}} - [(1-SS) * LEC_{GTPP}]}{SS} \tag{6}$$

4. Results and discussion

The simulation results of the present study are basis to figure out the optimal LFR field size that can be integrated to a trigeneration plant with different gas turbine sizes to cover the fixed plant steam side load. Figure 2 shows solar multiples for different areas of LFR solar field. The produced thermal powers by the integrated LFR gas turbine trigeneration power plant (LFR-GTPP) with different integration sizes is plotted in Figure 3.The results show that the obtained thermal power from solar system increases with LFR field area (solar multiple) till it reaches a maximum point for each capacity of the gas turbine. After that, it remains constant. This is due to the solar field over sizing, which means that the LFR field would be operated at part load after that point. The maximum thermal power required from the LFR system takes place when the burned fuel at the duct burners is extremely-small during the daytime. That means there is no room for integrating LFR technology after these specified points; this is because the thermal content in gas turbine exhaust gasses is enough for producing electricity, steam and chilled water at the plant steam side.

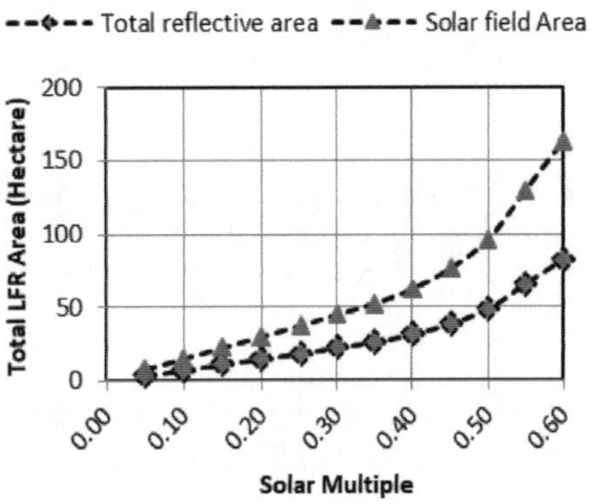

Figure 2. Solar multiples versus areas of the LFR field

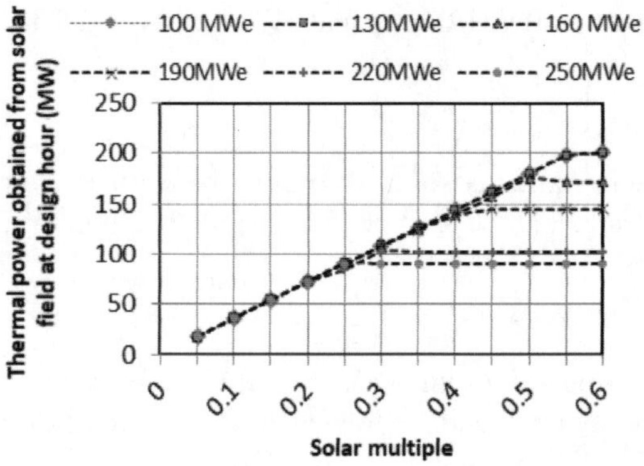

Figure 3. Obtained thermal power from the LFR field versus the solar multiple.

Figure 4 shows the variation in the instantaneous solar shear of the proposed plant (LFR-GTPP) for different values of gas turbine size and solar multiple. From the results, it is clear that an increase in the soar multiple consequently increase the instantaneous solar shear for each gas turbine size. The instantaneous solar shear reaches a maximum value for each gas turbine size. Beyond that, it remains constant. This is attributed to the area of LFR field, which is not fully used beyond that maximum value as mentioned earlier. The annual solar shear (the annual percent) of the energy produced by the LFR-GTPP due to solar integration has been plotted in Figure 5. According to the Fig, the plant with large gas turbine size, 250MWe, does not required much energy from the solar integration (about 4%). whereas, the plant with small gas turbine capacity, 100 MWe, has a small instantaneous solar shear value (about 12.2%).

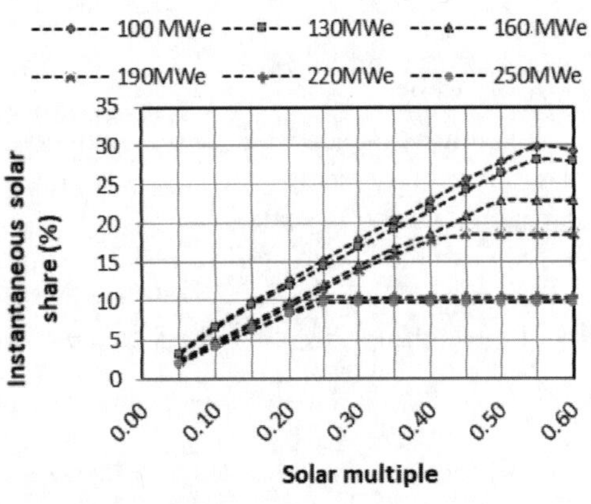

Figure 4. instantaneous solar share

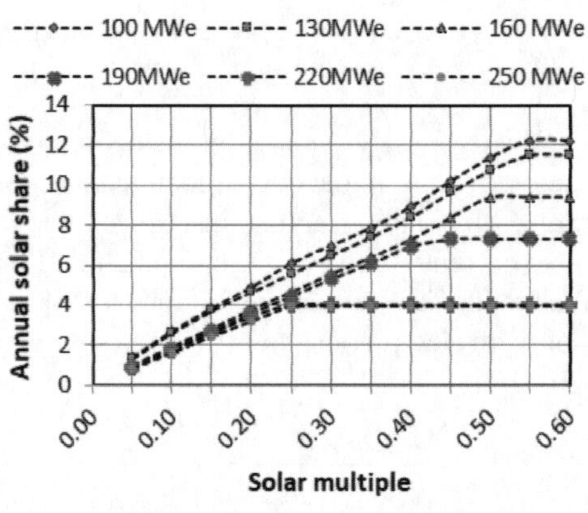

Figure 5. Annual solar share

The economic evaluation of the proposed power plants has been carried out by using LEC and solar SLEC. Figure 6 shows the LEC of the LFR-GTPP with different sizes of the solar field and gas turbine. As shown, the LEC of the plant with different gas turbine sizes slightly increased when solar multiple increases. Figure 7 shows how the SLEC of the plant with different gas turbine sizes vary in response to rise in solar multiple. From the Figure, it can be observed that SLEC reaches a minimum value for each gas turbine size. Beyond that, it increases. This is due to the over sizing of the LFR area, which is not usable after that value. The optimal solar integration occurs when the burned fuel at the duct firings is extremely small during the daytime. It implies that after the minimal value of SLEC, there is no room for integrating LFR system, where the recovered energy in the HRSG is enough for producing electricity, chilled water and steam. Optimum solar integration sizes (solar multiples) were determined at the minimum values of SLEC (Figure 7), which are 0.5, 0.5, 0.45, 0.40, 0.25, and 0.25 for gas turbine capacities 100, 130, 160, 190, 220, and 250 MWe respectively. Figure 8 presents LEC of both configurations of the GTPP and the LFR-GTPP with different gas turbine sizes. The results indicate that solar integration with trigeneration plant results in a minor increase in the LEC compared to that of the conventional trigeneration plant. This is because the simulation site is located in a high solar radiation level in a Sun Belt country (Yemen), which makes a specific size of LFR system producing maximum possible thermal energy. Since the solar integration of LFR to the trigeneration plant leads to a negligible increment in the LEC, it results in a significant avoidance in emissions of CO_2. Therefore, LFR-GTPP could be represented as a promotion solution to the environmental problems caused by emissions of CO_2. Figure 9 presents the annual CO_2 avoidance, which is the annual reduction of CO_2 emissions due to integrating LFR system. As shown, the maximum avoidance of the annual CO_2 emission takes place when the LFR integrated to 100 MWe gas turbine (110.34 k-tonne of CO_2), and this avoidance decreases while gas turbine capacity increases until it reaches 45 k-tonne of CO_2 at 250 MWe gas turbine capacity.

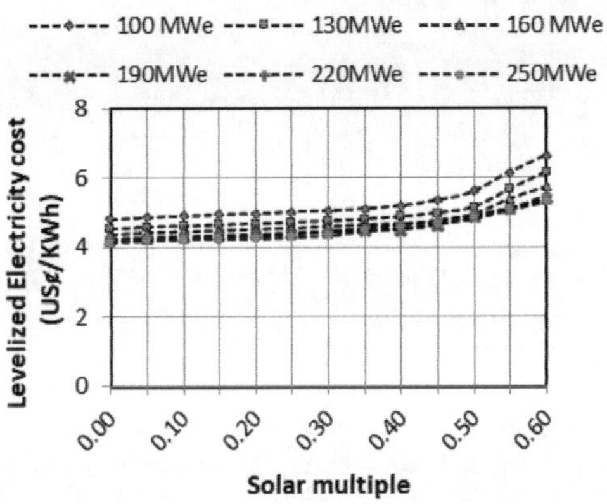

Figure 6. LEC for different integration sizes

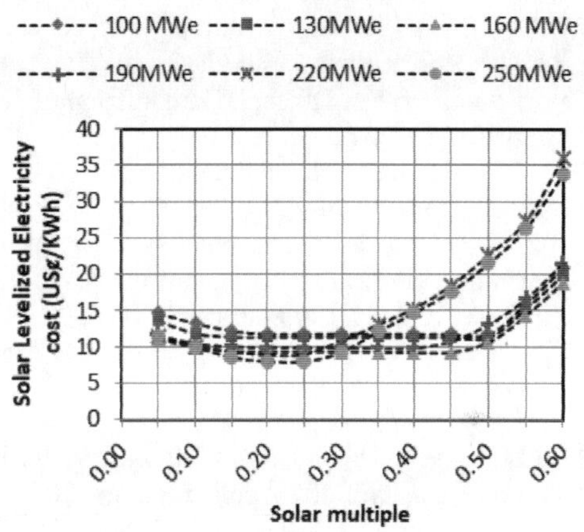

Figure 7. SLEC different integration sizes

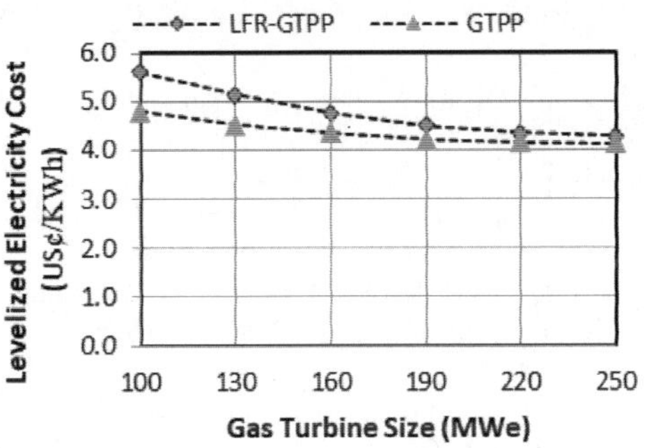

Figure 8. LEC for both configurations of GTPP and LFR-GTPP (at optimal integration).

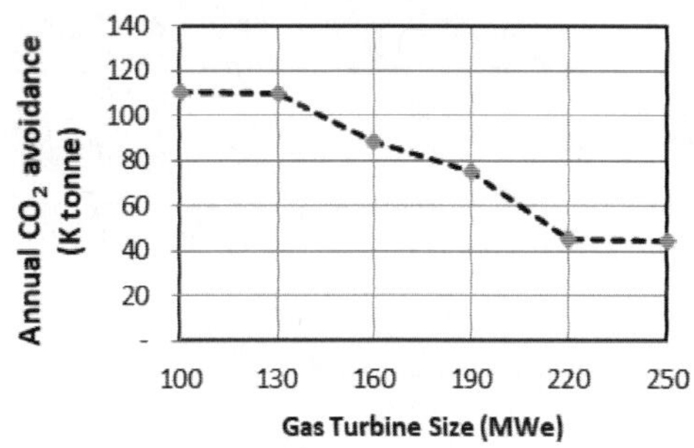

Figure 9. Annual CO_2 avoidance for utilizing LFR technology in the trigeneration power plant.

5. Conclusions

Solar energy is the most abundant source of energy on the earth and introducing integrated LFR with gas turbine trigeneration power plants offers stable and clean power plant for producing electricity, industrial process steam and chilled water. The annual and instantaneous solar shares as well as the annual reduction in CO_2 emissions are higher when the LFR technology integrated to the gas turbine trigeneration power plant with small gas turbine generation sizes (less than 190 MWe). Moreover, integrating LFR technology with a conventional gas turbine trigeneration power plant results in a minor increase in LEC compared to the conventional trigeneration plant. The LEC for the integrated plant varied between 4.28 US¢ and 5.6 US¢/ kWh, while this integration leads to a considerable reduction in CO_2 emissions compared to the conventional trigeneration plant.

Acknowledgments
This work has been supported financially by the National Science Foundation of China (51476159, 5171101721, and 51776193).

Reference

[1] Bishoyi D, Sudhakar K. Modeling and performance simulation of 100 MW LFR based solar thermal power plant in Udaipur India. Resource-Efficient Technologies. 2017.
[2] Mokhtar G, Boussad B, Noureddine S. A linear Fresnel reflector as a solar system for heating water: Theoretical and experimental study. Case Studies in Thermal Engineering. 2016;8:176-86.
[3] Bachelier C, Stieglitz R. Design and optimisation of linear Fresnel power plants based on the direct molten salt concept. Solar Energy. 2017.
[4] Dabwan YN, Mokheimer EMA. Optimal integration of linear Fresnel reflector with gas turbine cogeneration power plant. Energy Conversion and Management. 2017;148:830-43.
[5] Wang J, Yang Y. A hybrid operating strategy of combined cooling, heating and power system for multiple demands considering domestic hot water preferentially: A case study. Energy. 2017;122:444-57.
[6] Tora EA, El-Halwagi MM. Integrated conceptual design of solar-assisted trigeneration systems. Computers & Chemical Engineering. 2011;35:1807-14.
[7] Al-Sulaiman FA, Dincer I, Hamdullahpur F. Exergy modeling of a new solar driven trigeneration system. Solar Energy. 2011;85:2228-43.
[8] Al-Sulaiman FA, Hamdullahpur F, Dincer I. Performance assessment of a novel system using parabolic trough solar collectors for combined cooling, heating, and power production. Renewable Energy. 2012;48:161-72.
[9] Zhai H, Dai YJ, Wu JY, Wang RZ. Energy and exergy analyses on a novel hybrid solar heating, cooling and power generation system for remote areas. Applied Energy. 2009;86:1395-404.
[10] Zhang X, Li H, Liu L, Zeng R, Zhang G. Analysis of a feasible trigeneration system taking solar energy and biomass as co-feeds. Energy Conversion and Management. 2016;122:74-84.
[11] Baghernejad A, Yaghoubi M, Jafarpur K. Exergoeconomic comparison of three novel trigeneration systems using SOFC, biomass and solar energies. Applied Thermal Engineering. 2016;104:534-55.
[12] Meng X, Yang F, Bao Z, Deng J, Serge NN, Zhang Z. Theoretical study of a novel solar trigeneration system based on metal hydrides. Applied Energy. 2010;87:2050-61.
[13] Yao E, Wang H, Wang L, Xi G, Maréchal F. Multi-objective optimization and exergoeconomic analysis of a combined cooling, heating and power based compressed air energy storage system. Energy Conversion and Management. 2017;138:199-209.
[14] Jeffs E. Thermoflow integrates-THERMOFLEX and ST PRO with PEACE. TURBOMACHINERY INTERNATIONAL. 2003;44:20-3.
[15] Montes M, Abánades A, Martinez-Val J, Valdés M. Solar multiple optimization for a solar-only thermal power plant, using oil as heat transfer fluid in the parabolic trough collectors. Solar Energy. 2009;83:2165-76.
[16] Mokheimer EMA, Dabwan YN, Habib MA. Optimal integration of solar energy with fossil fuel gas turbine cogeneration plants using three different CSP technologies in Saudi Arabia. Applied Energy. 2017;185, Part 2:1268-80.
[17] Mokheimer EMA, Dabwan YN, Habib MA, Said SAM, Al-Sulaiman FA. Development and assessment of integrating parabolic trough collectors with steam generation side of gas turbine cogeneration systems in Saudi Arabia. Applied Energy. 2015;141:131-42.
[18] Montes M, Rovira A, Munoz M, Martinez-Val J. Performance analysis of an integrated solar combined cycle using direct steam generation in parabolic trough collectors. Applied energy. 2011;88:3228-38.
[19] Dabwan YN. Development and Assessment of Solar-Assisted Gas Turbine Cogeneration Systems in Saudi Arabia. King Fahd University of Petroleum and Minerals. 2013;MS.c Thesis.
[20] Schwarzbözl P, Buck R, Sugarmen C, Ring A, Marcos Crespo MJ, Altwegg P, et al. Solar gas turbine systems: design, cost and perspectives. Solar Energy. 2006;80:1231-40.

IOP Conf. Series: Materials Science and Engineering **556** (2019) 012024 doi:10.1088/1757-899X/556/1/012024

Airflow Organization in a Transformable Shelter

Liang Guo, Hong Ye*
Department of Thermal Science and Energy Engineering, University of Science and Technology of China, Hefei, PR China
E-mail: hye@ustc.edu.cn

Abstract. The environmental control system of a shelter can provide comfortable temperature, air velocity and fresh air supply for members indoor, protecting them from the extreme weathers or oxygen-poor environment in the field. A shelter with transformable walls will have a large available space including the core module and extended modules for diverse demands such as dining or sleeping. The airflows in such a limited space can be inevitably influenced by walls and furniture. In the study, a transformable shelter with top supply and bottom return air organization mode was taken as the research object. The air supply duct on the ceiling of the shelter was designed to uniform the supplying air quantity through each outlet by adding baffles in the duct. Baffles can also make the export winds nearly perpendicular to the duct wall. The indoor airflow velocity and temperature fields in the condition of people gathering for dinner were obtained by simulation, so as that in the condition of people scattering for sleeping. In the dining condition, the barriers to the uniform airflow and temperature are the crowd occupying the core module. In the sleeping condition, the dominating barriers are the walls between different modules. The airflows were uniformed between different modules by opening apertures across the partitions. In view of human comfort, a comprehensive evaluation of the air flow organization was given, which presents an essential reference for system design and optimization.

Keywords: transformable shelter, airflow organization, dining and sleeping condition, human comfort

1. Introduction

The environmental control system of a shelter can provide comfortable temperature, air velocity and fresh air supply for members indoor, protecting them from the extreme weathers or oxygen-poor environment in the field. The dimension of a vehicle-mounted shelter is limited by the vehicle. The expandable shelters with a small original volume and a high volume expansion ratio [1] have come into being. Common expandable shelters include origami-inspired shelters [2, 3] and rigid wall shelters.

Airflow organization includes optimization of air temperature and velocity distribution as well as the thermal comfort for personnel inside. Airflow organizations in an enclosed or semi-enclosed space such as accommodation [4], automobile cabin [5, 6] and highway sleeper coach [7] have been studied through experiments [8] and simulations [9, 10] comprehensively.

A shelter with transformable walls can stay on the vehicle and enlarge the space by moving the walls. The transformable shelter can provide spaces for diverse demands of personnel at different times and the requirements are also different for diverse use conditions. In this work, to cater to the different requirements of diverse using conditions, the airflow distribution inside has been simulated. The main factors impacting the airflow have been explored and several attempts have been made to optimize the airflow in the shelter.

2. Models and methods

The shelter before expansion can be hosted by a vehicle and are in the size of 6961 mm (*L*) × 2478 mm (*W*) × 2546 mm (*H*). When the shelter is transformed for dinner or sleeping, its size increases to 8866 mm (*L*) × 6336 mm (*W*) × 2546 mm (*H*). The transformed shelter consists of core module, left and right wing modules and front and back cells, as shown in Figure 1(a). The thickness of the shelter walls is 50 mm. An upper-supply and bottom-return airflow mode is used in the shelter. The air supply duct is located on the ceiling of the core module and front cell. The air return ducts are located on the bottom edges of the three modules. The width and height of the cross section of the air supply duct are 900 mm and 150 mm respectively. The closed tail of the air duct is 500 mm from the back cell. 9 pairs of air intakes are arranged on the vertical sides of the air duct, as shown in Figure 1(b). The air intakes are in the same size of 303 mm × 100 mm, and the distance between the adjacent intakes is 696 mm. The thickness of air duct wall is 5 mm.

When dining, members are gathering and standing in both sides of the long table in the closed space of the core module, seen in Figure 1(c). The area of the table is 6,000 mm × 600 mm and 750 mm from the floor. There are 10 members on each side of the table, each with a height of 1750 mm and a cross section size of 400 mm × 300 mm. When sleeping, members are scattering and lying on the beds hung on the walls of the core and wing modules, seen in Figure 1(d). There are 30 beds in total with 6 beds in core module and 12 beds in each wing module. The size of the bed is 1900 mm (*L*) × 700 mm (*W*) × 80 mm (*H*), and that of person is 1750 mm (*L*) × 600 mm (*W*) × 200 mm (*H*). The materials of the shelter walls and the duct walls are hardboard and stainless steel respectively. The beds and table are made of hardwood. Physical properties of persons are assumed to be the same as that of liquid water.

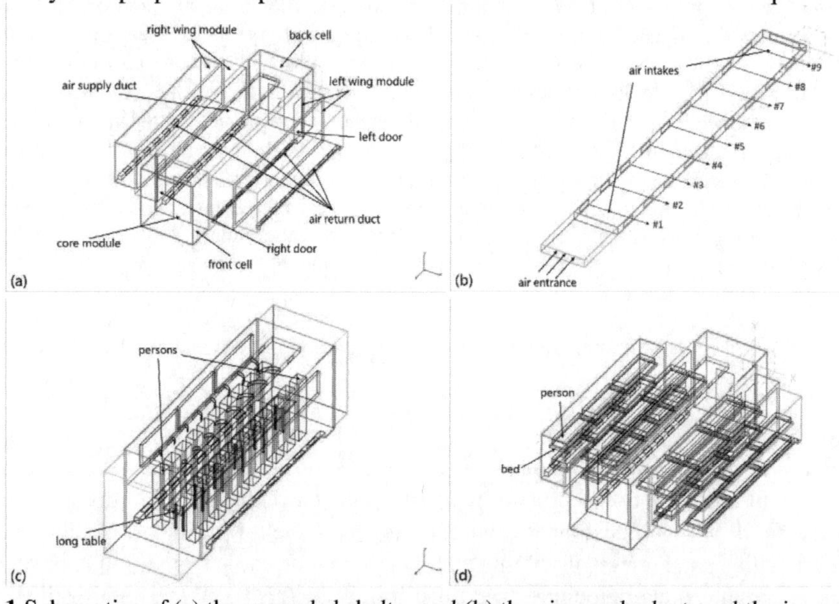

Figure 1 Schematics of (a) the expanded shelter and (b) the air supply duct, and the isometric views of (c) dining model and (d) sleeping model

Table 1 Physical properties of the related materials

object	material	Roughness, mm	Density, kg/m^3	thermal conductivity, W/(m·K)	specific heat capacity, J/(kg·K)
/	air [a]	/	1.207	0.026	1007
module wall	hardboard [b]	0.5	640	0.094	1170
duct wall	stainless steel[b]	0.15	7900	14.9	477
bed, table	hardwood [b]	0.5	720	0.16	1255
body	(water) [b]	2	1000	0.6	4180

a: build-in parameter of I-DEAS software.

b: appendix of *Fundamentals of Heat and Mass Transfer*, Sixth Edition.

I-DEAS software combines Computational Fluid Dynamics (CFD) and thermal solution technology with finite element method (FEM). It can be used to simulate the flow and temperature fields in the shelter. The calculation conditions of the models are given below. The air entrance of the air supply duct is a fan with a volumetric flow rate of 3500 m^3/h, and the air temperature is 17 °C. Each outlet of the return duct is a vent to the environment. The roughness of all the surfaces are listed in Table 1. The heat production power of each person is 900 W/m^3, and the surface temperature is assumed to be 33 °C. The shelter is located facing west at the Tuotuo River with an altitude of 4,500 m and an atmospheric pressure of 58,477 Pa. In the steady-state calculations, the ambient temperature is set as 40 °C.

To evaluate the performance of an airflow organization system, from the perspective of technicality, in addition to the averages of indoor temperature and velocity, the non-uniform coefficients of temperature (K_T) and velocity (K_v) are also needed. A non-uniform coefficient is the ratio of the root mean square (RMS) of a parameter to its average. Thus we have:

$$K_T = \frac{\sigma_T}{T_p},\ \sigma_T = \sqrt{\frac{\sum (T_p - T)^2}{n-1}},\ T_p = \frac{\sum T}{n} \tag{1}$$

$$K_v = \frac{\sigma_v}{v_p},\ \sigma_v = \sqrt{\frac{\sum (v_p - v)^2}{n-1}},\ v_p = \frac{\sum v}{n} \tag{2}$$

where T and v are the temperature and velocity, σ_T and σ_v are the RMS of temperature and velocity, T_p and v_p are the averages of the temperature and velocity, respectively, n is the number of measuring points. The smaller the non-uniform coefficient, the better the uniformity of temperature and velocity fields.

From the perspective of comfort, human comfort can be used as an evaluation indicator. The effective temperature difference θ, which can reflect the difference between perceived and actual temperatures, has an expression synthesizing the influences of temperature and airflow as

$$\theta=(T-T_n)-M(v-v_r) \tag{3}$$

where T_n is the given room temperature, usually the indoor average temperature T_p, v_r is the designing velocity of 0.15 m/s, M is the temperature value equivalent to the unit velocity effect with a value of 7.66 °C/(m·s^{-1}). If θ varies in the range from -1.7 to 1.1 °C, the members inside feel comfortable. The proportion of measuring points in this range is the air diffusion performance index (ADPI). When the ADPI is larger than 80%, this airflow organization can be considered satisfying.

3. Results and discussion

The airflow in the shelter needs to be organized and optimized for the uniform and appropriate air temperature and velocity, guaranteeing the human comfort indoor. In view of the complexity of the optimization process, only some significant results are given below.

The uniformities of the air temperature and velocity in the shelter depend on the uniformity of supplied air quantity through each intakes of the shelter, namely the outlets of the duct. The air supply should be optimized before the indoor airflow organization.

After entering the duct from entrance, the air is gradually diverted into the shelter from the outlets along the initial flow direction. To uniform the supplied air quantity, baffles are added at the downstream side of each outlet to control the airflow velocity and direction. The shape, length and direction of the baffles can be adjusted. Straight-shaped baffles can make the mass flow rate from each outlet nearly equal. However, the air from the outlet rushes into the rear space of the core module along the fixed direction of baffles, causing air stagnation in the front space, which is detrimental to the uniformity of air velocity. The arc-shaped baffles are optimized from the straight-shaped ones, whose shape is a quarter of a cylinder wall. The radius of the baffle at each outlet gradually increases, as shown in Figure 2(a). After multiple optimization calculation, the radius data in Table 2 are obtained, which can make the mass flow rate of each outlet nearly equal. The results in Table 2 show that the maximum deviation of the mass flow rate of each pair of outlets is 13%. As can be seen from the air velocity field in Figure 2(b), the air velocity direction from each air outlet has been deflected with respect to the direction of air entering. The outflow directions of the outlets have been optimized with the export winds nearly perpendicular to the duct walls. Therefore, the air supply duct with arc-shaped baffles is applied to subsequent calculations.

Table 2 The baffle radius and mass flux of each pair of outlets

	#1	#2	#3	#4	#5	#6	#7	#8	#9	average
Baffle radius /mm	100	135	175	220	250	300	360	400	/	/

mass flux /kg·(m²·s)⁻¹	1.27	1.25	1.20	1.16	1.10	1.20	1.28	1.28	1.40	1.24

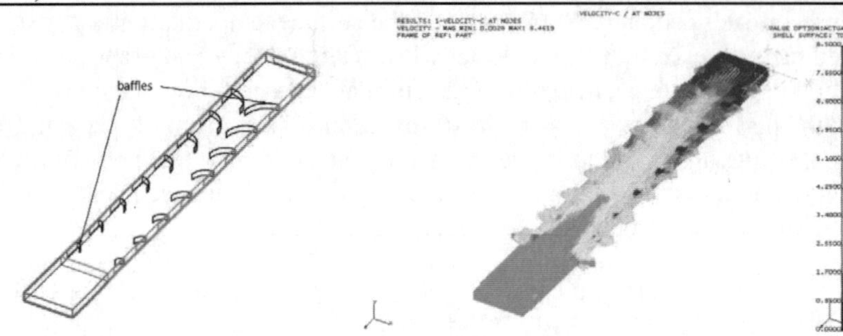

Figure 2 The schematic of air supply duct with arc-shaped baffles

The airflow velocity and temperature fields in the core module have been obtained by simulating the dining model in a steady-state condition. The distributions of air velocity in some typical cross sections are shown in Figure 3. As can be seen in top view (Figure 3(a)), there are regions with relative low velocity in the front and end parts in core module, so as in the top regions in lateral views (Figure 3(b)) and central regions in front views (Figure 3(c)). In most regions, the air velocity is lower than 0.63 m/s. Because the temperature of air is 17 °C and far lower than the ambient temperature of 40 °C, the temperature in the regions with high air velocity is lower than that with low velocity.

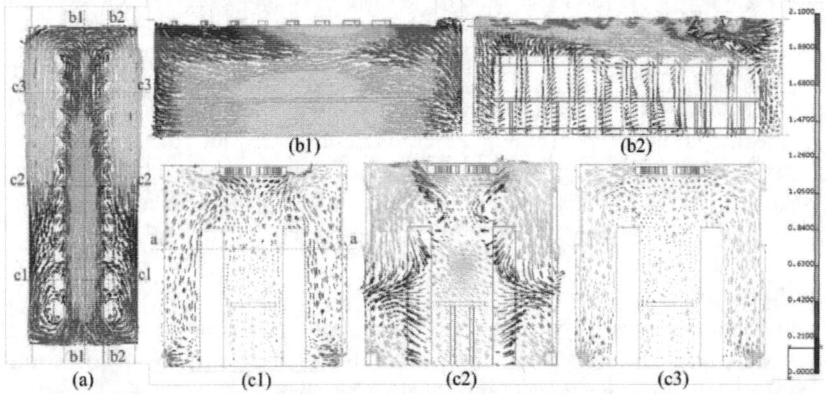

Figure 3 Air velocity distributions in the cross sections of the core module
(a) top view, (b) lateral views, (c) front views.

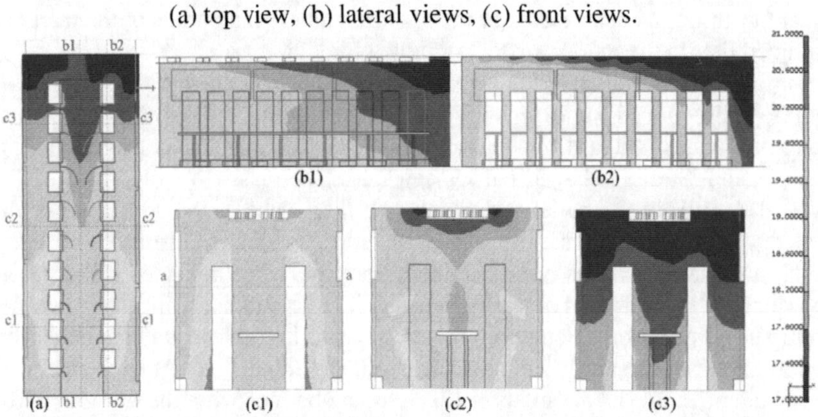

Figure 4 Air temperature distributions in the cross sections of the core module
(a) top view, (b) lateral views, (c) front views.

As shown in Figure 4, the highest air temperature is below 21 °C, which means that the refrigeration effect of the air supply of the environmental control system is powerful. The highest temperature occurs in the zones close to the feet of the persons standing in the front region of the module, and the air temperature shows a decreasing trend from the front to the back. Since the air enters the module backwards diagonally, there is more cold quantity obtained in back region than the front one in spite of the equivalent air mass fluxes, resulting in the temperature difference between front and rear spaces.

Table 3 shows the statistical results of the air velocity and temperature in the horizontal section near the neck of the human bodies (Figure 3(a) and 5(a)). It can be seen that the ADPI in this section is not satisfying and far lower than the comfort index of 80%. In dining condition, the barriers to the uniform airflow and temperature are the persons.

Table 3 Statistical results of air velocity and temperature in a horizontal section

air	average	RMS	non-uniform coefficient	ADPI
velocity	0.40 m/s	0.21 m/s	0.53	37.9%
temperature	18.7 °C	0.7 °C	0.04	

The airflow velocity field in shelter have been obtained by simulating the sleeping model in steady-state condition. The only passage from the core to the right wing module is the right door located on the wall of the front space, which is a dead zone for airflow and only a few fresh and cold air can be allocated in. The negative pressure occurred in the back of the right wing module leads to intrusion of the ambient air and affects the entire right wing module and front cell. Therefore, the situation of the air supply duct or the passages between the modules need to be optimized in order to obtain more uniform distribution of the air velocity and temperature.

The new air passages are carved on the walls of the core module. As shown in Figure 5(a), total four long narrow passages with a size of 200 mm × 5460 mm are carved 400 mm above bed rows. The airflow velocity fields of the improved sleeping model are shown in Figure 5(b). Compared with that of the original model, the airflow is fully developed and more uniform. The temperature fields shown in Figure 6 indicate that the temperature in each module is relative uniform and approximately 19 °C. The statistical results of the air velocity and temperature in the horizontal sections 100 mm above the top surfaces of human bodies (Figure 6) are listed in Table 4. The average velocities of two horizontal sections are 0.35 and 0.24 m/s, respectively. For upper section the ADPIs are lower than the comfort index of 80%, and for lower section, the ADPIs is higher than 80% except in left wing module.

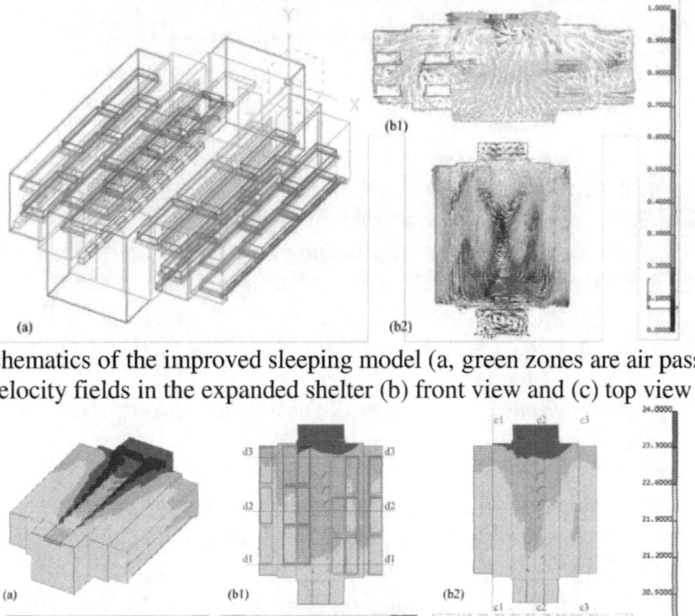

Figure 5 Schematics of the improved sleeping model (a, green zones are air passages) and velocity fields in the expanded shelter (b) front view and (c) top view

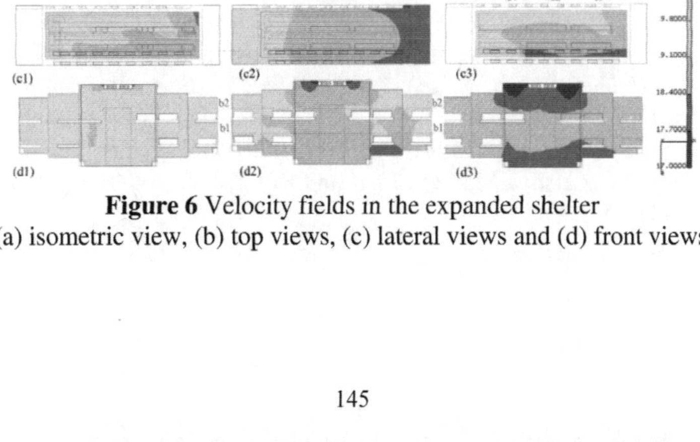

Figure 6 Velocity fields in the expanded shelter
(a) isometric view, (b) top views, (c) lateral views and (d) front views

Table 4 Statistical results of the air velocity and temperature in two horizontal sections

		air	average	RMS	non-uniform coefficient	ADPI
upper section	core module	velocity	0.37 m/s	0.18 m/s	0.48	53.8%
		temperature	19.0 °C	0.6 °C	0.03	
	left wing	velocity	0.31 m/s	0.17 m/s	0.54	57.8%
		temperature	19.5 °C	0.7 °C	0.04	
	right wing	velocity	0.35 m/s	0.16 m/s	0.45	57.9%
		temperature	19.4 °C	0.5 °C	0.02	
	all modules	velocity	0.35 m/s	0.17 m/s	0.49	53.9%
		temperature	19.3 °C	0.6°C	0.03	
lower section	core module	velocity	0.22 m/s	0.11 m/s	0.52	80.8%
		temperature	18.8 °C	0.3°C	0.02	
	left wing	velocity	0.29 m/s	0.12 m/s	0.41	54.8%
		temperature	19.2 °C	0.6 °C	0.03	
	right wing	velocity	0.25 m/s	0.08 m/s	0.32	88.5%
		temperature	19.5 °C	0.5 °C	0.03	
	all modules	velocity	0.24 m/s	0.11 m/s	0.48	78.7%
		temperature	19.1 °C	0.5 °C	0.03	

4. Conclusions

In this work, the temperature and velocity of the airflow in a transformable shelter are simulated in both steady and transient states. The air supply duct is optimized by adding arc-shaped baffles. In the dining model the dense crowd of the diners in the core module block the airflow, leading to a low human comfort, while in the sleeping model the dominating barriers are the walls between the modules.

Acknowledgment

This work was funded by the National Natural Science Foundation (No. 51576188).

References

[1] Martinez-Martin, F.J. and A.P. Thrall, *Honeycomb core sandwich panels for origami-inspired deployable shelters: Multi-objective optimization for minimum weight and maximum energy efficiency*. Engineering Structures, 2014. **69**: p. 158-167.

[2] Thrall, A.P. and C.P. Quaglia, *Accordion shelters: A historical review of origami-like deployable shelters developed by the US military*. Engineering Structures, 2014. **59**: p. 686-692.

[3] Quaglia, C.P., A.J. Dascanio, and A.P. Thrall, *Bascule shelters: A novel erection strategy for origami-inspired deployable structures*. Engineering Structures, 2014. **75**: p. 276-287.

[4] Kun, Z., L. Haixu, and W. Jinling, *CFD Simulation of Airflow Organization in the Ship Accommodation*. Applied Mechanics and Materials, 2015. **711**: p. 91-5.

[5] Neacsu, C., et al., *The evaluation of the overall thermal comfort inside a vehicle*. IOP Conference Series: Materials Science and Engineering, 2017. **252**.

[6] Chien, C.H., et al., *3-D numerical and experimental analysis for airflow within a passenger compartment*. International Journal of Automotive Technology, 2008. **9**(4): p. 437-445.

[7] Xie, X. and F. Shen, *Numerical Investigation on Thermal Environment inside the Passenger Compartment of A Highway Sleeper Coach, in Recent Trends in Materials and Mechanical Engineering Materials, Mechatronics and Automation, Pts 1-3*, Q. Luo, Editor. 2011. p. 215-218.

[8] Orzechowski, T. and Z. Skrobacki, *Evaluation of thermal conditions inside a vehicle cabin, in Efm15 - Experimental Fluid Mechanics 2015*, P. Dancova and M. Vesely, Editors. 2016.

[9] Sukri, M.F., et al., *simulation of vehicle cooling loads at different orientations*. Heat Transfer Research, 2016. **47**(4): p. 343-357.

[10] Hofstaedter, R.N., et al., *Heat capacity and heat transfer coefficient estimation for a dynamic thermal model of rail vehicles*. Mathematical and Computer Modelling of Dynamical Systems, 2017. **23**(5): p. 439-452.

Design of a small size PTC: computational model for the receiver tube and validation with heat loss test

M Salvestroni[1], G Pierucci[1], F Fagioli[1], A Pourreza[1], M Messeri[1], F Taddei[1], S Hosouli[1], H Rashidi[1] and M De Lucia[1]
[1] Department of Industrial Engineering of University Of Florence,
Via di Santa Marta 3, 50139 Florence – Italy
E-mail: michele.salvestroni@unifi.it

Abstract. In EU, the residential sector is responsible for the 40% of the total energy consumption. The integration of solar technologies in buildings is therefore necessary in order to reduce the use of fossil fuels. Concentrating collectors could offer an interesting solution but nowadays their application in buildings is rare due to high costs, large dimensions and complexity of the system. A novel small parabolic trough collector (PTC) has been conceived to overcome these limits and to allow the integration in buildings. The main features of realized prototype are the compact dimensions, strongly reduced compared to the PTC standards, and modularity. The paper deals with the numerical analysis necessary to design the concentrating collector properly and its validation with experimental results. An optical analysis has allowed to select the optimum values for the parameters of the parabola, aperture and rim angle. A thermo-fluid dynamics finite element model has been developed with Comsol Multiphysics, to analyse the relevant physical characteristics and to predict the performance of the receiver tube. The efficiency curve of the collector has been extracted. Successively a receiver tube has been built based on the indications of FEM model for what concerns geometry and materials. In order to evaluate the heat loss of the receiver and to validate the finite element model, a test bench has been realized. The results of off-sun heat loss tests on the receiver tube are reported for several temperatures. The computational model is in good agreement with experimental results and therefore it is validated.

Keywords: Small size parabolic trough, receiver tube, FEM model, buildings

1. Introduction

In EU, the residential sector is responsible for the 40% of the total energy consumption [1]. The integration of solar technologies in buildings is therefore necessary in order to reduce the use of fossil fuels and the emissions of greenhouse gases caused by buildings.

Regarding the production of heat for domestic heating, solar thermal collectors are a mature technology. Nevertheless, they still exhibit some problems, for instance a remarkable drop in efficiency when the fluid temperature grows; for this reason flat plane and evacuated solar collectors are not adequate for solar cooling. Concentrating collectors could offer an interesting solution but at the moment their application in buildings is strongly limited due to high costs, large dimensions and complexity of the system.

The cooling demand in EU, especially in Mediterranean area has been growing considerably in recent years; therefore, an open task is developing an economic and compact solar collector suitable to produce heat at medium temperature (up to 200°C) to exploit it in absorption cycle for refrigeration.

For these reasons, a novel small size parabolic trough collector (PTC) has been conceived. In order to realize a compact system suitable to be integrated on the roof, the dimensions of the prototype collector (mirror, receiver tube) have been strongly reduced compared to the PTC standards.

An accurate design phase has preceded the construction of a prototype. A numerical analysis from an optical and thermal point of view has been conducted.

The prototype of the receiver has been realized based on the indications of FEM (finite element method) model for what concerns geometry and materials.

2. Optical analysis and FEM model

2.1 Optical analysis

The aim of the project is to build a compact concentrating collector, suitable to be installed on the roof.

The design phase of the small size parabolic trough collector has started with an optical analysis that has allowed to choose the optimum values for the parameters of the parabola, aperture and rim angle.

In order to have a compact object the aperture should be equal or less than 1 m. The diameter of the receiver has been selected as a trade-off between concentration ratio and minimum acceptable dimensions. For these reasons, the aperture has been fixed to about 500 mm and the external diameter of absorber tube has been set to 10 mm.

The shape of parabola has been determined through an optical analysis with ray-tracing software. The optical errors (angular width of the sun, micro and macro imperfection of the parabolic shape, etc.) have been estimated and applied in the analysis. For a beam spread of about 0.9°, it has been found that the rim angle that maximizes the irradiance on receiver tube, is 110°. **Figure 1** shows the irradiance distribution on the receiver tube obtained. As can be seen, along the circumference of the absorber tube the irradiance is non uniform, since the upper part of the receiver (corresponding to the positions close to 0.15 and -0.15 on the horizontal axis in the plot) receives only direct radiation while the lower part gets the reflected radiation from the mirror.

2.2 FEM Model

The second stage of the design phase deals with the finite element model developed in order to analyse the relevant physical characteristics and to predict the performance of the receiver.

The model is a 3d thermofluid dynamic model of the receiver tube. It describes the dynamic of the fluid inside the absorber tube and the heat transfer between all the components of the receiver. Several numerical models are present in literature [2-8]. The flux obtained through a ray-tracing software is applied on the outer boundary of absorber tube.

A novel geometry solution is considered; two concentric tubes are inserted in an evacuated glass envelope, so that the fluid inlet and output are at the same side of One-End receiver. This configuration is simpler to build, performant and much more economic than the standard one. The inner absorber tube has an external diameter of 5 mm. The geometry and the domains of the computational model are shown in **Figure 2**.

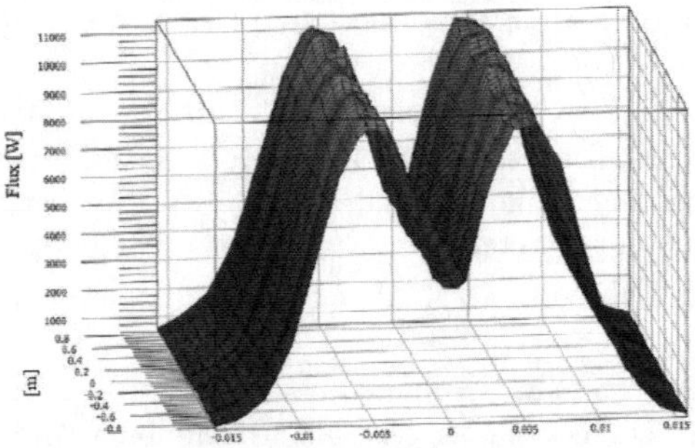

Figure 1 Solar flux distribution on the absorber tube obtained by Soltrace.

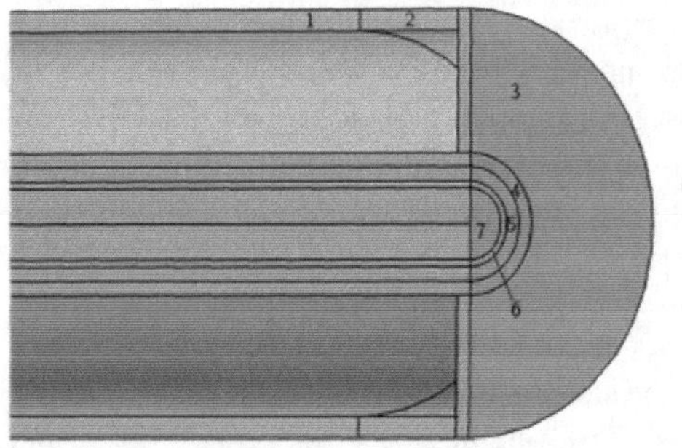

Figure 2 1) Glass; 2) Kovar; 3) Metal surface; 4) Outer absorber tube; 5) Fluid domain; 6) Inner absorber tube; 7) Fluid domain.

IOP Conf. Series: Materials Science and Engineering **556** (2019) 012025 doi:10.1088/1757-899X/556/1/012025

The model solves the continuity equation and the momentum equation in the fluid domain, the energy equations in all domains. A k-e turbulence model with wall function is used when the flow regime is turbulent.

The main boundary conditions are:

- the solar flux distribution obtained by the optical analysis is imposed on the outer absorber surface;
- radiation boundary condition on the inner boundary of glass envelope and on the outer boundary of the absorber; no convection inside the annulus since it is supposed to be under high vacuum, 10^{-4} mbar;
- convection and radiation on the external surface of the glass tube.

A parametric analysis has been conducted to investigate the performances of the receiver under different conditions for mass flow, inlet temperature, absorber tube materials, emissivity of the coating, irradiation, etc. Part of the results are reported below.

Figure 3 shows the results varying the inlet flow velocity. The optimal velocity has been selected as the one that maximize thermal efficiency and minimize pressure losses.

Figure 4 shows the results of thermal efficiency varying the emissivity of coating. Comparing the performances at 180°C of an optimum selective coating (5% of emissivity), and a medium selective coating (15% of emissivity), the deviation in efficiency between them is only 2%. Thus, it is possible to choose a more economic coating without a significant loss in thermal performances.

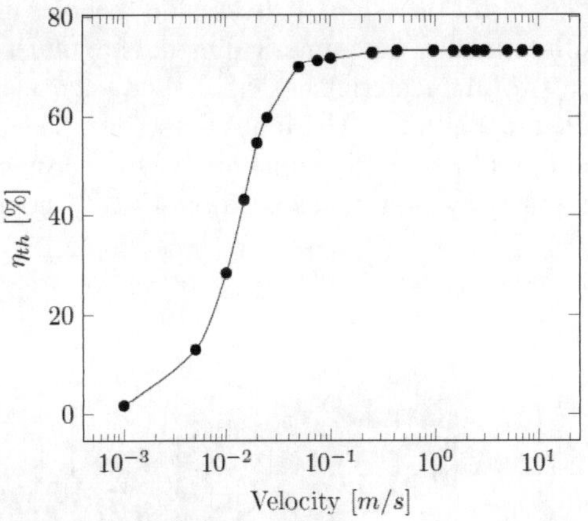

Figure 3 Thermal efficiency for different velocity of the flow.

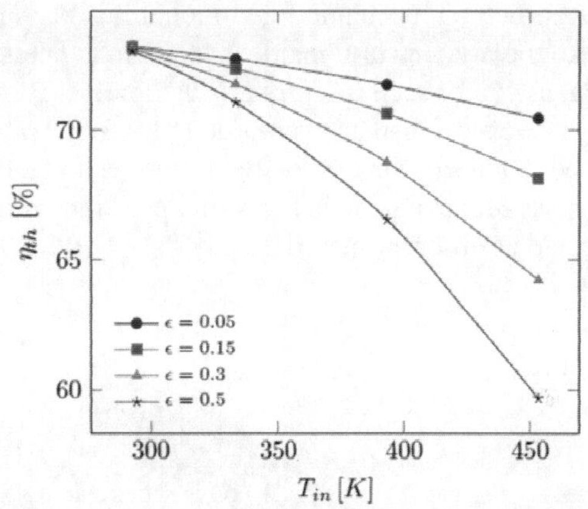

Figure 4 Sensitivity analysis: thermal efficiency vs coating emissivity.

Figure 5 shows the thermal efficiency versus fluid inlet temperature. The dot line represents the optical efficiency of receiver, calculated by tracing software, fixing the reflectivity of the mirror, the transmittance of the glass and the solar absorbance of the coating. The selective coating used for this simulation has emissivity equal to 0.15.

The optical efficiency of the small PTC is evaluated to be about 73%, due to the shape of the parabola and to the properties of materials. The latter are a trade-off between good performances and low costs. The drop in thermal efficiency at 180°C is only 5% thanks to vacuum in the annulus and to selective coating.

The increase of temperature inside the collector varying inlet fluid temperature for different conditions of solar irradiance reported in **Figure 6**. For an inlet fluid temperature of 25°C a rise of about 6 °C is predicted at 800 W/m^2.

.

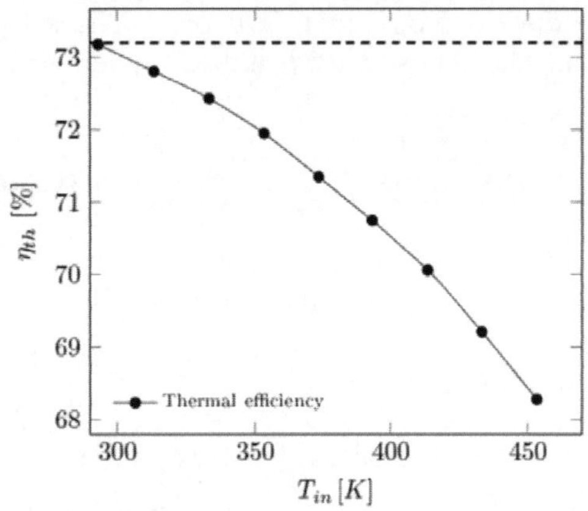

Figure 5 Thermal efficiency vs inlet flow temperature.

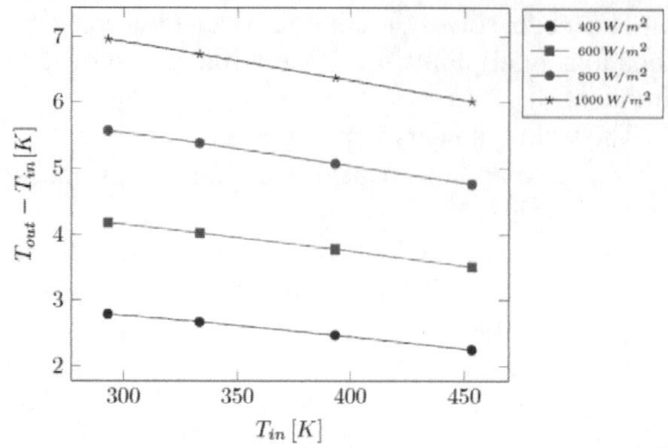

Figure 6 Temperature rise on the collector varying inlet fluid temperature for different irradiance conditions.

Steel, copper and aluminium have been investigated as absorber material; no so much thermal efficiency differences have been found. Therefore, the absorber material has been chosen based on structural and economic reasons.

A structural mechanic FEM model has been realized to verify the thermal stresses on receiver tube due to the non-uniform incident solar flux. The stresses evaluated by the numerical model are under 30 MPa, as can be seen in **Figure 7**, well below the yield stress of the materials tested.

The vertical displacement due to thermal stress reported in **Figure 8**. The inner absorber tube leans on the outer tube since it is free to move. The outer absorber tube properly constrained with springs to the glass tube, remains in the initial position without moving away from focus position. Finally, copper selected as absorber material.

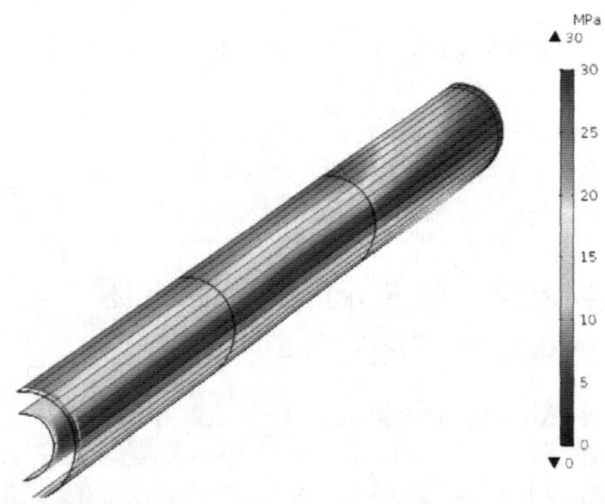

Figure 7 Von Mises stress on the outer absorber tube.

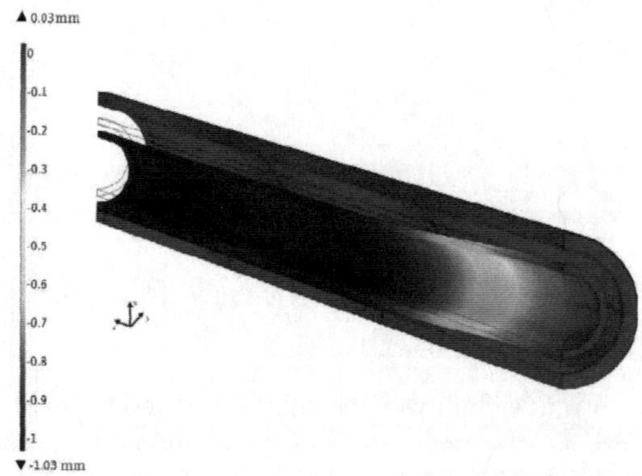

Figure 8 Vertical displacement of the outer and inner absorber tubes due to thermal stress.

Following the results of the numerical analysis, a prototype of the receiver and of the collector hasbeen built.

3. Prototype of the collector and validation of the FEM model

The small parabolic trough collector realized has 1800 mm of length with a chord of about 420 mm (**Figure 9** and **Figure 10**). The parabolic mirror is made of aluminium and it has a rim angle of 110°. The realized receiver tube is made by two concentric copper ducts where the heat transfer fluid flows, letting the flow enter and exit just by one side. The absorber is surrounded by a glass envelope coated by AR layer. In order to reduce heat loss, the annulus between glass and copper tube is evacuated. The absorber copper tube has a diameter of 10 mm (1 mm thickness), with a selective Cermet absorber (α=0.94 and ε=0.13 at ambient temperature). A smaller tube is used with an internal diameter of 5 mm (0.5 mm thickness) to feed the receiver with a counter current fluid. Four springs support both the pipes in glass cylinder and keep them aliened in the reflector focus.

Figure 9 The small size parabolic collector.

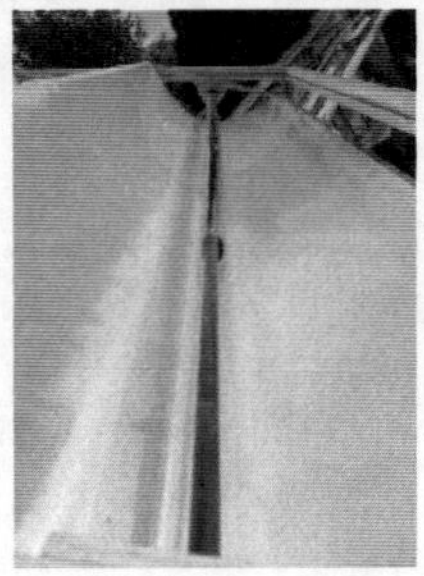

Figure 10 The receiver tube.

A test bench has been designed to measure the heat loss of the receiver tube and it has been described in a previous work [9].

The experimental results of off-sun heat loss test have been compared with the outcome of numerical simulation in order to validate the FEM model. The numerical model predicted well heat losses at low temperature but underestimated heat losses at high temperature. The reason was that emissivity had been considered constant in simulations, whereas the emissivity of coating grows with temperature. Therefore, a model calibration has been necessary. The value of emissivity, at maximum temperature, has been varied until the numerical heat loss has been equal to experimental one. Knowing the emissivity value at ambient temperature and the value at 200°C, it is possible to approximate the variation of emissivity with temperature with a linear fit. **Figure 11** shows the comparison between the numerical and experimental results.

All numerical results lie in the experimental error bars. The numerical model is therefore validated.

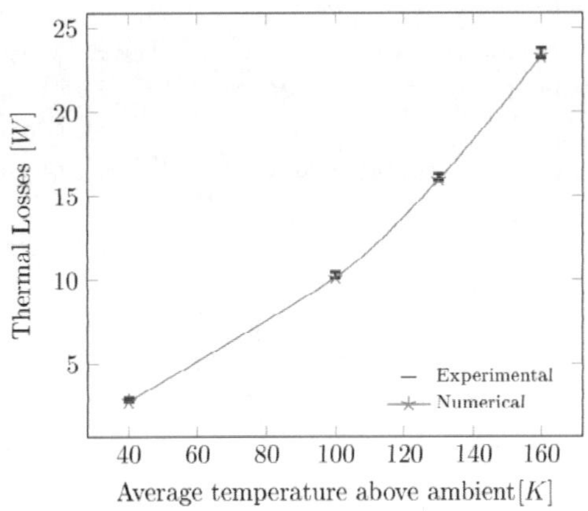

Figure 11 Comparison between experimental and numerical results.

4. Conclusions

A novel small parabolic trough, which can be easily installed on buildings roof, has been designed.

The optical design phase has allowed defining the main parameters of the optic components. The aperture of the parabola has been set to 500 mm in order to realize a compact collector suitable for residential application. The diameter of the absorber tube is 10 mm to intercept almost all the incoming radiation, keeping the concentration ratio high.

A FEM model has been realized in order to describe the heat transfer and the dynamic of the fluid inside the receiver tube. The solar flux obtained by ray tracing analysis has been imposed on the absorber surface in the FEM model. The performances of the receiver tube have been investigated deeply through a parametric analysis. The selective coating chosen has emissivity of 0.15, a trade-off between low thermal losses and costs. The expected optical efficiency is 73%. The drop in thermal efficiency with temperature due to heat losses is 4% at 180°C.

A structural FEM model was also carried out in order to check if thermal stress could cause the removal of the absorber tube from the focus position. The predicted Von Mises stress are well below the yield stress and thanks to a proper number of constraints, the outer absorber tube remains in the initial position.

The prototype of the collector has been realized following the indications coming from the numerical models. The numerical analysis has allowed to design and fix the main features of the concentrating collector: aperture and rim angle of the parabola, diameter of the receiver tube, properties of materials involved in order to reach the target of performance, the optimal mass flow, etc.

Off sun test have been carried out on the prototype of the receiver tube. This has allowed to validate the FEM model. The dependency of emissivity with temperature has been obtained calibrating the FEM model. Since the numerical results were in good agreement with the experimental measurements, the FEM model has been validated through the experimental results of heat loss test.

The collector developed could offer a promising solution in order to reduce strongly the consumption of fossil fuels in buildings. Indeed, the small size parabolic collector possesses features that are intermediate between a standard parabolic trough collector and a solar thermal collector. It is characterized by high performances like a PTC (although it works at lower temperatures) but it is basically different from a PTC because of the reduced dimensions and consequently, more similar to a solar thermal collector.

The high thermal efficiency of the collector at 180°C allows producing heat for solar cooling via an absorption cycle or for power production via ORC cycle, overcoming the temperature limitations of solar thermal collectors. Therefore, the small size parabolic trough might lead the way towards trigeneration in buildings. Finally, supposing to integrate the small PTC with other technologies like thermal energy storage and solar PV collectors, it could be possible to convert the building into a positive energy building (PEB).

[1] Economidou M, Atanasiu B, Despret C, Maio J, Nolte I and Rapf O 2011 Europe's buildings under the microscope A Country-by-country review of the energy performance of buildings *Building Performance Institute Europe (BPIE)*

[2] Cheng Z D, He Y L, Xiao J, Tao Y B and Xu R J 2010 Three-dimensional numerical study of heat transfer characteristics in the receiver tube of parabolic trough solar collector *International Communications in Heat and Mass Transfer* **37**(7): 782-7

[3] He Y L, Xiao J, Cheng Z D and Tao Y B 2011 A MCRT and FVM coupled simulation method for energy conversion process in parabolic trough solar collector *Renewable Energy* **36**(3) 976-985

[4] Padilla R V, Demirkaya G, Goswami D Y, Stefanakos E and Rahman M M 2011 Heat transfer analysis of parabolic trough solar receiver *Applied Energy* **88** 5097-5110

[5] Huang W, Hu P and Chen Z 2012 Performance simulation of a parabolic trough solar collector *Solar Energy* **86** 746-755

[6] Cheng Z D, He Y L, Cui F Q, Xu R J, Tao Y B 2012 Numerical simulation of a parabolic trough solar collector with nonuniform solar flux conditions by coupling FVM and MCRT method *Solar Energy* **86**(6) 1770-84.

[7] Hachicha A A, Rodríguez I, R. Capdevila R and Oliva 2013 A Heat transfer analysis and numerical simulation of a parabolic trough solar collector *Applied Energy* **111** 581-592

[8] Wu Z, Li S, Yuan G, Lei D and Wang Z 2014 Three-dimensional numerical study of heat transfer characteristics of parabolic trough receiver *Applied Energy* **113** 902-911

[9] Pierucci G, Hosouli S, Salvestroni M, Messeri M, Fagioli F, Taddei F and De Lucia M, 2017 Experimental methodology and thermal loss test on small size absorber tube for solar application *Energies* **11**(10) 2552

IOP Conf. Series: Materials Science and Engineering **556** (2019) 012026 doi:10.1088/1757-899X/556/1/012026

Design of a seasonal storage for a solar district heating in Florence

M Salvestroni[1], G Pierucci[1], F Fagioli[1], A Pourreza[1], M Messeri[1], F Taddei[1],S Hosouli[1], H Rashidi[1]and M De Lucia[1]
[1] Department of Industrial Engineering of University of Florence,
Via di Santa Marta 3, 50139 Florence – Italy
E-mail: michele.salvestroni@unifi.it

Abstract. The energy demand in residential sector constitutes an important fraction of the entire energy consumption (40% in EU). Solar district heating (SDH) is a key strategy to reduce use of fossil fuels in buildings. Inside the European project REPLICATE, financed by Horizon 2020 SCC1 Smart Cities and Communities, a SDH with a seasonal storage (STES) has been designed to be realized in the city of Florence. It is the first example of solar district heating with a STES in Italy. The design phase has aimed to size properly the extension of solar field and the volume of seasonal storage based on several parameters such as number of dwellings to feed, heat demand, solar resource, geology of the location and economic reasons. The paper deals with the model that has been realized through TRNSYS to describe the energy fluxes of heating plant andtheir optimization process. The computational model depicts the possible operating conditions and leads to define the control strategies of solar field and seasonal storage, integrated with commercial components that complete theplant such as gas boiler, heat pump and overall circuit. Hot water tank TES has been selected as the appropriate typology of storage for this application based on geological considerations. The thickness of insulation material and various layers have been determined. The numerical analysis fixes the volume of TES to be 3800 m^3 and a solar field of about 1000 m^2. The solar fraction expected by the district heating is 44%.

Keywords: Seasonal storage, solar district heating, TRNSYS

1. Introduction

The energy demand in residential sector constitutes an important fraction of the entire consumption of energy, counting for the 40% in EU [1].

In order to reduce the use of fossil fuels and the emissions of greenhouse gases caused by buildings, the integration of renewable energies, solar technologies in particular, plays an important role.

A very promising strategy is represented by the concept of solar district heating (SDH) in which the heat, provided by solar thermal collectors, is stocked in a seasonal thermal energy storage (STES or TES) during spring and summer to release it during autumn and winter [2]. In this way it is possible to increase considerably the amount of heat produced by solar technologies respected to the total heat demand, which is called solar fraction.

Many examples of SDH with STES are present in Europe. The countries that have invested more in this field are Denmark, Germany and Sweden [3-6]. In Italy few solar district heating have been developed in recent years but they are not associated with a seasonal storage.

The city of Florence is participating at the European project Replicate, financed inside the program Horizon 2020 SCC1 Smart Cities and Communities. Among the actions to undertake, it is planned the improvement of energy performances of two social housing (300 dwellings) and the realization of a

IOP Conf. Series: Materials Science and Engineering **556** (2019) 012026 doi:10.1088/1757-899X/556/1/012026

solar district heating with seasonal thermal storage (first case in Italy). In this context, a solar field and a tank seasonal storage have been designed.

A model has been realized through TRNSYS in order to evaluate the features of all the components in the heating plant (gas boiler, solar field, seasonal storage, etc.) and their interactions. The TRNSYS model highlighted the parameters that most influence the behavior of the system.

2. TES design

The design of a seasonal thermal energy storage deals with geological aspects, constructional features, energetic and economic issues.

First of all, the composition of the ground of area interested by the district heating project is not suitable for a borehole storage or an aquifer storage. Moreover, a comparison between pit storage and water tank storage has led to select the latter one as the appropriate for the project thanks to the best insulation performances that it guarantees.

In general a water tank STES is a complex structure composed by several layers of different materials. From the inside to the outside is as follows: a liner made by stainless steel or polymer to obtain water tightness, a concrete layer to support the structure, a vapor barrier against water permeation, and an insulation layer to reduce thermal losses [5-6]. In this project it has been decided to use a water tightness high density concrete without liner. Expanded glass granules have been selected as insulation material.

The design strategies are strongly affected by the project site. In this case, the Florence city rules forbid to build structures over the ground near the river Arno, due to flood problem, so that the STES must be almost fully underground. This fact limits the height of the storage because of the presence of aquifer 10 m below the ground level. There are no limitations for the width of the storage.

The TES needs to be designed carefully since its thermal performance, the amount of heat stored and its optimal dimensions are dependent and connected to many variables, such as: the radiation resource, the heat user demand, the size of solar field, the target solar fraction of the district heating, the supply and return temperature of the district heating, parameters related to its geometries (aspect ratio), constructive solutions (thermal properties of materials, layers width, etc.,), characteristics of district heating net (pipe length, insulation, etc.).

Due to complexity of the system, a numerical analysis is necessary to design properly a seasonal thermal energy storage. The solar field, from project, must guarantee 750 MWh/y minimum and available solar field area is limited by the roof buildings (**Error! Reference source not found.**).

Figure 1 Picture of the buildings and of the area that will be retrofitted. The sectors available for solar field are highlighted. Between the two buildings the area where TES will be built.

IOP Conf. Series: Materials Science and Engineering **556** (2019) 012026 doi:10.1088/1757-899X/556/1/012026

A numerical analysis, considering mutual shadowing and external shadows, has been conducted to find the configuration of the solar collectors that guarantees the maximum amount of heat collected. The inclination of collectors and the distance between them have been varied: in **Table 1** some configurations are reported.

Table 1 Different configurations for solar field. Temperature of inlet flow fixed at 80°C.

Inclination (°)	Net area of collectors (m^2)	Total energy (MWh)
25	504	354
15	792	544
15	864	585
10	1008	687

As can be seen, the maximum value for annual energy is reached filling the available area with the maximum number of collectors (i.e. net area of collectors), even if the tilt angle is not the theoretical optimum and shadows could occur between the rows of the solar field during the morning and the evening. This analysis has been used only to compare different configurations since the heat produced by solar field is strongly dependent from TES capacity and heat demand.

The performances of district heating (storage, solar field) strongly depend from the supply and return temperature of the heating plant. The return temperature from heating net fixes the minimum temperature of TES.

In this project, due to the presence of radiators in the dwellings, the supply and return temperature are quite high (66 °C and 51°C). Therefore, it would not be possible to fully discharge the storage and the temperature inside the TES would be limited between 95°C and 50°C, reducing strongly the storage capacity.

Following the international experience [7], a water-water heat pump, linked to TES, has been chosen to extract heat from TES and discharge it completely down to 25°C. In this way, the heat pump increases the capacity in producing heat by renewable energy source and the solar fraction of the district heating grows.

Figure 2 shows the layout scheme created in TRNSYS for the SDH. It can be seen it is divided in several loops that separate different hydraulic circuits. The solar field pump is switched on if solar radiation is available and circulates the water which is heated up to 120°C. When the temperature in solar field is higher than TES temperature, heat is transferred to the storage. The stored heat is then delivered to users. The gas boiler and the heat pump are inserted to supply heat in case of need. The heat pump uses TES as the source of heat (evaporator side), while the user is the sink (condenser side).

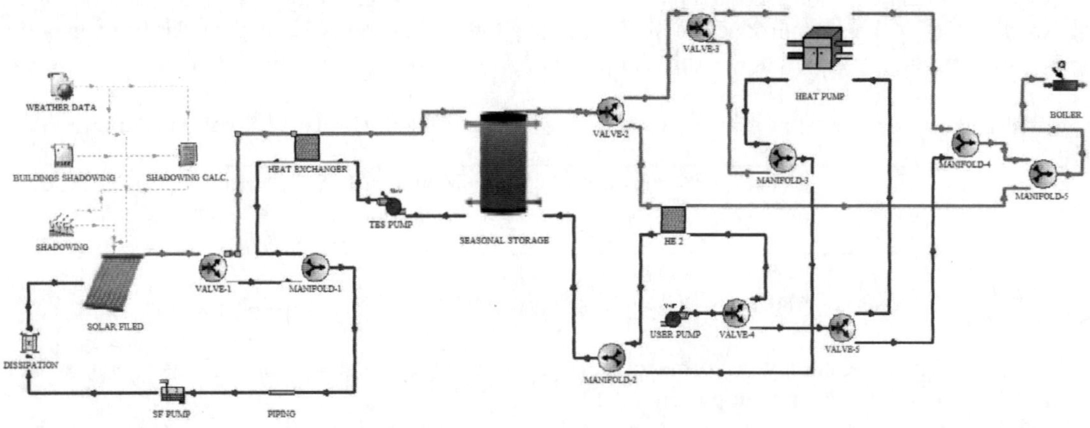

Figure 2 Layout scheme in TRNSYS.

In **Table 2** a part of the results of parametric analysis has been reported. The heat demand requested by the two buildings has been calculated to be 1522 MWh/y. The surface of the solar field is fixed to 1000 m^2. The shape of TES is supposed to be cylindrical. The coefficient of heat loss for the TES, has been fixed to 0.2 $W/(m^2K)$. This precautionary value has been selected considering that the measured heat losses in the thermal energy storages built in Europe have been considerably higher, up to 50-60% than expected [8].

In the first column of the table it is reported an important parameter for TES, the ratio between surface and volume (S/V), that represents the ratio between heat losses and thermal capacity of storage. Increasing the volume of the storage, the ratio decreases. Heat collected by solar field does not coincide with heat delivered by solar field to TES because of losses in the pipe that connects solar field and TES.

Table 2 Results of TRNSYS simulations for different volume of TES.

Volume (m^3)	S/V (m^{-1})	Heat collected by SF (MWh)	TES Losses (MWh)	Heat from TES to user (MWh)	Heat from TES to user directly (MWh)	Heat from TES to user by HP (MWh)	Maximum TES temperature (°C)	Solar Fraction (%)
1500	0.49	625	50	554	304	250	94	36
2500	0.43	701	69	609	322	287	94	40
3800	0.38	780	90	668	356	312	92	44
5000	0.37	808	105	682	355	327	82	45
6000	0.35	818	115	684	344	340	73	45

From **Table 2**, it is observed that the smaller volumes of TES does not allow to reach the target production of heat by solar field since TES capacity is small and TES is fully charged at the beginning of summer season. Solar fraction is 36-39%.

For larger volumes the energy produced by SF overcomes the target value, thanks to the higher capacity of the storage and the solar fraction grows up to 44-45%. The amount of useful heat delivered to the users from TES, passing from 3800 m^3 to 5000 m^3 rises weakly (only 15 MWh more) since heat losses increases although average temperature of TES decreases. Because of the external constraints (underground water) in this project case, the aspect ratio of TES, height to diameter, is far from the ideal condition. In order to increase the volume of the storage it is only possible to increase the width of the tank, lowering further the aspect ratio. This leads to the increase of the surface and of heat losses.

Finally, with the SF net aperture area and the height of TES fixed, there are no energy advantages to build a storage bigger than about 4000 m^3. The additional costs of excavations and of materials that will rise for a bigger storage are not justified.

The size of 3800 m^3 has been selected as the optimal volume for TES in this project. It is the trade-off between energetic performance and costs. The solar fraction expected for this configuration is 44%. It is worth to notice that the introduction of heat pump allows to rise remarkably the solar fraction of the solar district heating as reported in **Table 3**. For a TES volume of 3800 m^3, it passes from 35% to 44%.

Table 3 Comparison of performances in case without and with HP for TES volume 3800 m^3.

Hp	Heat by SF (MWh)	Heat from TES to user (MWh)	Minimum TES temperature (°C)	Solar Fraction (%)
No	686	535	51	35
Yes	780	668	23	44

The enhancement of energy performances is due to the better discharge of TES, as it can be seen from TES minimum temperature reported in table.

It is important to notice that the heat pump used in simulations is a commercial water-water heat pump. Selecting a heat pump with better performances, the solar fraction could be improved further lowering the minimum temperature to 10 °C. **Figure 3** shows the variation of average TES temperature during a year.

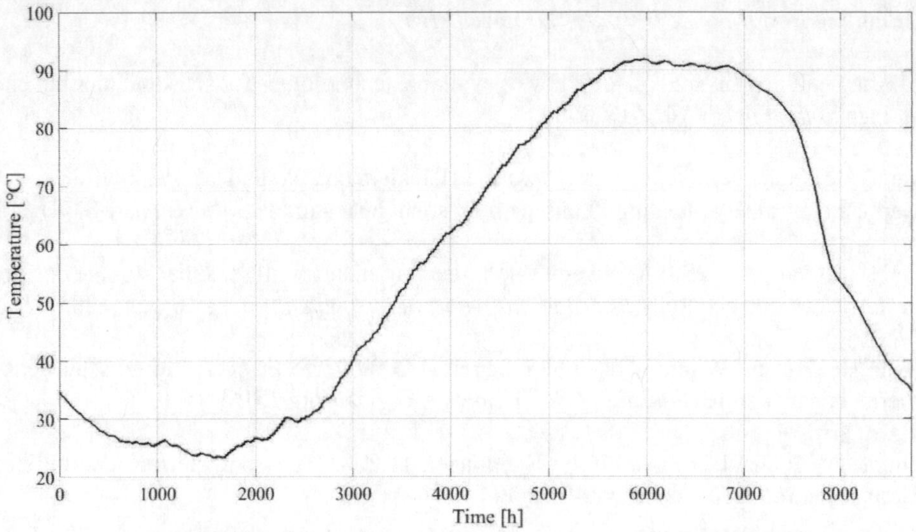

Figure 3 Temperature profile inside the seasonal storage.

The TRNSYS model has also allowed to establish the best control strategies for the district heating. The different operating conditions of the SDH can be explained through **Figure 3**. At the beginning of spring, TES is discharged and the SF starts to transfer heat to it. The charging process continues during summer time when irradiation from Sun is high and the user demand is limited to DHW. At the end of summer the storage is fully charged. In autumn time, the user demand grows and the TES begins to discharge providing heat directly for space heating and DHW. This process continues as long as TES temperature is higher than return temperature from user. When the temperature of TES drops below the return temperature, the heat pump is switched on. Heat pump extracts heat from TES in order to completely discharge the storage and to deliver heat to user. If the heat demand is higher than the heat provided by heat pump, the back-up boilers are turned on. At the end of the winter the thermal storage is discharged and the cycle is ended.

3. Conclusions

A TRNSYS model has been realized in order to design the seasonal thermal energy storage and the solar district heating that will be realized in Florence under the European project Replicate, financed inside the program Horizon 2020 SCC1 Smart Cities and Communities.

The model describes all the components of heating plant. It has led to recognize the most relevant parameters of the system. The model proves that a water-water heat pump is a key component for this project, necessary to increase considerably the performance of the system.

A wide parametric analysis has been conducted to select the optimal size of the solar field and the storage. The solar field net aperture area has been fixed to 1000 m^2 reaching a production of 780 MWh/year The TES volume of 3800 m^3 is the best trade-off between energetic performance and costs. The control strategies of the heating plant have been decided based on the best performances of TES and the expected solar fraction is 44%.

The concept of solar district heating associated with a seasonal thermal storage is a promising strategy to reduce the consumption of fossil fuels in buildings. Since district heating is widely present in Europe, the integration of solar collectors with a storage could be very interesting to supply the heat demand especially during summer months when the heat request is low and heat losses are high in percentage due to recirculation.

[1] Economidou M, Atanasiu B, Despret C, Maio J, Nolte I and Rapf O 2011 Europe's buildings under the microscope A Country-by-country review of the energy performance of buildings *Building Performance Institute Europe (BPIE)*

[2] Xu J, Wang R Z and Li Y 2014 A review of available technologies for seasonal thermal energy storage *Solar Energy* **103** 610-638

[3] Bauer D, Marx R, Nußbicker-Lux J, Ochs F, Heidemann, W and Müller-Steinhagen H 2010 German central solar heating plants with seasonal heat storage *Solar Energy* **84**(4) 612-623

[4] Dalenbäck JO and Jilar T 1985 Swedish solar heating plants with seasonal storage design, performance and economy *International Journal of Ambient Energy* **6** 123–128.

[5] Ochs F, Heidemann W and Müller-Steinhagen H 2009 Performance of large-scale seasonal thermal energy stores *Journal of Solar Energy Engineering* **131**(4) 041005

[6] Schmidt T, Mangold D, and Müller-Steinhagen H 2003 Seasonal thermal energy storage in Germany *ISES solar world congress* **Vol. 14**

[7] Hesaraki A, Holmberg S and Haghighat F Seasonal thermal energy storage with heat pumps and low temperatures in building projects—a comparative review 2015 *Renewable Sustainable Energy Reviews* **43** 1199–1213.

[8] Ochs F, Heidemann W and Muller-Steinhagen H 2008 Seasonal thermal energy storage: a challenging application for geosynthetics *EuroGeo4 4th. European Geosynthetics Conference*

IOP Conf. Series: Materials Science and Engineering **556** (2019) 012027 doi:10.1088/1757-899X/556/1/012027

Investigation on steady aerodynamic performance of a S-CO_2 compressor with different diffusers in solar power system

Qiuwan Du[1], **Di Zhang**[1,*] **and Yonghui Xie**[2]
[1] MOE Key Laboratory of Thermo-Fluid Science and Engineering, Xi'an Jiaotong University, Xi'an, China
[2] Shaanxi Engineering Laboratory of Turbomachinery and Power Equipment, School of Energy and Power Engineering, Xi'an Jiaotong University, Xi'an, China
E-mail: zhang_di@mail.xjtu.edu.cn

Abstract. Research on solar power system has become a hot spot recently. The supercritical carbon dioxide (S-CO_2) power cycle has been widely used in the solar power system with excellent advantages of high efficiency and small device size. Centrifugal compressor plays an important role in the S-CO_2 power cycle. The design of a centrifugal compressor is completed according to the requirements of thermal design in this paper. The differences of three kinds of diffusers on the steady aerodynamic performance are further investigated based on the designed model. Finally, the influence of diffuser blade number on the performance is discussed. The results show that the centrifugal compressor with wedge-shaped diffuser holds the best performance under the design and off-design conditions. The diffuser blade number has significant influence on the aerodynamic performance. The current design and performance analysis of centrifugal compressor could provide important reference value for further application of centrifugal compressor in solar power system.

Keywords: solar; centrifugal compressor; design; aerodynamic performance; diffuser

1. Introduction

As a new renewable energy, solar energy has the advantages of rich-reserved, clean, safety and so on. At present, many countries have strengthened the research and development of utilization of solar energy. Brayton cycle systems using S-CO_2 offers advantages such as smaller weight and volume, lower thermal mass and less complex power blocks and has been widely used in Concentrating Solar Power (CSP) plants[1].

Recently a large number of scholars have investigated S-CO_2 Brayton cycle in CSP. Ricardo[2] conducted exergy analysis of four different S-CO_2 Brayton cycle configurations in CSP and found that the thermal efficiency increases monotonically with the temperature. Dynamic characteristics of a S-CO_2 PCS in a solar thermal power plant were investigated by Rajinesh[3]. R. Chacartegui[4] used three CO_2 cycles in CSP and showed obvious benefits in terms of efficiency and cost. The influence of transient nature of the solar resource on solar thermal power generation was analyzed and a data set for stable S-CO_2 Brayton cycle was established by Iverson[5]. Garg[6] analyzed S-CO_2 Brayton cycle in CSP and selected thermal efficiency, specific work output and magnitude of irreversibility generation as performance indicators to compare with trans- and sub-critical operations.

Centrifugal compressor plays an important role in the S-CO_2 Brayton cycle. The flow characteristics analysis has become a hot pot. A centrifugal compressor required in a solar dish-Brayton system was designed and optimized by Wang[7] and the numerical simulation results showed a great performance. Rinaldi[8][9] numerically analyzed the performance map of a radial compressor operating with S-CO_2. The numerical results were consistent with the experimental data. Guo[10] investigated steady and unsteady performance of a S-CO_2 centrifugal compressor with splitters. A S-CO_2 centrifugal compressor with large input power was designed and numerically investigated by Wang[11] and the

IOP Conf. Series: Materials Science and Engineering **556** (2019) 012027 doi:10.1088/1757-899X/556/1/012027

unsteady flow characteristics were revealed in detail. Zhao[12] conducted the aerodynamic design and impeller aerodynamic optimization for a S-CO_2 centrifugal compressor, furthermore, the strength analysis was conducted to obtain the stress and deformation results.

In this paper, a centrifugal compressor is designed and the steady aerodynamic performance is analyzed in detail. The differences of three kinds of diffusers on the aerodynamic performance are further investigated based on the designed model. In addition, the influence of diffuser blade number on the performance is discussed. The current design of centrifugal compressor could provide important reference value for further application of centrifugal compressor in S-CO_2 Brayton cycle for solar power system.

2. Compressor design

The compressor design process is divided into three sub processes, namely 1D flow design, thermodynamic design and geometry design. The thermodynamic parameters of a centrifugal compressor are obtained by the combination of the real gas state equation, the continuity equation, the momentum equation and the energy equation. In addition, the geometry parameters of the impeller and diffuser are determined. The performance estimates and state parameters of the designed centrifugal compressor refer to Zhao's research[12], as listed in Table 1.

Table 1. Performance estimates and state parameters of the designed centrifugal compressor

Items	Parameters	Value
State parameters	Inlet temperature/K	306.7
	Inlet pressure/MPa	8.96
	Outlet pressure/MPa	18.94
	Rotating speed/($r \cdot min^{-1}$)	15000
Performance estimates	Input Power/ MW	5
	Mass flow rate/($kg \cdot s^{-1}$)	272.6
	Pressure ratio	2.1
	Efficiency/%	79.5

In this paper an impeller including 15 blades and three kinds of diffusers including 15 blades are designed, as shown in Figure 1. The geometry parameters are listed in Table 2.

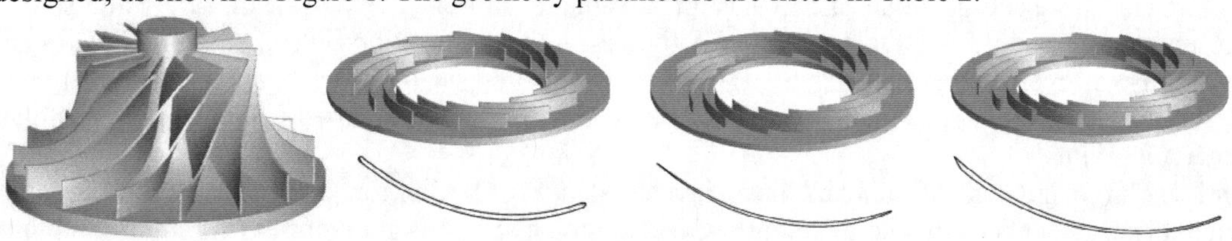

Figure 1. Geometry of the compressor

Table 2. Geometry parameters of the impeller and diffuser

Meridian plane	Parameters	Value
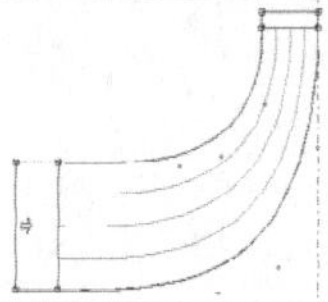	Blade height at outlet/mm	18.8
	Outer radius of impeller/mm	102
	Outer radius of inlet/mm	60
	Inner radius of inlet/mm	20
	Blade height /mm	18.8
	Outer radius of diffuser/mm	167
	Inner radius of diffuser /mm	112
	Outer radius of outlet /mm	202

3. Numerical method

3.1. Governing equations

The flow characteristics of the compressor were investigated by ANSYS CFX. The continuity, momentum and energy equations are selected as the governing equations. The general form of these equations can be expressed as[13]:

$$\frac{\partial(\rho\varphi)}{\partial t}+\mathrm{div}(\rho u\varphi)=\mathrm{div}(\Gamma\mathrm{grad}\varphi)+S \tag{1}$$

Where φ is the universal variable, and $\varphi=1$ indicates continuity equation. When φ is temperature, it represents energy equation, and when φ is set as the velocity of an arbitrary direction, Equation (1) corresponds to being a momentum equation. In addition, Γ is generalized diffusion coefficient and S is generalized source term.

In addition, the standard k-ε turbulence model was used for the analysis. The CO2RK database in ANSYS-CFX is adopted to describe the practical property of the working fluid SCO_2. It is based on the RK gas state equation[14] which is shown as:

$$P=\frac{RT}{V-b}-\frac{a}{T^{1/2}V(V+b)} \tag{2}$$

Where, $a=0.42748\frac{R^2T_c^{2.5}}{P_c}$, $b=0.08664\frac{RT_c}{P_c}$

Where, R - gas constant, T_c - critical temperature, P_c - critical pressure.

3.2. Mesh and boundary conditions

A single passage model is adopted for flow analysis. The mesh is shown in Figure 2. The hexahedral grid is used in order to improve the calculation accuracy. In addition, the O-H topology was set to the mesh topology. This can greatly reduce the number of grids and further improve the grid quality. The total number of grids for the three models is 201635, 203440 and 288674 respectively after grid independence validation.

Figure 2. Mesh of the compressor

The boundary conditions employed in this paper for steady aerodynamic performance are as follows: mass flow rate of the inlet (18.17kg/s), static temperature of the inlet (306.7K), and static pressure of the outlet (18.94MPa). Meanwhile, the rotating speed was 15000rpm.

4. Results and discussions

4.1. Steady aerodynamic performance comparison

The pressure contours at mid-span of blades for three models are shown in Figure 3. It can be seen that the pressure distributions of the three models are almost the same and the pressure reaches the minimum at the inlet of the impeller. Due to the centrifugal effect in impeller and the transformation from kinetic energy into pressure energy, the pressure enlarges along the flow passage. In addition, there exists a stagnation point at the leading edge of diffuser blades. Moreover, it is further obtained that the maximum pressure of the airfoil model is slightly larger than that of the other two models. However, the pressure distribution of wedge-shaped diffuser is more uniform.

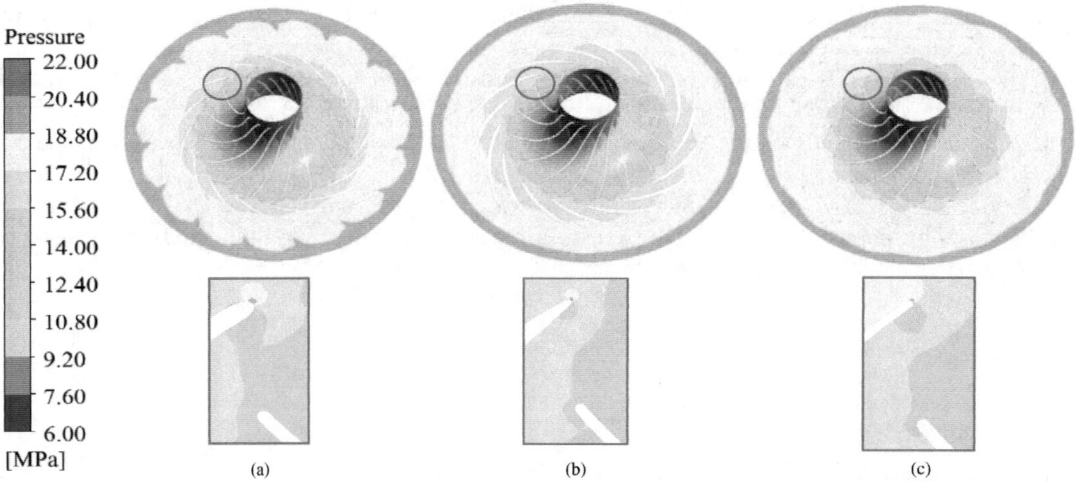

Figure 3. Pressure contours at mid-span of blades: (a) circle; (b) airfoil; (c) wedge-shaped

The temperature distributions are shown in Figure 4. It can be observed that the temperature distributions of the three models are almost the same and the minimum temperature occurs at the inlet of the impeller. The temperature gradually increases along the passage with the increase of pressure. The temperature reaches the maximum value at the leading edge of the diffuser blade. Moreover, it is further obtained that the maximum temperature of the circle model is slightly larger than that of the other two models.

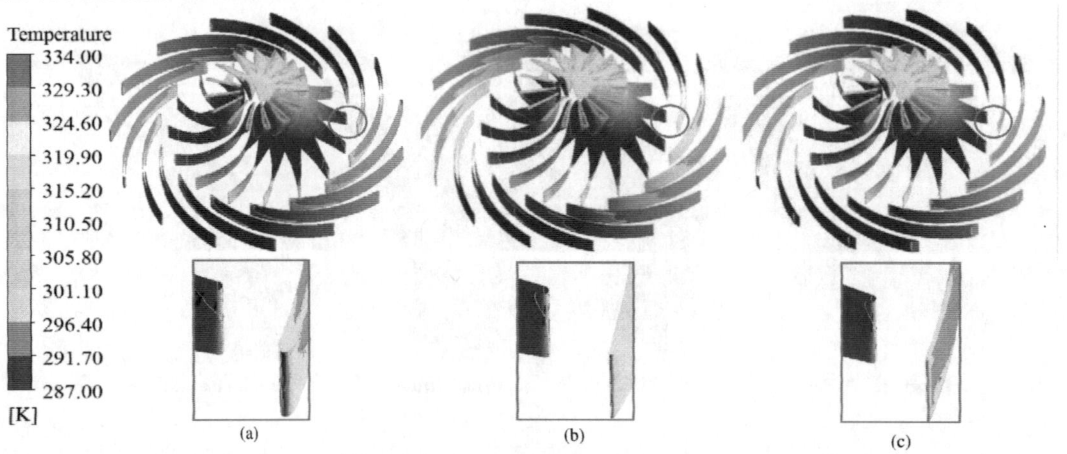

Figure 4.Temperature contours at mid-span of blades: (a) circle; (b) airfoil; (c) wedge-shaped

Figure 5 shows the streamline distributions of the three diffusers. As shown in Figure 5, the streamline distributions of the three diffusers remain almost identical. In addition, there is a velocity stagnation point located at the leading edge of the diffuser blade. The velocity gradually decreases through the diffuser passage resulting in the increase of pressure which is shown in Figure 3. Moreover, it is further obtained that the velocity of the wedge-shaped diffuser is larger than that of the two diffusers and the flow field is more uniform.

In addition, the performance parameters of the three models are also obtained. The powers of the circle model, airfoil model and wedge-shaped model are 4.98MW, 5.01MW and 5.11MW respectively. The pressure ratios are 2.134, 2.126 and 2.154 respectively. Combined with the calculation results of the pressure, temperature and flow field, it concludes that the wedge-shaped model holds the best aerodynamic performance under the design condition.

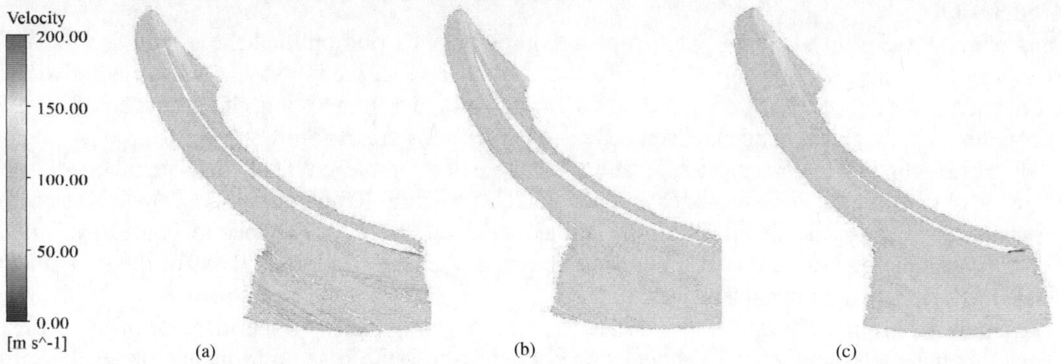

Figure 5. Streamline distributions of the three diffusers: (a) circle; (b) airfoil; (c) wedge-shaped

4.2. Off-design performance comparison

The off-design performance of the three models is investigated in this chapter by changing the mass flow rate which varies 17.80kg/s, 18.17 kg/s, 18.50 kg/s, 19.00 kg/s and 19.50 kg/s. The calculation results are listed in Table 3. It can be seen that the power of the three models increases as the mass flow rate increases while the pressure ratio decreases. It is worth mentioning that the calculation with the airfoil model is not convergent under the conditions with 17.80kg/s and 19.50kg/s, which indicates that the off-design performance of the airfoil model involved in this paper is the worst. Further comparison shows that the power and pressure ratio of the wedge-shaped model is higher than those of the other two models under the same working condition.

Table 3. The off-design performance parameters

Mass flow	Power/MW			Pressure ratio		
rate/(kg/s)	Circle	Airfoil	Wedge-shaped	Circle	Airfoil	Wedge-shaped
17.80	4.90	/	5.02	2.138	/	2.158
18.17	4.98	5.01	5.11	2.134	2.126	2.154
18.50	5.06	5.08	5.18	2.130	2.123	2.151
19.00	5.17	5.20	5.30	2.124	2.116	2.146
19.50	5.28	/	5.41	2.117	/	2.140

4.3. Influence of diffuser blade number

In order to further compare the differences between the three models, the influence of diffuser blade number on the performance is discussed in detail in this chapter. The numbers of diffuser blade are selected as 13, 14, 15, 16 and 17 respectively.

Table 4. The performance parameters with different diffuser blade number

Diffuser blade	Power/MW			Pressure ratio		
Number	Circle	Airfoil	Wedge-shaped	Circle	Airfoil	Wedge-shaped
13	4.96	4.95	5.02	2.100	2.102	2.124
14	4.94	4.99	5.07	2.120	2.115	2.142
15	4.98	5.01	5.11	2.134	2.126	2.154
16	5.02	4.96	5.06	2.144	2.134	2.158
17	5.04	4.97	5.15	2.155	2.139	2.164

The calculation results are listed in Table 4. It can be seen that the power and pressure ratio show an increase trend as a whole with the increase of the diffuser blade number. Further comparison shows that the power and pressure ratio of the wedge-shaped model are higher than those of the other two models under all working conditions.

5. Conclusions

In this paper, the design of S-CO_2 centrifugal compressor is carried out and the steady aerodynamic performance is analyzed in detail. The differences of three kinds of diffusers on the aerodynamic performance is further investigated based on the designed model. In addition, the influence of diffuser blade number on the performance is discussed. The conclusions are as follows:

(1) The distributions of pressure, temperature and velocity for the circle model, airfoil model and wedge-shaped model remain almost identical. The pressure, temperature and velocity reach the maximum value for the airfoil model, the circle model and the wedge-shaped model respectively.
(2) The power and pressure ratio of the wedge-shaped model are larger than those of the two models under the design and off-design conditions.
(3) The power and pressure ratio show an increase trend as a whole with the increase of the diffuser blade number. In addition, the wedge-shaped model show the best performance under the same conditions.

6. References

[1] Craig S T, Zhiwen M, Ty W N and Michael J W 2013 *J. Sol. Energy Eng* **135** 375-83
[2] Ricardo V P, Yen C S T, Regano B and Wes S 2015 *Appl Energ* **148** 348-65
[3] Rajinesh S, Sarah A M, Andrew S R and Peter A J 2013 *Energy* **50** 194-204
[4] R C, J M M D E, D S, B M and T S 2011 *Appl Therm Eng* **31** 872-79
[5] Brian D I, Thomas M C, James J P and Alan M K 2013 *Appl Energ* **111** 957-70
[6] Pardeep G, Pramod K and Kandadai S 2013 *J Supercrit Fluid* **76** 54-60
[7] Yongsheng W, Kai W, Zhiting T, Feng L, Chaoqun N and Abraham E 2013 *J Therm Sci* **22** 404-12
[8] Enrico R, Rene P and Piero C 2015 *J. Eng. Gas Turbines Power* **137** 072602
[9] Enrico R, Rene P and Piero C 2014 *ASME Turbo Expo 2014* GT2014-26966.
[10] Ding G, Dongbo S and Di Z 2017 *Therm Sci* **21** 185-92
[11] Yuqi W, Dongbo S, Di Z and Yonghui X 2017 *Appl Sci* **7** 310
[12] Hang Z, Qinghua D, Kuankuan Z, Hanzhen Z and Zhenping F 2014 *ASME Turbo Expo 2014* GT2014-26646
[13] Fujun W 2004 Analysis of computational fluid dynamics *Tsinghua University Press* pp 124-125
[14] O Redlich and JNS Kwong 1949 *Chem Rev* **44** 233-44

Acknowledgments

The authors gratefully acknowledge for the support by 111 project (Grant No.B16038).

Research on Operation Strategy of Solar Assisted Air Source Heat Pump System

Z Yuhan[1], W Jinshun[2*], L Mu[3], Z Xingxing[4], Z Lining[2], Z Weiya[2], Z Weijie[1], W Jun[2], X Liang[5], Z Weihong[6], W Xinru[5] and W Yixuan[5]

[1] Hebei university of engineering
[2] North China Institute of Science and Technology
[3]Nanjing Normal University
[4]Dalarna University
[5] The University of Nottingham Ningbo
[6] University of Science and Technology Liaoning
Eimail: Wu Jinshun. wujinshun2005@163.com

Abstract: Solar-assisted air source heat pump (SAASHP) has greatly improved the operating efficiency of the system. However, the operational performance of the key modules in the combined system is not clear. So a solar-assisted air source heat pump radiant heating system was established and a SAASHP was monitored. The key parameters of the solar collector outlet water inlet water temperature, and water tank temperature under different combined operation modes were calculated. The solar collector heat collection efficiency, system COP, and the energy consumption ratio under the corresponding mode were given. The solar collector heat collection efficiency, system COP, and the energy consumption ratio under the corresponding mode were given. Photo thermal conversion, evaporative heat transfer, and energy conversion between the condensers. By comparing the operating characteristics of each combination mode, the characteristics of SAASHP combined operation are finally obtained.

Keywords: Solar energy; Air source heat pump; Energy efficiency analysis; Operation strategy

1. Introduction

As the issue of energy and environmental protection becomes more prominent, the negative effects brought about by the use of non-renewable energy based on heating models are increasingly unable to meet the requirements of sustainable development in society [1-2]. Air source heat pumps (ASHP) have become one of the most widely used applications for a variety of heat pump types because of their low energy consumption and small environmental impact. However, when the temperature in winter is lower than the air dew point temperature and lower than 0°C, frost issues may occur on the outdoor heat exchanger coil surfaces [3-6].

To solve this problem many experts conducted research, and SAASHP greatly improved the system's operating efficiency [7-9]. Xiaolong Lv et. al.[8] studied a novel solar-assisted auto-cascade heat pump cycle model with the zeotropic mixture of R32/R290 and the results showed that it has 4.23-9.85% and 4.37-9.68 improvement compared to the ASHP. M. Shan [9] et al. researched an integrated active solar and ASHP water heating system with a passive house in cold climate. They confirmed the benefits of the ASHP system combined with solar system. However, there are few studies involving the operational performance of each module in the auxiliary operation of SAASHP. Therefore, this paper studies the SAASHP combined operation strategy, and calculates the heat collection efficiency of the collector, the system COP and the energy consumption ratio under the corresponding module, and then obtains the photothermal conversion, the evaporative heat transfer and the energy relationship between the condensers in the system. By comparing the special operations of each combination mode, the characteristics of the combined operation of the SAASHP are finally obtained.

2. Experimental system

2.1 System introduction

The system diagram is shown in Fig2-1. The experimental system is located in the air-conditioning energy-saving office on the eighth floor of an experimental building in North China. The laboratory area is 22 m², and the system is consisted of the solar collectors, thermal storage tank, Capillary circulation and heat pump system. The indoor cooling ends include capillary networks and air conditioners.

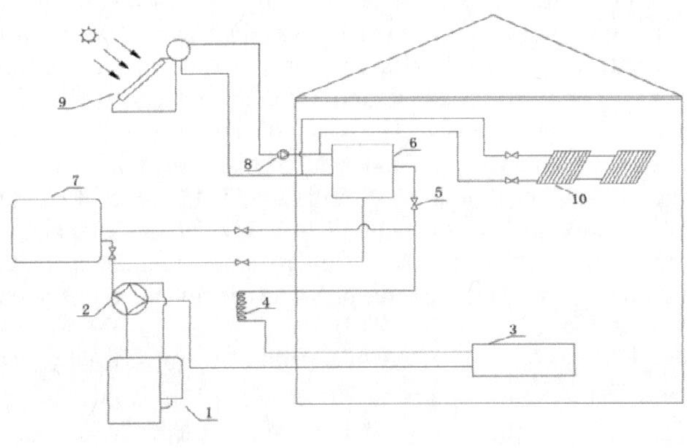

1 compressor, 2 4-way transfer valve, 3 indoor air conditioner, 4 throttle valves, 5 valves , 6 fluorine water plate heat exchanger, 7 outdoor heat exchangers,8 pumps, 9 solar collectors, 10 capillary network

Figure 1-1. The solar-assisted air source heat pump diagram

2.2 System operation mode

The system can generate multiple combinations of operating modes according to its own characteristics. This system mainly discusses the auxiliary forms of SAASHP. The specific operating strategy is shown in Table 2-1 below.

Table 2-1. Realizable operation modes

Mode	Solar thermal collector	Heat exchange	Evaporator	Air conditioner	Radiation Capillary
Mode 1	Daytime run	Nocturnal run	Daytime run	Whole day run	Whole day stop
Mode 2	Daytime run	Whole day stop	Daytime run	Daytime run	Nocturnal run
Mode 3	Daytime run	Daytime run	Nocturnal run	Whole day run	Whole day stop
Mode 4	Daytime run	Whole day stop	Nocturnal run	Nocturnal run	Daytime run

3. Comparison of system operation strategy

3.1. Combined Operation Mode Comparison

The single ASHP, SAASHP and the solar capillary radiant heating system can be combined to operate in the following four joint operation modes. The following is a comparative analysis of the differences in the operation of the four modes.

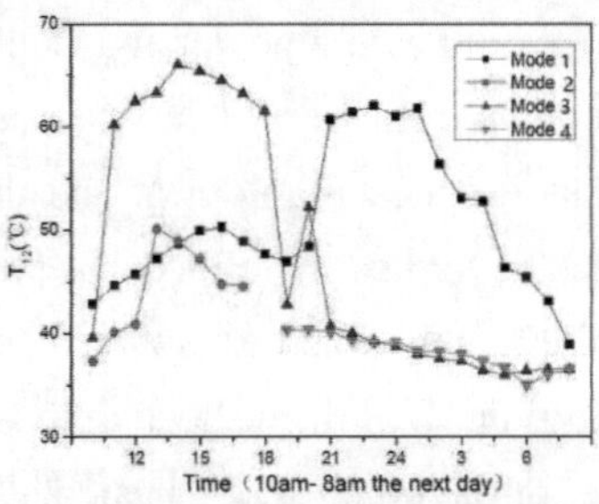

Figure 3-1. Comparison of T12 in four modes.

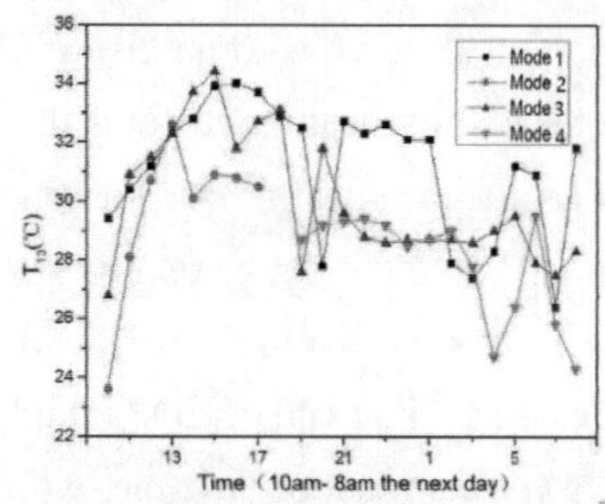

Figure 3-2. Comparison of T13 in four modes.

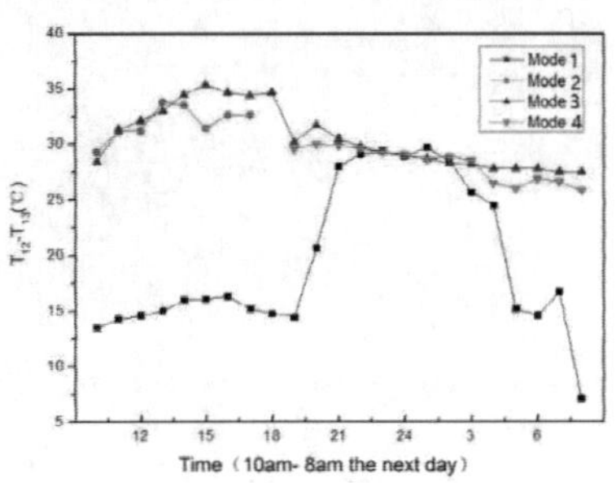

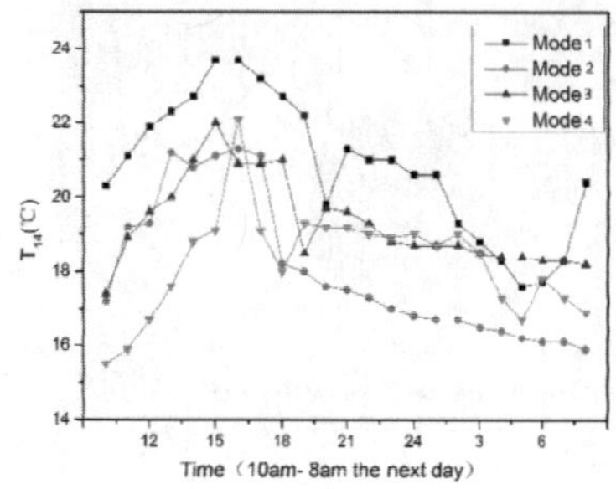

Figure 3-3. Comparison of T12 and T13 in four Modes. **Figure 3-4.** Comparison of T14 in four Modes.

The air conditioner refrigerant inlet and outlet temperature trend of four modes were shown as Figure.3-1, Figure.3-2 and Figure.3-3. It can be seen that the temperature in the combination of solar auxiliary mode in the daytime and night is highest, however the ASHP has a lower temperature in night compared to daytime. T12 is obviously different in different modes. In general, the air source heat pump is greatly affected by the day and night, thus resulting the T12 has a greater change in Mode 3 compared to Mode 1.

For the refrigerant outer temperature T13, there are little difference at daytime between ASHP and solar power assistance, however, the T13 is lower in ASHP compared to solar assisted heat pump. The temperature difference between T12 and T13 is basically maintained at about 20°C, but the temperature difference between Mode 1 is smaller.

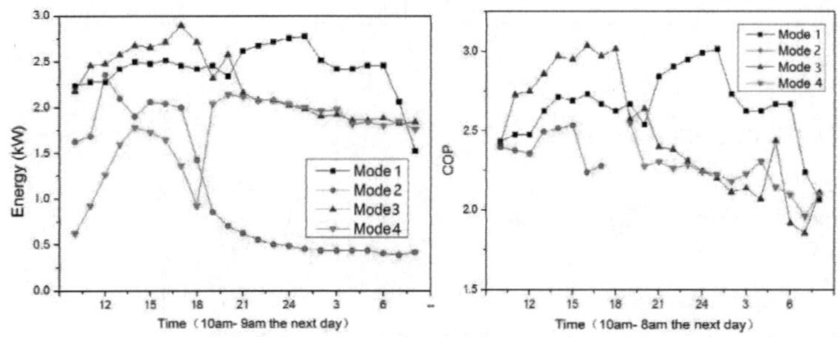

Figure 3-5. Comparison of heating power and COP in four modes.

The Figure 3-5 shows the comparison between the four modes of heating power and COP. It can be seen that the highest heating power is the solar assisted heat pump, followed by the ASHP, and finally the capillary radiant heating. In four mode, the Mode 1 has the most stable heat and the outdoor temperature during the day is relatively high. The ASHP has better heating effect. The water stored in the water tank is used to supply the auxiliary mode at night so that the heat source can be stable during the night. The trend of COP is similar to that of heating. The night-time heating COP of ASHP is much lower than that of daytime, which is more suitable for daytime use. The calculation results show that Mode 1 has the highest average COP.

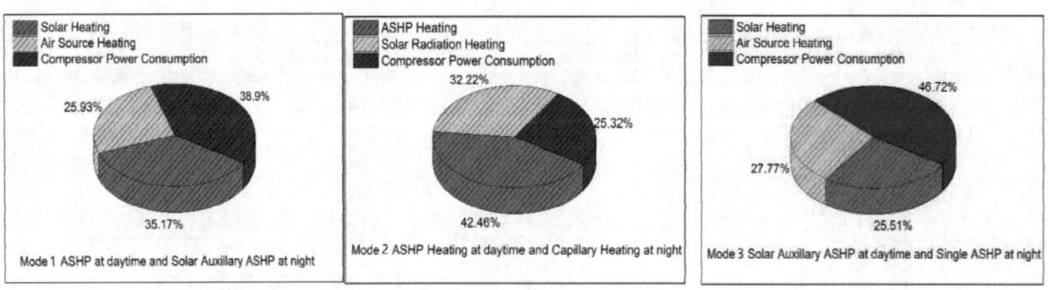

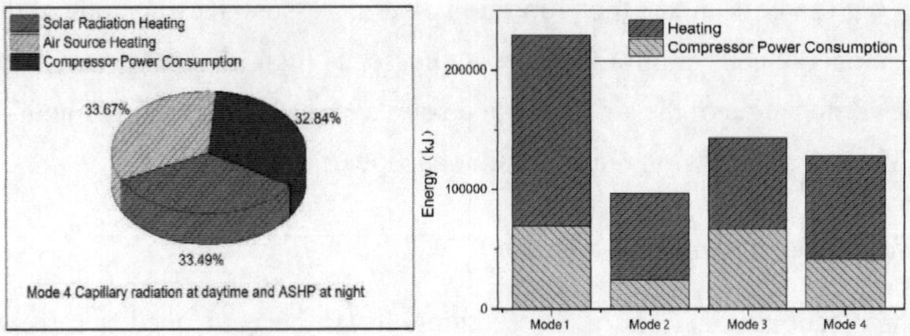

Figure3-6. Comparison of energy share heating and power consumption of compressor in four modes.

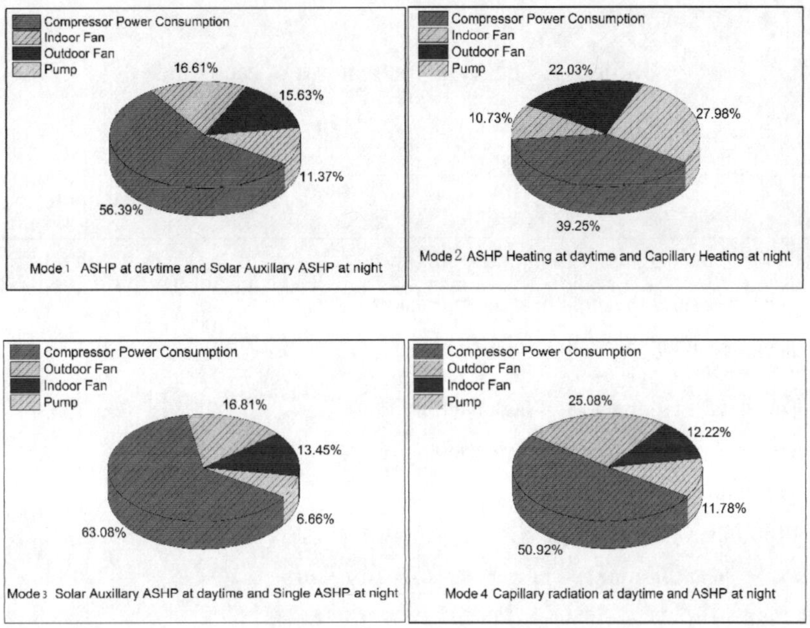

Figure 3-7. Comparison of share of power consumption in four modes.

By contrasting the energy ratio of the pie shown as Figure.3-6, Mode 2 has the smallest proportion of compressor, which is 25.32%. Mode 4 represents the largest proportion of energy consumption. This is mainly because solar radiation heating does not require the work of the compressor. Mode 2 uses capillary heating method, and capillary mode operation time is longer, energy consumption is reduced greatly. Similarly, the energy consumption of Mode 4 is lower than that. For Mode 3, solar energy auxiliary and air source heat pump require the work of the compressor to make heat, and the air source heat pump is less efficient at night causing less heat.

From the histogram in Figure3-6, it can be seen that although the proportions of energy consumption in Mode 2 and Mode 4 are smaller, but the heating capacity of one day in Mode 2 is the lowest among the four modes, while Mode 4 is the lowest. Mode 1 has the highest heat production per day, and its energy consumption is average in all four modes, and it is also a more energy-efficient mode.

IOP Conf. Series: Materials Science and Engineering **556** (2019) 012028 doi:10.1088/1757-899X/556/1/012028

From Figure3-7, it can be seen that the proportion of compressors in each mode of the compressor is the highest. The energy consumption of the compressor is high, thus resulting its proportion high. This makes the combination of the air source and solar energy assisted Mode 1 and Mode 3 consume more energy. Mode 2 capillary runs longer and it has lower overall power.

3.2 Portfolio mode operation strategy comparison.

Through the comparison and analysis of the sections, it can be concluded that the advantages and disadvantages of the eight operating modes of the ASHP, solar and capillary heating system, and also the corresponding architectural types. The following features of the eight operating modes are summarized, as shown in table 3-2.

Table 3-1. The operation features of each mode

Mode	advantage	Shortcoming
Mode 1	The indoor temperature is high throughout the day, which ensures the night heating and makes good use of air energy and solar energy.	The collector has low thermal efficiency and high energy consumption.
Mode 2	To meet the requirement of high daytime temperature and low nighttime demand, the temperature is high in the night, so that the capillary tube will not be heated in the morning and consume less power.	Less heat-collecting efficiency of Solar collector and less heating time at night.
Mode 3	The heating efficiency of collector is higher, and the indoor temperature can be guaranteed during the day.	The system energy consumption is higher, the temperature difference between night and night is large, the night system is not stable, and the indoor temperature is not guaranteed.
Mode 4	The efficiency of the collector is higher, and the indoor heating quantity is guaranteed. The indoor temperature difference between day and night is small, and the thermal comfort is relatively high.	Night air source heat pump system frequently defrost, operating instability.

4. Conclusion

This paper compares and analyses different modes of solar auxiliary ASHP radiant heating system through measurement and simulation, and presents different heating operation strategy for different buildings. The specific results are as follows:

In the solar auxiliary ASHP, solar hot water replaces the low temperature heat source of the heat pump cycle in outdoor evaporator, which can better utilize the solar energy collection heat capacity. Heat pump system can stable operation, the indoor temperature with small fluctuations basic keep at around 20°C; In low temperature capillary radiant heating system, capillary water temperature can meet the demand of indoor heating when it is higher than 28°C. Indoor temperature kept at around 18°C and the body feeling temperature is 20-21°C, where temperature field is even and the thermal comfort is higher. In the cycle of solar hot water participation, the system is generally energy-saving, stable, reliable and efficient. The operating modes of various combination systems have their own characteristics, so different combinations should be used for indoor heating for different buildings.

In the solar auxiliary ASHP, with the help of solar water, the ASHP system can stabilize the heating. The combined operation mode of ASHP solar capillary radiant heating system has advantages and disadvantages. Different heating strategies should be selected according to the non-function of buildings. But more energy efficient way is to as much as possible to heat by using low temperature capillary radiant heating system (water temperature greater than 28°C). When it cannot meet the requirements, the low temperature water can be used to heat the solar auxiliary ASHP system (Water temperature above 10°C). Finally, the single ASHP system is selected.

Acknowledgments

This research was financially supported by Innovative Research Groups of the Beijing Natural Science Foundation Project (3172041), Key R & D projects in Hebei (18394317D), and Hebei province graduate student innovation subsidy project (CXZZSS2018036).

Reference

[1] Yaning Zhang, Qin Ma, Bingxi Li, Xinmeng Fan, Zhongbin Fu, Application of an air source heat pump (ASHP) for heating in Harbin the coldest provincial capital of China. Int J Energy and Building 138(2007)96-103

[2] Xiao Wang, Jianlin Yu, Meibo Xing. Performance analysis of a new ejector enhanced vapor injection heat pump cycle. Int J Energy Conversion and Management 100(2015) 242-248

[3] Mengjie Song, Ailiu Chen, Ning Mao. An exoerimental study on defrosting performance of an air source heat pump unit with a multi-circuit outdoor coil at different frosting envennes values [J]. Applied Thermal Engineering, 94(2016) 331-340

[4] J.S. Byun, j. Lee, CD. Jeon, Frost retardation of an air source heat pump by the hot gas bypass method. Int, J. Refrig, 31(2)(2008):328-334

[5] Hwan-Jong Choi, Byung-Soon Kim, Donghoon Kang, Kyung Chun Ki.Defrosting method adopting dual hot gas bypass for an air-to-air heat pump. Int J Applied Energy 2011;88: 4544-4555

[6] F.Qin, Q.F.Xue, G.M.A. Velez,G.Y. Zhang, H.M.Zou, C.Q. Tian, Experimental investigation on heating performance of heat pump for electric vehicles at -20°C ambient temperature. Int J Energy Conversion and Management 102(2015)39-49

[7] Cai-hua Liang, Xiao-song Zhang, Xiu-wei Li, Xia Zhu. Study on the performance if a solar assisted air source heat pump system for building heating. Int J Energy and Building 43(2011) 2188-2196

[8] Martin Kegel, Justin Tamasauskas, Roberto Sunye, Antoine Langlois. Assessment of a solar assisted air source and a solar assisted water source heat pump system in a Canadian household. Int J Energy Procedia 31(2012)654-663

[9] M.Shan, T. Yu, X.Yang. Assessment of an intergrated active solar and air-source heat water heating system operated within a passive house in a cold climate zone. Int J Eenewable Energy 2015:1-8

Experimental Study on Heat Storage and Release Processes of Composite Phase Change Wall under Short Intermittent Heating Conditions

Yanru Li[1], Jin Li[1], Luting Xu[2], Yang Ming[1], and Enshen Long[1, 2, *]
[1]College of Architecture and Environment, Sichuan University, Chengdu, China
[2]Institute of Disaster Management and Reconstruction, Sichuan University-Hong Kong PolyU, China.

E-mail: longes2@163.com

Abstract: During intermittent heating, the short intermittent conditions are common in meeting rooms of office buildings and school classrooms. The thermal process of building envelope has coupling effects on the indoor thermal environment in the frequent start-stop of heating equipment. This paper integrated phase change materials (PCMs) with building envelope to improve its thermal performance. Measurements were carried out to study the dynamic thermal process of composite phase change wall (composite-PCW) in case 1 (heating 1h + suspension 1h) and case 2 (hating 2h + suspension 2h). Results showed that: 1) the indoor air temperature fluctuations were around 14°C and 18°C respectively for case 1 and case 2 when the indoor thermal processes reached relatively stable. 2) The heat storage of composite-PCW could maintain the inner surface temperature relatively constant, so that the inner surface temperature fluctuations were only 5°C and 6°C respectively for case 1 and case 2. 3) The heat storage utilization rates of the last cycles were 5.08% and 14.86% for case 1 and case 2. Thus, the heat transferred into the wall during the heating period was stored inside the wall to improve the indoor thermal environment by rational use to maintain the inner surface temperature during the cooling-down period.

Keywords: Phase change meterials; intermittent heating; heat storage; energy efficiency

1. Introduction

The indoor thermal environment in the intermittent heating process is different from that of continuous heating because of the frequent start-stop of the heating equipment, and the thermal process of building envelope has coupling effects on the indoor thermal environment [1]. Tsilingiris et al. [2] and Barrios et al. [3] studied the heat loss of different wall structures during intermittent heating. Wang et al. [4] investigated the floor heat storage and release during an intermittent in-slab floor heating process.

The thermal inertia of building envelope can affect the indoor environment variations [5], and the rational use of building heat storage and release processes help improve the indoor thermal environment. Ogoli et al [6] found that the wall constructed by heavy materials could reduce the indoor air temperature effectively. Cheng et al [7] comprehensively analyzed the effects of different building heat storage materials, different surface colours of exterior wall and different orientations on indoor temperature. During the intermittent heating process, the building energy consumption can be reduced through using building heat storage to optimal match the heating equipment operation.

Phase change materials (PCMs) are the ideal material for improving the heat capacity of building envelope due to the high thermal inertia in a relatively small temperature range [8]. Compared with the traditional building envelope, the building envelope integrated with PCMs is able to save energy consumption, improve the indoor thermal environment and transfer the electricity peak [9-10]. Zhang

IOP Conf. Series: Materials Science and Engineering **556** (2019) 012029 doi:10.1088/1757-899X/556/1/012029

et al. [11] found that the phase change wall could save the building energy consumption through increasing the fusion heat, reducing the thermal conductivity and selecting appropriate phase transition temperature. Castell [12] did experiments on two kinds of building envelope integrated with different PCMs. The results showed that the PCMs reduced t the air conditioning energy consumption by 15% in summer. During intermittent heating, the indoor air temperature changes severely when the heating is on/off, so that the phase changing process of PCMs in the building envelope is more obvious when indoor air temperature changes. Wang et al. [13] found that the phase change wall could reduce the heating load by 10-30% in the intermittent heating operation, and the PCMs could reduce 9-27% of the internal surface heat loss after turning off the heating equipment. However, that paper divided the intermittent heating process into the heating process and the cooling-down process, which neglects the lasting effect of building envelope heat storage on the indoor thermal environment and the heating energy consumption.

Therefore, the influence of building envelope thermal process on indoor thermal environment is different under different intermittent heating conditions [14]. Because the short-term use of space in office buildings and schools, the heating equipment starts and stops frequently. Thus, the heating time and heating energy consumption is able to be reduced through using the building envelope heat storage and release processes to improve the indoor thermal environment. In this paper, PCMs are integrated to the inner side of wall to realize the wall energy saving during intermittent heating. Measurements were carried out to study the dynamic thermal process of the composite phase change wall (composite-PCW) under two short intermittent heating conditions. And the energy efficiency and indoor thermal environment improvement were investigated. This work provides theoretical support for the application of composite-PCW in intermittent heated buildings.

2. Experimental set-up

The experiment was carried out in the wall dynamic test building in Sichuan University (Figure. 1(a)). The composite-PCW with size of 600mm×600mm×260mm consists of one layer of cement mortar of 20mm, one layer of PCMs of 20mm, one layer of foamed concrete of 200mm and one layer of cement mortar of 20mm from the inner side to outer side, and the schematic structures are presented in Figure 1(b). Table 1 shows the physical properties of each layer of material, and the phase-transition temperature range of the PCMs is from 18°C to 26°C, which has phase change latent heat of 178.5kJ/kg.

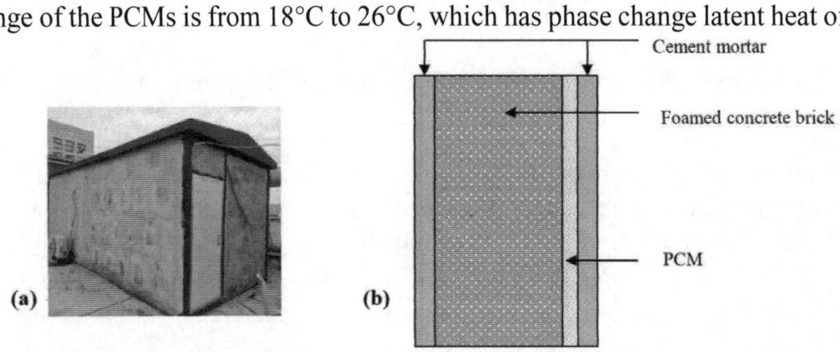

Figure 1. The experimental system ((a) the experiment test building; (b) The schematic structures of the composite-PCW).

Table 1. Physical properties of each layer.

Material	Density (kg/m^3)	Thermal conductivity (W/(m·°C))	Specific heat (kJ/kg·°C)
Cement mortar	1406	0.3505	1.05
Foamed concrete brick	500	0.1631	1.05
PCMs	1300	0.25(liquid), 0.5(solid)	1.785

T-type thermocouples with accuracy of ±0.5°C and heat flux meters with accuracy of 5% were selected, and all measurement data were recorded by a JTRG-II building thermal temperature automatic tester. Two short intermittent heating conditions were designed: a) case 1: heating 1h + suspension 1h, b) case 2: heating 2h + suspension 2h, which is shown in Figure 2. The measurements of case 1 and case

2 were carried out on 20th and 22th January 2016, and the first heating operation cycles both began at 9:00. The temperature and heat flux of the composite-PCW were measured. The heating thermostat setting was adjust to 30°C for winter space heating to ensure the phase change of PCMs.

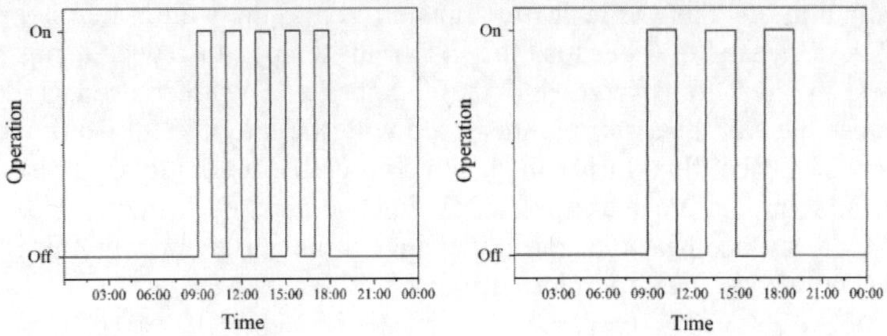

(a) Case 1 **(b) Case 2**

Figure 2. The short intermittent heating conditions.

3. Results and Discussion

3.1. Case 1: heating 1h + suspension 1h

Figure 3 presents the variations of the indoor and outdoor air temperatures in case 1. During the test period, the outdoor air temperature changed between 6.5-8.5°C. The indoor air temperature changed greatly with the heating equipment operation process in the five heating operation cycles. In the first operation cycle, the indoor air temperature was only around 11°C before heating, and rose to 29.95°C rapidly after heating for 1 hour while decreased to 14.62°C after stopping heating. Then the dynamic thermal response of the indoor air was accelerated because of the increasing heat storage of the room. The indoor air temperature changing rate increased after heating, and the minimum indoor air temperature of each cycle rose gradually when the heating was stopped. For the last four operation cycles, the indoor air temperature rose to around 31°C in the 1 hour heating process, and decreased to around 17°C in the 1 hour cooling-down process. The last heating process ended at 18:00, and the indoor air temperature decreased rapidly. The indoor air temperature was above 16°C while the cooling rate gradually decreased. After 2 hours cooling-down process (after 20:00), the cooling rate increased, and the indoor air temperature decreased to 9.8°C. Excepting for the first operation cycle, the indoor air temperature was always higher than 15°C during the intermittent heating, and the fluctuation range was about 14°C.

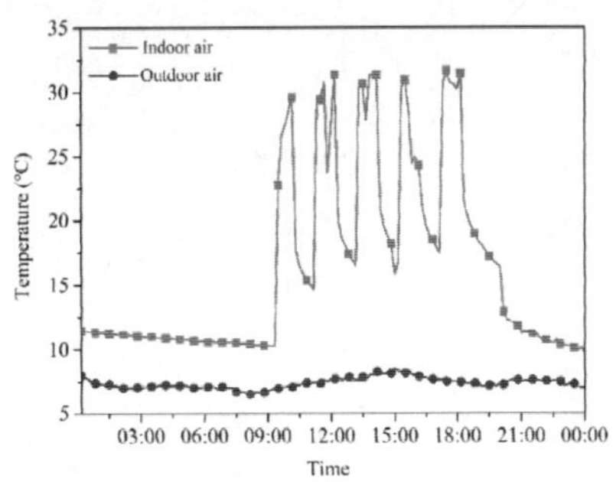

Figure 3. The variations of indoor and outdoor air temperatures in case 1.

Figure 4(a) presents the variation of surface temperatures of the composite-PCW in case 1. The variation of outer surface temperature was similar to the outdoor air temperature. And the outer surface temperature did not increase in the heating process of the five operation cycles, which indicates that the heat loss to outdoor through wall was less due to the wall thermal inertia. During the test period, the inner surface temperature of the composite-PCW changed greatly with the variation of indoor air temperature. And the wall stored heat during 1 hour heating process of each cycle to maintain the inner surface temperature constant, leading to the increase of the inner surface temperature. Thus, the inner surface temperature was always higher than 16°C in the last three cycles, and the temperature fluctuation

IOP Conf. Series: Materials Science and Engineering **556** (2019) 012029 doi:10.1088/1757-899X/556/1/012029

was only about 5°C. Figure 4(b) shows the variation of the inner surface heat flux. The value of inner surface heat flux remained around 0 from 0:00 to 9:00, indicating that there was no heat transfer between the indoor air and the composite-PCW. After starting heating at 9:00, the heat flux of the inner surface increased rapidly, which means that the heat was transferred into the wall. After stopping heating, the inner surface heat flux decreased to 0 because the indoor air temperature decreasing. In the following four operation cycles, the heat transferred into the composite-PCW decreased gradually during the heating process because the wall heat storage increased with the intermittent heating operation, so the increase of the inner surface heat flux during the heating processes gradually decreased. After stopping heating, the inner surface temperature decreased much slowly compared with the indoor air temperature, so that the value of inner surface heat flux did not change to negative until the third cycle, indicating that the composite-PCW released heat to indoor air.

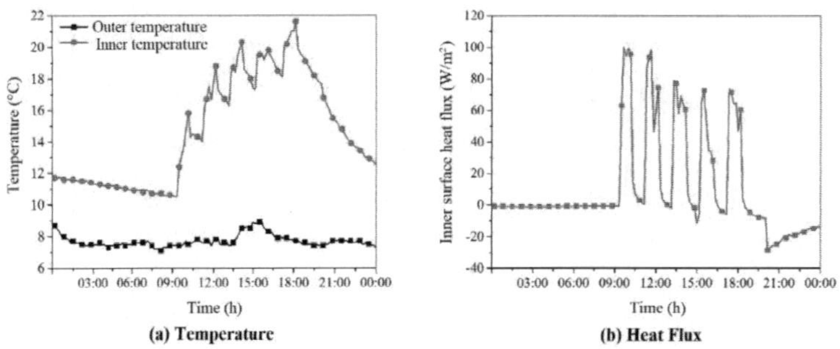

Figure 4. The variations of surface temperatures and inner surface heat flux of the composite-PCW in case 1.

3.2. Case 2: heating 2h + suspension 2h

Figure 5 presents the variations of the indoor and outdoor air temperatures in case 2. The outdoor air temperature was decreasing during the test period. In the three operation cycles, the indoor air temperature changed greatly. It was only around 8°C before the heating process of the first operation cycle, and increased rapidly after heating at 9:00. After 0.5 hours heating, the changing rate became slower, and finally the indoor air temperature rose to 31.46°C in the 2 hours heating process in the first cycle. After stopping heating, the indoor air temperature decreased to 13.52°C. For the three operation cycles, the indoor air temperature could remain above 15°C for more than 1.5h because of the wall heat storage and release. And the indoor air temperature fluctuation was around 18°C.

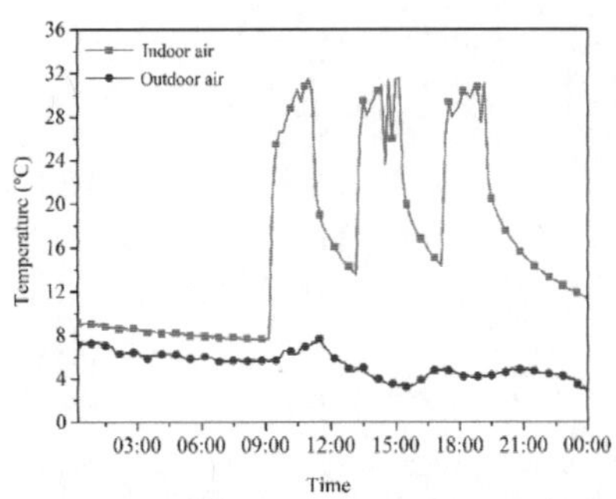

Figure 5. The variations of indoor and outdoor air temperatures in case 2.

The surface temperature variations of the composite-PCW was shown in Figure 6(a). During the test period, the variation of the outer surface temperatures were similar with the outer air temperature variation. The inner surface temperature of the composite-PCW changed greatly with the indoor air temperature variation, and showed a clear warming trend. The wall heat storage during the 2 hours heating process of each cycle maintained the inner surface temperature relatively constant, so that the fluctuation range was only about 6°C. Figure 6(b) shows the inner surface heat flux variation of the composite-PCW. From 0:00-9:00, the inner surface heat flux was maintained at around 0. After heating at 9:00, the heat was transferred into the wall inner surface, so that the inner surface heat flux rose rapidly. The inner surface heat flux decreased rapidly after stopping heating, and the wall released heat to indoor environment. In the last two operation cycles, the heat transferred into the composite-PCW decreased gradually because of the wall heat storage.

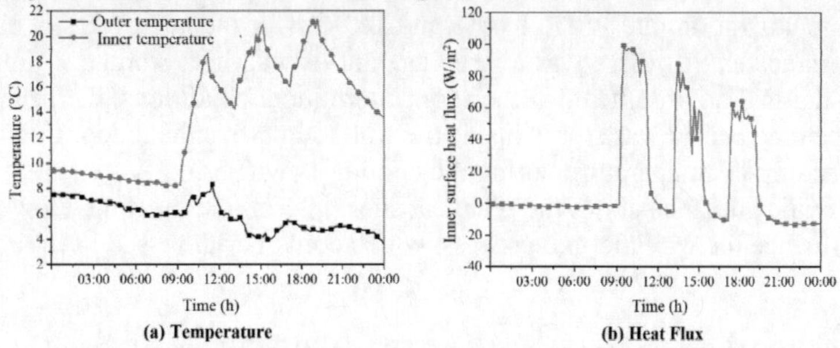

Figure 6. The variations of surface temperatures and inner surface heat flux of the composite-PCW in case 2.

4. The heat storage and release processes

In the whole cycle of intermittent heating (τ_0-τ_2), the wall begins store heat while the indoor air temperature rises after heating at τ_0. And the wall releases heat while the indoor air temperature decreases after stopping heating at τ_1, and the heat release process do not end until the end of the intermittent cycle at τ_2. The heat storage utilization rate η, the ratio of the effective heat release of the wall in the cooling-down period to the effective heat storage in the heating period, is given in Equation (1).

$$\eta = \frac{\int_{\tau_1}^{\tau_2} h_i (T_i - T_{in}) d\tau}{\int_{\tau_0}^{\tau_1} h_i (T_{in} - T_i) d\tau} \qquad (1)$$

Where, h_i is inside convective heat transfer coefficient, W/(m·°C); T_{in} are indoor air temperature, °C; T_i are inner surface temperatures, °C.

Table 2 presents the heat storage utilization of the composite-PCW in the last cycle of the two cases. For case 1, the heat transferred into the wall inner surface was 209.07 kJ/m^2 during the heating process. The heat lose to indoor environment was only 10.65 kJ/m^2 in the 1 hour cooling-down process, which indicated that there was still a large amount of heat stored in the composite-PCW. The inner surface temperature was higher than 18°C, and the wall had large heat release potential. For case 2, the η in the last operation cycle was 14.86%. The heat release of the composite-PCW in the cooling-down process helped maintain the indoor environment. The heating energy consumption can be reduced by stopping the heating equipment in advance while the indoor thermal comfort is not affected by rationally using the heat storage and release of the composite-PCW.

Table 2. The heat storage utilization of the composite-PCW in the two conditions.

Conditions	Heat storage (kJ/m^2)	Heat release (kJ/m^2)	η
Case1-cycle 5	209.07	10.65	5.09%
Case2- cycle 3	363.90	54.09	14.86%

5. Conclusion

This paper experimentally studied the dynamic thermal performance of building envelope. The following conclusions have obtained:

1) The indoor air temperature changed greatly during the short intermittent heating conditions. With the intermittent heating operation, the heat storage of the room increased, and the dynamic thermal response of the indoor air accelerated. The indoor air temperature fluctuations were around 14°C and 18°C respectively for case 1 and case 2 when the indoor thermal processes reached relatively stable

2) For case1, the wall thermal process became relatively stable in the last three cycles, and the inner surface temperature of the composite-PCW was always higher than 16°C. For case 2, the wall heat storage in the 2 hours heating remained the inner surface temperature relatively stable. The inner surface temperature fluctuations of composite-PCW were 5°C and 6°C respectively for case 1 and case 2。

3) The heat storage utilization rates were 5.09% and 14.86% for the last cycles of case 1 and case 2 respectively, which indicates that there was a large amount of wall heat storage which can be used to improve the indoor thermal environment. In the short intermittent conditions, the heat transferred into the wall during the heating period was stored inside the wall to improve the indoor thermal environment by maintain the inner surface temperature during the cooling-down period

This work can provide guidance about the practical short intermittent heating. However, the heating thermostat could influence the wall thermal process, which could be addressed in future research.

6. References

[1] Peeters L, Van der Veken J, Hens H, Helsen L, D'haeseleer W 2008 Control of heating systems in residential buildings: current practice *J. Energy and Buildings* 40 pp 1446-55

[2] P.T. Tsilingiris 2006 Wall heat loss from intermittently conditioned spaces – the dynamic influence of structural and operational parameters *J. Energy and Buildings* 38 pp 1022-1031

[3] G.Barrios, G.Huelsz, J.Rojas 2012 Thermal performance of envelope wall/roofs of intermittentair-conditioned rooms *J. Applied Thermal Engineering* 40 pp 1-7

[4] Dengjia Wang, Yanfeng Liu, Yingying Wang, Jiaping Liu 2014 Numerical and experimental analysis of floor heat storage and release during an intermittent in-slab floor heating process *J. Applied Thermal Engineering* 62 pp 398-406

[5] Zhe Wang, Borong Lin, Yingxin Zhu 2015 Modeling and measurement study on an intermittent heating system of a residence in Cambridgeshire *J. Building and Environment* 92 pp 380-386

[6] Ogoli David Mwale 2003 Predicting indoor temperatures in closed buildings with high thermal mass J. Energy and buildings 35 pp 851-862

[7] Cheng V., Ng E., Givoni B 2005 Effect of envelope color and thermal mass on indoor temperatures in hot humid climate *J. Solar Energy* 2005 78 pp 528-534

[8] Krupa I, Nógellová Z, Špitalsky′ Z, et al. 2014 Phase change materials based on high density polyethylene filled with microencapsulated paraffin wax *J. Energy Convers Manage* 87 pp 400-409

[9] Y. Dutil, D.R. Rousse, N.B. Salah, et al. 2011 A review on phase-change materials: mathematical modeling and simulations *J. Renewable and Sustainable Energy Reviews* 1 pp 112–130

[10] A.M. Khudhair, M.M. Farid 2005 A review on energy conservation in building applications with thermal storage by latent heat using phase change materials *J. Energy Conversion and Management* 2(45) pp 263–275

[11] Yinping Zhang, Kunping Lin, Yi Jiang, Guobing Zhou 2008 Thermal storage and nonlinear heat-transfer characteristics of PCM wallboard *J. Energy and Buildings* 40 pp 1771-1779

[12] A.Castell, I Martorell, M. Medrano, G. Perez, L.F. Cabeza 2010 Experimental study of using PCM in brick constructive solutions for passive cooling *J. Energy and Buildings* 42 pp 534-540

[13] Xu Wang, Hang Yu, Lu Li, Mei Zhao 2016 Experimental assessment on the use of phase change materials (PCMs)-bricks in the exterior wall of a full-scale room *J. Energy Conversion and Management* 120 pp 81-89

[14] Yanru Li, Mengwei Wang, Yin Zhang and Enshen Long 2018 The dynamic thermal process of indoor environment and building envelope during intermittent heating *J. Indoor and built environment*

Acknowledgments

This project is funded by the National Key R&D Program of China (2016YFC0700400), and the National Natural Science Foundation of China (51478280 and 51778382).

IOP Conf. Series: Materials Science and Engineering **556** (2019) 012030 doi:10.1088/1757-899X/556/1/012030

Photocatalytic activity enhanced by photo-thermal conversion with recyclable hollow Fe_3O_4@TiO_2 nanoparticles

Lei Shi[1], Yurong He[1*]

1 School of Energy Science and Engineering, Harbin Institute of Technology, Harbin, China, 1500014

E-mail: rong@hit.edu.cn

Abstract. Photocatalytic nanoparticles have been applied in the utilisation of solar energy because of their excellent harvesting of solar energy for wastewater treatment and chemical synthesis. In particular, TiO_2 nanoparticles exhibit unique photo-catalytic performance due to their high specific surface areas and distinctive optical properties. However, the presence of nanoparticles in solar energy systems can cause secondary pollution and severely limit large-scale application of the nanoparticles in solar technology. In the present work, recoverable hollow Fe_3O_4@TiO_2 nanoparticles were synthesized via a modified hydrothermal method. The prepared nanoparticles exhibit superior photocatalytic activity for degrading organic pollutants (Rhodamine B), which is 33.1% higher than that of P25, and the effect of photo-thermal conversion in enhancing the photo-catalytic activity of this nanoparticles was studied, which has not been investigated in most of previous reports. This study not only reveals that photo-thermal conversion plays an important role in enhancing photocatalytic activity, but also provides an approach to significantly reduce material consumption in the design of solar devices and facilitate the efficient utilization of solar energy applications.

Keywords: photo-thermal conversion; Fe_3O_4@TiO_2 nanoparticles; purification; magnetic separation

1. Introduction

Renewable energy resources are critical to the development of a globally sustainable society [1]. The development of sustainable energy technologies, especially solar energy, has become extremely important for securing future energy sources [2]. Solar energy is a sustainable and inexhaustible source of energy which has the potential to replace fossil fuels. Furthermore, there is significant potential for solar energy to be converted into other forms of energy, such as electric, thermal, and chemical energy [3]. Among them, photovoltaics is a promising commercial technology which may allow for the generation of electrical power for a wide range of applications [4]. Photo-thermal systems have potential in a range of industrial applications, such as vapour generation and seawater desalination. Besides, photo-thermal conversion has also been employed to selectively kill cancer cells in medical sciences [5]. Photochemical conversion has significant promising for photocatalysis, which is essential for a wide range of applications [6].

For the photochemical utilization of solar energy, the photocatalytic activities of nanoparticles (NPs) are largely dependent on their photo absorption and carrier-generation properties, which are usually affected by the materials and their structures [7]. In addition to photochemical processes, researchers have recently found that nanofluids, dispersions containing nanoparticles, exhibit excellent photo-thermal conversion properties in direct-absorption solar collectors (DASCs) [8]. In this application, carbon nano-materials, metallic oxides, and metallic nanoparticles were used to enhance solar light absorption and photothermal conversion. Enormous aforementioned literature has illustrated that TiO_2 exhibits excellent catalytic properties for water purification under the solar illumination [9,10]. This photochemical conversion process was accompanied by a photo-thermal conversion process which has an effect on photochemical conversion [11,12]. A number of researchers focused on the synthesis of nano-materials with excellent light absorption properties to improve the photo-thermal conversion efficiency of solar devices. Jin et al. [13] investigated the photo-thermal conversion mechanism of gold NPs-based solar volumetric receivers. Ni et al. [14] have reported a photo-thermal platform that has the potential to be scalable for a wide range of solar-based applications such as power generation, distillation, and sterilisation.

In this work, we reported the preparation of hollow Fe_3O_4@TiO_2 nanoparticles via a modified hydrothermal method, which combine the properties of TiO_2 and Fe_3O_4 to provide a material that facilitates recyclable purification and separation. Initially, the Fe_3O_4@TiO_2 NPs were characterised by SEM, TEM and magnetic property analysis. Subsequently, the photo-thermal conversion combined photocatalytic experiments were carried out to study the degradation performance compared to P25. We observed the photocatalytic performance of Rhodamine B (RhB) under visible light to investigate the photocatalytic disinfection activity of the Fe_3O_4@TiO_2 NPs.

2. Materials and Methods

2.1. Chemicals

Iron(III) chloride hexahydrate ($FeCl_3$•$6H_2O$), Ascorbic acid, Polyacrylamide (PAM), Urea (AR,99%), Titanium(IV) Fluoride and $C_6H_5Na_3O_7$•$2H_2O$ (Sodium citrate) were purchased from Aladdin Reagent (Shanghai, China). All the reagents were of the analytical reagent grade and were used as received. Deionized water was used through all the processes.

2.2. Synthesis of Fe_3O_4@TiO_2 NPs

1Mmol $FeCl_3$•$6H_2O$ was stirred vigorously in 80ml deionized water for 10 min. Then, the suspension was mixed with 2Mmol $C_6H_5Na_3O_7$•$2H_2O$ and 3Mmol urea. 0.8g PAM was added into the suspension and stirred vigorously 30 min. Then the mixed solution was transferred and sealed in a Teflon-sealed autoclave. The autoclave was kept at 200 °C for 12 h before being cooled in air naturally. Then the black precipitates Fe_3O_4 nanoparticles were washed with water and ethanol by magnetic attraction before oven drying at 50 °C for 12 h. In our synthesis system, citrate is used as a reducing reagent to provide the reducing atmospheres for the synthesis system as well as a coordination ligand to reduce the concentration of free iron ions in aqueous solution. The Fe_3O_4 nanoparticles were slowly mixed in the 8ml TiF_4 solution with the concentration of 0.04M and was stirred vigorously for 10 min. Later, the solution was transferred and sealed in a Teflon-sealed autoclave and add 70ml deionized water. The autoclave was kept at 180 °C for 48 h before being cooled in air naturally. Finally, the final Fe_3O_4@TiO_2 nanoparticles were separated from the reaction medium by magnetic separation.

2.3. Characterisation

Scanning electron microscopy (SEM) analysis was performed using a SUPRA 55 SAPPHIRE SEM (ZEISS, Germany) at an accelerating voltage of 20 kV. Transmission electron microscopy (TEM) patterns of the $Fe_3O_4@TiO_2$ nanoparticles were obtained using a field emission microscope (JEM-2010, Japan).

2.4. Experimental procedure

The experiments were conducted at room temperature (298 K) with the humidity of 25%. The experimental setup for the photocatalytic activity determination with photo-thermal conversion process (Fig. 1). The main experimental components of the device contain a cylindrical receiver (acrylic beaker), a solar simulator (CEL-NP2000, Beijing AuLight Ltd. Co., China) and an electric scale (Practum313-1CN, Sartorius, Göttingen, Germany). Meanwhile, the photocatalytic activities of the hollow $Fe_3O_4@TiO_2$ NPs were evaluated by the degradation of RhB in a suspension during 20 min.

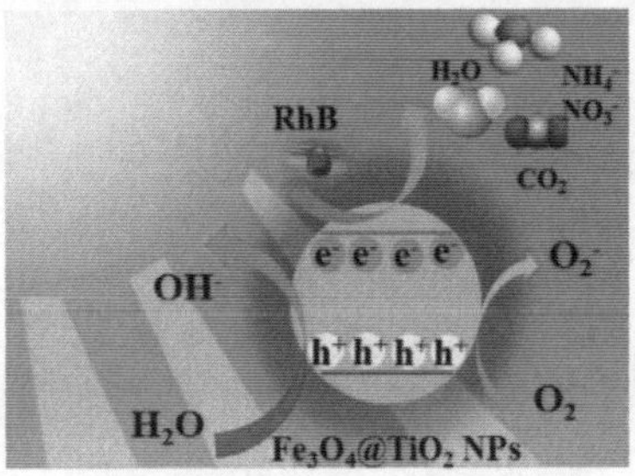

Fig. 1. Mechanism for photocatalytic activity enhanced by photo-thermal conversion.

3. Results and Discussion

3.1. Characterisation

The SEM image (Fig. 2a) and TEM (Fig. 2b) image showed that the hollow $Fe_3O_4@TiO_2$ NPs are of uniform size and morphology. The XRD pattern of Fe_3O_4 showed the characteristic peaks of magnetite (Fig. 2c), thus confirming its chemical composition. In the range of $20° < 2\theta < 80°$, the diffraction peak at 25.5°, 38.3° and 55.4° are assigned to the (1 0 1), (0 0 4) and (2 0 0) reflections of TiO2 (JCPDS No. 21-1272). The other diffraction peaks at the 2θ values of 30.5°, 35.4°, 42.8°, 53.6°, 57.2° and 62.5°, correspond well to the (2 2 0), (3 1 1), (4 0 0), (4 2 2), (5 1 1), and (4 4 0) reflections of the standard cubic phase of Fe_3O_4 (JCPDS No. 89-0691).

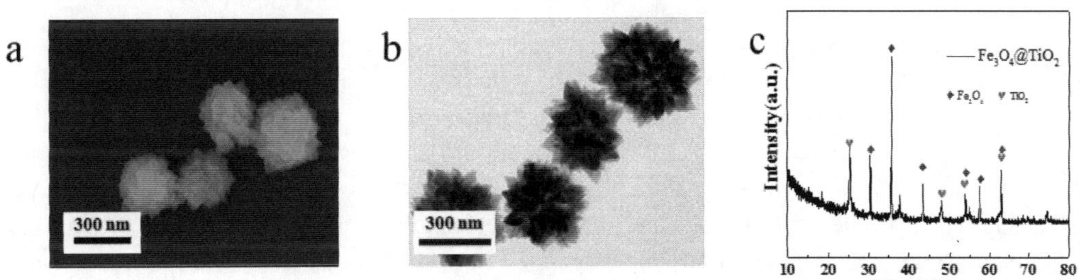

Fig.2. Morphology of the prepared hollow $Fe_3O_4@TiO_2$ nanoparticles: (a) SEM images; (b) TEM images; (c) XRD pattern of the prepared hollow $Fe_3O_4@TiO_2$ nanoparticles

3.2. Photocatalytic activity of the hollow Fe_3O_4@TiO_2 NPs

To investigate the photocatalytic properties of the prepared hollow Fe_3O_4@TiO_2 nanoparticles, hollow Fe_3O_4@TiO_2 nanoparticles was used to catalyze the degradation of RhB under simulated solar light compared to P25. Fig. 3a showed the UV-vis absorption spectra of the RhB solution after different solar-light illumination times, the intensity of the initial absorption spectrum decreases gradually with solar-light illumination duration. Following the Beer-Lambert law, the concentration of RhB is directly proportional to the intensity of the characteristic peak [15]. Hence intensity changes of the absorption peak were used to evaluate catalytic activity. Furthermore, the RhB degradation dynamics was expressed by plotting (TRC_f/TRC_0) as a function of time, as shown in Fig. 3b, the concentration of RhB decreases quickly as solar energy is converted into free carriers which promote the fast degradation of RhB. The photocatalysis kinetics of a semiconductor can be analyzed by the Langmuir-Hinshelwood first-order-reaction model [16]:

$$\ln(TRC_0/TRC_f) = kt \tag{1}$$

where TRC_0 is the initial concentration of RhB and TRC_f is the final concentration of RhB of the degraded RhB solution, k is the photo-degradation rate constant, and t is the solar irradiation time. Equation (4) affords the photo-degradation rate constant k through a plot of $\ln(C_0/C_t)$ versus solar irradiation time t. Fig. 3c illustrates the dependency of $\ln(C_0/C_t)$ on time, and the slope of the best fit line affords k. The total RhB concentration (TRC) of the degraded solution after 20 min of irradiation was calculated according to the Beer-Lambert law. The efficiency of RhB degradation (E_d) was calculated Equation:

$$E_d = \frac{TRC_0 - TRC_f}{TRC_0} \times 100\% \tag{2}$$

The reduction in the concentration of RhB is negligible under sun light radiation without photocatalyst.

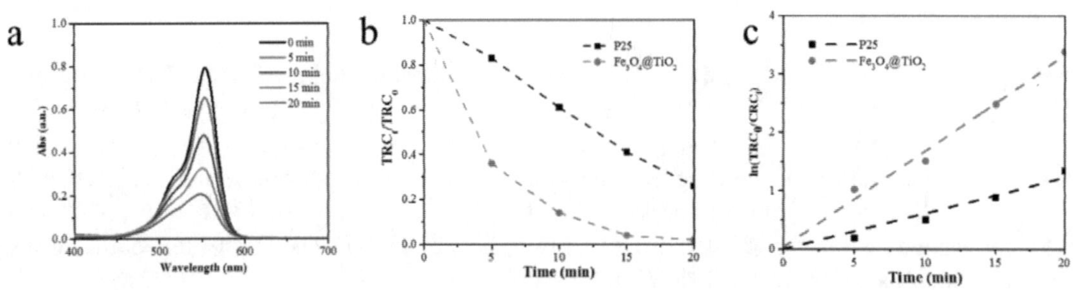

Fig.3. Photocatalytic activity of the hollow Fe_3O_4@TiO_2 nanoparticles: (a) Time dependence of the UV-vis absorption spectrum of the RhB suspension; (b) plot of TRC_f/TRC_0 vs. time; and (c) plot of $\ln(TRC_0/TRC_f)$ vs. irradiation time.

3.3. The effect of photothermal conversion on photocatalytic activity

Making the most efficient use of solar light is the main purpose of the research in photocatalysis. The solar power intensity has a significant effect on the temperature distribution in suspension, so varying the solar power could help to investigate the effect of the solar illumination intensity on the photo-thermal conversion efficiency of the device. In this study, direct solar absorption enabled by Fe_3O_4@TiO_2 nanofluids was experimentally investigated. Photo-thermal conversion experiments were conducted at a constant Fe_3O_4@TiO_2 concentration and the temperature rise of the suspension with the solar illumination intensity increased from 1 to 5 suns (1, 3, and 5 suns) were showed in Fig. 4a. Raising the solar power intensity cause an increase in the bulk temperature of the suspension, and the unconspicuous temperature gradient indicated that the penetration depth of sunlight in the suspension was more obvious, which may results in a temperature gradient in the suspension. The photo-degradation rate constant of hollow $Fe_3O_4@TiO_2$ nanoparticles under different solar illumination intensity was showed in Fig. 4b. The photo-degradation line have different slope under different solar illumination intensity and the slope is increasing with the increase of solar illumination. All above states indicated that the effect of photothermal conversion enhance the photocatalytic activity.

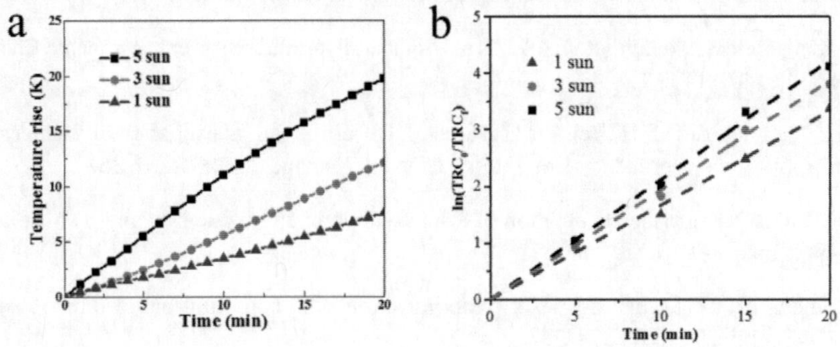

Fig.4. (a) Temperature rise in bulk fluid and (b) photocatalytic performance at different solar intensities (1, 3, and 5 sun)

4. Conclusion

In this study, we presented a convenient and efficient approach to synthesise hollow $Fe_3O_4@TiO_2$ NPs. The prepared nanoparticles exhibit superior photocatalytic activity for degrading organic pollutants (Rhodamine B), which was 33.1% higher than that of P25, and the effect of photo-thermal conversion in enhancing the photo-catalytic activity of this nanoparticles was studied, which has not been investigated in most of previous reports. This study reveals that photo-thermal conversion plays an important role in enhancing photocatalytic activity, moreover, it provides an approach to significantly reduce material consumption in the design of solar devices and facilitate the efficient utilization of solar energy applications.

Acknowledgements

This work is financially supported by the National Natural Science Foundation of China (Grant No. 51676060), the Natural Science Founds of Heilongjiang Province for Distinguished Young Scholars (Grant No. JC2016009), and the Science Creative Foundation for Distinguished Young Scholars in Harbin (Grant No. 2014RFYXJ004).

References

[1] Liu G, Zhou B, Liao S. Inverting methods for thermal reservoir evaluation of enhanced geothermal system. Renew Sustain Energy Rev 2018;82:471-6.

[2] Li Y, Chen H, Zhang X, Tan C, Ding Y. Renewable energy carriers: Hydrogen or liquid air/nitrogen? Appl Therm Eng 2010;30:1985-1900.

[3] Tong H, Ouyang S, Bi Y, Umezawa N, Oshikiri M, Ye J. Nano-photocatalytic materials: Possibilities and challenges. Adv Mater 2012;24:229-51.

[4] Bozkurt I, Karakilcik M. The effect of sunny area ratios on the thermal performance of solar ponds. Energy Convers Manage 2015;91:323-332.

[5] Wu W, Changzhong Jiang CJ, Roy VAL. Recent progress in magnetic iron oxide–semiconductor composite nanomaterials as promising photocatalysts. Nanoscale 2015;7:38-58.

[6] Chaudhary P, Rizwan M. Energy management supporting high penetration of solar photovoltaic generation for smart grid using solar forecasts and pumped hydro storage system. Renew Energy 2017;118:928-946.

[7] Keniar K, Ghali K, Ghaddar N. Study of solar regenerated membrane desiccant system to control humidity and decrease energy consumption in office spaces. Appl Energy 2015;138:121-32.

[8] Nafey A S, Abdelkader M, Abdelmotalip A, et al. Solar still productivity enhancement. Energy Convers Manage 2001;42(11):1401-1408.

[9] Bandarra Filho E P, Mendoza O S H, Beicker C L L, et al. Experimental investigation of a silver nanoparticle-based direct absorption solar thermal system. Energy Convers Manage 2014;84:261-267.

[10] Ma T, Yang H, Lu L. Performance evaluation of a stand-alone photovoltaic system on an isolated island in Hong Kong. Appl Energy 2013;112:663-672.

[11] Gorji T B, Ranjbar A A. Thermal and exergy optimization of a nanofluid-based direct absorption solar collector. Renew Energy 2017;106:274-287.

[12] Mohammad A, Alibakhsh K, Abtin A. Optimization of a combined solar chimney for desalination and power generation. Energy Convers Manage 2017;150:72-80.

[13] Jin H, Lin G, Bai L, et al. Steam generation in a nanoparticle-based solar receiver[J]. Nano Energy, 2016, 28: 397-406.

[14] Ni G, Miljkovic N, Ghasemi H, et al. Volumetric solar heating of nanofluids for direct vapor generation[J]. Nano Energy, 2015, 17: 290-301.

[15] Chehadi Z, Alkees N, Bruyant A, Toufailyb J, Girardonc J S, Capron M, et al. Plasmonic enhanced photocatalytic activity of semiconductors for the degradation of organic pollutants under visible light. Mater Sci Semicond Process 2016;42:81-84.

[16] Shi L, He Y, Huang Y, Jiang B. Recyclable Fe_3O_4@CNT nanoparticles for high-efficiency solar vapor generation. Energy Convers Manage 2017;149:401-408

Preliminary analysis of supercritical CO_2 Brayton cycle applied to the solar thermal plant

L Sun[1], YH Xie[1,*], D Zhang[2]
[1] Shaanxi Engineering Laboratory of Turbomachinery and Power Equipment, School of Energy and Power Engineering, Xi'an Jiaotong University, Xi'an, China
[2] MOE Key Laboratory of Thermo-Fluid Science and Engineering, School of Energy and Power Engineering, Xi'an Jiaotong University, Xi'an, China
E-mail: yhxie@mail.xjtu.edu.cn

Abstract. The use of a highly compacted and efficient supercritical CO_2 (S-CO_2) Brayton cycle has become an effective way in utilizing solar energy. This work performs a comprehensive parametric study for S-CO_2 Brayton cycle applied to the solar thermal plant. Several different power cycle layouts including basic regenerative cycle, recompression cycle, intercool cycle and reheat cycle are analysed by the computational models consist of various components in this paper. Then, the optimization of parameters including the cycle turbine inlet temperature and so on, is also performed based on genetic algorithm to achieve the best cycle performance. The most promising cycle configuration is selected by comparison of the compromise between cycle and recuperator performance. This study provides some reference for the related research of the S-CO_2 Brayton cycle integrated with solar thermal systems.

Keywords: S-CO2 Brayton cycle; genetic algorithm; power output

1. Introduction

Due to fossil fuel depletion, environmental and climate threats, Concentrating Solar Power (CSP) is playing a leading role in energy scenario which can providing carbon-free, renewable and dispatchable electric energy. more and more ways are used to convert solar heat into electricity to reduce the Levelized Cost of Electricity (LCOE) of this technology. Supercritical carbon dioxide (S-CO_2) cycle has been identified as a promising technology to utilize solar energy [1,2].

Supercritical fluid is a state between the gas state and the liquid state, with low compressibility near the liquid state and high viscosity near the gas state [3]. The critical temperature of S-CO_2 is 31.1°C, and the critical pressure is 7.38MPa [4]. The research on the application of S-CO_2 in Brayton cycle has been widely concerned and promoted, including system design and core components analysis [5-9].

Most of the above studies have considered so far sCO2 for nuclear power plants, while there is a growing interest in deploying them in CSP applications [10–12]. For example, S-CO_2 Brayton cycles have been compared with trans and sub-critical working conditions of the same fluid [13]. S-CO_2 cycles in CSP plants have been considered in indirect configuration [14]. The main challenges of S-CO_2 application, especially coupled to CSP plants which are heat exchanger design and material selection have been preliminarily studied [15].

However, most studies only compare fixed parameters when comparing thermodynamic cycle performance. It must be clear that the optimization parameters of each thermodynamic system are different after optimization, and the corresponding contrast results will also be different.

In this paper, several different power cycle layouts including basic regenerative cycle, recompression cycle, intercool cycle and reheat cycle are analysed by the computational models. Meanwhile, genetic algorithm is used to optimize the parameters of the cycles and a more impartial comparison is obtained.

2. Calculation model

The S-CO_2 Brayton cycle mainly consists of three kinds of equipment: turbines, compressors, and heat exchangers and the mathematical models are given as follows.

2.1. Turbine and compressor subroutine

The evaluation of the compression or expansion process starts from the machine inlet conditions that were specified in the input. The calculation procedure is the following:

$$\left(h_{c\_out} - h_{c\_in}\right) = \left(h_{c\_out\_id} - h_{c\_in}\right)/\eta_c \tag{1}$$

$$\left(h_{t\_out} - h_{t\_in}\right) = \left(h_{t\_out\_id} - h_{t\_in}\right)\times\eta_c \tag{2}$$

h, the enthalpy; η, the total to total efficiency; *c*, the compressor; *t*, the turbine; *in*, the inlet conditions; *out*, the outlet conditions; *id*, the ideal state.

Given the total to total efficiency and inlet conditions of the turbine and compressor work, the rest of the state points can be determined.

2.2. Recuperator subroutine

In this paper, printed circuit heat exchangers (PCHE) are utilized to calculate the recuperator. In order to improve the effectiveness of recuperator and take into account the minimum temperature difference of recuperator, the minimum temperature difference of recuperator r between the hot and cold side is set as 10°C. The overall heat transfer coefficient is calculated from [1]:

$$U_i = \left(\frac{1}{U_{i\_h}} + \frac{k}{t} + \frac{1}{U_{i\_c}}\right)^{-1} \tag{3}$$

U_{i_h} and U_{i_c} are the heat transfer coefficients on hot and cold sides respectively which are similarly to calculate. The parameter k is the thermal conductivity of the heat exchanger material and t is the plate thickness. U_{i_h} is calculated from:

$$U_{i\_h} = \left(\frac{\frac{f_h}{8}\square(\mathrm{Re}_h - 1000)\square\mathrm{Pr}_h}{1 + 12.7(\mathrm{Pr}_h^{2/3} - 1)\square\sqrt{\frac{f_h}{8}}\square(1 + \frac{PD}{L})^{2/3}}\right)\square\frac{L}{PD} \tag{4}$$

Where:

$$f_h = \left(\frac{1}{1.81\mathrm{g}(\mathrm{Re}_h) - 1.5}\right)^2 \tag{5}$$

$$PD = \frac{4\pi D^2/8}{\pi D/2 + D} \tag{6}$$

In the heat, the correlations above are used again to calculate the parameters of heat. Then, the required heat exchanger area can be calculated by the parameter D and L.

3. A brief introduction to the circulatory system

The circulatory system studied in this paper is shown in Figure 1. The basic regenerative cycle (Fig. 1, cycle (a)) comprises of recuperator, turbine, compressor, cooler and heater. The S-CO_2 after being cooled is compressed to an applicable pressure. Then, the cold S-CO_2 exchanges heat with the hot S-CO_2 exiting the turbine. The temperature of the pre-heated S-CO_2 is further elevated by the solar receiver before being expanded in the turbine.

IOP Conf. Series: Materials Science and Engineering **556** (2019) 012031 doi:10.1088/1757-899X/556/1/012031

The recompression cycle (Fig. 1, cycle (b)) has the same components as in the basic regenerative cycle but with different numbers and arrangements. Before the cooler, the S-CO_2 stream is splitted into two streams, one goes to the compressor directly while the other goes through the cooler before entering the compressor. Other arrangements are similarly to the basic regenerative cycle.
The intercool cycle (Fig. 1, cycle (c)) utilizes a second cooler which helps to reduce the compression work by reducing the temperature of the compressed S-CO_2. Besides, the arrangements can make a further bigger pressure ratio in the cycle. The reheat cycle (Fig. 1, cycle (d)) utilizes a second heater which helps to enhance the turbine work by adding heat after the expansion.

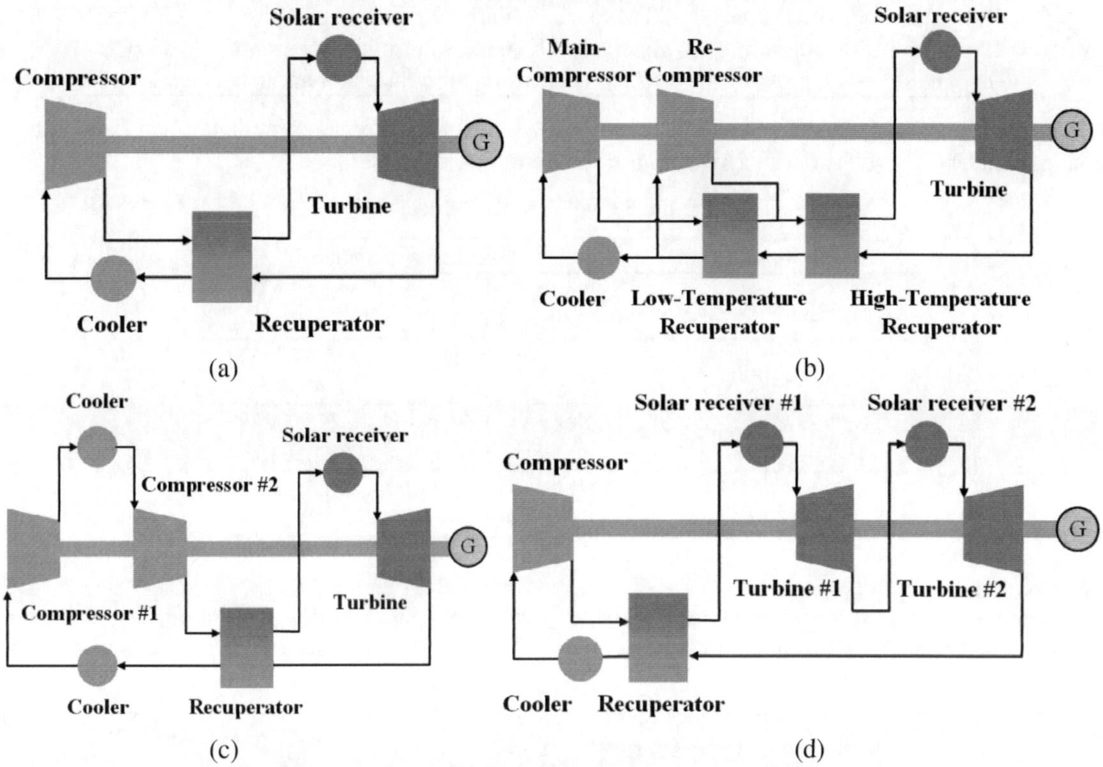

Figure 1. Supercritical CO_2 compound cycle layout

To study the above circulatory system a calculation model is built which including the main part, the compressor subroutine, the turbine subroutine and the heat exchanger subroutine. Carbon dioxide properties are interpolated directly from the existing physical data points. some assumptions are made to simplify the models:

1、All the circulatory system can reach a steady state in a short time.
2、The heat loss and the pressure drop in the circulatory system are neglected.
3、The CO_2 temperature at the outlet of the cooler can keep constant by changing the mass flow rate of the environmental water or air.

The mapping relation of the cost function from the input to the output can be obtained after the mathematical models above are built. Meanwhile, the efficiency of the circulatory system is taken as the output of the cost function, namely the optimization target. Table 1 summarizes the main design assumptions.

Table 1 Main assumptions for the power cycle simulation

fixed parameters	value
Solar thermal available	16 MW
Turbine isentropic efficiency	85%
Compressor isentropic efficiency	80%
Regenerator end temperature difference	10 °C

4. Results and discussion

4.1. Initial state and cycle results

After the establishment and construction of the equation above, the initial state parameters, which are not optimized, are calculated first and shown in Table 2.

Table 2. Parameter value table

Term	Value	Term	Value
Cycle maximum pressure	20 MPa	Turbine inlet temperature	600 °C
Cycle minimum pressure	8 MPa	Compressor inlet temperature	35 °C
Cycle intermediate pressure	12.62 MPa (utilized in intercool and reheat cycle)		

According to the fixed variables mentioned above, we get the system parameters that only use the initial state parameters before the optimization, as are shown in Table 3.

Table 3 Cycle results table

Circulatory system	Cycle performance	
basic regenerative cycle	Net power output	5.57 MW
	Exergy efficiency	37.16%
recompression cycle	Net power output	6.54 MW
	Exergy efficiency	43.64%
intercool cycle	Net power output	6.30 MW
	Exergy efficiency	42.01%
reheat cycle	Net power output	5.74 MW
	Exergy efficiency	38.23%

4.2. Optimization results

The genetic algorithm (GA) is used to optimize the circulatory system, with a population size of 50 and an iteration number of 50. As a result, the evolution curve of the best fitness value obtained by GA algorithm is shown in Figure 2.

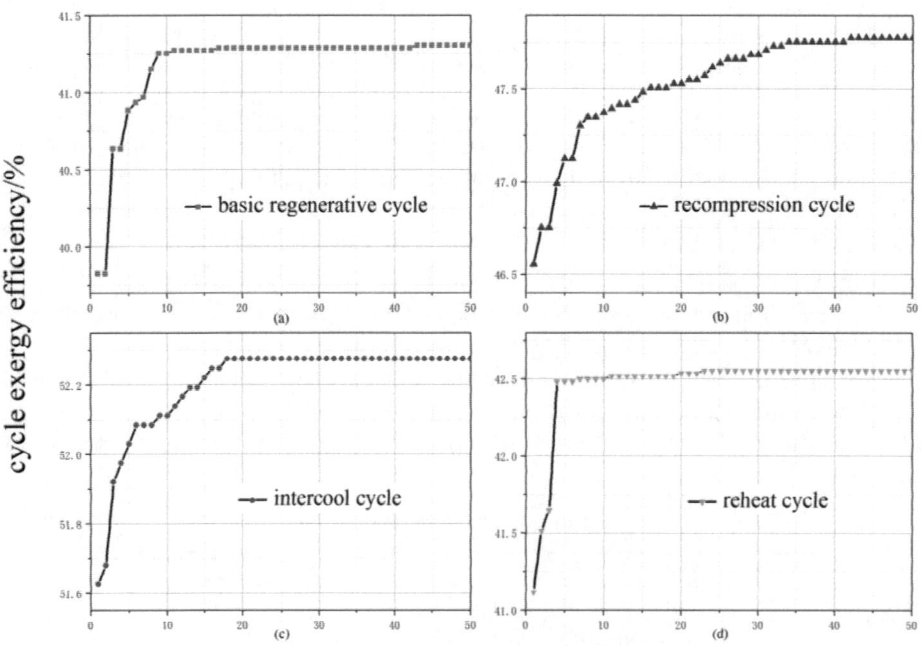

Figure 2 iterative convergence curve

The cycle efficiency of circulatory systems has reached 41.31%, 47.78%, 52.27% and 42.55% respectively. It can be found that the cycle efficiency of the circulatory system has been effectively improved. After the optimization, the cycle efficiency of the intercool cycle exceeds the recompression system, which indicates that it is necessary to make a certain optimization and then compare the cycle in the process of selecting the circulatory system. What's more, the change of the optimized input parameters is shown in Table 4.

Table 4 Optimization parameters table

Term and optimization interval	Cycle(a)	Cycle(b)	Cycle(c)	Cycle(d)
Cycle maximum pressure / (15,25)	24.95MPa	24.91 MPa	24.82 MPa	24.96 MPa
Cycle minimum pressure / (7.5,10)	7.64 MPa	7.87 MPa	9.42 MPa	7.61 MPa
Compressor Inlet temperature / (32,40)	32.37 °C	32.21 °C	32.07 °C	32.42 °C
Cycle efficiency after optimization	41.31%	47.78%	52.27%	42.55%

Then we calculate the required exchanging area of the circulatory system and the comparison of system efficiency and the required exchanging area among the above cycles is shown in Figure 3. It can be found that the demand for heat exchanging area in the recuperators increases while some structures have been added to the regenerative cycle layout. Among the above efficient cycles (a, b, c and d), the recompressed cycle (b) stands-out to be the excellent choice because it can achieve high thermal efficiency under acceptable demand for heat exchanging area. Obviously, the recompression cycle can utilize advantages in previous cycles by combining both intercool and reheat within the allowable cost range.

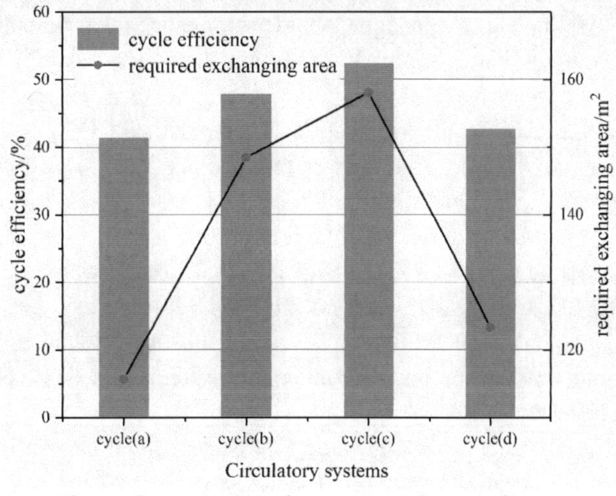

Figure 3 comparison between cycles graph

5. Conclusions

In this paper, several different power cycle layouts including basic regenerative cycle, recompression cycle, intercool cycle and reheat cycle are analysed by the computational models and compared after optimization using GA algorithm. Conclusions are as follows:

1. The cycle efficiency of the circulatory systems has been effectively improved (Efficiency increased by more than 4%) after the optimization which shows that the optimization algorithm used in this paper is effective.
2. It can be found that the closer the inlet temperature of compressor is, the higher the cycle efficiency is, and the best circulating pressure ratio may exist in some thermodynamic systems through the system optimization.
3. The demand for heat exchanging area in the recuperators increases while some structures have been added to the regenerative cycle layout because of the increase in the complexity of the system. Therefore, it is necessary to consider the relationship between system efficiency and

cost in actual selection.

4. After calculation, it can be found that the recompression cycle is the most suitable thermal system in the process of solar energy utilization and adding some structure on this can still bring about the improvement of system efficiency. In the actual process, we should consider the power level and cost limitation of solar power and rationally choose the thermal cycle structure to achieve better economic returns.

In conclusion, the recompression cycle is a suitable thermodynamic cycle for solar energy utilization, and it is feasible to increase part of the structure to improve cycle efficiency. This research can help the development of residual heat utilization and other engineering fields.

6. References

[1] Dostal, Vaclav. A supercritical carbon dioxide cycle for next generation nuclear reactors[J]. Massachusetts Institute of Technology, 2004, 154(3):265-282.

[2] Roux W G L, Bello-Ochende T, Meyer J P. A review on the thermodynamic optimisation and modelling of the solar thermal Brayton cycle[J]. Renewable & Sustainable Energy Reviews, 2013, 28(8):677-690.

[3] Hiroaki T, Niichi N, Masaru H, et al. Forced convection heat transfer to fluid near critical point flowing in circular tube[J]. International Journal of Heat & Mass Transfer, 1971, 14(6):739-750.

[4] Gil L, Otín S F, Embid J M, et al. Experimental setup to measure critical properties of pure and binary mixtures and their densities at different pressures and temperatures : Determination of the precision and uncertainty in the results[J]. Journal of Supercritical Fluids, 2008, 44(2):123-138.

[5] Dyreby J, Klein S, Nellis G, et al. Design Considerations for Supercritical Carbon Dioxide Brayton Cycles With Recompression[J]. Journal of Engineering for Gas Turbines & Power, 2014, 136(10):101701.

[6] Muto Y, Ishiyama S, Kato Y, et al. Application of supercritical CO2 gas turbine for the fossil fired thermal plant[J]. Journal of Energy and Power Engineering,2010, 4(9): 7-15.

[7] Bae S J, Lee J, Ahn Y. Preliminary studies of compact Bryton cycle performance for small modular high temperature gas-cooled reactor system[J]. Annals of Nuclear Energy,2015,75:11-19.

[8] Moullec Y L. Conceptual study of a high efficiency coal-fired power plant with CO2 capture using a supercritical CO2 Brayton cycle[J]. Energy, 2013,49: 32-46.

[9] Jeong W S, Lee J I, Jeong Y H. Potential improvements of supercritical recompression CO2 Brayton cycle by mixing other gases for power conversion system of a SFR. Nuclear Engineering and Design, 2011,241(6): 2128-2137.

[10] Crespi F, Gavagnin G, Sánchez D, Martínez GS. Supercritical carbon dioxide cycles for power generation: a review[J]. Appl Energy 2017; 195:152-83.

[11] Wang K, He Y-L, Zhu H-H. Integration between supercritical CO2 Brayton cycles and molten salt solar power towers: a review and a comprehensive comparison of different cycle layouts[J]. Apply Energy 2017; 195:819-36.

[12] Bauer ML, Vijaykumar R, Lausten M, Stekli J. Pathways to cost competitive concentrated solar power incorporating supercritical carbon dioxide power cycles. In: The 5th international symposium; 2016.

[13] Garg P, Kumar P, Srinivasan K. Supercritical carbon dioxide Brayton cycle for concentrated solar power[J]. Journal of Supercritical Fluids, 2013, 76(4):54-60.

[14] Iverson BD, Conboy TM, Pasch JJ, Kruizenga AM. Supercritical CO2 Brayton cycles for solar-thermal energy[J]. Apply Energy 2013;111:957–70.

[15] Cao G, Firouzdor V, Sridharan K, et al. Corrosion of austenitic alloys in high temperature supercritical carbon dioxide[J]. Corrosion Science, 2012, 60(7):246-255.

Study of Residential Thermal Transfer Values (RTTV) and Vertical Daylight Factor (VDF) for Hong Kong

Ernest K. W. Tsang [1] and Danny H. W Li [2]
[1] Department of Construction Technology and Engineering, Technological and Higher Education Institute of Hong Kong, 20A Tsing Yi Road, New Territories, Hong Kong SAR, China
E-mail: ernest_tsang@vtc.edu.hk
[2] Building Energy Research Group, Department of Architecture and Civil Engineering, City University of Hong Kong, Tat Chee Avenue, Kowloon, Hong Kong SAR, China

Abstract. In Hong Kong, over one-quarter of the total electricity use was consumed by residential buildings. Recently, the Hong Kong Government issued a practice note namely Design and Construction Requirements for Energy Efficiency of Residential Buildings. This practice note set out the Residential Thermal Transfer Values (RTTV) of building envelopes. To enhance energy efficiency of residential buildings, the RTTV of wall ($RTTV_{wall}$) and roof ($RTTV_{roof}$) should not exceed 14 and 4 W/m^2, respectively. Being one of the most densely populated cities in the world, most of building developments in Hong Kong are high-rise blocks located in densely built zones. The shading effects due to surrounding buildings could substantially restrict the diffuse light coming from the sky. In 2003, a practice note on lighting and ventilation requirements was issued. Referring to the performance-based approach, the note adopted the vertical daylight factor (VDF) to specify the daylighting performance of the building. The VDF should not be less than 8% for habitable rooms and 4% for domestic kitchens. This paper presents the energy performance of recently designed building fulfills the RTTV requirements. It found that VDF is a good indicator to shows the lighting elecricty consumption in buildings. For most of the bedroom, about 40% of air conditioning and 3 % lighting energy can be saved.

Keywords: RTTV, VDF, building façade designs

1. Introduction

Currently, buildings in Hong Kong accounted for more than 90% of the total electricity expenditure and over one-quarter of the total electricity use was consumed by residential buildings [1]. In most cities, electricity is mainly used to create thermally and visually comfortable built environments. Air-conditioning and electric lighting represent for the majority of the overall building energy use. In subtropical regions, residential building envelope's cooling load can contribute over one-third to two-thirds of the cooling requirement [2]. The principal objectives of fenestration designs is to limit the heat gain into the buildings. To control the heat transfer from outdoor to interior through the facades of a building, the Hong Kong Government recently issued a practice note namely Design and Construction Requirements for Energy Efficiency of Residential Buildings [3]. This practice note set out the Residential Thermal Transfer Values (RTTV) of building envelopes. To enhance energy efficiency of residential buildings, the RTTV of wall ($RTTV_{wall}$) and roof ($RTTV_{roof}$) should not exceed 14 and 4 W/m^2, respectively. Being one of the most densely populated cities in the world, most of building developments in Hong Kong are high-rise blocks located in densely built zones. The shading effects due to surrounding buildings could substantially restrict the diffuse light coming from the sky particularly for rooms at the lower floors [4]. In 2003, a practice note on lighting and ventilation requirements was issued [5]. Referring to the performance-based approach, the note adopted the vertical daylight factor (VDF) to specify the daylighting performance of the building. The VDF should not be less than 8% for habitable rooms and 4% for domestic kitchens. A high VDF would often mean a large RTTV and vice versa. It seems that there may be a contradiction on building façade designs for achieving the required RTTV and VDF. This paper studies the RTTV and VDF for typical domestic buildings in Hong Kong. The general correlation between RTTV and VDF was determined. Main features of the findings and building façade design implications are discussed.

2. Residential thermal transfer values and vertical daylight factor

2.1. Residential thermal transfer values (RTTV)

Residential buildings account for 27.7 % of total electricity in Hong Kong [6]. As standard of living increases during previous two decades, most of the residential buildings are installed with air-conditioning units. There are strong demands on governing the thermal performance in residential buildings. In 2014, the RTTV was adopted [3]. Basically, building with a lower RTTV uses less energy for space cooling. On the contrary, the higher the RTTV, more energy is used by air-conditioning system.
In most of the residential buildings, equipment is not operated continuously. During mid-season and non-peak operation, natural ventilation is sufficient for maintaining acceptable indoor environment. It is assumed that Air conditioners operate between April and October [3,7]. New residential buildings usually equipped with heat pump to provide both heating and cooling. In subtropical Hong Kong, heating is only required for not more than one month and hence there is no specific considerations for heating season in the RTTV.

2.2. Vertical daylight factor (VDF)

Hong Kong is one of the highest density cities in the world. Vast of high-rise buildings built closely together. Flats located on lower floors may not be able to admit sufficient daylight when facing heavy obstructions. In 2003, Buildings Department established a performance-based approach as an alternative path for buildings cannot fulfil the requirement stated in Building (Planning) Regulations [8] to obtain approval. This approach set minimum requirements of VDF for habitation rooms and kitchens. VDF is the amount of daylight fell on a vertical window compared to the illuminance level of an unobstructed CIE overcast sky. Under the same glazing type and internal layout, rooms with higher VDF have higher indoor illuminance levels.

3. Computer simulation approach

Computer-based building energy simulation has proved useful in assessing different daylighting schemes [10]. In this study, EnergyPlus was used to model the daylighting and energy performances of a residential building. It is capable to analyse a wide range of daylight design features and is able to calculate internally reflected component inside the thermal zones and specular & diffuse reflections from adjunct surfaces.

3.1. Building descriptions

A 31-floor residential tower which was designed and built recently was selected for this study. Windows are facing three major orientations includes southeast, southwest and northwest. The designers prevented to locate window on northeast direction as there is another high-rise hotel tower in this direction within 5m from the site boundary. On southwest direction, there are some old buildings which are relatively low compared to the building being studied. On southwest, there are few tall buildings mixed with some shorter one. They affect the daylight level at lowest floors. For the northwest direction, the obstruction is taller than other directions. Buildings in Hong Kong need to comply with the Building Energy Code (BEC) [10]. The government promote green buildings by granting an additional 10% gross floor area concessions for building completed green building assessment [11]. BEAM Plus [7] is the most commonly used green building assessment tool in Hong Kong. So, most of the new buildings are designed to conform the requirement in BEC and BEAM Plus. Table 1 shows the input parameters and building information while Figure 1 shows the building layout of the residential building.

Table 1. Descriptions of residential building

Building type and storey			31-Storey Residential building above 5-storey commercial podium
Floor area			Total Area: 14,371 m^2 Air-conditioned area: 9,719 m^2 (Living and dining room: 4,634m^2 Bedroom: 5,085 m^2)
Envelop	Opaque surface	U-value	Exterior Wall: 3.93 W/m^{2}K Roof: 0.36 W/m^{2}K
	Glazing	Shading Coefficient	0.57
		Visual transmittance	0.4
		U-value	5.71 W/m^{2}K
			Shading is provided to simulate occupant's behaviours
	Window to wall ratio		0.21
	$RTTV_{wall}$		13.29 W/m^{2}K
	$RTTV_{roof}$		0.98W/m^{2}K
Internal load	Occupancy		Living and dining room: 2 persons per room Bedroom: 2 persons per room
	Lighting		Living and dining room: 15 W/room Bedroom: 13 W/room LED dimmable lighting Design illuminance level: 150lx (living and dining room) 100 lx (bedroom) Living and dining room: 6:00 - 8:00; 13:00 - 14:00; 18:00 - 22:00 Bedroom: 5:00 - 9:00; 13:00 - 17:00; 18:00 - 24:00
	Equipment		Living and dining room: 142 W/room Bedroom: 45 W/room
HVAC	System		Split-type heat pump
	Coefficient of Performance		2.7 (Heating); 2.6(Cooling)
	Indoor Temperature		24°C
	Operation		Living and dining room: 13:00 - 22:00; April - October Bedroom: 0:00 - 7:00; 13:00 - 24:00; April - October

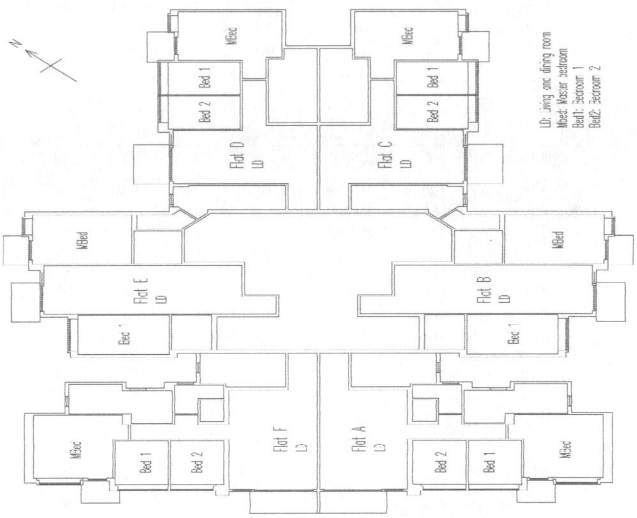

Figure 1. Floor plan for the residential building.

3.2. Weather file

EnergyPlus can conduct hourly and sub-hourly simulation by using 8,760-hour record of measured weather data. Weather varies from a year to another. It is essential to develop a weather database which is typical or representative of a particular geographical area so that different schemes can be readily compared. RTTV [3] adopted the typical mereological year by Chan et al. [12] for performing complex external shading calculation. This weather data was used in this study.

4. Result

The building comprises of 31 floors. Simulations were conducted for every five floors (i.e. 1st, 6th, 11th, 21st, 26th and 31st). The neighbourhood buildings were also included in the simulation model to simulate the shading effect of these surrounded buildings.

4.1. Visual Performance

To understand the energy saving opportunity offered by daylight-linked control system, total fraction of time that daylight can provide sufficient indoor illuminance is important. Table 2 reports the findings. Except master bedrooms and living and dining room, over 80% of daytime, daylight provides sufficient illuminance level. There are two reasons for bedrooms outperform other rooms. First, balconies provided for living and dining room and master bedroom usually block significantly the daylight entering the into the room. Second, it is found that living and dining room and master bedrooms are more sensitive to the floor level. In master bedroom and living and dining room, the differences between the bottom and the top floor may over 15%. Balconies blocks certain among of light from zenith which is the brightest part of sky under overcast sky condition. So, rooms can only admit small amount of light fluxes horizontally and therefore, more sensitive to external obstruction. For the remining rooms, the differences between topmost and lowest floor are around 3% only. The VDFs were calculated for all rooms as shown in Table 3. VDF has a good agreement with the fraction of time daylight exceeds the design illuminance level. In the living and dining room of Flat E, both VDF and the fraction of time daylight exceeded the design illuminance level change significantly with floor level. The daylight level has a constant increase from 1/F to 16/F with a becomes steady between 21/F and 31/F.

Table 2. Percentage of time indoor daylight illuminance exceed daylight level during daytime

		1/F	6/F	11/F	16/F	21/F	26/F	31/F
Flat A	Living and Dining room	67.1%	71.0%	74.7%	77.0%	77.5%	77.7%	77.8%
	Master Bedroom	63.3%	64.7%	68.5%	73.3%	73.4%	74.2%	73.7%
	Bedroom 1	81.6%	83.2%	84.1%	84.3%	83.7%	83.8%	83.8%
	Bedroom 2	80.8%	82.8%	83.9%	84.6%	84.0%	84.1%	84.1%
Flat B	Living and Dining room	74.1%	75.2%	76.4%	77.7%	78.8%	78.5%	77.5%
	Master Bedroom	78.7%	80.4%	80.0%	80.7%	80.7%	80.9%	81.0%
	Bedroom	81.1%	81.1%	81.4%	81.8%	82.2%	82.3%	82.3%
Flat C	Living and Dining room	60.2%	70.2%	72.5%	73.2%	73.5%	73.9%	74.2%
	Master Bedroom	77.3%	79.4%	79.9%	80.7%	79.8%	80.0%	80.1%
	Bedroom 1	82.2%	82.8%	83.1%	83.4%	83.1%	83.2%	83.3%
	Bedroom 2	81.0%	81.4%	81.7%	82.0%	82.3%	82.4%	82.5%
Flat D	Living and Dining room	58.4%	62.5%	65.7%	70.1%	70.9%	71.5%	71.8%
	Master Bedroom	73.9%	74.8%	76.3%	80.8%	81.3%	80.9%	80.5%
	Bedroom 1	82.6%	81.9%	84.0%	83.8%	84.2%	83.8%	83.8%
	Bedroom 2	81.4%	82.0%	82.4%	83.1%	83.4%	83.5%	83.6%
Flat E	Living and Dining room	55.7%	61.8%	65.5%	70.1%	73.6%	74.0%	75.3%
	Master Bedroom	79.2%	79.4%	81.3%	82.2%	83.3%	83.5%	83.0%
	Bedroom	77.4%	78.8%	80.8%	82.5%	83.3%	83.4%	83.5%
Flat F	Living and Dining room	66.1%	70.0%	74.2%	77.2%	77.4%	77.6%	77.7%
	Master Bedroom	64.3%	65.1%	69.2%	73.1%	73.1%	73.8%	73.3%
	Bedroom 1	81.9%	83.8%	84.2%	84.3%	83.7%	83.8%	83.8%
	Bedroom 2	81.6%	82.2%	84.0%	84.3%	84.6%	84.6%	84.7%

4.2. Energy

The energy consumptions for lighting and air-conditioner with and without daylighting control are shown in Figure 2. Bedrooms consume more electricity than living and dining rooms as bedroom has a longer operating hours of lighting during daytime. 40% of lighting electricity can be saved for bedroom by daylight-linked control system. Similar to the effect of daylight level, balconies reduce the amount of energy saved and increase the influence by obstruction/floor level.

The energy consumed by air-conditioning was also analysed. Instead of 31/F, 16/F has the highest energy consumption. 31/F is relatively free of obstruction and received more solar radiation. The fraction of time for windows shaded by internal shading increases. Hence, energy consumed by air-conditioner is slightly decreased for those top floors. With the implement of daylight control, most of the bedrooms can reduce 4% energy used by air conditioners. For living and dining rooms of Flat D has a very low VDF of 13.4% at 1/F. With the implement of daylight control, it helps to reduce the air-conditioning consumption by 6.5% which is the largest among all rooms. There are two plausible explanations. Firstly, this room is severely obstructed and hence, solar radiation is difficult to reach the room, so the total cooling load decreases. Therefore, the air-conditioning energy savings due to the reduction of lighting heat dissipation become significant. Secondly, about half the operation time of air conditioner for living and dining room is during daytime, so the effect of daylight-linked control on the air conditioning energy saving is more significant than that for bedrooms.

Table 3. VDF for each room

		1/F	6/F	11/F	16/F	21/F	26/F	31/F
Flat A	Living and Dining room	24.4	28.0	31.7	34.1	36.0	36.2	36.2
	Master Bedroom	18.7	22.9	26.8	29.2	30.9	31.1	31.1
	Bedroom 1	32.8	37.3	41.5	44.0	45.8	46.0	46.0
	Bedroom 2	33.3	37.6	41.5	43.9	45.8	46.0	46.0
Flat B	Living and Dining room	24.5	30.9	33.4	33.4	33.4	33.4	33.4
	Master Bedroom	25.8	33.3	35.8	35.8	35.8	35.8	35.9
	Bedroom	31.1	37.1	39.2	39.2	39.2	39.2	40.6
Flat C	Living and Dining room	13.6	14.0	14.0	14.0	14.0	14.0	18.4
	Master Bedroom	21.7	28.7	31.0	33.8	33.8	33.8	33.8
	Bedroom 1	25.6	32.0	34.6	35.9	36.0	36.1	38.0
	Bedroom 2	28.7	36.7	39.6	40.7	40.7	40.8	44.6
Flat D	Living and Dining room	13.4	14.2	16.8	21.3	22.8	22.9	25.3
	Master Bedroom	14.5	16.2	19.7	32.0	33.9	34.0	34.0
	Bedroom 1	25.4	27.2	31.8	41.6	43.0	43.2	45.3
	Bedroom 2	25.3	27.7	32.8	39.4	41.5	41.7	44.5
Flat E	Living and Dining room	23.1	24.2	26.7	30.2	33.3	33.6	33.7
	Master Bedroom	24.7	26.1	28.6	31.9	34.6	34.7	34.7
	Bedroom	24.0	26.0	28.6	32.0	34.8	35.0	39.1
Flat F	Living and Dining room	24.3	27.6	31.1	33.8	36.2	36.5	36.5
	Master Bedroom	11.8	14.4	17.1	19.5	21.6	21.9	21.9
	Bedroom 1	34.1	37.6	40.9	43.5	45.7	46.0	46.0
	Bedroom 2	33.6	37.3	40.8	43.4	45.7	46.0	46.0

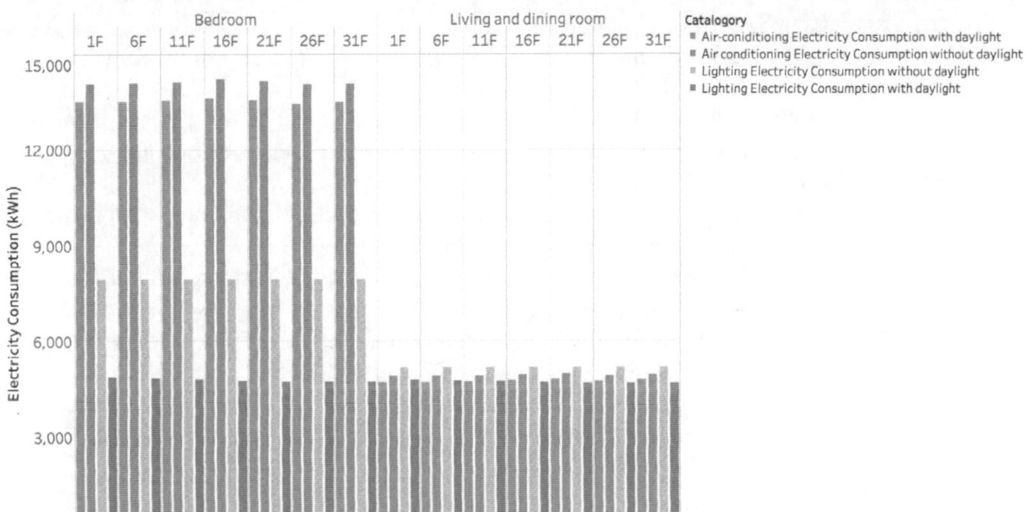

Figure 2. Electricity consumption for different floors of bedrooms and living rooms.

5. Conclusions

In Hong Kong, most of the residential buildings projects in the high-density zone are high-rise multi-storey blocks built closely together. The effects of obstruction on daylighting performance are of interested. Recently the HKSAR Government has introduced a new RTTV to govern the building envelop design. Simulations have been conducted for a newly design residential building which fulfils the daylight and thermal requirement by the government. It is found that VDF is closely related to the indoor illuminance level. Over 80% of daytime, daylight alone can provide sufficient illuminance for most of the bedrooms. For rooms with heavy obstruction, over 50% of daytime, the room is considered sunlit. For most of the bedrooms, 40% of lighting and 3% of air-conditioning energy consumptions can be reduced. It is believed that VDF is closely related to the energy consumptions and visual comfort aspect. This is a snapshot study on VDF and RTTV, more works are required to explore the relationships among VDF, RTTV and visual comfort to assist building designer to optimise their designs in early stage.

6. References

[1] Electrical and Mechanical Services Department 2017 *Hong Kong Energy End-use Data 2017.* (Hong Kong SAR Government) https://www.emsd.gov.hk/filemanager/en/content_762/HKEEUD2017.pdf
[2] Lam J C, Tsang C L, Li D H W and Cheung S O 2005 Residential building envelop heat gain and cooling energy requirements. *Energy* **30** 933-51.
[3] Buildings Department 2014 *Guidelines on Design and Construction Requirements for Energy Efficiency of Residential Buildings 2014* (Hong Kong SAR Government)
[4] Li D H W, Wong S L, Tsang C L and Cheung G H W 2006 A study of the daylighting performance and energy use in heavily obstructed residential buildings via computer simulation techniques *Energy and Buildings* **38** 1343-8.
[5] Buildings Department 2015 *PNAP APP-130 Lighting and Ventilation Requirements – Performance-based Approach* (Hong Kong SAR Government) https://www.bd.gov.hk/english/documents/pnap/APP/APP130.pdf
[6] Census and Statistics Department 2018 *Hong Kong energy statics annual* (Hong Kong SAR Government)
[7] BEAM Society Limited 2012 *BEAM Plus for new buildings version 1.2* (BEAM Society Limited, Hong Kong SAR)
[8] Hong Kong Government 1956 *Building (Planning) Regulations CAP 123F* (Hong Kong Government)
[9] John R, Sullivan R, Selkowitz S, Nozaki S and Conners C 1984 *Glazing energy performance and design optimization with daylighting LBL-15625* (Lawrence Berkeley Laboratory, University of California, USA)
[10] Electrical and Mechanical Services Department 2015 *Code of Practice for Energy Efficiency of Building Services Installation* (Hong Kong SAR Government)
[11] Buildings Department 2014 *Building Design to Foster a Quality and Sustainable Built Environment PNAP APP-151* (Hong Kong SAR Government)
[12] Chan A L S, Chow T T, Fong S K F and Lin J Z 2006 Generation of a Typical Meteorological Year for Hong Kong *Energy Conversion and Management* **47** 87–96

The Study On Anti-PID Performance Of High Efficiency Bifacial Cell Module

Xia Cai, Zhichun Ni*, Chengjin Chen, Po Ke, Henglei Chen, Haibo Cao, Qiangming Zhang
Suzhou Talesun Solar Technologies Co., Ltd. Changshu, Jiangsu, China
E-mail: paul@talesun.com

Abstract. This paper mainly studied the correlation between the PID attenuation of P/N type bifacial cell and the adhesive film. At present, with the development of glass-glass module technology and bifacial cell technology, glass-glass bifacial module is an important direction of industry development. Currently P type monocrystal PERC bifacial cell and N-type PERT bifacial cell are typical representatives of the bifacial technology. However, the PID problem of bifacial cell is more serious than single side cell. How to solve the bifacial cell PID problem has become a hot topic. We mainly studied the correlation between two different adhesive film (EVA&POE), and the anti-pid attenuation ability of P/N type bifacial cell. The influence of the water vapor transmission rate and resistivity of the adhesive film on PID attenuation of bifacial cell was studied. After PID 192h test, For P type, power decay of the front side is lower than rear side, while for N type cell, power dacy of the rear side is lower than front side. Both P&N type bifacial cell module with POE are much better than with EVA. For POE, recipe adjustment affects PID performance obviously.

Keywords: bifacial cell, PID, water vapor transmission rate, resistivity

1. Introduction

Potential induced degradation (PID) in photovoltaic (PV) module presents a great challenge for the solar industry, as it could lead to catastrophic failure of the devices [1]. It has previously been reported that certain PV modules suffered a 50% loss in maximum power after a six-month exposure to the outdoor environment due to PID [2]. Addressing PID is critical for ensuring 25-year lifetime of PV modules, and the development of PV [3]. Extensive research has been conducted in recent years to examine, analyze, and prevent PID in PV modules [4]–[14].The PID phenomenon exists not only in the conventional silicon module, but also in the bifacial cell double glass module. The main external factors affected the PID phenomenon of the module are the environment temperature, the environment humidity and the voltage potential. The internal factors are the cells, encapsulation film, and glass and so on.

The encapsulation film is the key material for protecting the module, which isolate cells with the atmosphere environment. The encapsulation film need tolerance all sorts of environment stress, and need have reliable insulation, water resistance and aging resistance. The encapsulation film is crucial to the outdoor reliability, power degradation and life time of the module.

The water-resistance of the encapsulation film determines the effect of the environment humidity on the module. And the resistivity of the encapsulated film determines the electron transfer rate in the module. We studied the correlation between the two different adhesive film (EVA&POE), and the anti-pid attenuation ability of P/N type bifacial cell. At the same time, the influence of the water vapor transmission rate and resistivity of the encapsulation film on the PID attenuation of bifacial cell was also studied.

IOP Conf. Series: Materials Science and Engineering **556** (2019) 012033 doi:10.1088/1757-899X/556/1/012033

2. Experimental method

Double glass module samples are choosed and PID tested in the conditions of 85℃, 85% relative humidity (RH) and -1500V bias voltage. The schematic diagram of the PID test is shown in Figure 1. The materials of the samples are same except the encapsulation film and cells. Schematic structure of N type bifacial silicon cell and P type bifacial silicon cell, as shown in Figure 2.

The effect of different adhesive film on the P/N type bifacial cell anti-pid performance was obtained by means of the crossover experiment of P/N type bifacial cell with EVA and POE. At the same time, the correlation of water vapor transmission rate and resistivity of the EVA/POE film with the PID attenuation of module samples is studied.

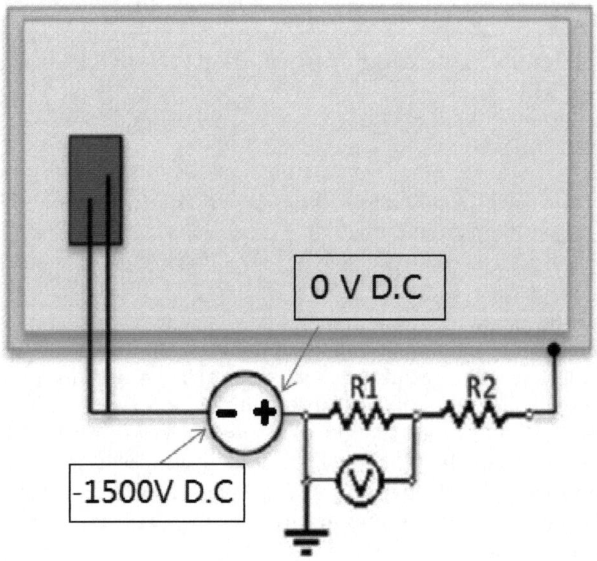

Figure 1. The schematic diagram of the PID test.

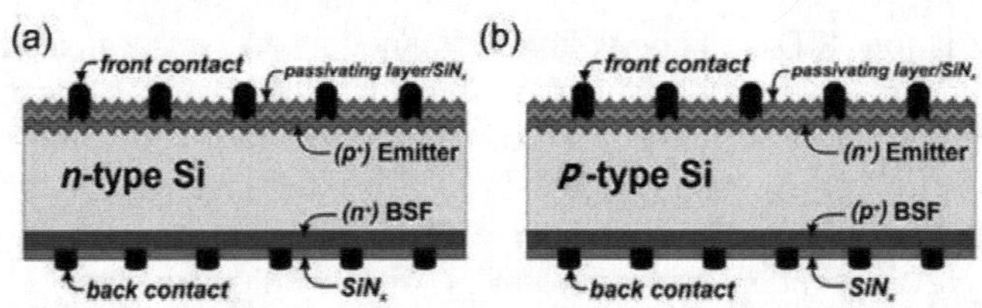

Figure 2. Schematic structure of N type bifacial silicon cell (a) and P type bifacial silicon cell (b)

3. Experimental results

In order to quickly analyse the anti-PID performance of P/N type cell bifacial with super anti-PID EVA and POE. We prepared module with one bifacial cell in the sample, and then PID tested in the conditions of 85℃, 85% relative humidity (RH) and -1500V bias voltage for 96h, 192h and 280h. The overall EL image after PID was shown in Figure 3. The EL test current remained at 4A. As the Figure 3 shown the N type bifacial cell with POE under PID tested 280h, the power attenuation is more obvious than P type bifacial cell.

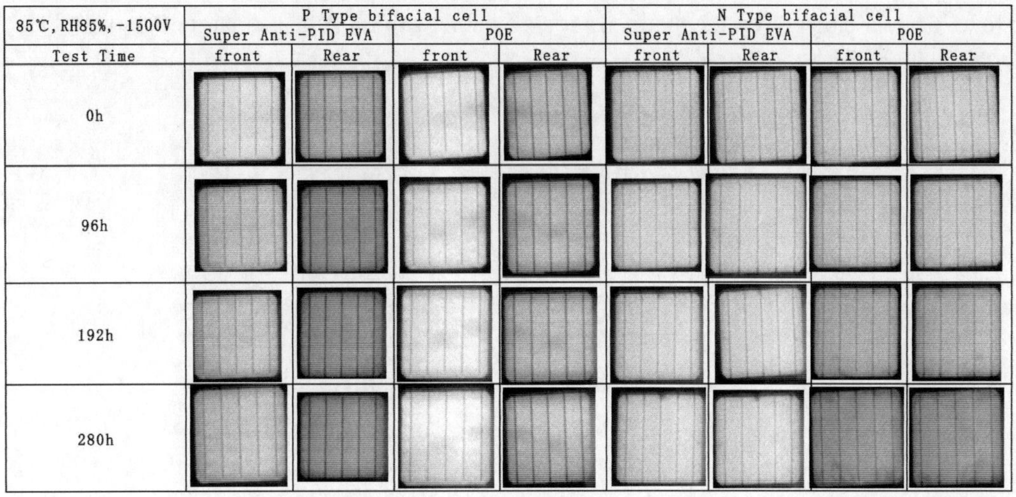

Figure 3. The EL image of P/N type bifacial cell with super anti-PID EVA/POE after PID test.

The PID test result of P type bifacial cell module and N type bifacial cell module with normal EVA under 192h is shown in Figure 4. The front side power degradation of N type bifacial cell module is much more serious than the P type cell module. And the front side power attenuation ratio gap between two type modules is 16.29%. The rear side power degradation of P type bifacial cell module is much more serious than the N type cell module. And the rear side power attenuation ratio gap between two type modules is 21.08%.

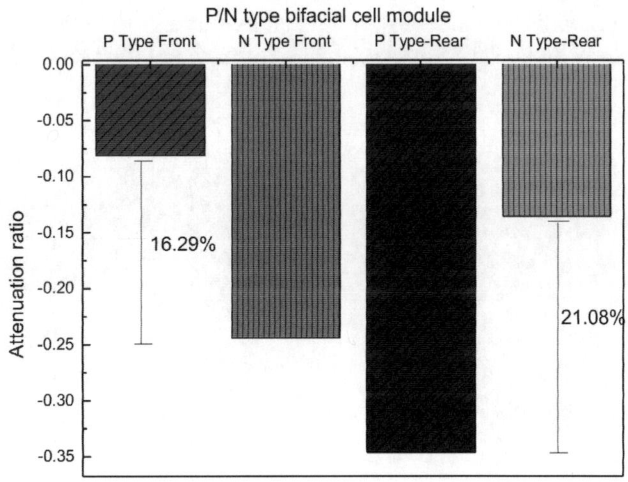

Figure 4. The PID test result of P type bifacial cell module and N type bifacial cell module with normal eva under 192h

The PID test result of P type bifacial cell module with normal EVA and poe under 192h PID test is shown in Figure 5. The power degradation of the module with POE is much better than with EVA, it can be seen that the POE exhibits better protection against PID than the PID-resistant EVA. The front side power attenuation ratio of module with EVA is larger than with POE, and the gap is 4.91%. At the

IOP Conf. Series: Materials Science and Engineering **556** (2019) 012033 doi:10.1088/1757-899X/556/1/012033

same time, the rear side power attenuation ratio of module with EVA is also larger than with POE, but the gap as high as 28.7%.

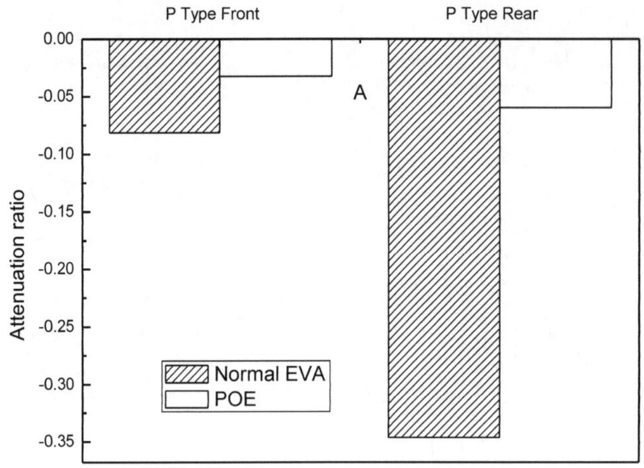

Figure 5. The P type bifacial cell module PID tested 192h with EVA and POE

The PID test result of N type bifacial cell module with normal EVA and poe under 192h is shown in Figure 6. The front side power attenuation ratio of module with EVA is larger than with POE, and the gap is 6.1%. At the same time, the rear side power attenuation ratio of module with EVA is also larger than with POE, and the gap as high as 5.36%. And the performance of the front side is much better than the rear side, whether it's with EVA or with POE.

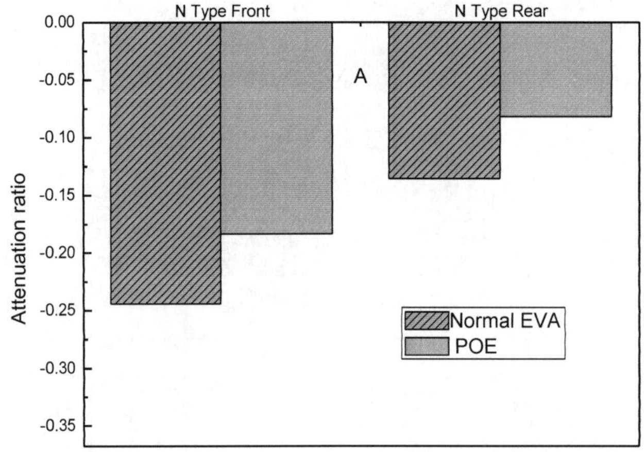

Figure 6. The N type bifacial cell module PID tested 192h with EVA and POE

According to Figure 5 and Figure 6, the power degradation of two types module with normal POE are greater than 5%. The next work, the P/N type bifacial cell modules with super anti-pid POE were further tested for 480h under the conditions of 85℃, 85% relative humidity (RH) and -1500V bias voltage, the

power degradation of two types module less than 5%.As shown in Figure 7. When the PID test time is 96h, the Anti-PID performance of N type bifacial cell module is better than P type bifacial cell module. However, the curve of N type bifacial module appears the turning point when the test time is 192h. The performance of N type bifacial module is poorer than P type bifacial module after 192h. And PID test process, the gap of the power degradation of this two modules grown larger.
Furthermore, the rear side power degradation of two types bifacial cell module after PID tested, as shown in Figure 8. When the PID test time is 192h, the Anti-PID performance of N type bifacial cell module is better than P type bifacial cell module. However, the curve of N type bifacial module drop seriously when the test time is 288h. The performance of N type bifacial module is poorer than P type bifacial module after 288h.

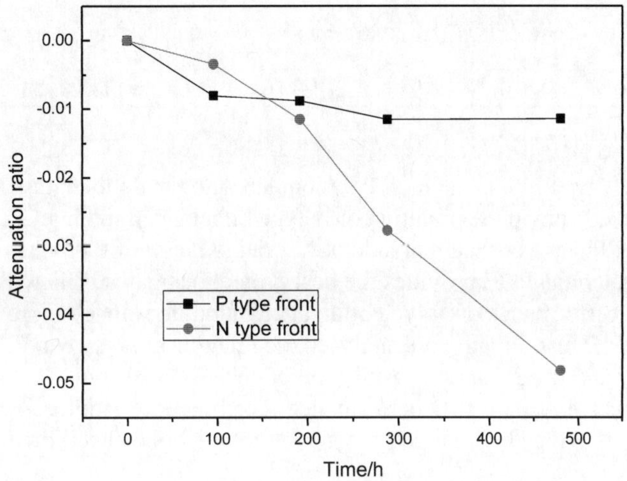

Figure 7. The front side of P/N type bifacial cell module performance after PID test

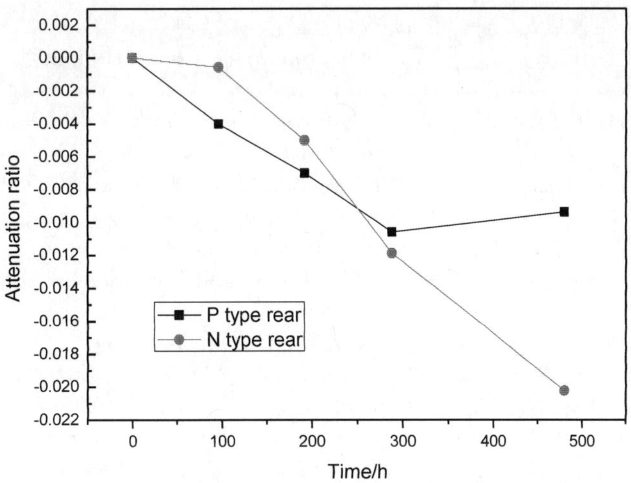

Figure 8. The front side of P/N type bifacial cell module' performance after PID test

4. Experimental analysis

Therefore, according to the results of previous experiments, the Anti-PID performance of P/N type bifacial cell module with POE is better than EVA. The relationship between PID performance of module with resistivity and water vapor transmission of encapsulated film was analysed.
The water vapor transmission rate and resistivity of the EVA and super anti-PID POE film were shown in Table 1. The water vapor transmission rate of super anti-PID POE is much smaller than EVA. The resistivity of super anti-PID POE is at least one order of magnitude higher than EVA. Therefore, the higher the water vapor transmission rate, the worse the anti-PID performance. The higher resistivity, the better anti-PID performance, whether it's P type bifacial cell module or N type bifacial cell module.

Table 1. The water vapor transmission rate and resistivity of the EVA and super anti-PID POE film.

	Unit	EVA	Super anti-PID POE	Reference Standard
Water Vapor Transmission Rate	g/m^2/day	28.3	3.2	ASTM F1249
Resistivity	Ω.cm	1.20E+16	1.50E+17	IEC 62788-1-2

5. Conclusions

In this study, PID in P/N type bifacial silicon PV modules with the double glass structure was investigated. The front side power degradation of N type bifacial cell module is much more serious than P type after 192h PID test, but the rear side power degradation of P type is much poorer. Among the two investigated materials, POE provides the best protection against PID with the slower degradation rates. The performance of N type bifacial cell module is much more serious than P type when under long time PID test. Finally, we analysed the relationship between PID performance of module with resistivity and water vapor transmission of encapsulated film. The higher the water vapor transmission rate, the worse the anti-PID performance. The higher resistivity, the better anti-PID performance, whether it's P type bifacial cell module or N type bifacial cell module.

6. References

[1] W.Luo et al., 2017 *Energy Environ. Sci.,* vol. 10, no. 1
[2] E. Schneller, N. S. Shiradkar, and N. G. Dhere, 2014 *Proc. 40th IEEE Photovolt. Spec. Conf.,* Denver,CO, USA, pp. 3216–3219
[3] W. Luo et al., 2018 *Sol. Energy Mater. Sol. Cells,* vol. 174
[4] S. Pingel *et al.*, 2010 *Proc. 35th IEEE Photovolt. Spec. Conf.*,, Honolulu, HI, USA, 2010, pp. 2817–2822
[5] P. Hacke *et al.*, 2011, *Proc. 37th IEEE Photovolt. Spec. Conf.*, Seattle, WA, USA, pp. 000814–000820
[6] M. B. Koentopp, M. Krober, and C. Taubitz, 2016. *IEEE J. Photovolt.*, vol. 6, no. 1, pp. 252–257
[7] P. Hacke *et al.*, Nov. 2015. *IEEE J. Photovolt.*, vol. 5, no. 6, pp. 1549–1553,
[8] V. Naumann, C. Hagendorf, S. Grosser, M. Werner, and J. Bagdahn, 2012. *Energy Procedia*, vol. 27, pp. 1–6
[9] W. Luo *et al.*, 2017. *IEEE J. Photovolt.*, vol. 7, no. 1, pp. 104–109
[10] K. Hara, S. Jonai, and A. Masuda, 2015. *Sol. Energy Mater. Sol. Cells*, vol. 140, pp. 361–365
[11] H. Nagel, P. Saint-Cast, M. Glatthaar, and S. Glunz, 2014. *Proc. 29th Eur. Photovolt. Sol. Energy Conf. Exhib.*, Amsterdam, The Netherlands, , pp. 2351–2355
[12] D. Lausch *et al.*, 2014. *IEEE J. Photovolt.*, vol. 4, no. 3, pp. 834–840
[13] S. Hoffmann and M. Koehl, 2014. *Prog. Photovolt., Res. Appl.*, vol. 22, no. 2, pp. 173–179
[14] D. Lausch *et al.*, 2014. *Energy Procedia*, vol. 55, pp. 486–493

Performance analysis of a high concentrating photovoltaic/thermal system with a water spray cooling device

Haifei Chen[1], Jie Yang[1], Nianyong Zhou[1,*], Jianfang Chen[1], Yang Zhang[2]
[1] Shool of Petroleum Engineering, Changzhou University, Changzhou, China
[2] School of Aeronautics and Astronautics, Purdue University, West Lafayette, IN, USA

E-mail:chfs@cczu.edu.cn

Abstract. A high concentrating photovoltaic/thermal system based on a water spray cooling device has been investigated in this paper. The system uses point-focus Fresnel lens as a concentrator with a concentration ratio over 1000. Both of the numerical thermal and electrical models have been developed to investigate the performance of the system. Moreover, the spray cooling has been used to solve the problem of high heat flux transfer and get good temperature uniformity, which is considered as an efficient method to develop the electrical efficiency and thermal efficiency. The results show that the overall efficiency is mainly influenced by water mass flow.

Keywords: solar energy; photovoltaic/thermal; concentrating system; spray cooling

1. Introduction

Facing the challenges of energy shortage and environmental pollution, photovoltaic power generation is an effective way to alleviate the problems. Photovoltaic technology has improved in the last ten years. However, photovoltaic system still has two urgent problems, including the high cost of photovoltaic systems and the low photoelectric conversion efficiency, which constrain the wide application of photovoltaic technology. Currently, concentration technology has been developed [1, 2], which can cut down the cell area to reduce the cost of the system. When sunlight reaches concentrating photovoltaic (CPV) system, the energy is not only converted into electric energy but also extra heat, which can cause the temperature rising of the cells. As the temperature of solar cells increases, the cell efficiency decreases. Therefore, a cooling system is used to cool the cells meanwhile the excess heat can be used as a heat source [3]. This concentrating photovoltaic thermal (CPV/T) system attracts a great deal of researchers' attention [4, 5], because the CPV/T system can produce electric energy using a low-cost concentrator but also supply hot water for thermal application. Schuetz designs a photovoltaic system based on compound parabolic concentrator, which is a 7 times concentration system using silicon photovoltaic cells with system power efficiency of 7.9% [6]. Joe S. Coventry introduces a concentrating photovoltaic/thermal solar collector with a concentration ratio of 37, which shows thermal efficiency around 58% and electrical efficiency about 11% [7]. In order to solve the low electrical efficiency, multi-junction cells have been researched, because of their high electrical conversion efficiency [8-10]. As MOVPE technology improves, GaAs cells have a large potential for CPV/T system. Li Ming analyses a trough concentrating photovoltaic/thermal system using GaAs cells, which are much better than silicon cells [11].

Compare with low concentrating PV/T system, the high concentrating PV/T (HCPV/T) system demands better tracking and cooling technologies. When the geometric concentration ratio reaches 500 or even 1000, it is hard to achieve a uniform radiation flux distribution and an acceptable temperature of solar cells, which severely affects the electrical efficiency of the HCPV/T system[12].

In order to solve the above problem, the paper presents a high concentrating photovoltaic/ thermal (HCPV/T) system with a water spray cooling device, which can provide both electrical energy and usable heat. The operation of the system is dynamic actually, thus a dynamic numerical model suitable for the system simulation is developed in this paper. It can provide the performance analysis of the system, including thermal gain, electrical gain and efficiencies. And the influence of the concentration ratio and the mass flow rate is also discussed in this model.

2. Description of the HCPV/T system

Figure 1 shows the structure of the HCPV/T system using Fresnel lens. The HCPV/T system mainly consists of 3 parts, including a concentrating device, a cooling device and a tracking device. It converges the sunlight into a small cell area through the concentrator. And GaAs cells are cooled by the cooling system using water as the working fluid, through which the heat is collected for thermal applications.

Generally, the tracking device should be adopted in the concentrating system. The system can only use direct solar radiation, when the concentration ratio reaches more than 10. In order to make sure that the system is accurately tracked, a two axes tracking device is applied in the HCPV/T system. The tracking device follows the location of the sun both in azimuth and elevation. And the tracking accuracy is about 0.2^{o}. A programmable logic controller device is designed to compute the sun position.

At present, many high concentrating PV/T systems use continuous parabolic structure. However, this structure has a non-uniform distribution of light intensity, which can cause a temperature gradient and overheating in solar cells. It will greatly reduce the power output of the cells. Therefore, the Fresnel lens array structure is applied in this HCPV/T system, which can achieve a more uniform light intensity distribution.

In the concentrating system, the temperature of solar cells will increase fast under high solar intensity. As the temperature of solar cells increases, the electrical efficiency decreases. Therefore, a cooling device must be applied to cool and protect solar cells. The HCPV/T system uses water spray cooling device to control the mass flow.

Figure 1. Photograph of the HCPV/T system using Fresnel lens.

3. Mathematical analysis

A dynamic model is developed to simulate both the thermal and electrical performance, which is based on the control volume, finite difference approach for the water cooling system. The system can be represented by four corresponding points through solving the transient energy balance equations. The point 'g' is for solar cells, the point 'c' is for spray cooling device, the point 'w' is for water, and the point 'i' is for the insulation material.

IOP Conf. Series: Materials Science and Engineering **556** (2019) 012034 doi:10.1088/1757-899X/556/1/012034

The direct irradiance absorbed by GaAs cell layer is given by

$$Q_g = \rho_c \alpha_g G - E_g \tag{1}$$

$$G = C_r R\, I_d A_g \tag{2}$$

Where ρ_c is the reflectivity of the concentrator, $\alpha_g$is the absorptance of cells, C_r is the concentration ratio of the system.
The electrical generation is given by

$$E_g = \rho_c r_g G \eta_{cell} \tag{3}$$

$$\eta_{cell} = \eta_r [1 - \beta_r (T_p - T_r)] \text{ [13]} \tag{4}$$

Where r_g is the ratio of GaAs cell area to aperture area, η_r is the reference GaAs cell efficiency at the reference temperature T_r, β_r is a temperature coefficient for GaAs.
The thermal resistances for the components of the system are given by

$$R_{ga} = 1/(h_{ga} A_{ga}) \tag{5}$$

$$R_{ge} = 1/(h_{ge} A_{pe}) \tag{6}$$

$$R_{cg} = 1/(h_{cg} A_{cg}) \tag{7}$$

$$R_{ci} = 1/(h_{ci} A_{ci}) \tag{8}$$

$$R_{cw} = 1/(h_{cw} A_{cw}) \tag{9}$$

$$R_{ai} = 1/(h_{ai} A_{ai}) \tag{10}$$

The heat transfer coefficients for the components of the system are given by

$$h_{ga} = 3.8 u_a + 5.7 \text{ [14]} \tag{11}$$

$$h_{ge} = \varepsilon_g \sigma (T_g^2 + T_e^2)(T_g + T_e) \tag{12}$$

$$h_{cp} = k_c / \delta_c \tag{13}$$

$$h_{ci} = k_i / \delta_i \tag{14}$$

$$h_{cw} = Nu_D k_w / D_h \tag{15}$$

$$Nu_D = 4.7 Re^{0.61} + Pr^{0.32} \text{ [15]} \tag{16}$$

Therefore, the thermal resistance equations are considered for the HCPV/T system:

$$\text{Solar cells: } M_g C_g \frac{\partial T_g}{\partial t} = k_g V_g \frac{\partial T_g}{\partial x^2} + \frac{T_a - T_g}{R_{ga}} + \frac{T_e - T_g}{R_{ge}} + \frac{T_c - T_g}{R_{cg}} + Q_g \tag{17}$$

$$\text{Spray cooling: } M_c C_c \frac{\partial T_c}{\partial t} = k_c V_c \frac{\partial T_c}{\partial x^2} + \frac{T_g - T_c}{R_{cg}} + \frac{T_i - T_c}{R_{ci}} + \frac{T_w - T_c}{R_{cw}} \tag{18}$$

Insulation: $$M_i C_i \frac{dT_i}{dt} = \frac{T_c - T_i}{R_{ci}} + \frac{T_c - T_i}{R_{ci}} \quad (19)$$

Water: $$M_w C_w \frac{\partial T_w}{\partial t} = \frac{T_c - T_w}{R_{cw}} - \dot{m}_w C_w \frac{\partial T_w}{\partial x} L \quad (20)$$

The thermal output is given by

$$Q_w = \dot{m}_w C_w (T_{out} - T_{in}) \quad (21)$$

Where T_{in} is inlet temperature and T_{out} is outlet temperature of water.
Then, the thermal efficiency is

$$\eta_t = Q_w/G = \dot{m}_w C_w (T_{out} - T_{in})/G \quad (22)$$

The electrical efficiency is

$$\eta_e = E_g/G \quad (23)$$

The total efficiency can be given by

$$\eta_o = \eta_t + \eta_e \quad (24)$$

4. Results and discussion

In order to know the performance of the HCPV/T system under different mass flow rate and concentration ratio, some parameters are assumed. The concentration ratio is 1000, direct irradiance is 800 w/m^2, and inlet temperature is 20 °C. The relevant parameter of the system is listed in Table 1.

Table 1. Mass, specific heat and conductivity of the system components.

	Specific heat ($Jkg^{-1}K^{-1}$)	Conductivity($Wm^{-1}K^{-1}$)
Solar cells	326	46
Cooling device	903	237
Insulation	795	0.034
Water	4181	0.63

Other data includes: $\varepsilon_p = 0.85$, $\alpha_p = 0.95$, u_a=3.5 m/s, L=0.14 m, $\eta_r = 0.366$, $\beta_r = 0.0026\ ^oC^{-1}$, $T_r = 25\ ^oC$.

As shown in Fig.2 and Fig.3, the performance of the PV/T changes rapidly in this model, when the mass flow rate is less than 1.0g/s. As the mass flow rate decreases from 1.0g/s to 0.1g/s, the electrical efficiency of GaAs cells will drop rapidly as shown in Fig.1; the temperature of GaAs cells increases fast as shown in Fig.3 If the mass flow rate is lower than 0.1g/s, GaAs cells will be damaged because of the high temperature. This is mainly because with the mass flow rate less than 1.0g/s, part of the water flow in the cooling device will evaporate and boil; therefore GaAs cells and the cooling device temperature rise, leading to rapid decline in the photoelectric conversion efficiency of GaAs cells. When the mass flow rate increases from 1.0g/s to 5.0 g/s, GaAs cells can be better cooled.

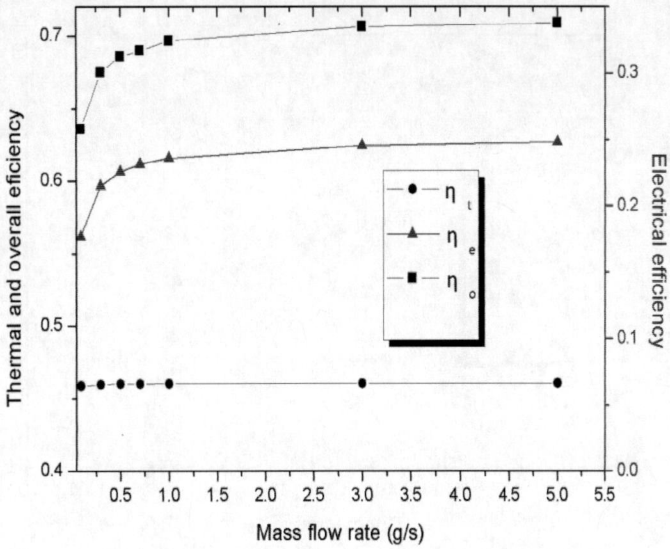

Fig. 2. Thermal and electrical efficiency of different mass flow rate.

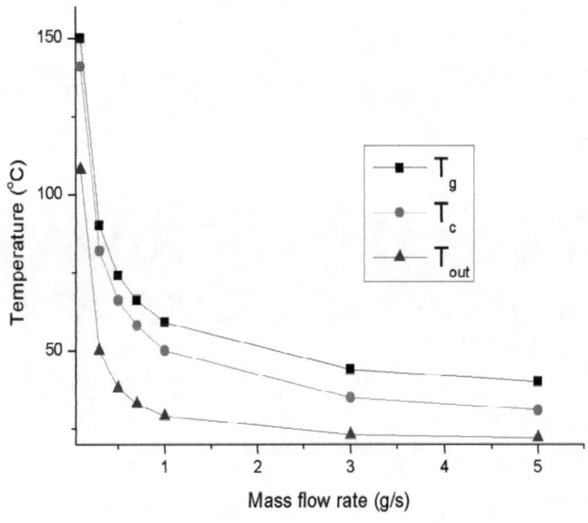

Fig. 3. Temperature of different mass flow rate

As shown in Fig.4, when the concentration ratio is less than 2000, the thermal efficiency has no significant change. However, the electrical efficiency decreases, if the concentration ratio continues to increase. It is mainly because solar radiation flux focusing on the surface of GaAs cells is too high. So part of the thermal energy can't be taken away, which causes the temperature of GaAs cells rising, as shown in Fig.5. The electrical efficiency will be lower than 20% with the concentration ratio at about 2000. However, the temperature of GaAs cells will be up to or more, which is not conducive to the long-term use of the cells. Therefore, the selection of concentration ratio should consider the operation of the system.

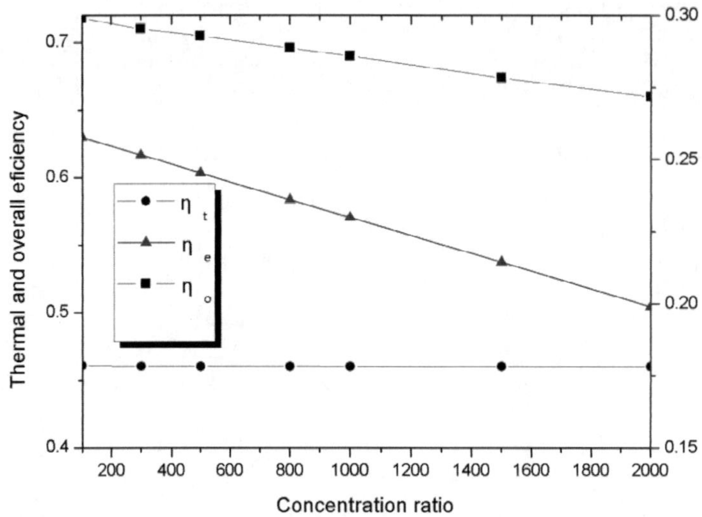

Fig. 4. Thermal and electrical efficiency of different concentration ratio

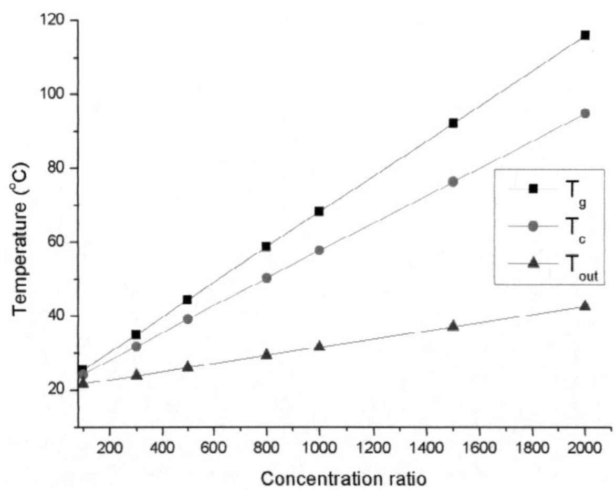

Fig. 5. Temperature of different concentration ratio

5. Conclusions

This paper presents a HCPV/T system using Fresnel lens, which can provide a uniform light intensity distribution. GaAs cells are cooled by water spray cooling device, through which the heat can be collected for thermal application. Hence, the total solar utilization efficiency is greatly increased.

In order to study the performance of HCPV/T system, a dynamic model is developed, which can perform complete energy flow analysis on the system. It shows instantaneous thermal efficiency around 47% and electrical efficiency around 20% ~ 25%. The simulation results also indicate that this system is strongly influenced by the concentration ratio and the mass flow rate. Then the mass fow rate and the concentration ratio selection should comprehensively consider the operation of the system.

Acknowledgements

This work is supported by the Natural Science Foundation of Jiangsu Educational committee (No. 18KJD480001) and No. YZGXYJS2018-KYCX-012.

References

[1].Desideri, U., F. Zepparelli, V. Morettini, et al.,2013. Comparative analysis of concentrating solar power and photovoltaic technologies: Technical and environmental evaluations. Applied Energy. 102:765-784.

[2].Grasso, G., A. Righetti, M.C. Ubaldi, et al., 2012. Competitiveness of stationary planar low concentration photovoltaic modules using silicon cells: A focus on concentrating optics. Solar Energy. 86(6) : 1725-1732.

[3].Kermani, E., S. Dessiatoun, A. Shooshtari, et al., 2009. Experimental investigation of Heat Transfer Performance of a Manifold Microchannel Heat Sink for Cooling of Concentrated Solar Cells. 2009 Ieee 59th Electronic Components and Technology Conference, Vols 1-4. New York: Ieee. 453-459.

[4].Helmers, H., A. Boos, F. Jetter, et al., 2011. Outdoor Test Setup for Concentrating Photovoltaic and Thermal (CPVT) Systems, in 7th International Conference on Concentrating Photovoltaic Systems, F. Dimroth, et al., Editors., Amer Inst Physics: Melville.

[5].Mittelman, G., A. Kribus and A. Dayan, 2007. Solar cooling with concentrating photovoltaic/thermal (CPVT) systems. Energy Conversion and Management. 48(9) : 2481-2490.

[6].Rosell, J.I., X. Vallverdu, M.A. Lechon, et al., 2005. Design and simulation of a low concentrating photovoltaic/thermal system. Energy Conversion and Management. 46(18-19) :3034-3046.

[7].Coventry, J.S., 2005. Performance of a concentrating photovoltaic/thermal solar collector. Solar Energy. 78(2) : 211-222.

[8].Cotal, H., C. Fetzer, J. Boisvert, et al., 2009. III-V multijunction solar cells for concentrating photovoltaics. Energy & Environmental Science. 2(2) : 174-192.

[9].Jones, R.K., J.H. Ermer, C.M. Fetzer, et al., 2012. Evolution of Multijunction Solar Cell Technology for Concentrating Photovoltaics. Japanese Journal of Applied Physics. 51(10).

[10].Zubi, G., J.L. Bernal-Agustin and G.V. Fracastoro, 2009. High concentration photovoltaic systems applying III-V cells. Renewable & Sustainable Energy Reviews. 13(9) : 2645-2652.

[11].Li, M., G.L. Li, X. Ji, et al.,2011. The performance analysis of the Trough Concentrating Solar Photovoltaic/Thermal system. Energy Conversion and Management. 52(6) : 2378-2383.

[12].Haifei Chen, Jie Ji, Gang Pei, et al., 2017. Experimental and numerical comparative investigation on a concentrating photovoltaic system, Journal of Cleaner Production, 174: 1288~1298

[13].Chow, T.T.,2003. Performance analysis of photovoltaic-thermal collector by explicit dynamic model. Solar Energy. 75(2) : 143-152.

[14].Kumar, S. and S.C. Mullick, 2010. Wind heat transfer coefficient in solar collectors in outdoor conditions. Solar Energy. 84(6) : 956-963.

[15].Rybicki J R, Mudawar I, 2006. Single-phase and two-phase cooling characteristics of upward-facing and downward-facing sprays[J]. International Journal of Heat & Mass Transfer, 49(1):5-16.

IOP Conf. Series: Materials Science and Engineering **556** (2019) 012035 doi:10.1088/1757-899X/556/1/012035

Solar photocatalytic degradation of methyl orange using PbS-ZnS/electrospun fiber composites

Jie Yang, Haiqun Chen, Haifei Chen*
School of Petroleum Engineering, Changzhou University, Changzhou, Jiangsu, China
E-mail: chfs@cczu.edu.cn

Abstract. A facile, efficient approach to fabricate PbS-ZnS heterojunction/electrospun fiber composites with micro-nano-sized structure was reported, involving the preparing of electrospun fibers by high-voltage electrospun technology and the synthesizing of chalcogenide semiconductors by low temperature hydrothermal method. The PbS-ZnS/electrospun fiber composites exhibited high photocatalytic activity for the degradation of methyl orange under ultraviolet (UV) light irradiation of solar light. Compared with titanium dioxide and the PbS-ZnS powder, PbS-ZnS/electrospun fiber composites exhibited excellent reusability with high photocatalytic efficiencies, indicative of its suitability for the solar photocatalytic applications. The residual mass of photodegradation of methyl orange in 300 min was 1.23%.

Keywords: Solar photocatalytic; Electrostatic spinning; Degradation; PbS-ZnS/electrospun fiber composites

1. Introduction

Since the discovery of photocatalytic degradation pollutants in water using TiO_2 by Fujishima in the year of 1972[1], photocatalytic degradation pollutants had been considered as one of the most promising avenue solar energy conversion methods. However, pure semiconducting photocatalyst powders are facing the challenge of large-scale application due to their low degradation rate because of the fast recombination of photo-generated electron/hole pairs and the aggregation of semiconducting particles[2, 3]. Therefore, semiconducting photocatalyst powders had been widely grafted onto organic [4-6] or inorganic carriers [7-9]. It was proved that the carriers could improve the photocatalytic activity as well as the stability of semiconducting phtocatalysts[10].

Inorganic carriers still have some shortcomings such as poor flexibility, low light transmittance, difficulty of loading and processing, etc[11]. Polymers have good flexibility, good film forming properties, high light transmittance and are easy to process compared with Inorganic carriers[12]. Therefore, the use of polymer as a photocatalyst carrier has become a hot topic of current research[13]. Fluorocarbon polymers have good photodegradation resistance, chemical resistance, heat resistance and dielectric properties. It is a good photocatalyst carrier that has attracted more and more attention. Studies have shown that composite materials have good degradation resistance and repeatability [6, 12] .

In this paper, PbS-ZnS/electrospun fiber composites were prepared by supporting PbS-ZnS heterojunction on the FVDF/SMA polymer fibers in the way of hydrothermal synthesis. The photocatalytic performance of the composites was studied by the mercury lamp which simulating ultraviolet (UV) light irradiation of solar light.

2. Experimental

2.1. Materials

Polyvinylidene Fluoride (PVDF) was supplied from Shanghai 3F New Material Co., Styrene-maleic anhydride (SMA) was purchased from Shanghai SpecTek petrochemical high tech Co., Ltd. N, N-dimethylacetamide(DMAC) was purchased from Shanghai Pilot Chemical Co., Ltd. Thiourea, lead nitrate and zinc nitrate were obtained from Sinopharm Shanghai Chemical Reagent Co., Ltd, acetone was brought from Shanghai Su Yi Chemical Reagent Co., Ltd. All chemicals were used as received.

2.2. Electrospinning of PVDF/SMA polymer fiber

A typical electrospinning process prepared the PVDF/SMA polymer fiber. 0.41g SMA and 3.6g PVDF were added into the mixture of acetone and DMAC in a stirred three-necked flask with N2 protection. Afier stirring 24 hours, the solution was electrospun at 24 KV positive voltage, 16 cm working distance, and flow 1.2 ml per hour. All process steps were operated at room temperature. The mat was cut to 5.0cm×5.0cm in dimension for the follow-up experiments.

2.3. Preparation of PbS-ZnS/electrospun fiber composite

The as-prepared electrospun fibers were put into 25 ml, 0.25 mol/L zinc nitrate for 12 h to form the complex between the fiber and the Zn^{2+}. Then the fibers were removed from the solution, washed with distilled water for three times, and dried in a vacuum for 2 h at 60 °C. Subsequently, the fibers were put into a 50 mL sealed Teflon-lined stainless steel autoclave, which include desired concentration of thiourea solution. The autoclave was placed in an oven for 12 h at 120°C, and then it was cooled to room temperature. The ZnS/electrospun fiber composites were washed with distilled water under ultrasonic vibrations in order to remove the byproducts and unreacted precursor, and were dried in vacuum for 3 h at 60 °C.

The dried fibers were immersed in 25ml, 0.25 mol/L lead nitrate aqueous solution at room temperature for 24h. Then the fibers were removed from the solution, washed with distilled water for three times, and dried in a vacuum for 2 h at 60 °C. Subsequently, the fibers were put into a 50 mL sealed Teflon-lined stainless steel autoclave, which include desired concentration of thiourea solution. After being cooled, washed and dried, the PbS-ZnS/electrospun fiber composites were prepared.

As a comparative experiment, the PbS-ZnS powders were prepared by the similar conditions, while the latter was not loaded on the electrospun fiber mats.

2.4. Phtocatalytic reaction

Photocatalytic activities of the as-prepared samples for degradation of methyl orange were carried in SGY-IB multi-purpose chemical reactor from Nanjing Hoop Technologies Co., Ltd., the light source was mercury lamp (300W, 225nm), and the system temperature was kept at 25±1°C using flowing cool water to remove the heat generated by mercury lamp. The electrospun fiber composites (0.29g containing 0.15g PbS-ZnS) were added in an aqueous solution(250mL) containing 0.005 g of methyl orange, while the PbS-ZnS powers(0.15g) and the P25-TiO_2 powders were dispersed in the same solution with a magnetic stirrer separately. The system was initially purged by N_2 in a dark environment to exclude the oxygen. The concentration of methyl orange was measured by UV/VIS spectrophotometer after photocatalytic degradation.

2.5. Characterization·

Scanning electron microscopy was carried out with a JSM-6490LV, which was bought from JEOL Co., Ltd. The X-ray diffraction (XRD)(Rigaku D/max-β B, Rigaku Corporation, Japan) measurement was carried out using a Cu-Kα radiation source. X-ray photoelectron spectroscopy (XPS) spectra was detected by the ESCALAB-MK-II. Ultraviolet-visible (UV-Vis) absorption spectra were obtained on a Shimadzu Solidepec-3700 DUV spectrophotometer at room temperature. The thermo-gravimetic analysis (TGA) was applied to evaluate the content of ZnS-PbS of PbS-ZnS/electrospun fiber composites on a Netzsch TG-209-F3. The high-voltage electrostatic generator was bought from Jinan Loufang Co., Ltd.

3. Results and discussion

IOP Conf. Series: Materials Science and Engineering **556** (2019) 012035 doi:10.1088/1757-899X/556/1/012035

3.1. Structure and morphology of composites

The SEM images of ZnS/ electrospun fiber composites, PbS-ZnS /electrospun fiber composites were shown in Fig. 1(a) and Fig. 1(b) respectively. The spherical ZnS particles with the diameter of about 600nm grew on the surface of the fibers. Fig. 1(b) showed that many PbS-ZnS particles grew on the surface of electrospun fibers. ZnS/electrospun fiber composites which were immersed in thiourea solution, there are numerous ZnS particles after reacting with thiourea. The size of PbS particles on the surface is to the point, and the particles distributed evenly.

Figure 1. SEM images of (a) ZnS/ electrospun fiber composites, (b) PbS-ZnS /electrospun fiber composites.

Fig.2 showed the XRD patterns of ZnS/PVDF electrospun fiber composites and PbS-ZnS/PVDF electrospun fiber composites. The diffraction peaks located at 2θ = 28.47 °, 47.58 °, 56.22 °corresponded to the (111), (220) and (311) planes of the sphalerite structure of ZnS (ICDDPDF NO.05—0566). The diffraction peak appearing in Fig. 2(b) corresponded to the (111), (200), (220), (311), (222), and (440) planes of the galena crystal structure of PbS (JCPDS 5-0592), the narrow and sharp peaks indicate that the resulting product had a good crystalline state. The absorption peaks of ZnS and PVDF/SMA electrospun fibers were also shown in Fig. 2(b), which indicated that PbS-ZnS/electrospun fiber composites had been formed on the electrospun fibers.

XPS spectra of PbS-ZnS/electrospun fiber composites was showed in Fig.3. The spectra demonstrated that PbS-ZnS/electrospun fiber composites were composed of C, O, F, Zn, Pb and S elements, which were corresponding to the composites.

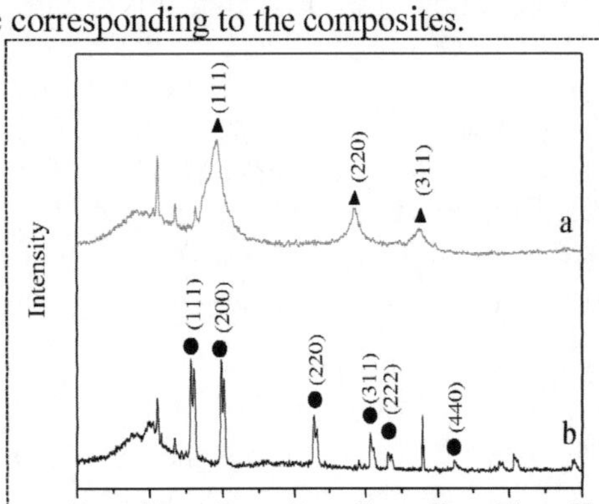

Figure 2. XRD patterns of (a) ZnS/ electrospun fiber composites (b)PbS-ZnS/ electrospun fiber composites

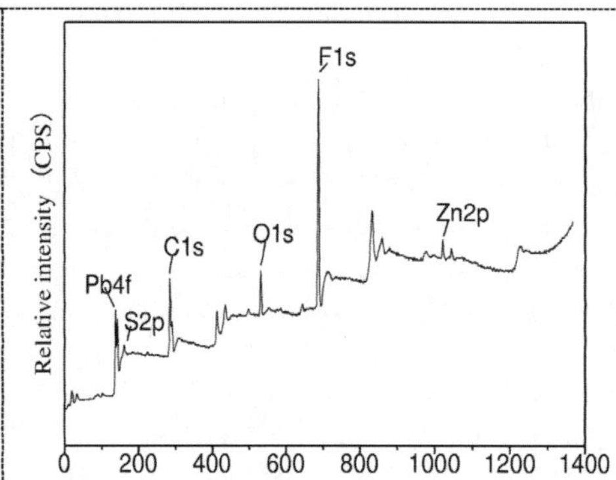

Figure 3. XPS spectra of PbS-ZnS/ electrospun fiber composites

3.2. Ultraviolet-Visible absorption spectra

Fig. 4.5 (a) shows the absorption of PbS-ZnS heterojunction/electrospun fiber composites at 350nm

IOP Conf. Series: Materials Science and Engineering **556** (2019) 012035 doi:10.1088/1757-899X/556/1/012035

had a maximum absorption peak, indicating that the photocatalyst can absorb ultraviolet light with a wavelength less than 360nm, which corresponded to the band gap value of sphalerite ZnS-PbS. In Fig. 4.5 (b), the unsupported electrospun fibers have no evident absorption between 300 and 800 nm, while composites have absorption peak sat round 280 nm, revealing that the conducting electrospun fibers do not interfere with the light absorption during the photocatalytic reaction.
Fig.5. shows the TGA curve of PbS-ZnS / electrospun fiber composites. The final mass loss of PVDF / SMA electrospun fibers is 93% by TGA, while the final weight loss of PbS-ZnS heterojunction/ electrospun fiber composites after loading is 45.23%, so the mass fraction of catalyst is about 51.36%.

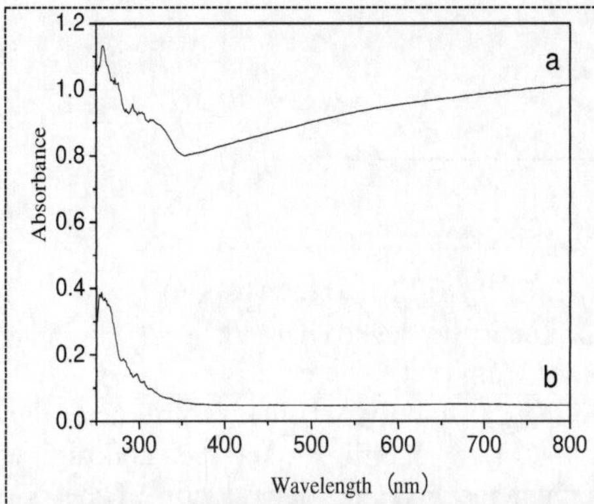

Figure 4. UV/VIS absorption spectra of (a) PbS-ZnS/ electrospun fiber composites (b) PVDF/SMA electrospun fiber

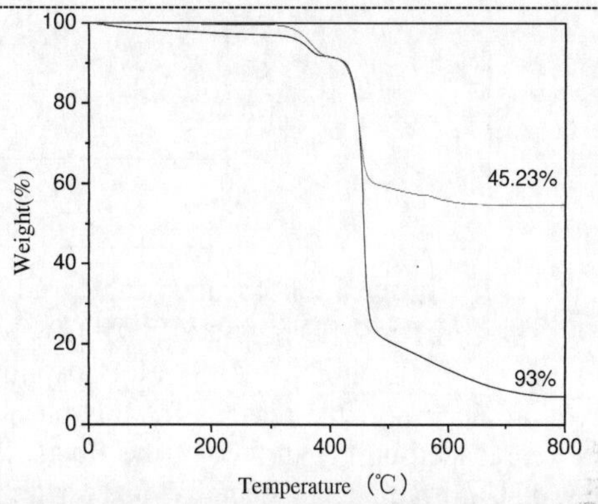

Figure 5. TGA curve of PbS-ZnS/ electrospun fiber composites

3.3. Photocatalytic activity measurements
The photocatalytic performance of PbS-ZnS / electrospun fiber composites was evaluated by degradation of methyl orange under ultraviolet (UV) light irradiation of solar light. The degree of the photocatalysis is presented by the ratio of the concentration C/C_0 under various times. Fig.6 shows the trend of photocatalytic degradation. It can be clearly observed that the residual mass fraction of methyl orange was 1.23% after PbS-ZnS/electrospinning fiber composites photo-degradation for 300 min, while the removal efficiency of methyl orange reaches to 32.4% by PbS-ZnS powders and 44.2% by P25-TiO_2 powders. This shows that the photocatalytic degradation of PbS-ZnS/electrospinning fiber composite is much more efficient than PbS-ZnS heterojunction powders and P25-TiO_2 powders, which is due to the the PbS-ZnS heterojunction loaded on the surface of the electrospun fiber have a larger surface area and higher light utilization. In addition, the PVDF/SMA electrospinning fiber has strong adsorption capacity, which makes it easy for methyl orange to be absorbed and migrated to the surface of inorganic particles of composite materials, performs photocatalytic degradation of methyl orange and forms an adsorption-migration-photo degradation chain. However, the adsorption capacity of PbS-ZnS heterojunction powders is so weak that the process of the entry of -OH reactive groups into the solution becomes the key factor of photodegradation rate. In addition, as can be seen from Fig. 6, the degradation rate of PbS-ZnS heterojunction powders is significantly higher than that of Degussa P25-TiO_2, which is due to the special band structure and carrier transport characteristics of semiconductor heterojunctions.

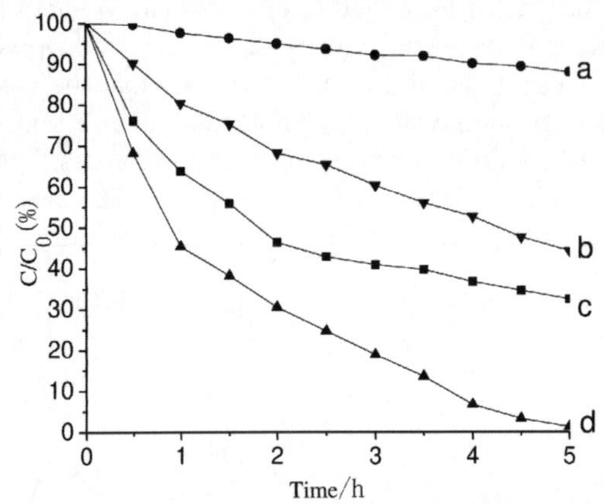

Figure 6. Photocatalytic degradation by (a) PVDF/SMA electrospun fiber (b) Degussa P25 TiO2 (c) The PbS-ZnS powder (d) The PbS-ZnS / electrospunning fiber composite

There are three possible reasons for this phenomenon (Fig.7). Firstly, when PbS-ZnS heterojunction particals were loaded on the surface of the fibers, they have a large specific surface area and high-energy efficiency of UV irradiation. Secondly, the carrier can prevent the loss and degradation of the powdery catalyst. Thirdly, the photo-generated eletron-hole pairs within the PbS-ZnS heterojunction particles were produced in the light, the electron migrate to the surface of the PbS and degraded methyl orange in water. The holes which migrate to the surface of the electrospun fiber and react with the sacrificial agents were adsorbed on the surface of the fiber.

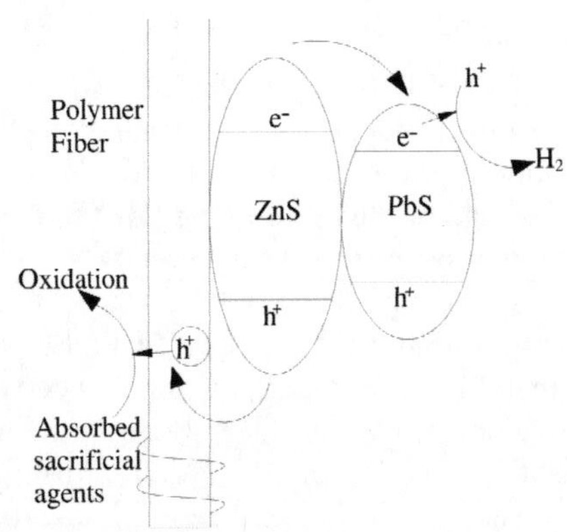

Figure7. Structure and catalytic mechanism diagram of PbS-ZnS/electrospun fiber composites

4. Conclusion

In summary, novel composite photocatalysts- PbS-ZnS/electrospun fiber composites were successfully synthesized and had an obvious effect on enhancing the photocatalytic efficiency of the photocatalysts. The UV-Vis absorption spectra of PbS-ZnS / electrospun fiber composites shows that the materials have good absorption under visible light, and electrospun fiber absorption peak in the mercury lamp simulated sunlight catalytic hydrogen water does not affect the light absorption of the composite. The residual mass fraction of methyl orange was 1.23% after PbS-ZnS/electrospinning fiber composites photo-degradation for 300 min, which is better than PbS-ZnS heterojunction powders and P25-TiO_2 powders.

Acknowledgements

This work is supported by the Natural Science Foundation of Jiangsu Educational committee (No. 18KJD480001) and No. YZGXYJS2018-KYCX-012.

References

[1].Fujishima, A.; Honda, K., 1972, Electrochemical photolysis of water at a semiconductor electrode. *Nature* 238, (5358), 37-8.
[2].Gaya, U. I.; Abdullah, A. H., 2008, Heterogeneous photocatalytic degradation of organic contaminants over titanium dioxide: A review of fundamentals, progress and problems. *Journal of Photochemistry and Photobiology C-Photochemistry Reviews* 9, (1), 1-12.
[3].Li, Z.; Qiu, N.; Yang, G., 2009, Effects of synthesis parameters on the microstructure and phase structure of porous 316L stainless steel supported TiO2 membranes. *J. Membr. Sci.* 326, (2), 533-538.
[4].Du, M.; Du, Y.; Feng, Y.; Yang, K.; Lv, X.; Jiang, N.; Liu, Y., 2018, Facile preparation of BiOBr/cellulose composites by in situ synthesis and its enhanced photocatalytic activity under visible-light. *Carbohydr. Polym.* 195, 393-400.
[5].Fan, W.-J.; Zhou, Z.-F.; Xu, W.-B.; Shi, Z.-F.; Ren, F.-M.; Ma, H.-H.; Huang, S.-W., 2010, Preparation of ZnIn(2)S(4)/fluoropolymer fiber composites and its photocatalytic H(2) evolution from splitting of water using Xe lamp irradiation. *Int. J. Hydrogen Energy* 35, (13), 6525-6530.
[6].He, T.; Zhou, Z.; Xu, W.; Ren, F.; Ma, H.; Wang, J., 2009, Preparation and photocatalysis of TiO2-fluoropolymer electrospun fiber nanocomposites. *Poly* 50, (13), 3031-3036.
[7].Grilla, E.; Petala, A.; Frontistis, Z.; Konstantinou, I. K.; Kondarides, D. I.; Mantzavinos, D., 2018, Solar photocatalytic abatement of sulfamethoxazole over Ag3PO4/WO3 composites. *Applied Catalysis B-Environmental* 231, 73-81.
[8].Zheng, J.; Zhang, L., 2018, Rational design and fabrication of multifunctional catalyzer CO2SnO4-SnO2/GC for catalysis applications: Photocatalytic degradation/catalytic reduction of organic pollutants. *Applied Catalysis B-Environmental* 231, 34-42.
[9].Zhou, X.; Shi, T.; Zhou, H., 2012, Hydrothermal preparation of ZnO-reduced graphene oxide hybrid with high performance in photocatalytic degradation. *Appl. Surf. Sci.* 258, (17), 6204-6211.
[10].Pelaez, M.; Nolan, N. T.; Pillai, S. C.; Seery, M. K.; Falaras, P.; Kontos, A. G.; Dunlop, P. S. M.; Hamilton, J. W. J.; Byrne, J. A.; O'Shea, K.; Entezari, M. H.; Dionysiou, D. D., 2012, A review on the visible light active titanium dioxide photocatalysts for environmental applications. *Applied Catalysis B-Environmental* 125, 331-349.
[11].Tong, H.; Ouyang, S.; Bi, Y.; Umezawa, N.; Oshikiri, M.; Ye, J., 2012, Nano-photocatalytic Materials: Possibilities and Challenges. *Adv. Mater.* 24, (2), 229-251.
[12].Guo, X.; Zhou, X.; Li, X.; Shao, C.; Han, C.; Li, X.; Liu, Y., 2018, Bismuth oxychloride (BiOCl)/copper phthalocyanine (CuTNPc) heterostructures immobilized on electrospun polyacrylonitrile nanofibers with enhanced activity for floating photocatalysis. *J. Colloid Interface Sci.* 525, 187-195.
[13].Konstantinou, I. K.; Albanis, T. A., 2004, TiO2-assisted photocatalytic degradation of azo dyes in aqueous solution: kinetic and mechanistic investigations - A review. *Applied Catalysis B-Environmental* 49, (1), 1-14.

IOP Conf. Series: Materials Science and Engineering **556** (2019) 012036 doi:10.1088/1757-899X/556/1/012036

Study on aluminum honeycomb plate solar air collector and its building heating potential in western Sichuan

T Meng[1], X Meng[1,2], Y Zhao[1] and C Jing[1]

[1] College of Architecture & Environment, Sichuan University, Chengdu, China

[2] Kyushu University, Kyushu, Japan

Abstract. In this paper, the efficiency of aluminum honeycomb plate solar air collector has been analyzed through both experimental and theoretical methods, followed by a study a on the applicability of using honeycomb plate solar air collector in western Sichuan. Solar energy on building vertical wall has been investigated firstly, then the solar heating potential of building facades in three representative regions of western Sichuan has been computed. Lumped parameter method was used to simulate dynamic heating capacity of the collector. The results indicated that aluminum honeycomb plate solar air collector has a relatively large heating potential in western Sichuan.

Keywords: honeycomb plate solar air collector; western Sichuan solar resources; solar heating potential

1. Introduction

According to the statistics, in 2015 the domestic building energy consumption has accounted for 27.5% of total social energy consumption [1]. Heating energy consumption will continue to grow rapidly along with the improvement of living standard and the rising heating demand. Solar air heating technology is a convenient alternative of tradition heating system since building heating requires low-grade thermal energy and solar energy can exactly provide this type of power [2]. Solar air collector, as the main form of solar air heating technology, is simple in structure, easy to install, high in efficiency and therefore worth investigating and promoting. A new type of solar air collector system that introduced a honeycomb plate for enhancing heat transfer is studied, its efficiency is estimated through both experiments and numerical modeling. Its dynamic performance is examined under weather condition of western Sichuan for the purpose of analyzing heating potential of the aluminum honeycomb plate solar air collector.

2. Experimental methods

The experiment was carried out in Chengdu, on the roof of a residential building in Wuhou District. The structure of the collector used in this experiment is shown in Figure 1. The collector is composed of four sections, thermal insulation layer, wooden structural box, aluminum honeycomb core and transparent cover plate (the PC sun board). The aluminum honeycomb is fixed in the middle position of the channel

IOP Conf. Series: Materials Science and Engineering **556** (2019) 012036 doi:10.1088/1757-899X/556/1/012036

as a part of the flow-path, the aperture of the honeycomb hole is 6mm, and the depth of the honeycomb hole is 10mm. The collector has an air inlet and an outlet on the back of the box, which are used for connecting the collector with indoor space and hence circulating air through the collector to heat the room. A radiometer is placed in parallel with the collector panel to measure solar radiation intensity. In order to record airflow velocity, a hot wire anemometer is placed in the center of the inlet and outlet, with its probe in the direction of airflow. Air circulation is mechanically driven by a fan placed at the inlet, providing wind speed that increases from 2m/s to 5m/s by an interval of 0.5m/s. During the experimental period, wind speed of the outdoor environment ranges from 0.15m/s to 1.3m/s, the average is 0.4m/s. Ambient temperature varies form 35°C to 38°C, the relative humidity is approximately 55%, and solar radiation intensity is 700-800w/m^2.

In order to study the actual working condition of the collector, temperature distribution and variation of the airflow are measured and recorded by using several thermocouples. Temperature sensing points are arranged both inside the inlet and outlet air duct and at the center of the air channel. The efficiency of aluminum honeycomb collector can be calculated once the inlet-outlet temperature difference is confirmed.

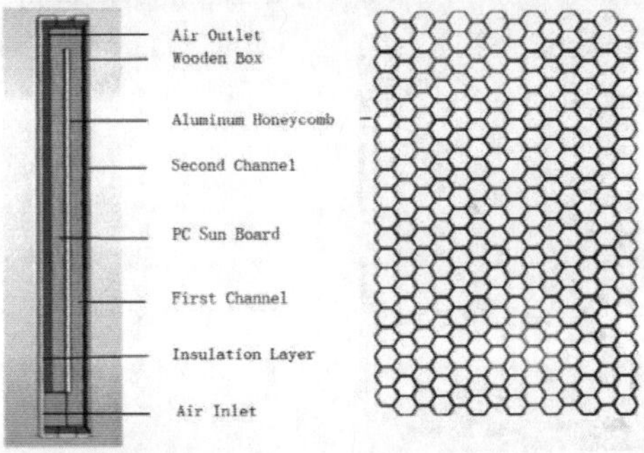

Figure 1. Structure of collector.

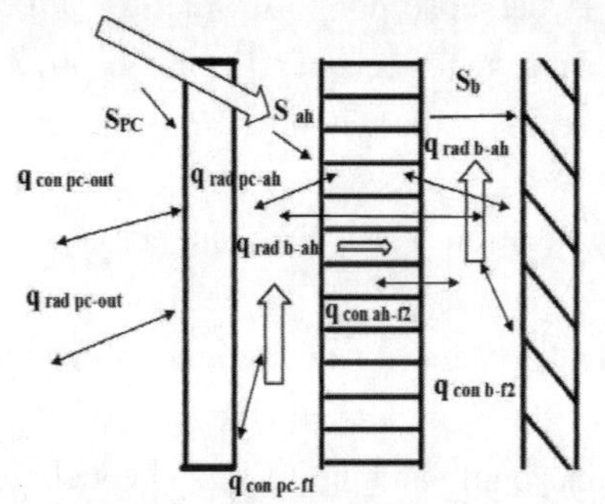

Figure 2. Heat transfer process illustration.

3. Mathematical modeling

Mathematical model of the collector is built on the basis of energy conservation during heat transfer process. As shown in Figure 2, the mathematical model of aluminum honeycomb solar air collector is mainly divided into three parts: thermal insulation, aluminum honeycomb core and transparent cover. The heat transfer process inside the collector is discussed by identifying absorbed and released heat of each section individually (detailed formulas are not presented in this paper). Next, heat balance equations are established to summarize the general process and hence to build up the mathematical model for simulating working condition of the collector. Assume that the heat transfer process between various parts of the collector is steady, the air is incompressible Newton viscous fluid and airflow is uniform and stable. Airflow velocity through each honeycomb hole is assumed to be consistent. Since temperature variation range of the collector is small, it is assumed that the physical parameters of each part are constant.

Heat balance equations of PC sun board, first channel, aluminum honeycomb core, second channel and thermal insulation backboard are established respectively and listed below.

Heat balance equation of PC sun board:

$$S_{pc} = q_{rad\ pc-ah} + q_{rad\ pc-b} + q_{rad\ pc-out} + q_{con\ pc-f1} + q_{con\ pc-out} \tag{1}$$

where S_{pc} is the solar radiation absorbed by the PC sun board, $q_{rad\ pc-ah}$ is the radiation heat transfer between PC sun board and aluminum honeycomb; $q_{rad\ pc-b}$ is the radiation heat transfer between PC sun board and thermal insulation backboard; $q_{rad\ pc-out}$ is the radiation heat transfer between PC sun board and external environment; $q_{con\ pc-f1}$ is the convection heat transfer between PC sun board and the first channel; $q_{con\ pc-out}$ is the convection heat transfer between PC sun board and external environment.

Heat balance equation of first channel:

$$q_{f1} = q_{con\ pc-f1} \tag{2}$$

where q_{f1} is the heat gain of the first channel; $q_{con\ pc-f1}$ is the convection heat transfer between PC sun board and the first channel

Heat balance equation of aluminum honeycomb core:

$$S_{ah} = -q_{rad\ pc-ah} - q_{rad\ b-ah} + q_{con\ ah-f2} \tag{3}$$

where S_{ah} is the absorption of solar radiation by aluminum honeycomb core; $q_{con\ ah-f2}$ is the convection heat transfer between PC sun board and the second channel; $q_{rad\ b-ah}$ is the radiation heat transfer between aluminum honeycomb core and thermal insulation backboard.

Heat balance equation of second channel:

$$q_{f2} = q_{con\ b-f2} + q_{con\ ah-f2} \tag{4}$$

where q_{f2} is the heat gain of the second channel; $q_{con\ b-f2}$ is the convection heat transfer between thermal insulation backboard and the second channel; $q_{con\ ah-f2}$ is the convection heat transfer between aluminum honeycomb and the second channel.

Heat balance equation of thermal insulation backplate:

$$S_b = q_{rad\ b-pc} + q_{rad\ b-ah} + q_{con\ b-f2} + q_{con\ b-out} \tag{5}$$

where S_b is the solar radiation absorbed by the thermal insulation back plate; $q_{rad\ b-pc}$ is the radiation heat transfer between thermal insulation backboard and PC sun board;$q_{rad\ b-ah}$ is the radiation heat transfer between thermal insulation backboard and aluminum honeycomb;$q_{con\ b-f2}$ is the convection heat transfer between thermal insulation backboard and second channel; $q_{con\ b-out}$ is the convection heat transfer between PC thermal insulation backboard and external environment.

4. Experimental and mathematical results

Experimental results are shown below in Figure 3 and Figure 4. It can be seen in Figure 3 that the temperature of the inlet keeps relatively stable, while the temperature of the outlet presents a decreasing trend as the flow velocity increases. As a result, the efficiency of the collector increases with the rising of wind speed. When wind speed is 2m/s, the efficiency is 62%. When the speed is 5m/s, the efficiency reaches as high as 79%.

IOP Conf. Series: Materials Science and Engineering **556** (2019) 012036 doi:10.1088/1757-899X/556/1/012036

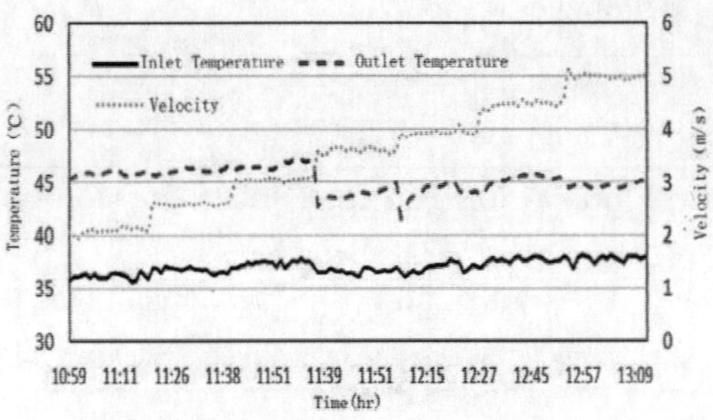

Figure 3. Experimental results.

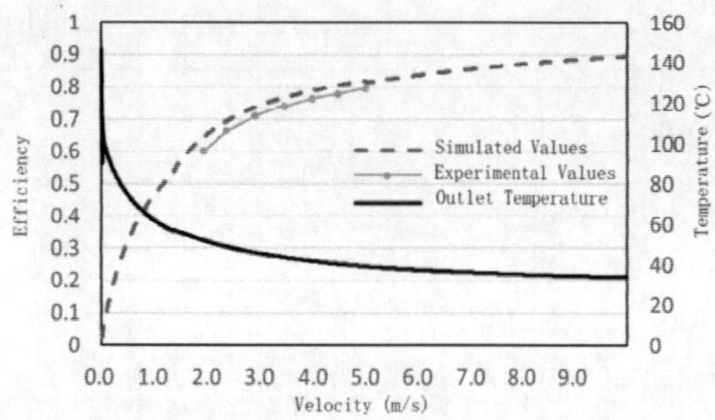

Figure 4. Mathematical vs experimental results.

In order to get the numerical simulation results more accurately, the initial conditions of the simulation are taken from the experimental measurements. Solar radiation intensity is 700W/m^2 and the ambient wind speed is 0.4m/s. Similar to the experiment, this paper takes the efficiency of collector and the air temperature of the outlet as the research object in the simulation, and carries out the computation of the mathematical model with flow velocity 0-10m/s through MATLAB programming.

Figure 4 shows the simulation results, it can be seen that the outlet temperature of the collector reduces, and the efficiency increases continuously with the rising of flow rate. Maximum efficiency is 83% and the minimum outlet temperature is 40°C at 10m/s. Simulated efficiency agrees well with the value obtained from experiment results, with an average error of about 3%. It is proved that the mathematical efficiency calculation model of the collector proposed in this paper is effective and reliable.

5. Heating potential of aluminum honeycomb collector in western Sichuan

In the plateau regions of Sichuan, which are mainly distributed in western part, solar radiation level is much higher than the basin regions due to less clouds blocking. Furthermore, the sunshine in autumn and winter is relatively stronger than that in spring and summer, and the total radiation in most western Sichuan areas is as much as 4500 MJ/m^2~6400 MJ/m^2[3]. The weather condition provides a suitable environment of applying solar air collectors for building heating purpose. Therefore the application performance of aluminum honeycomb solar air collector is examined under these conditions.

We select three typical cities of western Sichuan, Xichang, Jiulong, Ganzi, and build a room model for simulating the actual performance of introducing aluminum honeycomb solar air collector as its heating system. Assume the collector operates at the attainable highest efficiency from the study above.

5.1 Physical model

Assume that the room is in standard heating floor, heating load is only caused by south exterior wall and windows and air infiltration , while the solar energy source of the collector also only comes from south exterior wall (window excluded). The outer layer of the wall is made of clay bricks with cement mortar plastering on both sides, the floor is reinforced concrete. In addition, the external wall is equipped with a polyethylene foam insulation layer, and a hollow double-layer structure is used for the outer window. All interior walls of the room and upper and lower floors are treated as an insulated boundary along their center lines. Structural dimensions of the building envelope are shown in Table 1.

Table 1. Structural dimensions of building envelope.

Name	Width (m)	Height (m)	Thickness (mm)
External wall	4	3	20(Mortar)+40(EPS)+240(Brick)+20(Mortar)
Interior wall	5	3	240(Brick)
Floor	4	5	100 (Reinforced concrete)
External window	1.2	1.8	6(Glass)+10(Air)+6(Glass)

5.2 Solar air collector heating potential dynamic analysis

Dynamic analysis is conducted on the basis of lumped parameter method, i.e. all building envelope and indoor air are treated as a whole, its overall temperature dynamically increase or decrease along with the solar heat penetration or building envelope heat dissipation. Calculation is processed according to the following simplified heat balance formula.

According to F Tang [4], total southward vertical radiation intensity is:

$$I' = I_{dn} + I_{dv} + I_R \tag{6}$$

Where I_{dn} is the diffuse solar radiation intensity, I_{dv} is Vertical diffuse solar radiation intensity, I_R is ground diffuse solar radiation.

Heat transmitted from building envelope to the room is:

$$Q_s = I' \times A \times \eta_1 \tag{7}$$

where A is the area of solar air collector plate, the program selected $1m^2$, $2m^2$, $3m^2$, $\eta_1$is the efficiency of solar air collector, selected 80%.

Wall and floor heat loss ΔU:

$$\Delta U = \sum (V_i\rho_i C_i + V_f\rho_f C_f)(t_i^\tau - t_i^{\tau-1}) \tag{8}$$

where V_i is the volume of the wall,ρ_i is the density of the wall, C_i is heat capacity of wall, t_i^τ is indoor temperature at time τ, $t_i^{\tau-1}$is indoor temperature at time $\tau - 1$, V_f is Floor volume,ρ_f is density of the floor, C_f is Floor heat capacity.

Time-varying heat consumption:

$$Q_1' = k_1 A_1 (t_i^\tau - t_e^\tau) \tag{9}$$

$$Q_2' = 0.287 C_p V \rho_w (t_i^\tau - t_e^\tau) \tag{10}$$

$$Q_3' = I' A_2 \sigma (t_i^\tau - t_e^\tau) \tag{11}$$

where Q_1' is the basic time-varying heat consumption of the envelope,Q_2' is the time-varying air penetration heat consumption of door and window,Q_3' is the time-varying heat consumption of the window, t_e^τ is outdoor temperature at time τ, A_1 is outdoor structure area, K_1 is the total heat transfer coefficient of building envelope,C_p is the specific heat capacity of dry air,ρ_w is the density of air, V is the volume of room,A_2 is the area of window, σ is the transmissivity of glass.

According to heat balance principle, (7)- (11) is written together as:

$$Q_1' + Q_2' + Q_3' + \Delta U = Q_s \tag{12}$$

5.3. Results of solar air collector heating potential dynamic analysis

From November 1st to March 15th is assumed to be the heating period in this section in order to discuss the influence of the heat collection area on the indoor temperature. In final analysis, suitable area of solar air collectors that satisfy the residential indoor heating design condition in "Public Building Energy Efficiency Design Standards is discussed consequently. The design condition requires room temperature keeps above 18°C (comfortable heating temperature). It can be seen from Figure 5, Figure 6, and Figure 7 that 3m^2 of heat collector can meet the heating needs of the three districts of Sichuan.

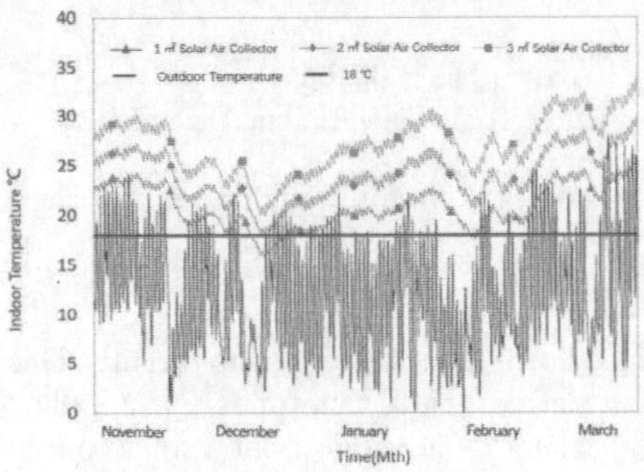

Figure 5. Xichang's heating dynamicanalysis.

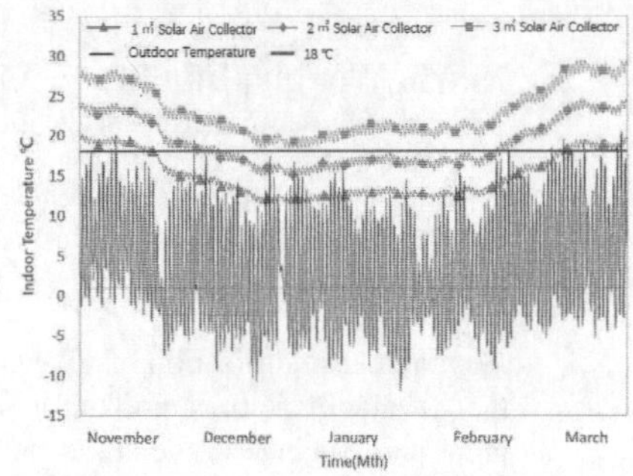

Figure 6. Jiulong's heating dynamic analysis.

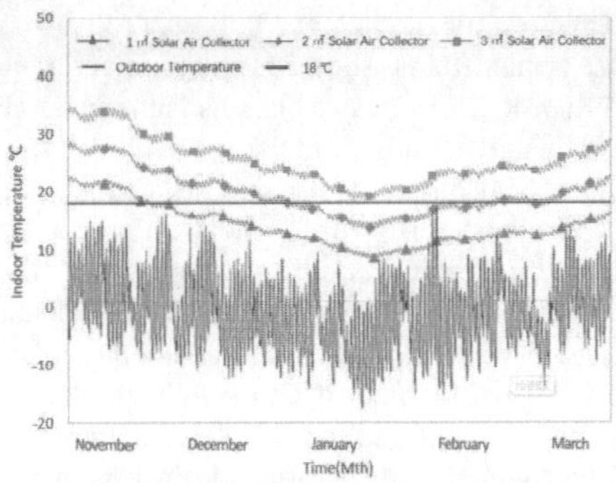

Figure 7. Ganzi's heating dynamic analysis.

6. Conclusion

The new type of aluminum honeycomb plate solar air collector has been tested both experimentally and numerically, and the result verifies its high efficiency as up to 80%. For the actual room case studied in this paper, the collector shows good performance in energy conservation through dynamic heating potential analysis. With the continuous optimization and improvement of the solar collector technics, aluminum honeycomb plate solar air collector has revealed its considerable potential in the application aspect in western Sichuan.

7. References

[1] L Hou 2015 *J Huazhong Architecture* 33(12) *Research on energy consumption of Chinese building*, p94-100.

[2] X Ge and H Ye 2003 *J solar energy* 1(1) Solar air heating system, p13-14.

[3] S Yang, Z Zhan, and X Fan 2007 *J plateau mountain meteorological research* 27(2) Distribution characteristics of solar energy resources in Sichuan province and its utilization suggestions p15-17.

[4] F Tang 1998 *J Building Science Research of Sichuan* (2) Solar radiation intensity on the vertical plane, p53-55.

IOP Conf. Series: Materials Science and Engineering **556** (2019) 012037 doi:10.1088/1757-899X/556/1/012037

Graphene oxide stabilized carbon nanotube-water nanofluids for direct absorption solar collectors

Xinzhi Wang[1,2], Dan Liu[1,2], Xing Liu[1,2] and Yurong He[1,2*]
[1] School of Energy Science and Engineering, Harbin Institute of Technology, Harbin, Heilongjiang 150001, People's Republic of China
[2] Heilongjiang Key laboratory of New Energy Storage Materials and Processes, Harbin, Heilongjiang 150001, People's Republic of China
E-mail: rong@hit.edu.cn

Abstract. The utilization of solar thermal energy has shown remarkable growth recently. Due to the development of nanomaterials, suspensions of nanoparticles (nanofluids) have exhibited great performance in the solar thermal systems, especially in direct absorption solar collectors (DASC). The fundamental advantage of DASC by using nanofluids is the minimizing of solar energy transfer steps and reducing thermal losses in converting sunlight, as nanoparticles could harvest the solar energy directly and achieve good photothermal conversion properties. Carbon nanotube (CNT) based nanofluids showed great potential as a working fluid in DASC. Nevertheless, the stability of CNTs in suspensions is the main obstacle for the large use of CNT nanofluids. In this work, small amount of graphene oxide (GO) was introduced to stabilize CNT-water nanofluids without using any other organic surfactants. The GO stabilized CNT nanofluids exhibited long-term and high-temperature stability. The π-π interactions between the GO and CNT played a significant role for the good dispersion and stability of CNT in water. Furthermore, optical characterizations showed that the GO stabilized CNT nanofluids have widely absorbing over the solar spectrum which enabled highly efficient solar energy collections. Eventually, photothermal conversion performance of GO stabilized CNT nanofluids was tested under an irradiation of simulated solar light and the nanofluid demonstrated a high efficiency in DASC. The long-term stability coupled with broadband absorption properties of GO stabilized CNT nanofluids make them ideal candidates as photothermal conversion media for direct solar thermal collectors.

Keywords: solar thermal utilization; photothermal conversion; nanofluids; carbon nanotubes; graphene oxide; direct absorption solar collectors

1. Introduction

As a kind of clean energy source with huge reserves, solar energy has broad applications. Two most popular approaches of solar utilization, photovoltaic and photothermal utilization can meet the principal energy demands in domestic life and industrial manufacture. The solar-thermal utilization is one of the most direct and straightforward approaches in which absorbed solar radiation is converted to heat for further use, such as power generation, water heating, solar stills, and so on. Hence, the absorption of solar light and the photothermal conversion process are the key issues associated with the solar-thermal method [1].

For traditional solar thermal collectors, the solar energy is absorbed by the black painted surface and converted into thermal energy. Then, the heat is transferred to the heat transfer fluid via thermal conduction and thermal convection. There would be inevitable thermal losses from the absorber to the environments due to the high temperature at the absorption surface. Therefore, a kind of volumetric solar energy harvesting technology was developed, which was called direct absorption solar collectors

(DASC). Usually, solar light absorption media is dispersed into the fluid in a DASC to realize direct solar absorption by working fluid. By using this method, the photo-thermal conversion is achieved in the fluid media and it achieves less thermal losses and high efficient solar energy collection.

The DASC was firstly proposed by Minardi and Chaung by using a black fluid (water with 3.0 g/L, India ink) to absorb the solar radiation. Recently, nanofluids (suspensions of nanoparticles) have attracted much attentions of scientists due to its unique thermophysical properties including high thermal conductivities [2], high thermal diffusivities [3] and high heat-transfer coefficients [4]. In 2009, Tyagi et al. [5] investigated theoretically the performance of a DASC containing an Al_2O_3-water nanofluid. Otanicar et al. [6] experimentally studied the effects of three kinds of nanofluid (carbon nanotubes, graphite and silver-based) in a direct solar absorption system. Wang et al. [7] experimentally investigated the solar-collection performance of single-wall carbon-nanotube-based nanofluids for direct and volumetric solar steam generation.

Carbon based nanomaterials have great advantages in the preparing of nanofluids for DASC due to their high light absorption properties. For instance, carbon nanotubes (CNT) have extremely high capability of absorbing solar spectrum and acceptable stability. One of the major problems for CNT nanofluids applications in DASC is the suspension instability. Due to the bad dispersity of CNT, a large amount of organic surfactant was required for the dispersing of CNT into water and the method is considered as a common but easy-handling, time-saving and cost-effective choice for mass production. However, it might reduce the other thermophysical properties of CNT nanofluid, such as rheological properties. Furthermore, the stability of CNT nanofluids under long time solar irradiation is underlying.

Graphite oxide sheet, now called graphene oxide (GO), is the product of chemical exfoliation of graphite and has been known for more than a century. GO has been known to disperse well in water since its first discovery over a century ago and thus has been routinely described as hydrophilic in the literature. GO's water dispersity and hydrophilicity have been mainly attributed to the ionizable edge-COOH groups. However, its basal plane is essentially a network of hydrophobic polyaromatic islands of unoxidized benzene rings. Therefore, GO should be viewed as an amphiphile with a largely hydrophobic basal plane and hydrophilic edges. Recently, it has been investigated as a new kind of colloidal surfactant. Since GO has many π-conjugated aromatic domains in its basal plane, it should be able to strongly interact with the surface of graphite and CNTs through π-π attractions [8]. In this work, a kind of GO stabilized CNT nanofluids was prepared and the photothermal performance of CNT/GO nanofluids was investigated under concentrated solar light. In addition, the effects of solar light intensities and mass concentrations of CNT in nanofluids were discussed. Moreover, the stability of CNT/GO nanofluid for a long time and after solar light irradiation was demonstrated.

2. Experimental section

2.1. Materials

Graphite powder, CNT powder, ammonia water, $NaNO_3$, $KMnO_4$, H_2SO_4, and H_2O_2 were supplied by Aladdin Chemical Co., Ltd.. Double-deionized water (DDI water, Sartorius Water Purification System, arium® mini; 18.2 MΩ) was used in these experiments. Graphene oxide (GO) was synthesized via a chemical exfoliation of the graphite powder by a modified Hummers method [9] as follows: 46 mL concentrated H_2SO_4 was slowly added into a mixture of 2 g graphite powder and 1 g $NaNO_3$ at 0 °C. Then, 6 g $KMnO_4$ was added with stirring while the temperature of the mixture was maintained below 20 °C using an ice bath. Afterward, the ice bath was removed, and the temperature of the mixture was raised to 35 °C and stirred for 30 min. Then, 92 mL deionized water was added to the mixture with stirring for 15 min, followed by the addition of 60 mL H_2O_2 solution at 60 °C. After the suspension turned brown, it was washed with deionized water and centrifuged for several times. Finally, the GO powder was obtained by freeze-drying method and kept for further use.

2.2. Preparation of nanofluids

A two-step method was used to prepare the GO stabilized CNT nanofluids. Firstly, the GO was dispersed into DDI water with a high concentration of 8 mg/mL and stored for the further use. Then CNT powder was added in to the GO-water dispersion at a mass ration of 1:1 (CNT/GO). The initial concentration of CNT in the solution was 100 ppm (0.1 mg/L). Subsequently, the mixture of CNT/GO-water was sonicated by a high-power tip ultrasonic oscillator for 1 hour to make sure CNTs well dispersed in GO-water. After sonication, the supernatant was carefully collected after the sample aged for 1 day to remove undispersed chunks. The well dispersed supernatant was GO stabilized CNT nanofluids. Furthermore, the CNT/GO nanofluids were diluted for the photothermal collections.

2.3. Solar thermal conversion experiments

Figure 1 shows the experimental setup for direct solar energy collections. Solar light was generated by a solar simulator (CEL HXF300, CEAULIGHT, Beijing, China). The light was generated by a 300W Xenon lamp, which could realize the exportation of collimated light with high energy. The Xenon lamp is one of the most common types of lamp for continuous solar simulators. It offers high intensities and an unfiltered spectrum matching reasonably well to sunlight. An acrylic tube filled with working fluid was placed on the electric balance and was used as a solar collector. The temperature was measured and recorded using thermocouples (TT-T-40-SLE, Omega, US) and a data-acquisition system (34972A, Agilent Technology, Santa Clara, CA, US). Due to the open system of collectors, there were evaporations due to the high temperature of the nanofluids. Therefore, the mass loss due to the evaporation was also recorded to make a better evaluation of solar thermal collection performance of the prepared nanofluids. Mass change due to evaporation was measured using an electric balance (Practum313–1CN, Sartorius, Göttingen, Germany). The inner height and diameter of the solar collector were 60 mm and 35 mm, respectively. The size and weight of the designed solar collector tube could be matched with the size of the solar light simulator and weighing range of the electric balance in the experiments. Five T-type thermocouples were inserted into the solar collector at different heights (H = 10, 20, 30, 40, and 50 mm).

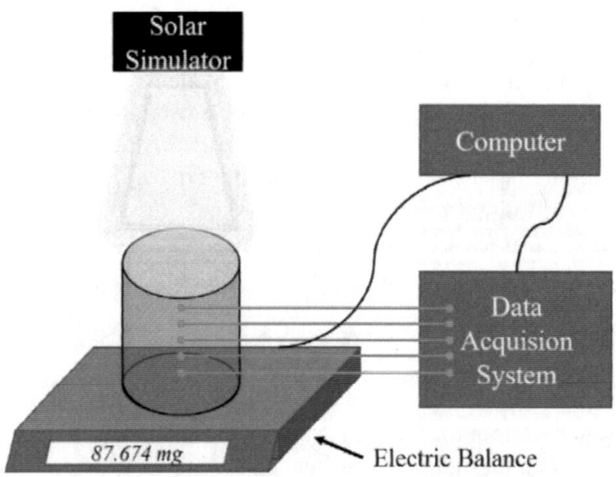

Figure 1. The diagram of experimental setup for direct solar energy.

To investigate the effect of solar illumination intensity, the solar collector was placed under different solar illumination intensities (1, 3, 5, and 7 sun, where 1 sun = 1 kW m^{-2}). The CNT/GO nanofluids with different concentrations (6.25, 12.5, 25, 50, and 100 ppm) were prepared to investigate the effect of CNT concentrations on the photothermal collection performance. Usually, the photo-thermal conversion efficiency η of water and the three nanofluids was calculated according to the following formula,

$$\eta = \frac{mc_p(T_s - T_i)}{IA\Delta t} \quad (1)$$

where m and C_p are the mass and specific heat of the nanofluids, T_i is the initial temperature, T_s is instantaneous temperature, A is the top surface area of the receiver, I is the heat flux of the incident light, and Δt is the time exposed to the light radiation. Considering the heat losses by the evaporation of the water, a modified photo-thermal conversion efficiency η model was proposed as follow,

$$\eta = \frac{(m_i - \Delta m)c_p(T_s - T_i) + \Delta mh}{IA\Delta t} \quad (2)$$

where m_i is the initial mass of the nanofluids, Δm is the weight losses of the nanofluids due to evaporation, h is the latent heat of vaporization for water.

2.4. Characterization

The morphologies of the HP-Au particles were examined by field emission scanning electron microscopy (FESEM, SUPRA 55, ZEISS, Germany). High-power tip ultrasonoscope (Biosafer 900-92, 900W, Biosafer, Beijing, China) was used to disperse the GO and CNT into water for preparing nanofluids. Ultraviolet-visible (UV-Vis) spectras were recorded on a two-beam UV-Vis spectrometer (TU1901, China). During the UV-Vis spectra test, the fluids are placed in a quartz cuvette with optical length of 1 mm and the DDI water is used as blank reference. All samples were 4 times diluted for the UV-Vis spectra.

3. Results and discussion

A tremendous amount of efforts have been devoted to making CNTs water processable through wrapping by water-soluble materials. Since many surfactants for dispersing CNTs have polyaromatic components, GO should be able to adhere to CNTs and disperse them in water as well. The photograph and microstructures of the CNT/GO-water nanofluids and CNT dispersion was shown in Figure 2. The left vial in Figure 2a shows that CNTs indeed dispersed well in GO-water with a 1:1 mass ratio after sonication. While the CNT can not disperse in water alone, as shown in the right vial of Figure 2a. Microscopy analysis revealed that the initial CNTs samples were heavily entangled (Figure 2b), which remained largely unaffected by sonication in water (Figure 2c). In contrast, CNTs sonicated in GO water were completely disentangled. Extensive microscopy observations by SEM (Figure 2d) revealed that almost all the CNTs in the sample were adhered to GO (Figure 2d).

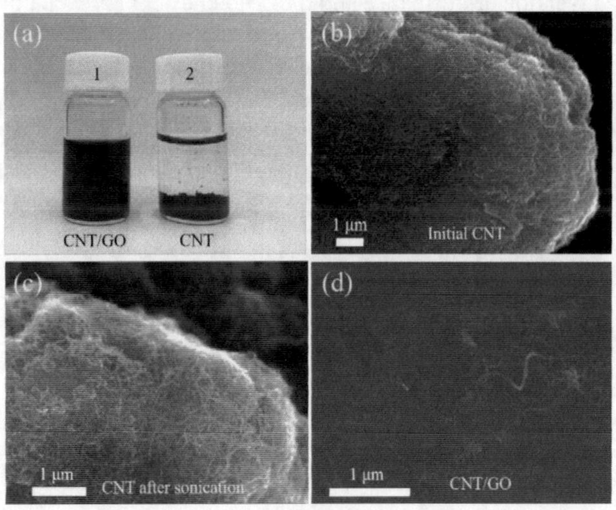

Figure 2. Dispersion of CNTs in GO-water: (a) photograph of CNT/GO dispersion and CNT dispersion. SEM images of (b) initial CNT, (c) CNT after sonication and (d) CNT dispersed in GO-water.

Furthermore, the stability of the CNT/GO nanofluids were monitored by its optical absorbance over a period of 10 hours (Figure 3), which remained nearly constant after sonication. Figure 3a shows the spectra of CNT/GO nanofluids for 10 hours and Figure 3b shows the changes of the values at the peak wavelength. The results demonstrated that CNT/GO nanofluids could keep good stability for a long time.

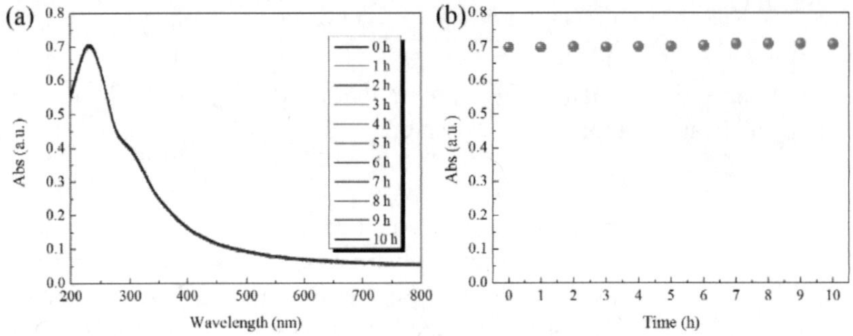

Figure 3. (a) UV-Vis spectra of CNT/GO nanofluids for 10 hours and (b) the variation of values at peak wavelength.

For the photothermal conversion test of CNT/GO nanofluids, the nanofluids with different concentrations were characterized under a solar light intensity of 7 suns (1 sun = 1kW/m^2). Figure 4 shows the photothermal conversion performance of CNT/GO nanofluids. It can be seen from figure 4a that the mass losses due to evaporation increased as the mass concentration of CNT increased from 6.25 ppm to 100 ppm. For the concentrations above 12.5 ppm, the mass losses of water evaporation did not increase greatly. The mass losses were around 5 mg for the CNT/GO nanofluids with concentrations form 12.5 ppm to 100 ppm. The average temperatures of the bulk fluids increased as the solar light irradiation absorbed by nanofluids as show in Figure 4b. While the average temperatures of the bulk fluids decreased as the increase of CNT concentrations. It means that the nanofluids with low concentration make more contribution in sensible heat storage and the nanofluids with high concentration make more contribution in latent heat for evaporation. Furthermore, the photo-thermal conversion efficiency of CNT/GO nanofluids with different concentrations was calculated, as demonstrated in Figure 4c. The photo-thermal conversion efficiency of the nanofluids increased firstly and then decreased as the increasing of fluid temperature. The mass concentration of 12.5 ppm exhibited the best photo-thermal conversion performance and the efficiency could reach up to ~90% at the working temperature around 35 °C. It is ascribed to the both relative high evaporation rate and average temperature increase for the CNT/GO nanofluids with concentration of 12.5 ppm.

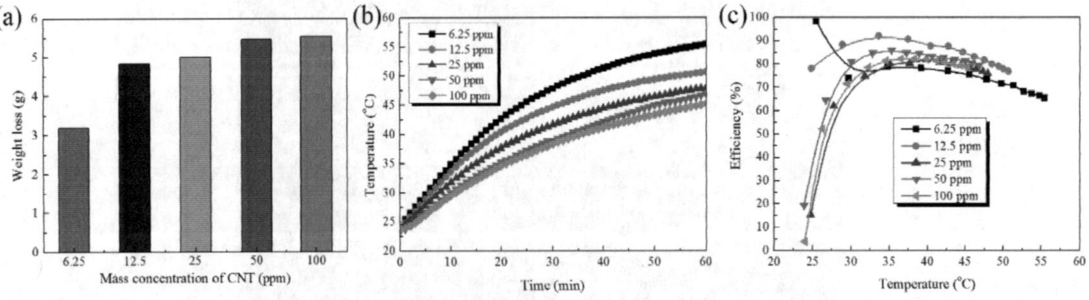

Figure 4. Solar thermal conversion performance of CNT/GO nanofluids with different concentrations: (a) weight losses Vs time, (b) average temperature Vs time, (c) photo-thermal conversion efficiency.

Eventually, the stability of the CNT/GO nanofluids before and after photothermal conversion experiments was investigated. The UV-Vis spectra of the CNT/GO nanofluids with different concentrations before and after solar light irradiation for 1 hour under 7sun is shown in Figure 5a. It can be seen that the absorption intensities slightly increased at low concentrations and increased obviously at high concentration after irradiation. In addition, the photographs of the sample before and after the solar light irradiation is shown in Figure 5b and 5c, respectively. The CNT/GO nanofluids still retained high dispersion rate and no sediment was observed. While the colour of the sample got darker after the solar irradiation, which could also be confirmed from the UV-Vis spectra in Figure 5a. The improved light absorption properties of CNT/GO nanofluids might be ascribe to the reduction of GO.

IOP Conf. Series: Materials Science and Engineering **556** (2019) 012037 doi:10.1088/1757-899X/556/1/012037

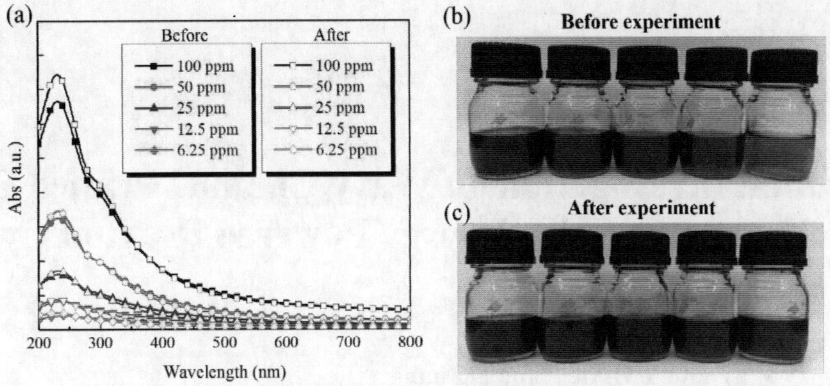

Figure 5. (a) UV-Vis spectra of CNT/GO nanofluids before and after the experiments, photographs of CNT/GO nanofluids (b) before and (c) after the experiments.

4. Conclusion

In a summary, small amount of graphene oxide (GO) was introduced to stabilize CNT-water nanofluids without using any other organic surfactants in this work. The results demonstrated that CNT/GO nanofluids could stable for over a long time. In addition, the CNT/GO nanofluids could retained high dispersivity even after a concentrated solar light irradiation (7 kw/m^2). The π-π interactions between the GO and CNT played a significant role for the good dispersion and stability of CNT in water. Eventually, the photothermal conversion properties of CNT/GO nanofluids in a DASC was characterized. The CNT/GO nanofluids exhibited excellent photothermal performance. The solar thermal collection efficiency could reach up to ~80%. CNT/GO nanofluids is a promising alternative for the future solar thermal utilizations.

5. References

[1] Tang B, Wang Y, Qiu M, and Zhang S 2014 *Sol Energy Mater Sol Cells* **123** 7–12.
[2] Wen D, and Ding Y 2004 *J Thermophys Heat Transf* **18** 481–5.
[3] Zhang X, Gu H, and Fujii M 2006 *Int J Thermophys* **27** 569–580.
[4] Qi C, Wang G, Yang L, Wan Y, and Rao Z 2017 *Int J Heat Mass Transf* **105** 664–72.
[5] Tyagi H, Phelan P, and Prasher R 2009 *J Sol Energy Eng* **131** 041004.
[6] Otanicar TP, Phelan PE, Prasher RS, Rosengarten G, and Taylor RA 2010 *J Renew Sustain Energy* **2** 033102.
[7] Wang X, He Y, Cheng G, Shi L, Liu X, and Zhu J 2016 *Energy Convers Manag* **130** 176–83.
[8] Chang X, Li Z, Zhai X, Sun S, Gu D, and Dong L 2016 *Mater Des* **98** 324–32.
[9] Guo P, Chen P, and Liu M. 2013 *ACS Appl Mater Interfaces* **5** 5336–45.

Acknowledgments
This research work has been financially supported by the National Natural Science Foundation of China (Grant No. 51676060), the Natural Science Funds of Heilongjiang Province for Distinguished Young Scholar (Grant No. JC2016009), the Science Creative Foundation for Distinguished Young Scholars in Harbin (Grant No. 2014RFYXJ004).

IOP Conf. Series: Materials Science and Engineering **556** (2019) 012038 doi:10.1088/1757-899X/556/1/012038

Optimization Investigation on Wavy Channel Printed Circuit Heat Exchanger for Solar Energy Powered Brayton Cycle

Y Z Li[1] and Y H Xie[1,*] and D Zhang[2]

[1] Shaanxi Engineering Laboratory of Turbomachinery and Power Equipment, School of Energy and Power Engineering, Xi'an Jiaotong University, Xi'an, China

[2] Key Laboratory of Thermo-Fluid Science and Engineering, Ministry, School of Energy and Power Engineering, Xi'an Jiaotong University, Xi'an, China

E-mail: yhxie@mail.xjtu.edu.cn

Abstract. The Printed Circuit Heat Exchanger(PCHE) has received widely attention for their high effectiveness and compact size and they appear promising to be employed as the heat exchanger in solar thermal system. In this research, the thermal and hydraulic performance of the wavy channels for solar energy were investigated detailed in different amplitudes, inclined angles and channel diameters. The three-dimensional numerical analysis was performed by computational fluid dynamics(CFD) tool and the model is validated by comparing with previous experimental data. Besides, it was difficult to get the universal correlations of Nusselt number and friction factor for various geometrical parameters. In this article, the Kriging response surface method is applied to predict the friction factor and *j* factor for PCHE wavy channels in different geometrical parameters based on the CFD simulation results. The thermal-hydraulic characters predicted by Kriging response surface showed overall good agreement with the CFD results. Moreover, one of the most popular multi-objective genetic algorithms NSGA-II was adopted in optimization. to maximize *j* factor, while minimize the fanning friction factor.

Keywords: solar thermal system; printed circuit heat exchanger(PCHE); wavy channels; Kriging response surface; non-dominated sorting genetic algorithms (NSGA-II)

1. Introduction

The global energy demand dramatic increase while the fossil resource is keeping in continuous consumption. Solar and other new clean energy are all expected alternatives to replace the conventional fossil resource. In the past three decades, large studies around the applications of solar energy have been done especially for the solar energy power for electricity generation and heat collection/refrigeration thermal cycle. Yamaguchi et al.[1, 2] proposed a solar energy powered Rankine cycle for combined heat and power. Their designed and constructed an experimental prototype in Kyoto, Japan, the experimental results showed that the solar energy powered Rankine cycle using CO_2 operate stably with electricity and heat generated. Fahad et al.[3] conducted five different SCO_2 Brayton cycles based on solar power tower and they found that the recompression cycle has the highest thermal efficiency. Pardeep et al.[4] investigate the SCO_2 Brayton cycle for possible concentrated solar power applications and compared it with trans- and sub-critical operations of the same fluid for thermal efficiency, specific work output and magnitude of irreversibility generation. The results show that the SCO_2 Brayton cycle has a comparatively better performance. Francesco et al.[5] reviewed all the works published in the SCO_2 cycle including the cycle powered by solar energy.

The PCHEs are more and more popular in many areas, including nuclear energy, electronic cooling and ocean engineering. Compared with the conventional shell and tube heat exchanger, the PCHEs can reduce 85% weight and only take place one third room. Besides, the diffusion bonding technique allows the PCHEs enduring a wide range of temperature and pressure. Due to plenty of advantages, researchers pay much attentions on PCHE and make abundant studies around the thermal-hydraulic performance. Justin et al. have investigated the thermal and hydraulic performance for straight channel PCHEs in Very High Temperature Reactors(VHTRs). Ting Ma et al.[6]identifies the local thermal-hydraulic performance of zigzag channel PCHE at high temperature of 900°C. Jeon et al.[7] outlined an innovative type of PCHE, considered the effect of channel sizes, the spacing between channels and the channel cross-sectional shape on the thermal-hydraulic performance. Aneesh et al.[8] clarified the heat transfer rate and thermo-hydraulic performance of trapezoidal, sinusoidal and triangular channel PCHE. The helium test loop and numerical simulations were conducted by In Hun Kim et al.[9], they obtained the fanning-friction factor and Nusselt number correlations with the operation condition parameters.

The academic community has extensively explored the thermal performance for solar Brayton cycle with SCO_2. However, little research has been conducted of the recuperators. The study explores thermal-hydraulic performance of sinusoidal channel PCHE used in solar Brayton cycle. We then model and optimize the recuperators with the goals of minimizing the friction factor and maximizing the *j* factor.

2. Calculation model and numerical method

2.1. System description

The schematic of SCO_2 Brayton cycle is shown as figure 1. The important components are turbine, compressor, recuperator, air cooled pre-cooler and receiver. The receiver absorbs the concentrated solar energy to heat the fluid, the hot fluid goes through the turbine and then the hot fluid heats the cold fluid from the compressor in the recuperator which consists of several channels and multiple plates.

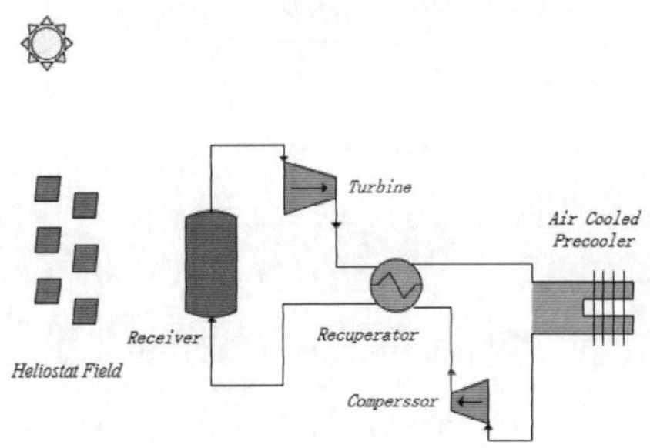

Figure 1. Schematic of solar energy powered SCO_2 Brayton cycle

2.2. Calculation model

The PCHE system as illustrated in figure 2 right contains a stack of photochemical etching planets with many semicircular fluid passages. The sinusoidal channel PCHEs are investigated and the reduced models are used in three-dimensional simulations. In this work, the radium of the hot channel R_h and cold channel R_c change from 0.5mm to 1.5mm, the pitch length L_p ranges from 10mm to 30mm and the angle α varies from 5° to 45° as illustrated in figure 2 left. The cold and hot SCO_2 are utilized as the working fluid, the inlet temperature and the outlet pressure in cold fluid are set as 450K and 20MPa, while those in hot fluid are set as 630K and 7.4MPa respectively.

The ANSYS FLUENT is used to numerical simulate the PCHEs and the finite volume method is taken to describe the governing equations. The RNG k-ε model is adopted for modelling the

IOP Conf. Series: Materials Science and Engineering **556** (2019) 012038 doi:10.1088/1757-899X/556/1/012038

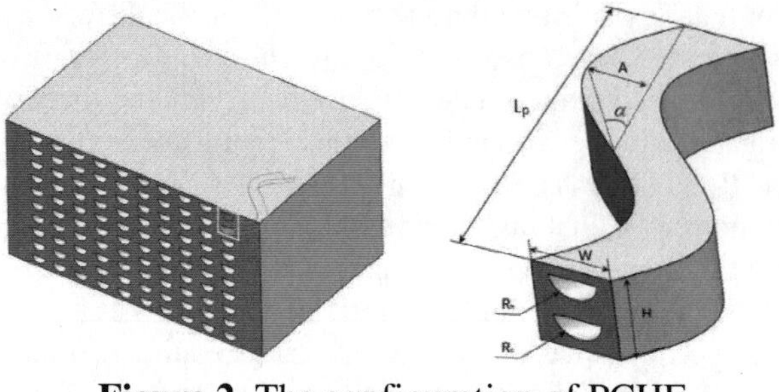

Figure 2. The configuration of PCHE

turbulence. The mass flow inlet and pressure outlet boundary conditions are specified on the cold channel and hot channel correspondingly. The periodic boundary condition is applied on the up and down walls and the adiabatic boundary condition is set for the front, back walls and side walls. The NIST real gas model is applied to estimate the thermal properties of the SCO2. The thermal-hydraulic characteristic considered include the heat transfer coefficient, Nusselt number and fanning friction factor as illustrated in equations (1-4). The *j* factor is shown in equation (5).
The hydraulic diameter of the channel is

$$D_h = \frac{\pi R}{(\pi / 2 + 1)} \tag{1}$$

where the R is the channel diameter. The heat transfer coefficient is described as followed

$$h = \frac{q}{T_b - T_w} \tag{2}$$

where the T_b is the volume average temperature of fluid, T_w is the area weighted temperature of the channel walls. The Nusselt number is shown as followed

$$Nu = \frac{hD_h}{\lambda} \tag{3}$$

where the λ is the heat conductivity of fluid. The fanning friction factor is expressed as followed

$$f = \frac{\Delta p \cdot D_{\mathrm{h}}}{2\rho L u_{\mathrm{in}}^2} \tag{4}$$

where the Δp is the pressure different between the inlet and outlet., ρ is the fluid density, L is the total length of channels and the u_{in} is the inlet velocity. The *j* factor is expressed as followed

$$j = \frac{D_{\mathrm{h}}}{4L} Pr^{2/3} N \tag{5}$$

where the *N* is the heat transfer units number of the whole PCHE.

3. Optimization methods

In this study, the Kriging response surface and multi-objective genetic algorithm NSGA-IIare combined to optimize the wavy PCHEs. As for the recuperators, the thermal and hydraulic performance should be considered. The geometric parameters include the radium of cold and hot channel, the width and height of single unit, the length of single pitch and the angle of the wavy PCHE.

*3.1. Kriging response surface and Non-dominated Sorted Genetic Algorithm-**II***

The Kriging response surface is a spatially optimal linear unbiased estimation method based on the spatial distribution of spatial attributes. The semi-variogram is used to determine the weights of the points to be interpolated in order to realize the estimation of the attributes of the points to be interpolated. The method takes into account the spatial structure and randomness of the variables and statistics by simulating the correlation and variability of the spatial distribution of geographical

phenomena. The Kriging response surface provide an approximation of the input parameters of the values of objectives, which allows a faster optimization operation. Firstly, only limited cases selected by space-filling designs should be calculated of three dimensions to fit the model. Secondly, during the optimization the population is simulated by the Kriging models. Therefore, the Kriging response surface is used as the surrogate model based on the simulation results in this paper.
Over the past decade, plenty of multi-objective evolutionary algorithms(MOEAs) have been proposed. The MOEAs should find multiple Pareto-optimal solutions in a single simulation run, and the NSGA is the first algorithm achieving the goal. However, the NSGA algorithm has high computational complexity, lacks elitism, needs the user to specify sharing parameters. To overcome those weak points, the improved algorithm NSGA-**II** was proposed by Kalyanmoy et al.

3.2. Objective functions and design parameters
In this study, the *j* factor and the fanning friction factor were considered as the objective functions. Besides, the geometric parameters such as the incline degree, the pitch length, the radium of channel, the height and the width were set as the design parameters. The Kriging metamodel fit was created by using one hundred design points which were generated by Latin Hypercube sampling. Latin hypercube and randomly selects survey units without any grouping, classification and queuing. Each unit of the sample is completely independent, and there is no certain correlation between them.

4. Results and discussions

4.1. Sensitivity analysis of thermal and hydraulic performance
The effects of the radium on the two channels, the inclined angle and the pitch length on the fanning friction factor and the j factor are shown in figure 3. In this section, the radium of cold/hot channel ranges from 1.5mm to 0.5mm in step of 0.1mm, the inclined angle ranges from 45° to 5° in step of 4°, the pitch length ranges from 30mm to 10mm in step of 2mm.
figure 3(a) shows the effects of the design parameters on the fanning friction factor. The results prove that the cold fluid fanning friction factor decrease with all the variables except the inclined angle. It is obvious that the bigger the radium is the lower the velocity is, so that the pressure drop is relatively lower and the fanning friction factor declines. The pressure drop is much bigger with a longer channel, while the pressure drop will be bigger when the inclined angle increase.
The effects of design variables on the *j* factor is shown in figure 3(b). The results reveal that the j factor increases with the radium of hot channels. The cases with smaller hot channel has a larger Nusselt number but a much larger Prandtl number, which causes a higher j factor. However, the j factor decreases with the decrease of the inclined angle and the pitch length which because of the lower heat transfer coefficient. The j factor decreases with the radium of cold fluid declining. The smaller cold channel leads to higher heat transfer coefficient, but the influence of the radium reduction on the Nusselt number dominants.
As indicated in figure 3(c), the trends of hot fluid fanning friction factor are similar to the cold one. The radium of channels and the pitch length tend to decrease the fanning friction factor because of the lower pressure drop.
Regard of the j factor of hot fluid, the trends of hot fluid fanning friction factor expect that of the radium of hot channel are almost similar to the cold one as shown in figure 3(d). The j factor decreases with the decrease of the inclined angle and the pitch length which because of the lower heat transfer coefficient. The j factor decreases with the radium of cold fluid declining because the reduction of Nusselt number. However, the j factor decreases first and then increases with the radium of hot channel. It is observed that the Prandtl number and the Nusselt number both increase with the radium of hot channel, but the Prandtl number increasement is more rapid first. In addition, the values of the hot fluid j factor are a little higher that the cold fluid j factor.

4.2. Optimization results
In this section, multi-objective optimization using NSGA-IIis conducted to trade-off between the fanning friction factor and the j factor for them are mutually conflicting where no single solution can

IOP Conf. Series: Materials Science and Engineering **556** (2019) 012038 doi:10.1088/1757-899X/556/1/012038

fully satisfy together. In fact, a set of design points displayed as a Pareto optimal solutions curve will be selected. figure 4 shows the results of Pareto optimal curve with the corresponding design parameters. It is obvious that the fanning friction factor increases with the j factor. Moreover, ten of the selected optimal parameters with a less fanning friction factor and a higher j factor are shown in table 1. From the table, it can receive that the radium tends to the two extremes of 1.5mm and 0.5mm, the larger inclined angle prefers a smaller pitch length.

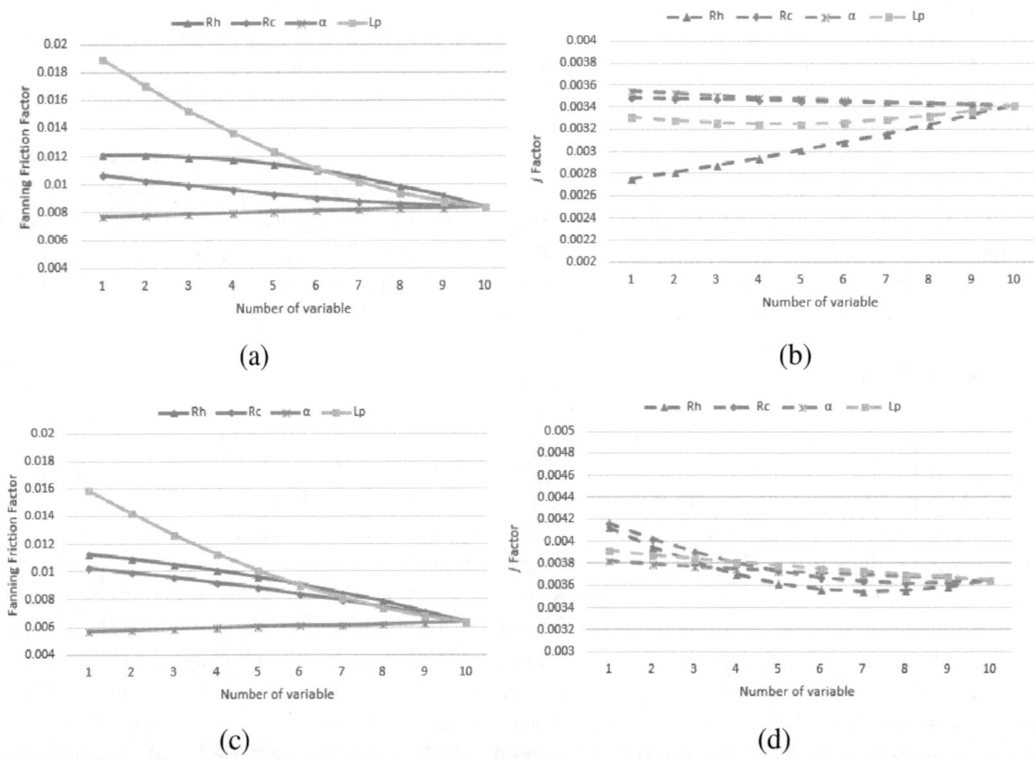

Figure 3. The effect of radium of hot channel, the radium of cold channel, the inclined angle and the pitch length on (a)fanning friction factor of cold fluid (b) *j* factor of cold fluid (c) fanning friction factor of hot fluid (d) *j* factor of hot fluid

5. Conclusions

In this paper, the Kriging response surface and the Non-dominated Sorted Genetic Algorithm-II were combined for the optimization of wavy channel PCHEs. The main conclusions are listed as below:

(1) The sensitivity analysis of the objective functions with respect to the design parameters are analyzed from qualitative to qualitative which is significant to guide the wavy channel PCHEs of the Brayton cycle powered by solar energy.

(2) The obtained Pareto optimal points reveal that the radium tends to the two extremes of 1.5mm and 0.5mm, the larger inclined angle prefers a smaller pitch length.

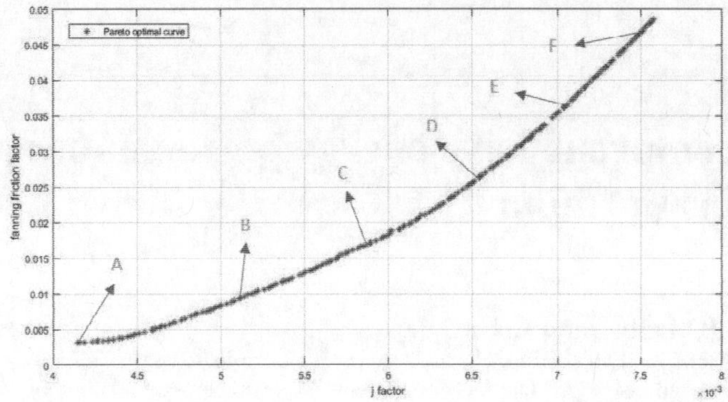

Figure 4. Pareto optimal curve of fanning friction factor and j factor

Table 1. The selected Pareto optimal values

Points	R_h/mm	R_c/mm	α/°	L_p/mm	f	j
A	1.50	0.55	7.96	28.73	0.0043	0.0041
B	1.38	0.66	12.77	19.69	0.0086	0.0051
C	0.50	1.23	38.80	10.30	0.0180	0.0052
D	0.50	1.41	42.25	12.0	0.0262	0.0065
E	0.5	1.49	44.11	11.95	0.0372	0.0072
F	1.5	1.5	45	10	0.0468	0.00746

Reference

[1] Yamaguchi H, Zhang X R, Fujima K, Enomoto M and Sawada N Solar energy powered Rankine cycle using supercritical CO2 2006 *Applied Thermal Engineering* **26** 2345-2354

[2] Zhang X R, Yamaguchi H, Uneno D, Fujima K, Enomoto M and Sawada N Analysis of a novel solar energy-powered Rankine cycle for combined power and heat generation using supercritical carbon dioxide 2006 *Renewable Energy* **31** 1839-1854

[3] Al-Sulaiman F A and Atif M Performance comparison of different supercritical carbon dioxide Brayton cycles integrated with a solar power tower 2015 *Energy* **82** 61-71

[4] Garg P, Kumar P and Srinivasan K Supercritical carbon dioxide Brayton cycle for concentrated solar power 2013 *The Journal of Supercritical Fluids* **76** 54-60

[5] Crespi F, Gavagnin G, Sánchez D and Martínez G S Supercritical carbon dioxide cycles for power generation: A review 2017 *Applied Energy* **195** 152-183

[6] Ma T, Li L, Xu X-Y, Chen Y-T and Wang Q-W Study on local thermal–hydraulic performance and optimization of zigzag-type printed circuit heat exchanger at high temperature 2015 *Energy Conversion and Management* **104** 55-66

[7] Jeon S, Baik Y-J, Byon C and Kim W Thermal performance of heterogeneous PCHE for supercritical CO 2 energy cycle 2016 *International Journal of Heat and Mass Transfer* **102** 867-876

[8] Aneesh A M, Sharma A, Srivastava A and Chaudhury P Effects of wavy channel configurations on thermal-hydraulic characteristics of Printed Circuit Heat Exchanger (PCHE) 2018 *International Journal of Heat and Mass Transfer* **118** 304-315

[9] Kim I H and No H C Thermal hydraulic performance analysis of a printed circuit heat exchanger using a helium–water test loop and numerical simulations 2011 *Applied Thermal Engineering* **31** 4064-4073

IOP Conf. Series: Materials Science and Engineering **556** (2019) 012039 doi:10.1088/1757-899X/556/1/012039

Study on the comprehensive energy performance of different shading systems in China

X Li[1], JQ Peng[1*] and C Curcija[2]
[1] College of Civil Engineering, Hunan University, Changsha 410081, China
[2] Windows and Envelope Materials Group, Lawrence Berkeley National Laboratory, United States
E-mail: jqpeng@hnu.edu.cn

Abstract. This study evaluates the impacts of different shading systems on the comprehensive energy performance of buildings in China. Firstly, simulation models for four shading systems were developed using THERM, WINDOW and EnergyPlus. Then, four shading systems with different qualities, setting modes and baseline window systems were defined and simulated. Two cities, Beijing and Guangzhou, were selected as representative cities for heating-dominated and cooling-dominated regions in this study. Finally, the comprehensive energy performance, including cooling electricity consumption, heating electricity consumption and lighting electricity consumption for buildings with different shading systems were analysed. The results show that in Beijing, the most suitable shading system is roller shade with D quality (the combination of performance parameters), H (half shading) setting mode and low-e glazing system, while in Guangzhou, the most suitable shading system is roller shade with A quality, F (Full shading) setting mode and low-e glazing system. Besides, buildings in Beijing with suitable shading systems can achieve at least 20% energy saving potential (including air conditioning and lighting electricity consumption) compared with buildings without shading systems and the corresponding figure for Guangzhou is 50%. The conclusions of this study provide a reference for design and selection of shading systems in different regions in China.

Keywords: shading systems, comprehensive energy performance, qualities, setting modes

1. Introduction

The energy consumption for civil buildings accounts for about 20% of the total energy consumption of our society in China [1~3]. With the rapid development of modern construction industry and significant improvement of people's living standard, the proportion of building energy consumption to the total energy consumption of our society will continue to increase. Therefore, building energy conservation has become an important issue in the strategy of sustainable development in China.

Windows are the most energy-consuming part among the enclosure structures of buildings [4, 5], however, the installation of suitable shading systems can effectively reduce the heat gain of external windows in summer, thereby reducing the cooling load of air conditioning, which is of great significance for building energy conservation [6]. At present, the traditional shading systems for residential buildings in China mainly consist of cellular shades, fixed awnings, roller shades, horizontal louvered blinds and so on. The shading systems will have a significant impact on buildings' cooling load, heating load and indoor lighting distribution. Therefore, this paper analyses the impacts of different shading systems on buildings' comprehensive energy performance, including cooling electricity consumption in summer, heating electricity consumption in winter and lighting electricity consumption for the whole year.

2. Methodology

2.1. Simulation Model

The simulation model established in EnergyPlus was based on a typical residential apartment in China. The whole apartment was divided into four zones viz. living room, bedroom, kitchen and bathroom. There were two windows on each orientation of north, east and west while the south side of the apartment was set with three windows. The size of each window is 1.6m (height) * 1.5m (width) and the material of the window frame is aluminum alloy. Moreover, the shading systems were mounted on the upside of each window in the east, south and west. Taking the fixed awnings as an example, the model established in EnergyPlus is shown in figure 1.

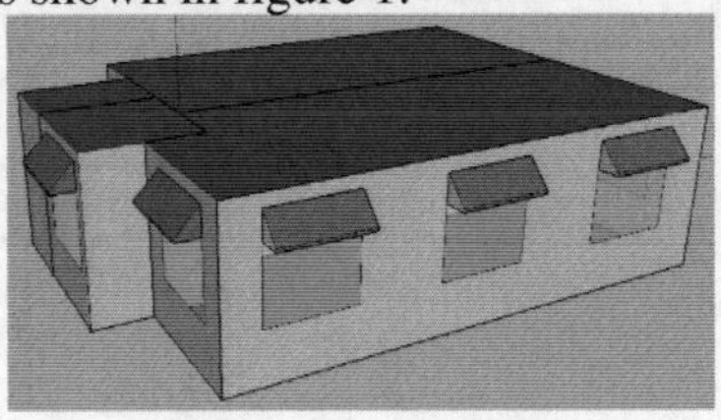

Figure 1. The fixed awnings model in EnergyPlus

The detailed process for establishing different shading systems is seen in figure 2. The design and heat transfer calculation of the aluminum alloy frame was conducted in THERM. Then, the files generated by THERM were imported in WINDOW. And the IDF file of the whole window, containing the data of the aluminum alloy frame, the baseline window system and the shading system, was derived from WINDOW. Finally, the IDF file of the whole window was integrated with the whole building's IDF file and run in EnergyPlus.

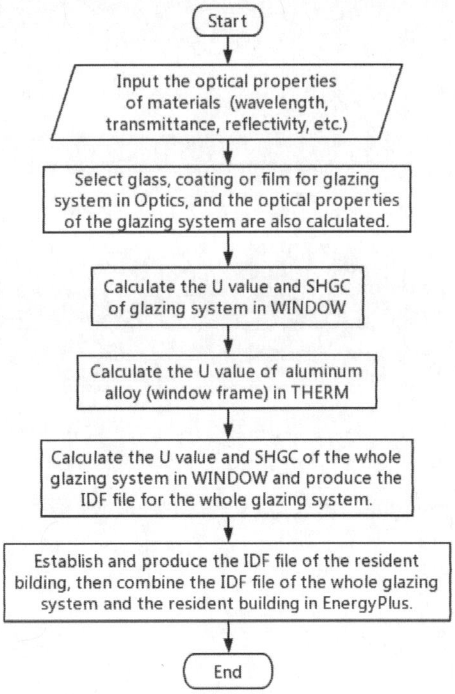

Figure 2. Modelling and integrating of traditional shading systems for buildings

The heat transfer model was employed to simulate the hourly heating and cooling load and the daylighting model was used for simulating the electricity consumption of artificial lighting, besides, the lighting control points were set in the middle of each area with a height of 0.75 m. The illumination level and lighting density were set to be 300 lux and 9W/m^2, respectively.

2.2. Qualities and setting modes for shading systems

The shading systems selected in this paper mainly consist of cellular shades, roller shades, fixed awnings and horizontal louvered blinds. Three baseline windows (double clear glazing system, low-e glazing system and single clear glazing system) were selected to compare the energy-saving potential of different shading systems under various glazing systems. In addition, four "qualities" were defined to represent a wide variety of heat transfer and daylighting performances. Although the parameters for each quality are developed with the intent that quality A has the best performance and qualities B, C, and D are continuously "worse". It is recognized that the overall energy outcomes are not necessarily going to be "Best" to "Worst". The definition of qualities can be seen in table 1.

Table 1. Definition of qualities and setting modes for shading systems

Shading type	Quality	Emissivity	Reflectance	Transmittance	Openness	Conductivity, k, (W/m·K)	Gap (mm)	Cell Side length (mm)	Thick (mm)	Tilt angle (°)	Remark
Fixed awnings	A	0.1	0.8	0.0							F
	B	0.9	0.6	0.0							F
	C	0.9	0.4	0.05							F
	D	0.9	0.02	0.50							F
Roller shades	A	0.1	0.8	0.0	0	0.10	0				F H
	B	0.9	0.7	0.1	0.02	0.10	0				F H
	C	0.9	0.5	0.2	0.05	0.15	3				F H
	D	0.9	0.05	0.5	0.3	0.5	12				F H
Cellular shades	A	0.1	0.8	0			0	19			F H
	B	0.9	0.65	0.2			3	19			F H
	C	0.9	0.45	0.25			3	15			F H
	D	0.9	0.1	0.5			12	10			F H
Horizontal louvered blinds	A	0.2	0.9	0		0.08			3	0	F / 0 ° slats H / 0 ° slats
										45	F / 45 ° slats H / 45 ° slats
										-45	F / -45 ° slats H / -45 ° slats
										90	F / 90 ° slats H / 90 ° slats
	B	0.9	0.75	0		0.2			0.2	0	F / 0 ° slats H 0 ° slats
										45	F / 45 ° slats H / 45 ° slats
										-45	F / -45 ° slats H / -45 ° slats
										90	F / 90 ° slats H / 90 ° slats
	C	0.9	0.6	0.6		0.2			0.2	0	F / 0 ° slats H / 0 ° slats
										45	F / 45 ° slats H / 45 ° slats
										-45	F / -45 ° slats H / -45 ° slats
										90	F / 90 ° slats H / 90 ° slats
	D	0.9	0.1	0		160			0.2	0	F / 0 ° slats H / 0 ° slats
										45	F / 45 ° slats H / 45 ° slats
										-45	F / -45 ° slats H / -45 ° slats
										90	F / 90 ° slats H/ 90 ° slats

As EnergyPlus cannot simulate the energy performance of half shading systems, the whole window was divided into two small and equal windows to analyse the energy performance of half shading systems. The upper half of the window was defined as window1 while the lower half of the window was defined as window2 and the size of the whole window is 1.6m (length) * 1.5m (width), thus the sizes of the divided windows are all 1.6m (length) * 0.75m (width). The shading systems were modelled in their three deployment states: when window1 and window2 were not deployed shading systems, this mode was defined as O (Open); When window1 was deployed shading system, window2 was not deployed shading system, this mode was defined as H (Half shading); When Windows1 and window2 were all deployed shading systems, this mode was defined as F (Full shading). For fixed awnings, F and O setting modes were adopted, while for cellular shades, roller shades and horizontal louvered blinds, H, F and O setting modes were adopted. Therefore, for fixed awnings, the combination of qualities and setting modes are AF, BF, CF, DF, while for cellular shades and roller shades, the combination of qualities and setting modes are AH, AF, BH, BF, CH, CF, DH, DF. For horizontal louvered blinds, the slat angle also should be considered, and defined 0° (perpendicular to the wall), 45° (the upper half of the window facing out), -45° (the upper half of the window facing in), 90° (parallel to the wall) as 1, 2, 3, 4. Therefore, the horizontal louvered blinds possess 32 layout methods, which are A1H, A1F, A2H, A2F, A3H, A3F, A4H, A4F, B1H, B1F, B2H, B2F, B3H, B3F, B4H, B4F, C1H, C1F, C2H, C2F, C3H, C3F, C4H, C4F, D1H, D1F, D2H, D2F, D3H D3F, D4H, D4F.

3. Results and discussion

Figure 3 to figure 5 compare the comprehensive energy performance of four traditional shading systems with different setting modes, qualities and glazing systems in Beijing and Guangzhou. It can be seen that the shading systems with suitable qualities and setting modes will show better energy performance while the shading systems with bad qualities and setting modes will consume more electricity compared with the O setting mode. For three baseline window systems in two regions, the heating electricity consumptions and lighting electricity consumptions of shading systems with H setting mode are all less than that with F setting mode, while the cooling electricity consumptions of shading systems with H setting mode are all more than that with F setting mode.

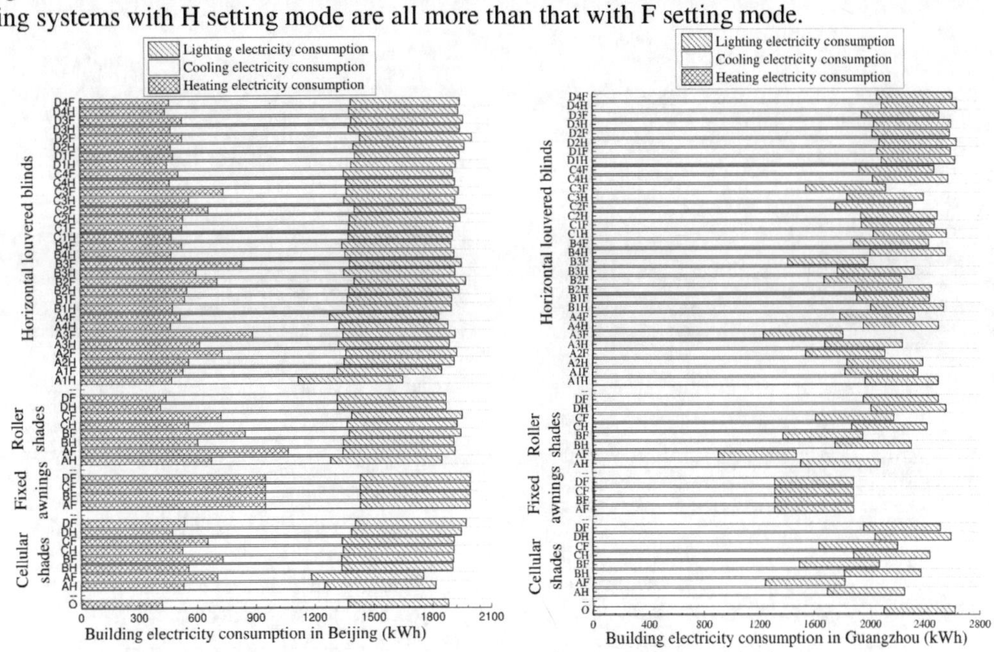

Figure 3. Building energy consumption for double clear glazing system

Besides, In Beijing, the total electricity consumptions of shading systems with F setting mode are all generally less than that with F setting mode for double clear glazing system and low-e glazing system, while for single clear glazing system, the total electricity consumptions of shading systems with F setting mode are all generally more than that with F setting mode. However, in Guangzhou, the total electricity consumptions of shading systems with F setting mode are all generally more than that with F setting mode for three baseline window systems. Besides, the impacts of setting modes and qualities on the energy performance of fixed awnings in two regions are not very obvious while this impact is pretty sensitive for cellular shades, roller shades and horizontal louvered blinds. And for low-e glazing system, the impacts of setting modes and qualities on energy performance of different shading systems are less significant than that for double clear glazing system and single clear glazing system.

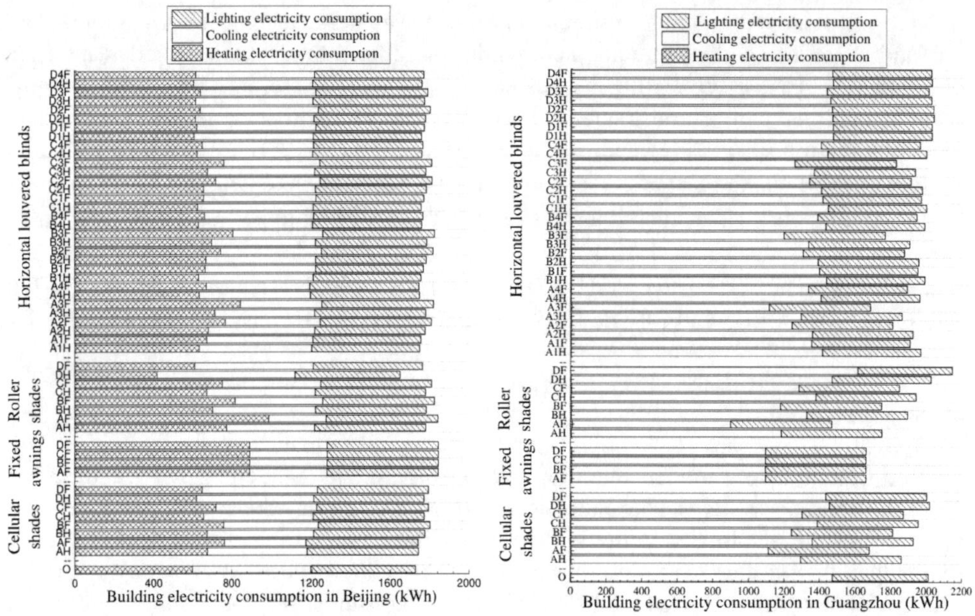

Figure 4. Building energy consumption for low-e glazing system

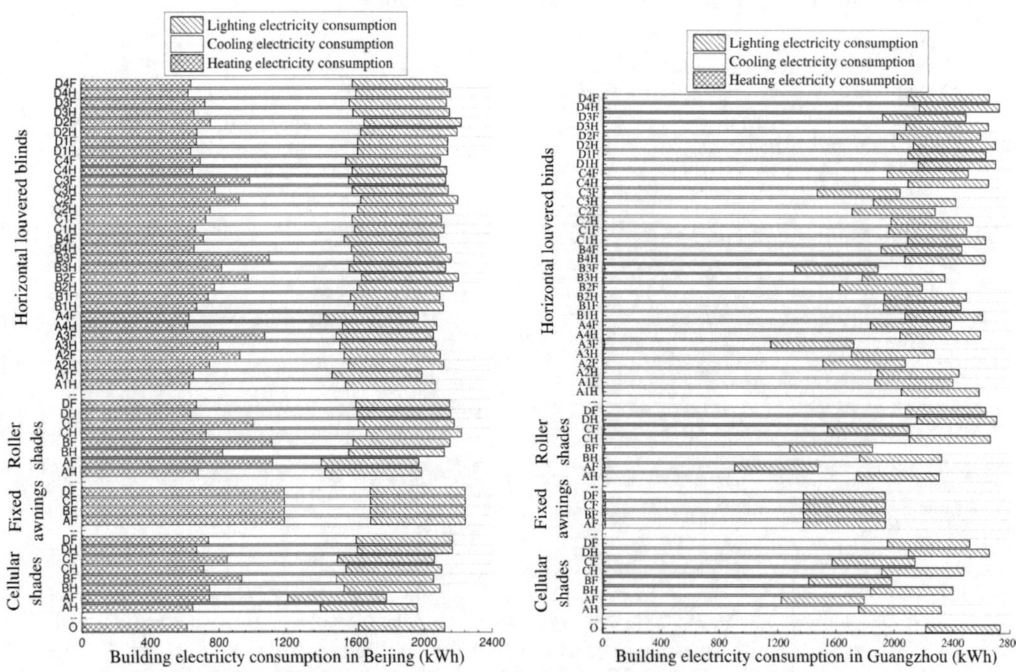

Figure 5. Building energy consumption for single clear glazing system

From figure 3, it is seen that for double clear glazing system in Beijing, the most suitable shading system is horizontal louvered blinds with A quality, H setting mode and 0° slat tilt, the corresponding heating electricity consumption is 473.28kWh, cooling electricity consumption is 645.51kWh, lighting electricity consumption is 530.4kWh and the total electricity consumption is 1649.19kWh. Compared with buildings without shading systems, the energy saving potential for buildings with horizontal louvered blinds is 15.54%; In Guangzhou, the most suitable shading system is roller shades with A quality and F setting mode, the corresponding heating electricity consumption is 4.07kWh, cooling

IOP Conf. Series: Materials Science and Engineering **556** (2019) 012039 doi:10.1088/1757-899X/556/1/012039

electricity consumption is 1500.75kWh, lighting electricity consumption is 569.32kWh and the total electricity consumption is 1472.25kWh. Compared with buildings without shading systems, the energy saving potential for buildings with roller shades is 50.43%.

Figure 4 shows that for low-e glazing system in Beijing, the most suitable shading system is roller shades with D quality and H setting mode and compared with buildings without shading systems, the energy saving potential for buildings with roller shades is 24.67%; In Guangzhou, the most suitable shading system is roller shades with A quality and F setting mode and compared with buildings without shading systems, the energy saving potential for buildings with roller shades is 50.60%.

Figure 5 shows for single clear glazing system in Beijing, the most suitable shading system is cellular shades with A quality and F setting mode and compared with buildings without shading systems, the energy saving potential for buildings with cellular shades is 18.59%; In Guangzhou, the most suitable shading system is roller shades with A quality and F setting mode and compared with buildings without shading systems, the energy saving potential for buildings with roller shades is 50.33%.

4. Conclusions

In this paper, the comprehensive energy performance of four types of shading systems with different qualities, setting modes and baseline window systems were analysed, and the main conclusions obtained were listed as follows.

1. The shading systems with suitable qualities and setting modes showed excellent energy-saving potential. Besides, the impacts of setting modes and qualities on the energy performance of fixed awnings were not obvious while this impact is pretty sensitive for cellular shades, roller shades and horizontal louvered blinds.
2. In Beijing, the most suitable shading system is roller shades with D quality, H setting mode, and low-e glazing system while in Guangzhou, the most suitable shading system is roller shades with A quality, F setting mode and low-e glazing system.
3. Compared with buildings without shading systems, buildings with suitable shading systems can save at least 20% and 50% energy (including air conditioning and lighting electricity consumption) in Beijing and Guangzhou, respectively.

References

[1] Price L et al. 2011 Assessment of China's energy-saving and emission-reduction accomplishments and opportunities during the 11th Five Year Plan *Energy Policy.* **39** 2165—2178

[2] Fridley D G. 2008 *Estimating total energy consumption and emissions of China's commercial and office buildings* (Lawrence Berkeley National Laboratory) p 1—39

[3] Building energy conservation research center, Tsinghua university 2017 *Research Report on the annual development of China's building energy conservation* (Beijing: China architecture and building press) p 8—10

[4] Misiopecki C, Bouquin M, Gustavsen A and Jelle B P 2018 Thermal modelling and investigation of the most energy efficient window position *Energy Build.* **158** 1079-1086

[5] Lee J W, Jung H J, Park J Y, Lee J B and Yoon Y 2013 Optimization of building window system in Asian regions by analyzing solar heat gain and daylighting elements *Renew. Energy.* **50** 522-531

[6] Hart R 2018 Numerical and experimental validation for the thermal transmittance of windows with cellular shades *Energy Build.* **166** 358-371

Acknowledgments

The authors appreciate the financial supports provided by the National Natural Science Foundation of China (Project No. 51608185), the Collaborative Innovation Centre of Building Energy Conservation & Environmental Control and the Fundamental Research Funds for the Central Universities (Hunan University, Project No. 531107040899).

IOP Conf. Series: Materials Science and Engineering **556** (2019) 012040 doi:10.1088/1757-899X/556/1/012040

The Assembly of Embedded Systems for Integrated Photovoltaic windows in Rural Buildings (E-IPB)

K Shanks[1], Hasan Baig[1], Ashley Knowles[2], Adam Brierley[3], Henry Orr[3], Jayesh Patel[4], Neil Foster[4], Stephen Ball[4], Yanyi Sun[5], Yupeng Wu[5], Sundaram Senthilarasu[1], Tapas Mallick[1]

[1]Environmental and Sustainability Institute, University of Exeter Penryn Campus, Penryn, UK
[2]Yorkshire Photonics, Yorkshire, UK.
[3] Brinell Vision, Dundee, UK.
[4] Couch Perry Wilkes, Birmingham, UK.
[5] Nottingham University, Nottingham, UK.
E-mail: k.shanks2@exeter.ac.uk

Abstract. Embedded PV systems are required to help improve the synergy of renewable energy and smart buildings. A novel concentrator photovoltaic (CPV) system embedded as a window for integration into buildings is presented. The system is made up of crossed compound parabolic concentrators (CCPC), Talesun Silicon solar cells and glass panes. The materials and manufacturing methods are presented along with their advantages and disidvantages. Two sized prototypes are presented, 7cm by 7cm and 20cm by 20cm. The maximum power obtained was 3.53 Watts for the larger prototype. The glass CCPC optics produce more power but plastic alternatives are only 44% the weight of the glass optics. The best performing plastic optics were made of Topaz and injection moulded. The cell soldering and alignment method is explored and in particular silver tracks printed on glass are analysed for their resistance, aesthetics and benefit to the assembly process. Using plastic optics for CPV technology is a relatively new area of research, and the combination of silver tracks printed on glass could make this innovative design revolutionary in its field.

Keywords: solar; energy; optics; concentrator photovoltaics; building integration;

1. Introduction

Concentrator photovoltaic (CPV) systems reduce the requirement of photovoltaic materials needed in a system installation [1]. This in turn reduces the carbon footprint of making this type of solar energy technology and can also reduce the cost of the system whilst maintaining or even increasing the solar cell efficiency. For example, the record solar cell efficiency at present is 46%, held by Fraunhofer and Soitec for a four-junction solar cell under high concentrated sunlight [2]. Despite these advantages, CPV costs limit the technologies competitiveness with flat plate silicon panels. Applications where space is limited and energy demand is high, such as for domestic use, are where CPV technology may contend, especially if more interesting, aesthetically pleasing designs are required. CPV technology also has the advantage of being flexible in its size, shape and transparency. Recent market research suggests that for CPV to succeed alongside standard flat plate PV, niche applications such as building integration, using embedded systems, need to be developed [3,4]. There is a lot of potential for smart integrated solar windows [10,11] with many designs, materials and manufacturing processes in need of investigation. Thus we present an embedded crossed compound parabolic concentrator (CCPC) plastic optic, arrayed within double glazed windows as a means to provide both electricity and natural sunlight to a building (Figure 1).

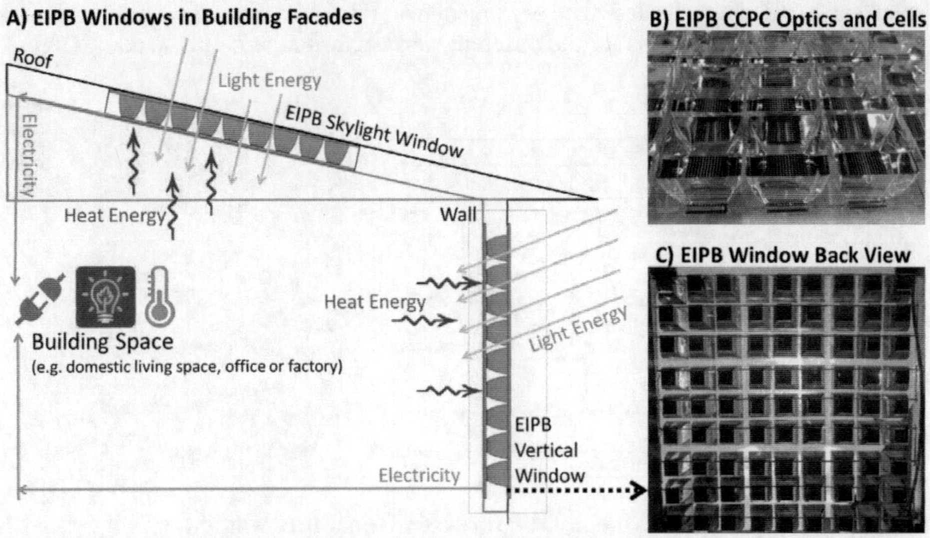

Figure 1. A) Example of roof and wall installations of EIPB window and the transfer of light, heat and electricity for a building space. B) Glass Crossed Compound Parabolic Concentrator optics optically coupled to tabbed silicon cells. C) Back view of 20cm by 20cm prototype showing light passing through module.

The weight of such a design however should be minimised to aid application to any kind of building façade, especially those on roofs and skyscraper buildings. Glass, although a very reliable and high performing material, adds weight to devices and in a window, is already being used for the front and back sheets. Plastics such as polymethylmethacrylate (PMMA) and Polycarbonate (PC) are popular alternative materials in CPV's with lower weight but also lower transmittance values [5–7]. For photovoltaic technology embedded into building facets and windows, plastic materials are an evident alternative to reducing weight, cost and handling difficulty [8,9] however more materials need to be investigated. The proposed embedded system as an integrated photovoltaic window in buildings (E-IPB) utilizes highly transparent Topaz plastic for the CCPC optics inside the windows (Figure 2). This improves the power to weight ratio of the window in comparison to glass optics as can be seen from figure 2.

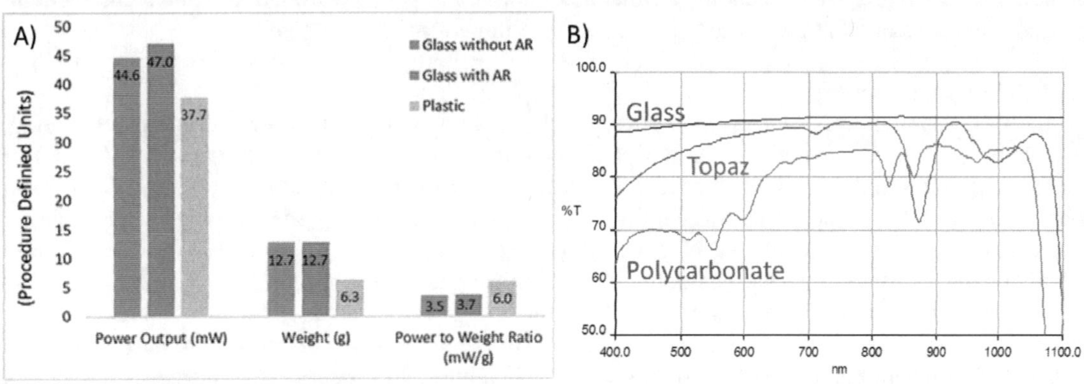

Figure 2. A) Power and weight comparison for glass and plastic CCPC's. B) Transparency spectra of Glass, Topaz and Polycarbonate.

2. Design components and manufacturing methods

One of the key challenges in this type of embedded system, as an integrated photovoltaic for building structures, is the manufacturing and assembly process. Cost effective large scale manufacturing of arrayed solar cells and optics sealed with wiring in window units requires optimization. Here we present the requirements and options for the different subsystems in the E-IPB window (figure 3a).

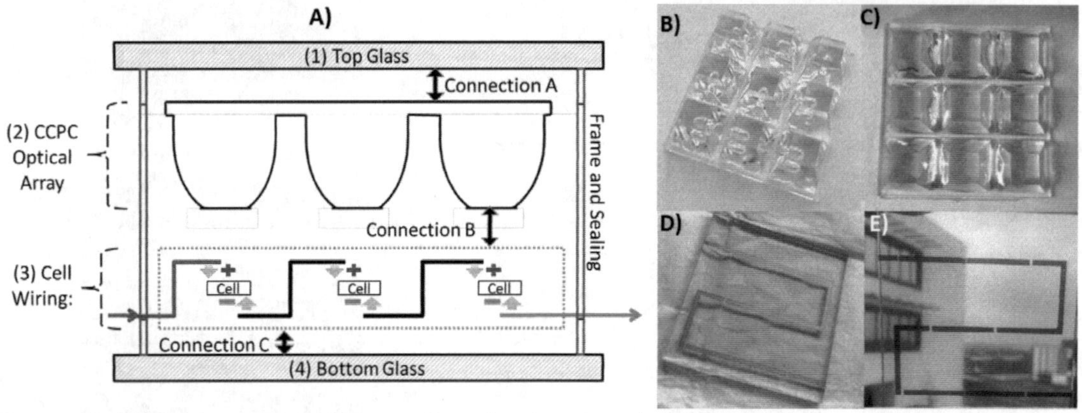

Figure 3. A) Diagrammatic representation of subsystems in EIPB window prototype. B) Injection moulded plastic 3 by 3 optical array with voids and shrinkage. C) Minimized flaws in injection moulded optic by oven cooling after injection. D) Top view of silver tracks printed on glass. E) Back view of silver tracks on glass showing gaps for series connection of cells.

2.1. Top Glass

The top cover glass, which will be the outer pane of the window when installed, should ideally have a high transmittance and an anti-reflective coating which covers most of the solar cell working range (300-1100nm for silicon) for maximum electrical performance of the device. As this is a window required to also be thermally insulating or cooling (location dependent) and perhaps provide certain light filtering (UV or glare for offices with computer screens) these requirements could compromise the exact best coating and material of the top glass. Most coatings are also fragile and cannot be handled or cleaned by ordinary measures which limits these options further for the outer surface which should be able to be installed by non-specialists and cleaned as a regular window. Coatings on the inner surface of the Top and Bottom Glass (1 and 4 in figure 3a) are however a good option but again may be compromised if connection A between the top glass and the CCPC optics is filled with a refractive index coupling material. Optically joining the top glass to the CCPC optics would reduce light loss due to Fresnel reflections as the incoming light passes through the different medium interfaces but adds another processing stage and hence more cost to the assembly.

2.2. CCPC Optical Array

For the CCPC optical array, arrays of optics which are cast as one part allow faster manufacturing, neat placement and alignment. This however also requires the cells to match the spacing. Alignment between the optics and the cells is crucial for maximum performance. Injection molding was the most appropriate, and for some plastics the only option, for manufacturing the designed CCPC plastic optics. Previously, in house casting of plastic crystal clear optics was investigated but the injection moulded parts had superior surface smoothness and optical quality as expected.

However, due to the thickness of the CCPC optics, the bulk material in the center of these optics was found to cool at a much slower temperature than the outer parts of the optics during material injection. This meant the inside of the optics was still in a hot liquid state which contracted and shrunk as it cooled, causing voids (bubbles) or pulling the outer shape inwards (figure 3b and c). This effect could be minimized (figure 3c) by either leaving the optics in the mold until they had fully cooled -a very time consuming and hence costly method in terms of manufacturing large quantities- or by quickly moving the optics into an oven to cool at a slower rate. Ideally this process could be done on a customized lehr (conveyer belt) set up for large scale manufacturing.

Despite this flaw in the optics the Topaz 3 by 3 arrayed optics still achieved a good short circuit current as shown in figure 4c. Glass however maintains the lowest absorbance and hence is still most efficient material, but as seen from figure 2 has the lowest power to weight ratio.

2.3. Cells and electrical connections

The solar cells are Silicon solar cells provided by Talesun following our own metallization and cutting pattern. A single cell (1cm x 1cm) under 1 sun provides ~20 mW of power with an efficiency of 19-20% following an I-V curve as shown in figure 4. The cells have a top contact and bottom contact (component no. 3 in figure 3a) and to be connected in series require tabbing connections as shown in figure 3. This is the most time consuming part of the assembly unless automated tabbing machines can be utilised. Another option for the connection and correct spacing of the cells is to use printed silver tracks on glass (Figure 3d and e). Conductive tracks can be printed onto the bottom glass with the correct spacing for the cells to be soldered onto (Connection C in figure 3a is hence also achieved). Some tabbing would still be required to connect the top contact of the cells to the next conductive track on the glass for our prototyping. Bottom contact cells would benefit most from this type of wiring. One of the disadvantages observed however in these prototyping investigations was the increased chance of misconnections as both the cell and the glass are inflexible allowing slight tilts and uneven soldering to cause incomplete connections along the full length of the cells' bottom contact.

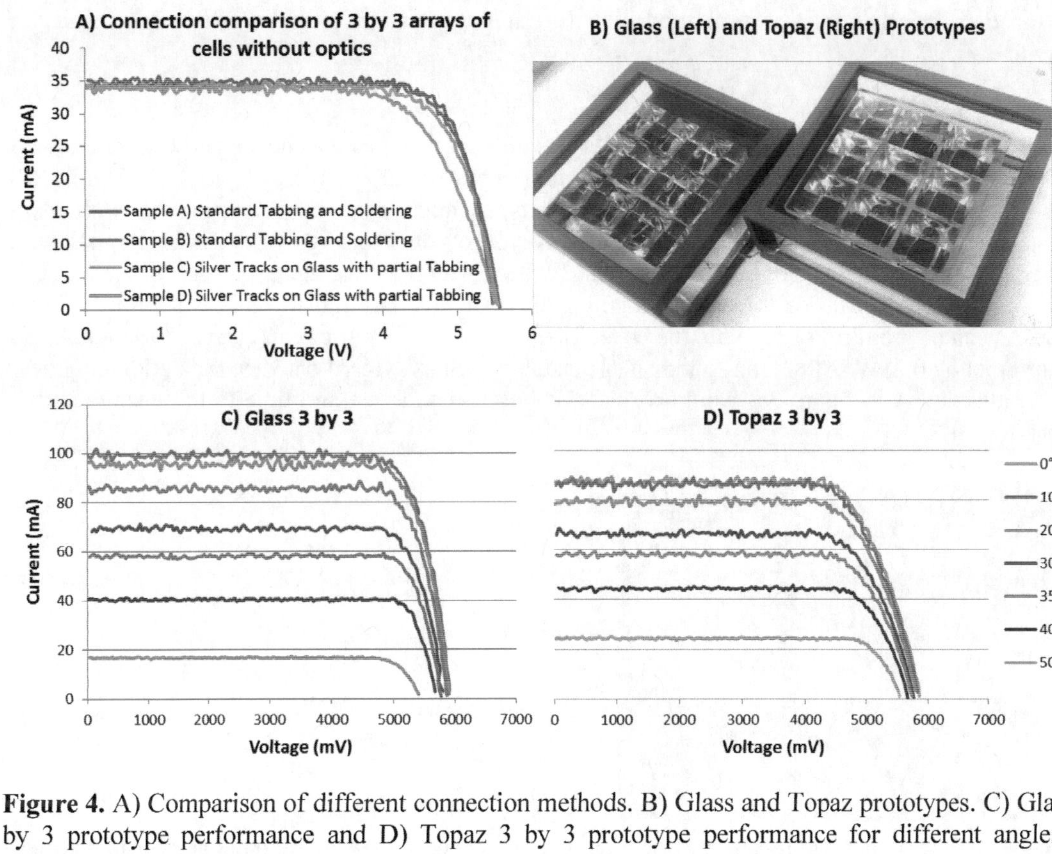

Figure 4. A) Comparison of different connection methods. B) Glass and Topaz prototypes. C) Glass 3 by 3 prototype performance and D) Topaz 3 by 3 prototype performance for different angles of incidence.

2.4. Bottom Glass

As described above, this could be printed with conductive tracks for the cell connections. This also opens up the potential for patterned windows or specific designs which also connects the cells but may be more aesthetically pleasing. Another disadvantage of this however is that it cannot coincide with any special coatings on the inner surface of the bottom glass. For thermal insulation some double glazed windows have low emissivity coatings on the inner surface of the bottom glass which can allow light energy into a room but reduce the heat energy out of the room through that glass (figure 1a). Another consideration is if these printed conductive tracks have increased resistance in comparison to current standard soldering and tabbing methods.

3. Prototype performance

3.1. Conductive tracks on glass
As can be seen from figure 4a, there is negligible difference between the different cell connection methods though as already stated it was more likely to be a fault obtained with the silver tracks on glass and partial tabbing connection method, giving a result such as sample C in figure 4a. With scaled up processes and automated pick and place soldering the chance of misconnections may be minimised.

3.2. Topaz vs. Glass CCPC optics
3 by 3 prototypes utilising glass optics and Topaz optics were made and tested first as shown in figures 4b-d. As expected the glass optics achieved a higher current and fill factor (figure 4c) than the Topaz counterpart (figure 4d). Interestingly the Topaz optics performed better at wider angles of light incidence but this is expected to simply be due to the topaz optics being joined in an array as opposed to the glass optics being individual and separate. The layer of Topaz which joins the 3 by 3 optical units together appears to act as a method of light trapping, but this is to be understood clearer in further investigations.

3.3. 9 by 9 arrayed Prototype (20cm by 20cm)
A larger 20cm by 20cm module consisting of 9 rows of 9 glass optics was developed (figure 5a) next but with standard tabbed cell connections. The maximum power output obtained was 3.53W with a fill factor of 0.722 as shown in figure 5b. These results although good can be improved with more accurate automated machinery and customised conveyer systems to align the cells and optics with repeated accuracy as opposed to the in house manual placement done for this prototype. Since the cells were connected in series, one lower functioning cell or slightly misaligned optic will limit the total output. As can be seen from figure 5b, the cover glass reduces the short circuit current by 8.55mA and power output by 0.28W. AR coatings and an encapsulate such as sylgard between the optics and cover glass (connection A in figure 3a) would reduce this but also add cost and time to the manufacturing process.

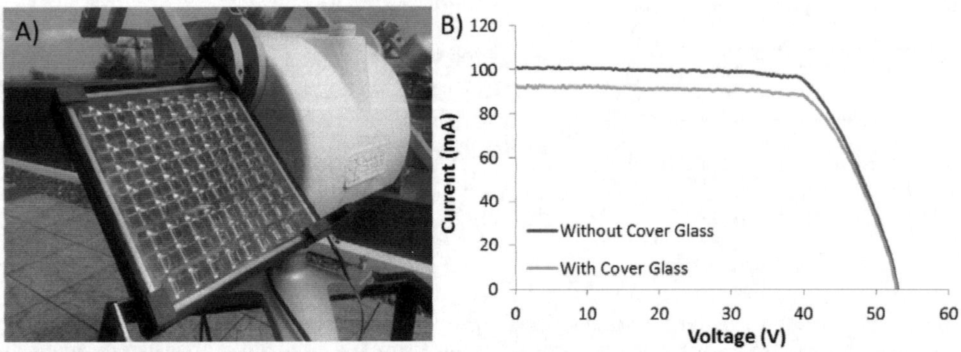

Figure 5. 20cm by 20cm prototype outside mounted on tracker at the Environment and Sustainability Institute, University of Exeter, Penryn, UK. B) I-V Outputs of prototype with and without cover glass.

The 20cm by 20cm prototype is the thickness of a standard double glazed window, framed and sealed by Cornwall Glass in Penryn, UK. This means integrations of this type of system into a wide selection of commercially available window frames should be easily achieved but consideration of the wiring and electrical connections needs to be carried out and will be investigated next as part of the EIPB project.

4. Conclusion

Plastic optics made of such materials as Topaz have serious potential for replacing glass optics in concentrator photovoltaic products to improve their power to weight ratios beyond glass and expand their applications for building integrated skylights and windows. The durability, aesthetics and electrical connection of these windows within a building have to be investigated further but the findings presented here confirm feasibility and assembly optimisation utilising Topaz injection moulded optics and silver tracks printed on glass cell connections. The Topaz optics achieve a power to weight ratio almost double that achieved by the glass optics and are best manufactured using a lehr such that the newly injection moulded arrays can be cooled slowly to minimise air voids and surface shrink back flaws. Silver printed on glass conductive tracks is more aesthetic, scalable and does not add extra resistive losses to the system. AR coatings and extra optical coupling between the optics and glass cover would reduce Fresnel losses but may also reduce the light trapping effect seen by the Topaz 3 by 3 arrays and add another stage and cost to the manufacturing system.

5. References

[1] Perez-Higueras P and F. Fernandez E 2015 *High Concentrator Photovoltaics: Fundamentals, Engineering and Power Plants* ed P Perez-Higueras and E F Fernandez (Jaen: Springer International Publishing)

[2] National Renewable Energy Laboratory (NREL) - National Center for Photovoltaics 2017 Research Cell Record Efficiency Chart

[3] Morgan J P 2017 Why Hasn't CPV Taken Off Yet and Will the Future Look Different *AIP-13th International Conference on Conentrator Photovoltaic Systems (CPV-13)* (Ottowa)

[4] Ekins-Daukes N 2017 Cost Requirements for High Volume CPV Manucaturing *AIP-13th International Conference on Conentrator Photovoltaic Systems (CPV-13)* (Ottowa)

[5] Sansoni P, Fontani D, Francini F, Mercatelli L, Jafrancesco D, Sani E and Ferruzzi D 2010 Internal Lighting by Solar Collectors and Optical Fibres *Solar Collectros and Panels, Theory and Applications* ed R Manyala (InTech) pp 1–30

[6] Schulz U and Kaiser N 2006 Vacuum coating of plastic optics *Prog. Surf. Sci.* **81** 387–401

[7] Shanks K, Senthilarasu S and Mallick T K 2016 Optics for concentrating photovoltaics: Trends, limits and opportunities for materials and design *Renew. Sustain. Energy Rev.* **60** 394–407

[8] Tripathy M, Sadhu P K and Panda S K 2016 A critical review on building integrated photovoltaic products and their applications *Renew. Sustain. Energy Rev.* **61** 451–65

[9] Shanks K, Baig H, Singh N P, Senthilarasu S, Reddy K S and Mallick T K 2017 Prototype fabrication and experimental investigation of a conjugate refractive reflective homogeniser in a cassegrain concentrator *Sol. Energy* **142** 97–108

[10] Wu Y, Connelly K, Liu Y, Gu X, Gao Y and Chen G Z 2016 Smart solar concentrators for building integrated photovoltaic façades *Sol. Energy* **133** 111–8

[11] Connelly K, Wu Y, Chen J and Lei Y 2016 Design and development of a reflective membrane for a novel Building Integrated Concentrating Photovoltaic (BICPV) "Smart Window" system *Appl. Energy* **182** 331–9

Acknowledgments

This work is funded by Innovate UK through the E-IPB project. Authors acknowledge all funding agencies for the support. In support of open access research all underlying article materials (such as data, samples or models) can be accessed upon request via email to the corresponding author.

Thermal Performance Analysis of Concrete Small Hollow Block

Bi-chao YE and Hao ZHOU
Changzhou Vocational Institute of Engineering, Changzhou, China
E-mail: 8000000619@czie.edu.cn

Abstract: In this paper, the thermal insulation coefficient and thermal inertia index of small hollow concrete blocks with different hole types are calculated by theoretical method and numerical method, and the block structure with the best thermal insulation property and heat-shielding performance is respectively determined. The relationships between thermal insulation property and heat-shielding performance and block structure parameters are analyzed. The conclusion obtained provides a good reference value for the manufacturer of the block.

Keywords: concrete hollow block; hole type; hole ratio; thermal insulation property; heat-shielding performance

1. Introduction

The concrete small hollow block has the advantages of wide source of raw materials, avoiding destruction of land and fired bricks, low production energy consumption, less environmental pollution, and consumption of some industrial waste, etc. It is one of the new wall materials promoted by China's wall reform. In recent years, it has been widely used in residential and other buildings. In this paper, measures to improve the thermal insulation performance and thermal insulation performance of small concrete hollow blocks are studied.

At present, the main specification of concrete small hollow block is 390mm×190mm×190mm, presented in fig.1. The density of the concrete material used in the concrete hollow block studied in the paper is 1400kg/m^3 and the thermal conductivity is 0.53W/m•°C. Since the minimum outer wall thickness of ordinary concrete hollow blocks should not be less than 30mm, the minimum rib thickness should not be less than 25mm, and the hole rate should not be less than 25%[1]. Under the above conditions, the thermal insulation performance and heat-shielding performance of small hollow blocks were analyzed.

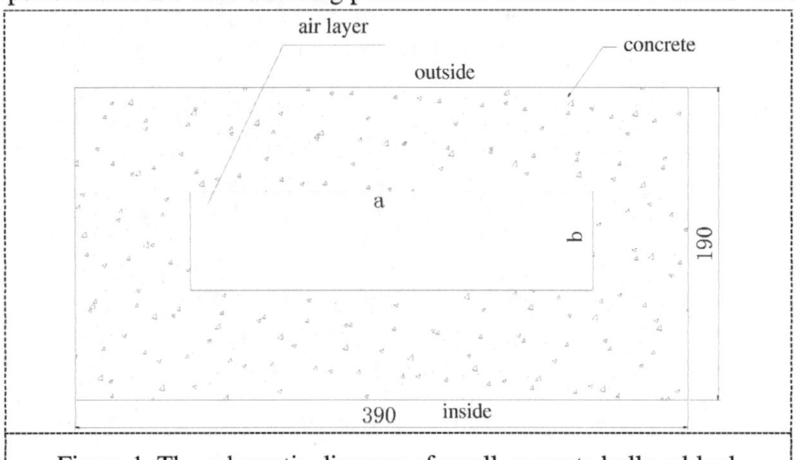

Figure 1. The schematic diagram of small concrete hollow block

2. Analysis of block thermal insulation property

IOP Conf. Series: Materials Science and Engineering **556** (2019) 012041 doi:10.1088/1757-899X/556/1/012041

The thermal insulation coefficient is an important parameter reflecting the heat insulation performance of the envelope structure. The higher the thermal insulation coefficient is, the better the insulation performance is. The heat insulation coefficient of the block is obtained when the hole shape is rectangular and the hole rate is 25%, 30%, 35%, 40%, and 45% respectively, according to the calculation method of the average thermal insulation coefficient of non-homogeneous building envelops. The calculation results are shown in Table 1.

Table 1. The heat insulation coefficient of the block(the hole rate is 25%)

Number	Block configurations	Hole rows	Max.	Max. a×b	Min.	Min. a×b
1		One	0.5466	330.0×56.1	0.4653	142.5×130.0
2		One	0.5374	152.5×60.5	0.4654	71.0×130.0
3		One	0.5273	93.0×66.5	0.4653	47.5×130.0
4		Two	0.6273	330.0×28.0	0.5511	176.5×52.5
5		Three	0.7134	330.0×18.7	0.6292	247.0×25.0
6		Two	0.6192	152.5×30.4	0.5431	88.2×52.5
7		Two	0.5820	152.5×40.5	0.5449	117.5×52.5
8		Two	0.5503	93.3×49.6	0.5473	88.2×52.5
9		Four	0.7778	330.0×14.0	nonexistent	nonexistent

In Table 1, when the hole rate is constant, by comparing all the data horizontally, the thermal insulation coefficient of each type of block increases with the increase of a and the decrease of b. If the a and b of the air layer in the block are constant, changing the thickness of the ribs and the outer wall does not affect the thermal insulation coefficient of the block, and staggering the air layers does not change the value of the thermal insulation coefficient. After exchanging the upper hole and the lower hole of the 7th and 8th blocks in Table 1, it is found that the value of the thermal insulation coefficient did not change.

When the hole rate is fixed, the thermal insulation coefficient of the block with four rows of holes is the largest, which is obtained by longitudinal comparison of all the data. If a is the same, the thermal insulation coefficient decreases as b increases. If b is the same and the total length of the air layer of the blocks is equal, then the thermal insulation coefficient is also equal. for example, b is 130mm in Table 1, the minimum of the thermal insulation coefficient of the 1st, 2nd, and 3rd blocks is equal. If a is the same, dividing an air layer into multiple layers can increase the thermal insulation coefficient of the block by comparing the maximum values of the 1st, 4th, 5th, and 9th block in Table 1. The variation law of the calculated results is the same as the above law when the hole rate is 30% ~ 45%. For the same structure, with the increase of the hole rate, the change law of the thermal insulation coefficient is shown in Table 2 for the 4th, 5th, and 9th blocks in Table 1. In the table 2, a is 330mm, and the value in parentheses is the b value at the corresponding hollow ratio.

Table 2. Maximum value of thermal insulation coefficient($m^2·K/$)and corresponding b(mm) value

Hole rows	Hole rate（%）				
	25	30	35	40	45
Two	0.6228 (28.0)	0.6234 (33.7)	0.6247 (39.3)	0.6052 (44.9)	0.5892 (50.5)
Three	0.7134 (18.7)	0.7044 (22.5)	0.6902 (26.2)	0.6926 (29.9)	0.6840 (33.7)
Four	0.7778 (14.0)	0.7784 (16.8)	0.7793 (19.7)	0.7678 (22.5)	0.7688 (25.3)

Observing the data in Table 2, it can be seen that with the increase of the hole rate, the variation law of the thermal insulation coefficient is not fixed, and the two rows and the four rows of holes increase first and then decrease, and the three rows of holes decrease continuously. For the block in this paper, in order to obtain a better thermal insulation effect, b of the two rows of holes should be about 40 mm, b of the three rows of holes should be about 19 mm, and b of the four rows of holes should be about 20 mm. This is because the thermal conductivity of air is smaller than the thermal conductivity of other materials constituting the envelope structure, so the air layer can play a role in blocking heat transfer within a certain thickness. When the air layer reaches a certain thickness, if its thickness is further increased, the thermal insulation coefficient of the air layer hardly increases due to the increase in the heat transfer space. At this time, the total thermal conductivity of the air layer is closer to the thermal conductivity of other materials constituting the envelope structure. This not only does not hinder the heat transfer effect, but also promotes heat transfer.

If the shape of the air layer is circular, it can be converted into a square shape with the same area, and then calculated according to the above method. From the above analysis, it can be seen that the thermal insulation coefficient of the block with a circular hole shape is at between the maximum and minimum values. When the shape of the air layer is elliptical, parallelogram, and triangular, the related literature does not describe how to determine the heat insulation coefficient. In this paper, the thermal insulation coefficients of all the blocks in Table 1 are calculated by numerical calculation method. The error between numerical calculation results and theoretical calculation results does not exceed 5%. Therefore, the numerical calculation method is used to calculate the thermal insulation coefficient of a block whose shape is elliptical and other shape. Take 1 of them as shown in Table 3, the shape of the block is shown in Fig. 1.In this case, a and b are 330 and 56.1 mm, respectively, and the hole rate is 25%.

Table 3. Comparison of thermal insulation coefficient of the block(the hole rate is 25%)

Block configurations	a×b	Rectangle	circular fillet1	circular fillet2	oval	Triangle	parallelogram	circle
	330×56.1	0.5523 (0.5456)	0.5522	0.5517	0.5403	0.5208	0.5151	0.4710 (0.4608)

When the hole type in Table 3 is rectangular, the calculated thermal insulation coefficient of the block is 0.5523 $m^2 \cdot K/W$, the theoretical calculated value is 0.5456 $m^2 \cdot K/W$, the error between the two values is 1.20%. The four corners of the rectangular hole are changed to rounded corners, the calculated values at the corner radius of 10 and 20 mm are 0.5522$m^2 \cdot K/W$ and 0.5517$m^2 \cdot K/W$, respectively, which are slightly reduced when compared with the rectangular hole. The calculated thermal insulation coefficient of the circular hole block is 0.4710 $m^2 \cdot K/W$, the theoretical calculation is 0.4608 $m^2 \cdot K/W$, and the error between the two values is 2.17%. The other hole patterns have smaller values than the rectangular hole patterns. The data in table 3 is the data when the hole rate is 25%. When the hole rate is other values, the law is same.

In summary, the hole-type rectangular four-hole block d has the best insulation performance when the hole rate is 35%, a is 330mm and b=19.7mm.

3. Analysis of block heat shielding performance

The thermal insulation coefficient is one of the performance parameters describing the heat transfer capacity of the envelope, and is the evaluation index of the building envelope under steady-state heat transfer conditions. Under non-steady-state heat transfer conditions, the thermal performance of the envelope structure should also be evaluated using the thermal inertia index D. According to the calculation method of D given in the literature[2], the blocks in the paper are calculated. The analysis of the calculation results reveals that if both the a and b values change, that is, if a increases, b decreases or a decreases, b increases, and finally the average thermal inertia index D of the mixed layer decreases when the hole rate is constant. Because a increases, the average heat storage coefficient and the average thermal insulation coefficient of the mixed layer composed of the air space and the concrete will be reduced, while the decrease of b will increase the thermal insulation coefficient of the pure concrete layer, and finally increase the thermal inertia index D of the pure concrete layer. How the thermal inertia index D changes depends on the algebraic sum of the above two parts.

D value is smaller than the D value when the shape of the air gap is rectangular When the shape of the air layer is ellipse, circle, parallelogram and triangle, respectively. For each type of block in Table 1, because the air heat storage coefficient is 0, the thermal inertia index D of the block decreases as the hole rate increases. Based on the above conclusions, the D value of the 5th block in Table 1 is the largest (2.41), when the hole type is rectangular, when the hole rate is 25%, a=10, b=20mm. That is, the heat shielding performance is best, and the thermal insulation coefficient at this time is 0.6184 $m^2 \cdot K/W$.

4. Conclusions

The impact of the hole types on the thermal insulation property and heat shielding performance of small hollow concrete block was analyzed in this paper. It can be concluded that:

(1) With a certain hole rate, increasing hole rows of the air layer can improve the thermal insulation property of the block.

(2) When the hole rate is fixed, the thermal insulation property when the hole type is rectangular is better than the insulation performance when the hole type is elliptical, circular, parallelogram and triangle.

(3) Reducing the hollow ratio can enhance thermal insulation performance. When the hole rate is constant, When the hole rate is constant, the three-hole or four-row hole block has the best heat shielding performance.

(4) The best thermal insulation performance and the best heat shielding performance of the hollow blocks do not occur at the same time. The maximum thermal insulation coefficient of 0.7793$m^2 \bullet K/W$ appears in: the hole shape is rectangular, the hole rate is 35%, a=330mm, b=19.7mm, at this time D=2.13; the maximum value of thermal inertia coefficient 2.41 appears When the hole type is rectangular, the hole rate is 25%, a=231.3mm, and b=26.7mm, the thermal insulation coefficient at this time is 0.6184$m^2 \bullet K/W$.

Besides, If combined with the mechanical properties of hollow blocks such as compressive and shear resistance, the choice of block structure will be more comprehensive. The mechanical properties of hollow blocks are the direction of future research.

References

[1] Ministry of Housing and Urban-Rural Development of the People's Republic of China 2014 *Normal Concrete Small Hollow Block* (BeiJing: China Planning Press)

[2] Yao qing Lu 2007 *Practical Heating and Air Conditioning Design Manual* ed E (BeiJing: China Building Industry Press) P252

[3] Y. Zhang, K. Du, J.P. He, L. Yang, Y.J. Li and S.H. Li 2014 Impact factors analysis on the thermal performance of hollow block wall *Energy and Buildings* **75** 330-341
[4] Yuan Zhang and Qian Wang 2017 Influence of Hollow Block's Structural configuration on the thermal characteristics of hollow block wall *Procedia Engineering* **205** 2341–2348

IOP Conf. Series: Materials Science and Engineering **556** (2019) 012042 doi:10.1088/1757-899X/556/1/012042

Dynamic performance analysis of façade-integrated tri-functional photovoltaic/thermal solar collector

Jinwei Ma, Xue Kou, Shuang Liu, Weiwei Tong and Tingyong Fang
Anhui Jianzhu University, Hefei, China
E-mail: majw@mail.ustc.edu.cn

Abstract. Research on building-integrated photovoltaic/thermal (BiPV/T) technology has been conducted for years because the system can provide electric power and thermal energy simultaneously. A façade-integrated tri-functional PV/T collector system is presented and studied in this paper. The collector can work in PV/water-heating mode or PV/air-heating mode. Besides, it can also work in the composite mode that air and water be heated simultaneously. So optimum work mode of the collecor can be chosen for different energy demands. The collector can be combined with the southern wall of building, then it operates as PV-Trombe wall for passive room heating and electricity generation in winter, and for domestic water heating and electricity generation in other seasons. For the different working modes, theoretical research is carried out and dynamic model of the collector and building are developed. Based on the models, the performance of collector and the thermal behavior of the room are investigated under different climates conditions.

Keywords: photovoltaic/thermal collector; tri-functional; composite heating

1. Introduction

Passive air heating system can provide heat for a building interior by air natural circulation and has been researched and used widely due to the advantage of easy operation, easily combining with the building and its low cost [1]. The passive heating solar system combined with buildings has various forms, such as the traditional Trombe wall [2], PV-Trombe wall [3], Barra-Costantini wall improved by adopting metal panel [4] and Trombe wall system with solar air collectors [5] et al. Passive air heating system can reduce energy consumption for space heating in winter, but it has no significant effect in other seasons or even causes the buildings overheat during summer days [6]. It was found that after water tubes were attached to the rear surface of the absorber, the solar energy absorbed by the absorber plane could be removed by the water flow in the tubes and then the building overheating problems can be alleviated or even solved [7]. A dual-function solar collector can work in both air and water heating modes has been mentioned, and the thermal performance under different operating modes has been investigated [8].

Now a novel tri-functional PV/T solar collector is presented in the present study. The design of this collector is based on the sheet-and-tube solar water collector and single pass flat plat PV/T air collector. In order to apply this novel tri-functional PV/T solar collector to real life, what we going to do is simulation to confirm the feasibility of this collector in the building and how to integrate the tri-functional PV/T solar collector with building.

2. The introduction of theoretical model

2.1. Structure of the system

A virtual building that the indoor temperature can be control based on the general practical architectural features was designed. The room is of the dimension of 3.6 m×4.8 m×2.8 m. The structure of the tri-functional PV/T solar collector consists of the following components: glass cover, Polycrystalline silicon photovoltaic cells, Aluminium plate, Copper tubes, air inlet, air outlet, and the insulation layer. The illustrative diagrams of the building -integrated dual-function solar collector system is presented in figure 1. The PV cells are packaged and attached to the absorber plate with a packing factor of 0.59. The effective irradiation-collector area of the collector is 1.95 m × 0.91 m and the cell area is 1.05 m2. The PV panel consists of 72 pieces of cells and the rest of the absorber plate is covered by black painted surface.

Between the glass cover and the PV cells, there is an air gap to reduce the heat loss from upper surface. Copper tubes are welded at the bottom of the absorber plate. The air flow is of the dimension of 194 cm (length) × 93 cm (width) ×2 cm (height). The backboard is composed of 2 cm glass fibre along with iron plates for both inside and outside of the channel sides. Two opening vents with dimension of 36 cm × 12 cm are positioned at the top and bottom of the blackboard of the module for the air heating mode of the collector. In PV/air-heating mode, the collector is connected to a ventilation system and both ends of the copper tubes are closed. While in PV/water-heating mode, the inlet and outlet of the air flow channel are closed and a water tank is connected.

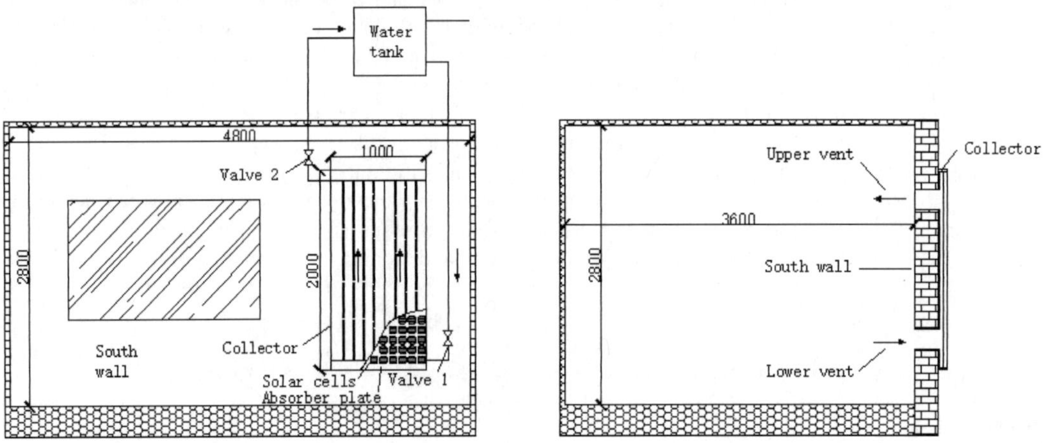

Figure 1. Illustrative diagrams of the building-integrated solar collector system.

2.2. Theoretical model of the building room

The envelop of the building is consisted of brick, and the thickness is 0.24 m, thermal conductivity is 0.84 W/(mK), specific heat is 835 J/kg K, density is 1920 kg/m^3. Absorptivity of the outside surface of the southern wall is 0.58. There is a window on the south wall which is of the dimension of 2.0 m (length) ×1.2 m (height). The solar radiation transmittance of the curtain is set to 0.3. A passive tri-functional photovoltaic/thermal solar collector module is installed on the outside of the southern wall. The depth of the air channel in the collector is 0.04m. The indoor temperatures in winter and summer are set to 18℃and 23℃. In the winter heating mode, indoor temperature is controlled as not lower than 18℃.

The heat flux through walls of the building envelope is determined by one-dimensional along its thickness, so the energy balance equation of the wall is given as

$$\rho_w c_w \frac{\partial T_w}{\partial t} = k_w \frac{\partial^2 T_w}{\partial x^2} \tag{1}$$

2.3. The theoretical model of the PV /passive air-heating mode

The essential components of the collector are the glass cover, backboard and the absorber plate. Under the assumption that the temperatures of the glass cover, absorber, and the back plate vary only in the direction of air flow, the energy equations were established. The temperatures of the air stream and essential components can be obtained by solving the equations. The scheme of heat transfer of the components of the collector is shown in Figure 2.

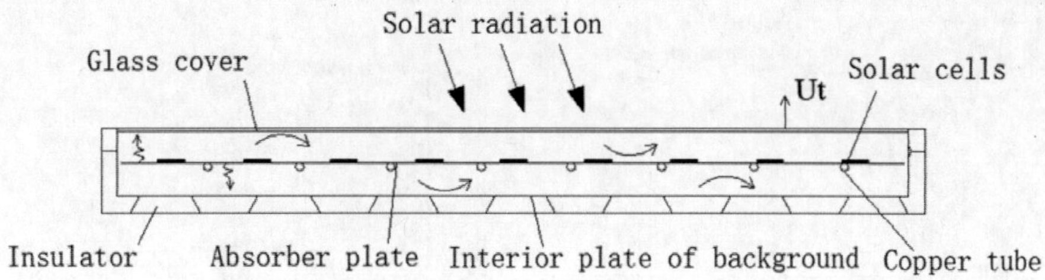

Figure 2. The schematic diagram of heat transfer process of PV/air heating.

The energy equation of absorber plate:

$$\rho_p c_p d_p \frac{\partial T_p}{\partial t} = k_p d_p \frac{\partial^2 T_p}{\partial z^2} - (hc_{pg} + hr_{pg})(T_p - T_g) - hc_{pf}(T_p - T_f) - (T_p - T_{pv}) + \alpha_p \tau_g S - \xi E_{pv} \quad (2)$$

The energy equation of air in the flow channel:

$$\rho_f c_f d \frac{\partial T_f}{\partial t} = \rho_f c_f u d \frac{dT_f}{dz} + hc_{pf}(T_p - T_f) + hc_{bf}(T_b - T_f) \quad (1)$$

2.4. The theoretical model of the PV /water-heating mode

The energy equation of water in the flow channel:

$$\frac{\dot{m} C_p}{w} \frac{dT_f}{dx} = hc_{pf}(T_p - T_f) + hc_{bf}(T_b - T_f) \quad (4)$$

2.5. Numerical Approaches

As a result of finite difference method, the energy equations are transformed into algebraic equations, which are solved by iteration method. In the solution procedure the initial condition and the instantaneous boundary conditions, such as solar radiation and ambient temperature, are set based on the experimental data. In this study, numerical simulation under unsteady state is run with FORTRAN.

3. Result and discussion

3.1. Study on winter operation performance of the system

Figure 3 shows the ambient temperature and solar radiation received by the vertical south surface on March 3th. The maximum solar irradiation is 617.8 W/m^2, and the total amount of radiation on the south wall throughout the day is 15.46 MJ/m^2. The maximum ambient temperature is 13.5℃ and the lowest is 1.5℃, the average ambient temperature is 6.3 ℃ throughout the day. Since the collector is in the passive heating mode, the heat quantity fluctuates greatly with the change of the solar irradiation and room temperature is difficult to reach a constant value, so the room temperature is set to no less than 18 ℃during the simulation. Figure 4 shows the Photovoltaic power generation. It can be seen that the power generation starts to rise at 9 am, drops to zero at 17 pm. The maximum output power can reach 48.3 W at about 13 pm of the day.

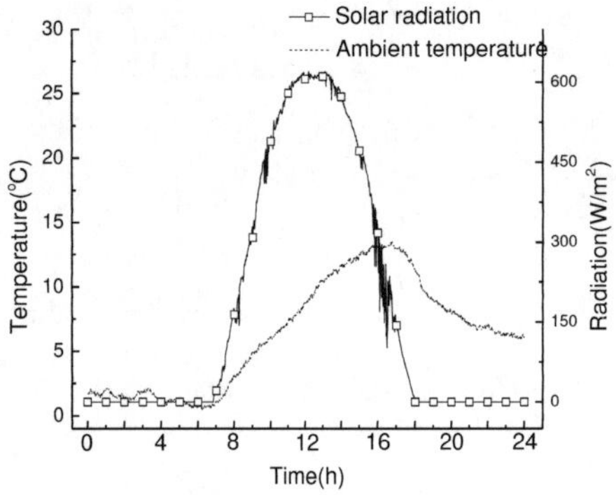

Figure 3. The curve of ambient temperature.

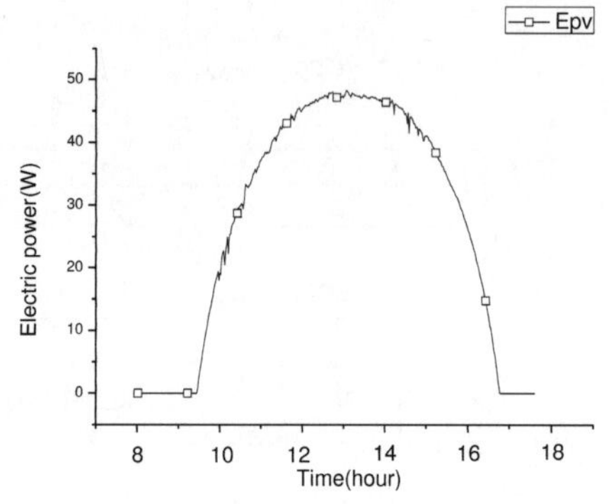

Figure 4. The curve of PV power.

Figure 5 shows the variation of indoor temperature of the building integrated with collector. It can be seen that the temperature is maintained at 18 ℃ from 0 am to 9:52 am and from 16:47 pm to 24 pm. The average indoor temperature is higher than 18 ℃ during the time from 9:52 am to 16:47 pm, and the maximum temperature reaches 21.2 ℃, which means that the room could keep the indoor air temperature by the use of the collector without additional power consumption. The air-conditioning system is started to keep the room at 18℃.

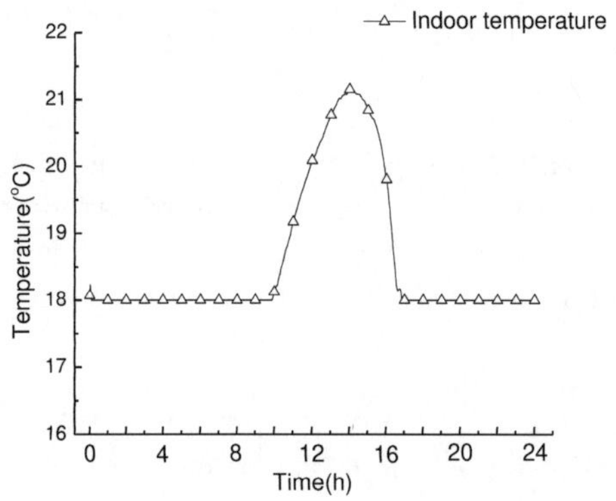

Figure 5. The average temperature of indoor air of the room.

3.2. Thermal behavior of the system in PV/water heating mode

Figure 6 shows the ambient temperature and solar radiation received by the south-facing surface of September 1st. The maximum solar radiation is 419.5 W/m^2 and the total amount of radiation on the south wall is 9.9 MJ/m^2. The maximum ambient temperature is 34.9°C and the minimum is 25.5°C. The daily average ambient temperature is 28.8°C. Figure7 shows the PV electric power generation of the day. It can be seen that the power generation increases from 8 am and the maximum output power can reach 48 W at about 13 pm. The solar radiation in summer is greater than that in winter, so the electric power gain is earlier than winter.

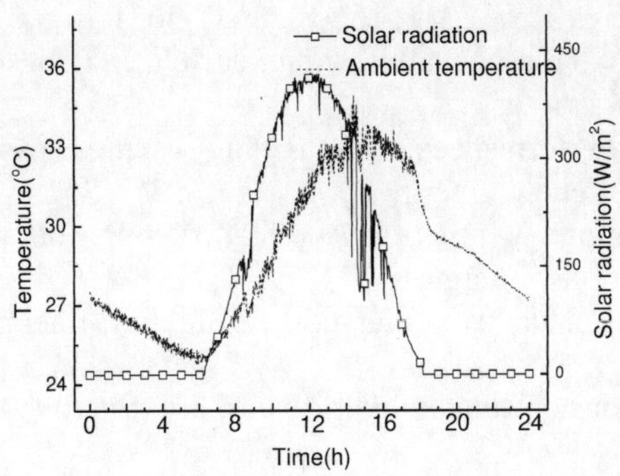

Figure 6. Solar radiation intensity and outdoor air temperature.

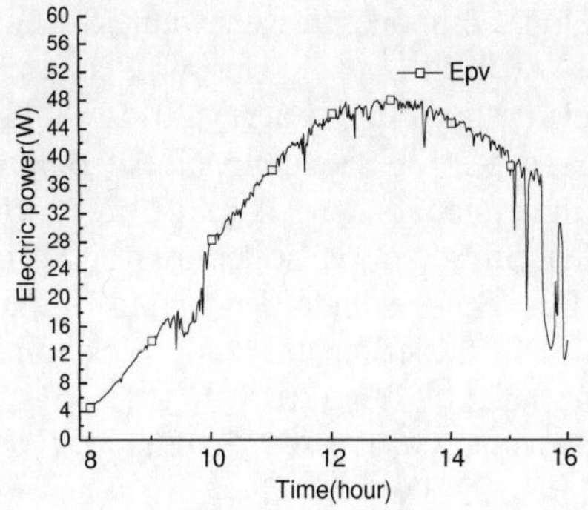

Figure 7. The curve of PV electric power generation.

Figure 8 shows the average water temperature in the tank from 8 am to 16 pm of the day. It can be seen that the temperature increases from 25.4 °C to 38.4 °C as a result of natural circulation. Therefore, the collector can not only generate electric power but also heat the water effectively when operating in the PV/water heating mode in summer.

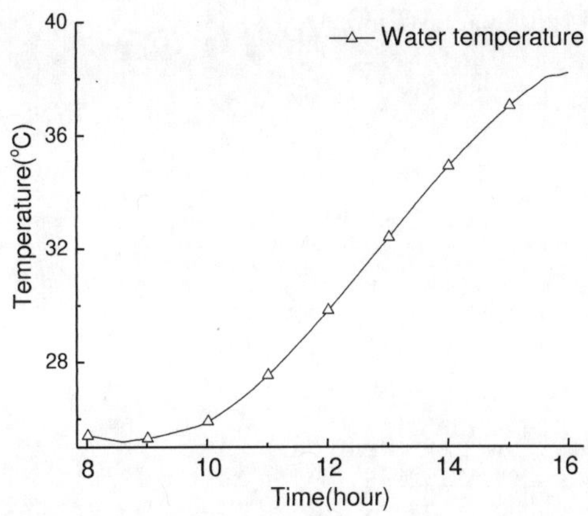

Figure 8. The water temperature in the water tank.

4. Conclusion

This paper designs a virtual building room integrated with a tri-function solar collector on the southern wall and establishes a theoretical model of the system. The theoretical model contains PV/passive air heating mode and the PV/water heating mode. The electric generation performance and thermal behavior of the two working modes have been theoretically studied.

1. The average indoor temperature is higher than 18 ℃ from 9:52 am to 16:47 pm in winter, which means that the room could keep the indoor air temperature by the use of the collector during this time without additional power consumption. In addition, electric power can be generated, the maximum output power can reach 48.3 W.

2. When the PV/water heating mode is operated in summer, the water in the tank can be heated up to 38.4 °C. The efficiency of photovoltaic power generation is about 10%.

It can be seen that the façade-integrated tri-functional photovoltaic/thermal solar collector works well both in winter and summer.

5. References

[1] Kuang, Y.; Wang, R., Applications and prospects of solar thermal technology in energy efficiency in buildings of china. Refrigeration & Air Conditioning 2001, 1, (4), 27-34.

[2] Sadineni, S. B.; Madala, S.; Boehm, R. F., Passive building energy savings: A review of building envelope components. Renewable Sustainable Energy Rev. 2011, 15, (8), 3617-3631.

[3] Sun, W.; Ji, J.; Luo, C.; He, W., Performance of PV-Trombe wall in winter correlated with south façade design. Appl. Energy 2011, 88, (1), 224-231.

[4] Imessad, K.; Messaoudene, N. A.; Belhamel, M., Performances of the Barra–Costantini passive heating system under Algerian climate conditions. Renew. Energy 2004, 29, (3), 357-367.

[5] Budea, S., Solar Air Collectors for Space Heating and Ventilation Applications—Performance and Case Studies under Romanian Climatic Conditions. Energies 2014, 7, 3781-3792.

[6] Wang X., Experimental study on cooling method of passive solar house. Building Energy & Environment 2009, 1, 018.

[7] Rockendorf, G.; Janssen, S.; Felten, H., Transparently insulated hybrid wall. Sol. Energy 1996, 58, (1-3), 33-38.

[8] Ma, J.; Sun, W.; Ji, J.; Zhang, Y.; Zhang, A.; Fan, W., Experimental and theoretical study of the efficiency of a dual-function solar collector. Appl. Therm. Eng. 2011, 31, (10), 1751-1756.

Acknowledgments

This work was supported by the National Natural Science Foundation of China (No. 51606002), the Natural Science Foundation of Anhui Province (No. 1608085QE108), the National Key Technology R&D Program (No. 2017YFC0702605), the International Scientific and Technological Cooperation Program of Anhui Province (No. 1503062012) and the Natural Science Research Project of Anhui Province Education Department (No. KJ2016JD05).

Estimating hourly global solar irradiance using artificial neural networks – A case study of Hong Kong

Wenqiang Chen [1*], Danny H.W. Li [1], Shuyang Li[1], Joseph C Lam[1]
[1]Building Energy Research Group, Department of Architecture and Civil Engineering, City University of Hong Kong, Tat Chee Avenue, Kowloon, Hong Kong SAR, China
E-mail: chenwenqiang123@sina.cn

Abstract. Solar energy is the most popular resource for power generation among the various available renewable energy alternatives. Solar radiation data are important for solar photovoltaic (PV) systems and passive energy-efficient building designs. Due to the unavailability of measurement in rural locations, solar radiation prediction models are required. In recent years, the Artificial Neural Networks (ANN) were successfully used for predicting solar radiation. However, previous works indicated that the ANN techniques were mainly focusing on prediction of monthly average or daily solar radiation, a few of them were modelled for predicting solar irradiance in hourly basis. In this study, prediction models of global solar irradiance on a horizontal surface will be developed based on neural-network techniques. Hourly meteorological variables between 2012 and 2015 acquired from the measurements made by local meteorological station were used for the study. To consider the effectiveness of individual predictors, different combinations of input variables were analysed using the Levenberg-Marquardt (LM) algorithm. Finally, equations were modelled by regression based on the important predictors. The developed models were used to estimate the global irradiance and assessed against measurement results.

Keywords: Global horizontal irradiance, ANN, Sensitivity analysis, Hong Kong

1. Introduction

Solar radiation data are important to active solar energy systems [1] and passive energy-efficient building designs [2]. It is a widespread concern to estimate the amount of such data at the specific location. Long-term ground measurement is the most effective and accurate way of setting up the databases of the required climatic parameters. However, many places do not offer measured solar radiation data [3]. Without direct measurement, the data can be predicted from empirical models based on geographical parameters and meteorological variables. Empirical prediction models tend to involve regression techniques, the principal merit of which is the simple format of mathematical expression [4]. The issue is that these simple models underperform if the systems being modelled are nonlinear. Recently, various Machine Learning (ML) algorithms, including Artificial Neural Network (ANN) and Support Vector Regression (SVR), have been used for estimating solar radiation [5, 6]. Models developed using these techniques tended to be more complex, less interpretable [7], and yet more accurate than regression equations [8]. More importantly, ML algorithm can be adopted to identify the contribution of individual input variables in the estimation of output in terms of mean-bias-error (MBE), root-mean-square-error (RMSE) and coefficient of determination (R^2) [9]. Such approaches have been used of many problems with large datasets, large quantities of relevant variables and complex interrelationships between input and output variables [10]. This paper proposes an approach for calculating the horizontal global solar radiation. Sensitivity analysis of individual meteorological parameters in determining the global solar radiation was analyzed using ML techniques and regression equations were modelled based on the important parameters. The findings and building design implications are discussed.

2. Solar irradiance and meteorological data

The horizontal global solar irradiance (GHI) levels and fluctuations are influenced considerably by the meteorological data. Available climatic parameters could be used as predictors to estimate hourly global solar irradiance. Different combinations of climatic parameters and geographical parameters were used in previous empirical models. The choice depends on the availability, accuracy and suitability of these climatic data.

Solar irradiance and meteorological data acquired from the measurements made by the Hong Kong Observatory (HKO) was used for the study. The three years hourly data from 2012 to 2014 were analyzed for the ANN models and regression models. It is inevitable that there are some short periods of missing data or some inaccurate data for various reasons. To eliminate spurious data and inaccurate measurements, quality-control tests based on the CIE guidance [11] were adopted. Table 1 summarizes the tests and results. Under Level 0 test, 1390 hourly irradiance data with a solar altitude of less than 4° or the GHI less than 20 W/m^2 were rejected. Then, 4 and 1224 hourly data were excluded under Level 1 and Level 2 tests respectively based on the conversion relationship between the three aspects of irradiance. After all the tests, 10475 hourly solar irradiance and meteorological datasets were retained for the analysis. Table 1 summarizes the number of accepted and rejected data for each process.

Table 1. Data quantity control process

CIE test level	Criterion	Number of data record (Hour)	
		Rejected	Accepted
Total	Solar altitude $\alpha > 0$		13226
Level 0	Solar altitude $\alpha > 4$; GHI > 20 W/m^2	1390	11836
Level 1	0 < GHI < 1.2 *GHE; 0 < DHI < 0.8 *GHE; 0 ≤ DNI < GHE	4	11832
Level 2	0.9 * (GHI – DNI* sin α) < GHI < 1.1 * (GHI – DNI* sin α); DHI ≤ GHI	1224	10608
Level 3	Effective record of meteorological parameters	133	10475

GHI: global horizontal irradiance; GHE: extraterrestrial solar irradiance; DNI: direct normal irradiance; DHI: diffuse horizontal irradiance;

3. ANN model description

ANN-based models are considered more appropriate in predicting the solar irradiance compared to the linear regression models. However, the accuracy of the ANN models is sensitive to the architecture of the neural networks. The process of selecting the optimum number of hidden neurons in an ANN model is still a challenge since it is normally based on the trial and error approach.

3.1 ANN architecture

To evaluate the hourly global solar irradiance in terms of various climatic parameters, a 3-layer feed forward neural network with the default number of neurons in hidden layer and 1 output neuron was developed initially. For ANN training, the network architecture with Levenberg–Marquardt learning algorithm and hyperbolic tangent sigmoid transfer function was used in the developed model. The three years hourly data obtained from HKO was used for the ANN development. The proportions of the data for training, validation and test sets were 75%, 15% and 15%. The elements of three subsets were chosen randomly to guarantee the representativeness of all climate patterns. The input variables (predictors) to the ANN were the solar altitude and metrological parameters, all the variables were normalized to the range of -1 to 1 before training. The target output was the hourly clearness index K_t (ratio of global horizontal irradiance to extraterrestrial horizontal irradiance) which is the most commonly used indicators in both global solar radiation models and direct/diffuse solar irradiance prediction models.

The most common approach of estimating K_t is to correlate it with other meteorological data used as predictors in simple regression equation. Table 2 summarizes all the predictors of hourly clearness index K_t for the ANN models. All variables were evolved in the initial neuron networks which consists of 10 input neurons and 1 output neuron. The crucial issue is how to determine the most appropriate number of neurons in the hidden layer. If the number of neurons is too small, the ANN may not capture the contributions of each predictor and if it is too large, the network has the tendency to overfit the data and requires more computations.

Table 2. Predictors of hourly clearness index K_t for ANN models.

Variable	Notation	Unit	Variable	Notation	Unit
Sine of solar altitude angle	*sin (α)*	-	Air temperature	*Temp*	°C
Sunshine duration	*Sh*	hour	Wet bulb temperature	*Wet*	°C
Cloud cover amount	*Cld*	octas	Dew-point temperature	*Dew*	°C
Wind speed	*Spd*	m/s	Relative humidity	*Rh*	(%)
Visibility	*Vis*	km	Mean sea level air pressure	*Pre*	kPa

To find the optimum number of hidden neurons, the networks were built up with increasing number of hidden neurons from 2 to 50 to control the complexity of the neural network. Each proposed ANN configuration has been trained 50 times to stabilize the weight initialization process and delivered the best accuracy in the shortest processing time. The network performed well in terms of the root mean square error. To further evaluate the performance of each configuration, the R (correlation coefficient) of all datasets were examined during the process. Fig. 1 presents the average, minimum and maximum value of R for the networks which have been trained 50 times with the different number of hidden neurons. It can be found that the average value of R increases smoothly from 2 to 21 neurons in hidden layer and reaches 0.924 representing that the performance of the networks increases with the complexity of the networks. However, it decreases at 22 neurons, and after that, the R does not increase significantly. Thus, the optimum range of neurons was set to 20 to 40 for the ANN model and for the next analysis.

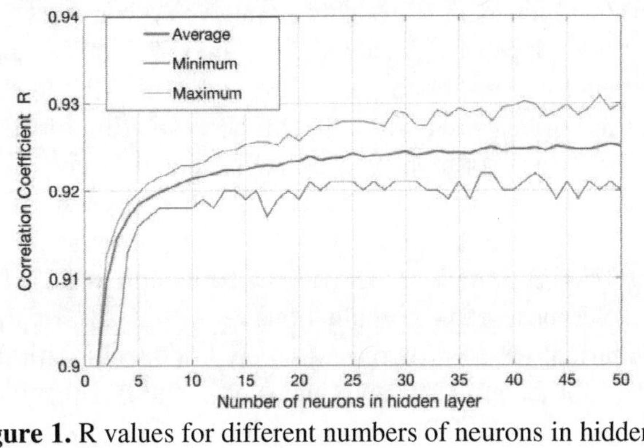

Figure 1. R values for different numbers of neurons in hidden layer

3.2 Sensitivity analysis of each parameter

To study the importance of each input variables on the predicted values of the target clearness index K_t and select appropriate predictors for the regression models, different configurations of the ANN model were established. To consider the effectiveness of individual predictors, models excluding (drop) one predictor from all the input variables were developed and the performance (R) of the corresponding neural network was examined. Each neural network was trained, validated and tested for 100 times with random number (20 to 40) of neurons in hidden layer following the results of the analysis of the optimum range of hidden neuron numbers.

The developed ANN with all the 10 predictors was found to give maximum R of 0.925 of predicted hourly global irradiance. When *sin (α)* was dropped from the networks, the R declined to 0.820 indicating the importance of α for estimating global solar irradiance. The performance of other models with different parameters being dropped is shown in the boxplot in Fig. 2. As shown in the figure, the importance of each predictor can be identified by the descend range from the R value of models with all predictors. Except for the *sin (α)*, the most important predictors identified by the ANN models were sunshine duration (*Sh*), amount of cloud cover (*Cld*), mean sea level pressure (*Pre*) and visibility (*Vis*).

IOP Conf. Series: Materials Science and Engineering **556** (2019) 012043 doi:10.1088/1757-899X/556/1/012043

Among these predictors, *Sh* (expressed in fractions of an hour over a 60 min interval) shows the strongest correlation with GHI, and it is mostly used in empirical models for predicting monthly or daily average global solar irradiance. Through the sensitivity analysis, these important predictors were selected to examine the various correlations with K_t as used by other regression techniques. Several different combinations of these predictors were used in the regression analysis and the mathematical expressions of the clearness index K_t were given.

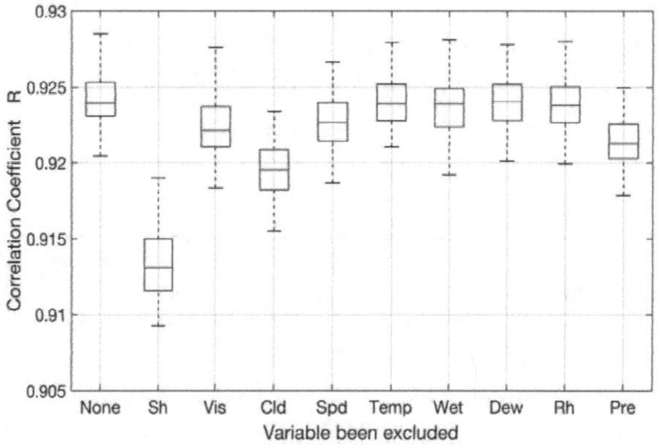

Figure 2. Performance of the neural networks when a single variable is excluded

4. Regression and results

Although models developed using ML techniques tend to be more accurate, most of them are too complex and less interpretable which cannot be easily adopted by other users. What's more, some of the predictors of these models such as meteorological variables are not always available in rural areas where the measurements are limited by some conditions. However, the ML algorithm can be adopted to identify the contribution of individual input variables in the estimation of output (i.e. K_t) as shown in Section 3. Thus, the relevant predictors identified by ANN models were selected to create empirical models by regression techniques. It should be emphasized that regression models represent a good statistical tool with both the simplicity and the flexibility to capture the relationship between K_t and various climate parameters. This section gives the mathematical expressions of the empirical models for predicting K_t with several different variable combinations.

To establish different empirical models, regressions were conducted with the different combination of 6 predictors namely solar altitude-*sin (α)*, day number of a year-*n*, sunshine duration-*Sh*, amount of cloud cover-*Cld*, mean sea level pressure-*Pre* and visibility-*Vis*. The *sin (α)* and *n* were used in all the equations since *α* is always available at a specific location and *n* can reflect the seasonal fluctuations of GHI. The choice of suitable models under specific circumstance depends on the availability of these meteorological data and the accuracy of the models. The models developed by regression for estimating horizontal GHI with their coefficient of determination (R^2) are shown in Eqs. (1) to (15). Polynomial or exponential expressions of the predictors are commonly used in modeling the equations, the differences are the number of input parameters and their combinations.

- **Model with 2 predictors**:

$$K_t = \exp[8.231\, sin(\alpha) - 7.479 sin^2(\alpha) + 2.714 sin^3(\alpha) - 0.001\, sin(n) - 0.226\, sin^2(n) - 0.416\, sin^3(n) - 4.095],\ R^2 = 0.485. \quad (1)$$

- **Models with 3 predictors**:

$$K_t = sin(\alpha)[0.119 + 0.120\, sin(\alpha) - 0.061\, sin(n) + 0.889(Sh) - 0.467(Sh)^2)] - 0.002,\ R^2 = 0.801. \quad (2)$$

$$K_t = sin(\alpha)[0.507 + 0.218\, sin(\alpha) - 0.072\, sin(n) + 0.395\left(\frac{Cld}{8}\right) - 0.909\left(\frac{Cld}{8}\right)^2)] - 0.011,\ R^2 = 0.736. \quad (3)$$

$$K_t = sin(\alpha)[0.33 + 0.176\, sin(\alpha) - 0.137 sin(n) + 0.028(Pre - 101.325) + 0.029(Pre - 101.325)^2)] - 0.015,\ R^2 = 0.473. \quad (4)$$

$$K_t = sin(\alpha)[0.596 - 0.054\, sin(\alpha) - 0.097\, sin(n) + 0.390\left(\frac{Vis}{16.5} - 1\right) - 0.485\left(\frac{Vis}{16.5} - 1\right)^2)],\ R^2 = 0.542. \quad (5)$$

$$K_t = sin(\alpha)[0.095 + 0.125\, sin(\alpha) - 0.059\, sin(n) + 1.698(Sh) - 2.801(Sh)^2 + 1.556(Sh)^3)] - 0.0012,\ R^2 = 0.804. \quad (6)$$

- **Models with 4 predictors:**

$$K_t = sin(\alpha)[0.210 + 0.224\, sin(\alpha) - 0.070\, sin(n) + 0.329(Sh) - 0.172\left(\frac{Cld}{8}\right)] + 0.012,\ R^2 = 0.791. \quad (7)$$

$$K_t = sin(\alpha)[0.074 + 0.198\, sin(\alpha) - 0.072\, sin(n) + 0.410(Sh) + 0.012(Pre - 101.325)] + 0.013,\ R^2 = 0.782. \quad (8)$$

$$K_t = sin(\alpha)[0.122 + 0.104\,sin(\alpha) - 0.062\,sin(n) + 0.390(Sh) + 0.120\left(\frac{Vis}{16.5} - 1\right)] - 0.005\,,\ R^2 = 0.789. \quad (9)$$

$$K_t = sin(\alpha)[0.645 + 0.244\,sin(\alpha) - 0.084\,sin(n) - 0.568\left(\frac{Cld}{8}\right) - 0.036(Pre - 101.325)] - 0.007,\ R^2 = 0.705. \quad (10)$$

$$K_t = sin(\alpha)[0.565 + 0.171\,sin(\alpha) - 0.071\,sin(n) - 0.530\left(\frac{Cld}{8}\right) + 0.279\left(\frac{Vis}{16.5} - 1\right)] - 0.021,\ R^2 = 0.746. \quad (11)$$

$$K_t = sin(\alpha)[0.288 + 0.170\,sin(\alpha) - 0.057\,sin(n) + 0.860(Sh) - 0.578(Sh)^2 - 0.052\left(\frac{Cld}{8}\right) - 0.209\left(\frac{Cld}{8}\right)^2] - 0.004,\ R^2 = 0.822. \quad (12)$$

- **Models with 5 predictors:**

$$K_t = sin(\alpha)[0.062 + 0.16\,sin(\alpha) - 0.067\,sin(n) + 0.385(Sh) + 0.028(Pre - 101.325) + 0.0142\left(\frac{Vis}{16.5} - 1\right)] + 0.005,\ R^2 = 0.791. \quad (13)$$

$$K_t = sin(\alpha)[0.569 + 0.16\,8sin(\alpha) - 0.071\,sin(n) - 0.531\left(\frac{Cld}{8}\right) - 0.002(Pre - 101.325) + 0278\left(\frac{Vis}{16.5} - 1\right)] - 0.021,\ R^2 = 0.746. \quad (14)$$

- **Models with 6 predictors**

$$K_t = sin(\alpha)[0.23 + 0.175\,sin(\alpha) - 0.063\,sin(n) + 0.285(Sh) - 0.205\left(\frac{Cld}{8}\right) + 0.013(Pre - 101.325) + 0.165\left(\frac{Vis}{16.5} - 1\right)],\ R^2 = 0.804. \quad (15)$$

The performance of each model can be evaluated by R^2 between measurement values and predicted values. The R^2s are various based on the number of parameters and complexity of the models with the minimum value of 0.473 (Eq.4) and maximum value of 0.822 (Eq.12). As shown in Eq. (1), without any measured climate parameters, K_t can be roughly estimated by solar altitude although the performance of the model is relatively poor. When an additional predictor was added for modeling, there was a general rise in R^2. Specifically, the R^2 increased significantly when *Sh* or *Cld* was included, while there was no increase or only slightly increase when *Pre* or *Vis* was added. The R^2 of the model with only *sin (α)*, *n* and *Sh* reached to 0.801. In this connection, Eq. (6) was created by adding the cubic *Sh* to the polynomial and the R^2 increased to 0.804. *Sh* is a widely available climatic variable measured in many meteorological stations. Eq. (6) is a simple regression model that takes into consideration only *Sh* effects but still offers widespread applicability and good performance for predicting hourly GHI. When the number of predictors was increased to 4, 5 and 6, there was a general improvement of R^2 comparing with those equations with 3 predictors. Eqs. (7) and (9) shown that the contribution of *Cld* and *Vis* to K_t is similar when they were added to *Sh* based model. Generally, when *Sh* is included in the models, the accuracy of these models tends to be higher than those with other predictors.

Table 3. Summary of the performance of regression models

Models	Input parameters	MBE (W/m^2)	%MBE (%)	RMSE (W/m^2)	%RMSE (%)	R^2
Eq.1	$sin(\alpha)$, n	+11.04	+0.13	213.8	59.4	0.485
Eq.2	$sin(\alpha)$, *n*, *Sh*	+0.64	-1.73	127.5	35.4	0.801
Eq.3	$sin(\alpha)$, *n*, *Cld*	+2.35	-0.66	138.3	38.4	0.736
Eq.4	$sin(\alpha)$, *n*, *Pre*	+16.88	+0.63	214.3	59.5	0.473
Eq.5	$sin(\alpha)$, *n*, *Vis*	-68.05	-0.28	236.2	65.6	0.542
Eq.6	$sin(\alpha)$, *n*, *Sh*	+1.21	+0.3	124.5	34.6	0.804
Eq.7	$sin(\alpha)$, *n*, *Sh*, *Cld*	-0.36	-0.1	127.7	35.5	0.791
Eq.8	$sin(\alpha)$, *n*, *Sh*, *Pre*	+2.35	+0.7	133.5	37.1	0.782
Eq.9	$sin(\alpha)$, *n*, *Sh*, *Vis*	-54.15	-15.0	140.4	39.0	0.789
Eq.10	$sin(\alpha)$, *n*, *Cld*, *Pre*	-2.40	-0.7	150.4	41.8	0.705
Eq.11	$sin(\alpha)$, *n*, *Cld*, *Vis*	-129.7	-36.0	205.6	57.1	0.746
Eq.12	$sin(\alpha)$, *n*, *Sh*, *Cld*	**-1.65**	**-0.5**	**117.3**	**32.6**	**0.822**
Eq.13	$sin(\alpha)$, *n*, *Sh*, *Pre*, *Vis*,	-62.1	-17.2	144.9	40.2	0.791
Eq.14	$sin(\alpha)$, *n*, *Cld*, *Pre*, *Vis*,	-129.3	-35.9	205.2	57.0	0.746
Eq.15	$sin(\alpha)$, *n*, *Sh*, *Cld*, *Pre*, *Vis*	-76.4	-21.2	148.2	41.2	0.804

To further evaluate the performance of the developed empirical models, measured solar irradiance and climate data of 2015 was used to test the models. The performance indices in terms of MBE, %MBE, RMSE and %RMSE were calculated and the results are summarized in Table 3. It can be seen that most of the MBEs are negative which means the models mainly underestimate the GHI. The RMSEs were basically smaller with the increase of R^2, the smallest and maximum value was found to be 117.3W/m^2 (32.6%) and 236.2 W/m^2 (65.6%). In general, the performance of the regression models increases with the increase of the number of predictors and the complexity of the expressions. The increase of polynomial power could further enhance the performance of the models. Eq. (12) has the highest R^2 and lowest RMSE when the quadratic expressions of *Sh* and *Cld* were used in the polynomial function. It indicates that *Sh* and *Cld* have the largest contribution to the estimation of hourly horizontal GHI. The findings will be helpful for users to select appropriate models according to the availability of measured parameters.

5. Conclusions

This study proposes an approach for estimating the horizontal global solar irradiance from meteorological parameters. Sensitivity analysis of individual parameters was conducted using ANN techniques and the importance of each predictor was identified. Four climate variables along with solar altitude and day number were selected to create the prediction models by regression techniques according to the results of sensitivity analysis. Hourly meteorological and solar irradiance data of Hong Kong between 2012 and 2015 were used for the analysis. The equations of various models with different combinations of predictors were given and their performance was tested by several indices. The estimation accuracy of the regression models depends on the number of predictors and the complexity of the mathematical expressions. More long-term measured data for different locations will be analyzed in the future work of this study, more combinations of the predictors and various forms of the equations should also be evaluated. Nonetheless, the empirical models developed by this study would be helpful for the estimation of hourly GHI when less meteorological data is available. The findings are important for solar energy system and energy efficient building designs.

References

[1] Li DHW, Cheung KL, Lam TNT, Chan W. A study of grid-connected photovoltaic (PV) system in Hong Kong. Appl Energy 2012;90:122-7.
[2] Li DHW, Yang L, Lam JC. Zero energy buildings and sustainable development implications – A review. Energy 2013;54:1-10.
[3] Muneer T. Solar radiation and daylight models. Oxford: Elsevier; 2004.
[4] Jiang Y. Prediction of monthly mean daily diffuse solar radiation using artificial neural networks and comparison with other empirical models. Energy Policy. 2008;36:3833-7.
[5] Celik AN, Muneer T. Neural network based method for conversion of solar radiation data. Energy Conversion and Management. 2013;67:117-24.
[6] Chen J-L, Li G-S. Evaluation of support vector machine for estimation of solar radiation from measured meteorological variables. Theoretical and Applied Climatology. 2014;115:627-38.
[7] Hastie T, Tibshirani R, Friedman J. The Elements of Statistical Learning. New York Springer; 2009.
[8] Kaytez F, Taplamacioglu MC, Cam E, Hardalac F. Forecasting electricity consumption: A comparison of regression analysis, neural networks and least squares support vector machines. International Journal of Electrical Power & Energy Systems. 2015;67:431-8.
[9] Zou L, Wang L, Lin A, Zhu H, Peng Y, Zhao Z. Estimation of global solar radiation using an artificial neural network based on an interpolation technique in southeast China. Journal of Atmospheric and Solar-Terrestrial Physics. 2016;146:110-22.
[10] Chen J-L, Liu H-B, Wu W, Xie D-T. Estimation of monthly solar radiation from measured temperatures using support vector machines – A case study. Renewable Energy. 2011;36:413-20.
[11] CIE. Guide to recommended practice of daylight measurement, Technical report. Vienna, Austria: Central Bureau of the CIE, 1994. Report No. CIE 108- 1994.

Acknowledgements

Work described was fully supported by a Strategic Research Grant from the City University of Hong Kong (Project no. 7004629). Wenqiang Chen was supported by a City University of Hong Kong Postgraduate Studentship.

IOP Conf. Series: Materials Science and Engineering **556** (2019) 012044 doi:10.1088/1757-899X/556/1/012044

Analysis of Point Daylight Factor (PDF) Average Daylight Factor (ADF) and Vertical Daylight Factor (VDF) under various unobstructed CIE Standard Skies

Danny H W Li[1], Shuyang Li[1], Wenqiang Chen[1*], Siwei Lou[2]
[1] Building Energy Research Group, Department of Architecture and Civil Engineering, City University of Hong Kong, Tat Chee Avenue, Kowloon, Hong Kong SAR, China
[2]School of Civil Engineering, Guangzhou University, 230 GuangZhou University Town Outer Ring Road, Panyu District, Guangzhou, P.R. China, 510006
E-mail: chenwenqiang123@sina.cn

Abstract. The usual method for computing natural illuminance for a given location in a building is the daylight factor (DF) approach, which is based on the traditional overcast sky without sunlight, as defined by the International Commission on Illumination (CIE). In 2003, the CIE adopted a range of 15 skies as international standard models that represent the actual skies of many places and cover the whole probable spectrum of skies found in nature. Previously, we developed a number of calculation tools to predict daylight illuminance for rooms facing different orientations under various CIE Standard skies. This paper presents the work on the calculation of the point DF (PDF), average DF (ADF) and vertical DF (VDF) under various overcast and non-overcast skies. The strong correlations between these three DF types can help to estimate the required DF values for architectural and daylighting designs and evaluations.

Keywords: Point Daylight Factor; Average Daylight Factor; Vertical Daylight Factor; CIE Standard Skies; Split-flux theory

1. Introduction

Estimation of the daylight level for any point within an interior space is essential to daylighting designs and assessments [1]. The usual method for computing natural illuminance for a given location in a building is the daylight factor (DF) approach, which is based on the traditional overcast sky without sunlight, as defined by the International Commission on Illumination (CIE). The DF approach can be analytically calculated using simple design aids [2]; annual daylight data are not required. The point DF (PDF) has shown the effectiveness of daylight-linked controls in saving energy on electric lighting [3]. The average DF (ADF), which is directly related to window area and less detailed input data than those provided by the PDF, could be a useful daylighting criterion [4]. The vertical DF (VDF) representing light coming directly from the sky and daylight reflected from surrounding buildings and the ground onto a vertical surface has been appropriately used for compact urban areas with large exterior obstructions, such as Hong Kong [5]. Such DF approaches have gained favour due to their simplicity. Practitioners have become accustomed to using the daylight criteria that are usually expressed in the DF approach, which has been adopted in many countries [6]. However, such DF approaches are not flexible enough to predict the dynamic daylight variations caused by the movement of the sun when the sky is not overcast [7]. Therefore, the daylight metric based on the DF lacks realism. In 2003, the CIE adopted a range of 15 skies as international standard models [8] that represent the skies of many places and cover the whole probable spectrum of skies found in nature [9]. Previously, we developed a number of calculation tools to predict daylight illuminance for rooms facing different orientations under various CIE Standard skies [10]. This paper presents the work on the calculation of the PDF, ADF and VDF under various overcast and non-overcast skies. Calculation procedures are demonstrated. Characteristics of the findings and design implications are discussed.

2. Mathematical Expressions for calculating Daylight Factors

For flats at the top floors of high-rise buildings or rooms of houses located in low densely-built zones, the approaches for estimating DF should be computed under unobstructed skies. The DF for an internal space is often split into three components, viz. the sky component (SC), the externally reflected component (ERC) and internally reflected component (IRC). Under unobstructed skies, there is no outdoor illuminance reflected from opposing external vertical surfaces to the interior point (i.e. no ERC). Previously, a nomograph and a Waldram diagram for calculating the SC were established [11]. Fig. 1 presents the nomograph consisting of a chart with two scales A and B for calculating the SC under CIE standard clear sky (Sky 12). Similar nomgraphs can also be plotted for the other CIE Standard Skies. Based on a number of procedures, the required SC under various unobstructed CIE Standard Skies can be determined. The computation of the IRC can be based on the theory of the split-flux principle. Mathematically, average IRC can be given as:

- $$\mathrm{IRC} = \mathrm{tW}\left(\frac{CR_{fw}+DR_{cw}}{A(1-R)}\right) \quad (1)$$

where A is the total area of all the interior surfaces, m²; C and D are the configuration factors of the daylight flux incident on the mid-height of the window pane from above and below the horizon, respectively, dimensionless; R is the average reflectance of all the interior surfaces, dimensionless; R_{cw} is the average reflectance of the ceiling and upper walls above the mid-height of the window, excluding the window wall, dimensionless; R_{fw} is the average reflectance of the floor and lower parts of the walls below the mid-height of the window, excluding the window wall, dimensionless; t is the visual transmittance of the window, dimensionless; W is window area, m²

The important issue is to determine the configuration factors C and D under various sky conditions. For an unobstructed sky, C is the vertical sky component (VSC) which is the ratio of the vertical (E_{VD}) to the horizontal (E_{HD}) sky-diffuse illuminance available at the same point. Constant VSC values of 40, 46 and 50% representing, respectively, the Standard Skies 1, 3 and 5 were obtained. For other 12 CIE Standard Skies, the VSC substantially relies on the solar location and the VSC could be identified by the scattering angle (χ). To provide precise VSC values for subsequent study, the template of Gaussian functions [12] that depict the symmetric bell curve shape were used to correlate VSC with the χ for the 12 CIE Standard Skies as shown in Eq. 2.

- $$\mathrm{VSC} = A_1 exp\left\{-\left[\frac{\chi_{ref}-B_1}{C_1}\right]^2\right\} + A_2 exp\left\{-\left[\frac{\chi_{ref}-B_2}{C_2}\right]^2\right\} \quad (2)$$

where A_1, B_1, C_1, A_2, B_2 and C_2 are the regression coefficients

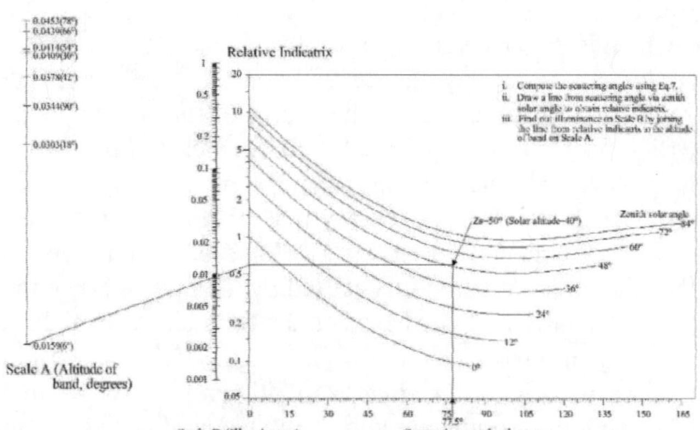

Figure 1. Nomograph calculating daylight iluminance for the CIE standard clear sky.

Table 1 summarizes the regression coefficients for Eq. 2 under the 12 CIE Standard Skies other than Skies 1, 3 and 5 [12]. Accordingly, the VSC can be computed. For non-overcast skies (i.e. Skies 7 to 15) the ground reflected component due to direct sunlight (E_{HB}) should be included.

Table 1. Coefficients A1, B1, C1, A2, B2 and C2 (for the Gaussian Function with 2 terms).

Sky No.	A_1	B_1,	C_1	A_2	B_2	C_2
2	0.528	10.9	106	0.306	198	70.7
4	0.626	5.93	95.7	0.424	222	107
6	0.688	2.41	91.9	0.549	257	146
7	0.82	6.76	75.1	0.958	336	178
8	1	7.3	66.7	0.745	294	150
9	0.891	−2.5	95.1	0.624	243	142
10	1.04	2.33	75.9	2.01	461	246
11	1.3	−2.48	73.3	2	427	211
12	1.45	−1.5	68.1	2.60	510	267
13	1.86	−13.1	71.8	7	801	372
14	2.39	−24.9	78.3	1.8	499	278
15	3.91	−56.3	93.6	1.31	445	222

Mathematically, D can be written as:

- $D = \frac{(E_{HD}+E_{HB})\rho_g}{2E_{HD}} (with\ E_{HB})$ (3)
- $D = 0.5\rho_g (without\ E_{HB})$ (4)
- $E_{HB} = E_{NB} \operatorname{Sin} \alpha_s$ (5)

where E_{NB} is the direct normal sunlight (lux); α_s is the solar altitude (degrees); ρ_g is the ground reflectance (dimensionless)

The approximate E_{HD} and E_{NB} may be either measured or obtained from a number of articles [13]. Under non-overcast skies, the sun can cast a large and well defined shadow in front of the window façade. To cater such effect, it can be assumed that only diffuse component was considered for those shaded areas (using Eq. 4) [14]. The study can be extended to include the determination of the VDF and ADF for unobstructed skies [15]. VDF is the sum of C (VSC under unobstructed skies) and D. Referring to the Longmore equation [16], the additional term is the ratio of C to the area of floor and lower parts of the walls below the mid-height of the window (A_{fw}). Knowing the C and D values, the IRC, VDF and ADF for various unobstructed CIE Standard Skies can then be estimated accordingly.

3. Model for estimating DF

The PDF, ADF and VDF under the 15 unobstructed CIE Standard Skies were estimated based on the following case studies. The vertical window area of 4m (W) x 2m (H) with the sill height of 1m off the floor was set. Figure 2 presents the layout of the room and the reference points for estimating the DF. The centre of the vertical window faces south (i.e. sun-shading vertical surface) on 21 December (winter solstice) solar noon (α_s=45° and ϕ_s=180°) with façade reflectance 0.6 and ground reflectance 0.2 under 15 CIE unobstructed skies.

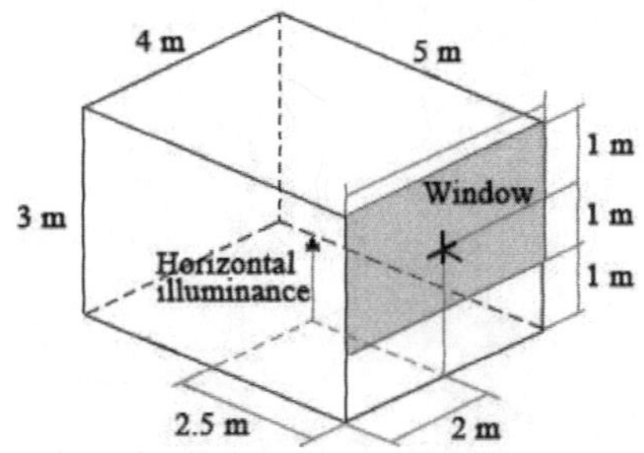

Figure 2. The room layout and the reference points.

4. Results Analysis and Design Implications

Accordingly, the C, D, SC and IRC for ADF and PDF were computed and analysed. Figure 3 presents the ADF, PDF and VDF at various scattering angle (χ) under Skies 1, 8 and 13 representing overcast, partly cloudy and clear skies, respectively. The VDF is higher than the corresponding PDF and ADF. Sky 13 has the maximum ADF of 29%, PDF (centre point) of 45% and VDF of 154%. Under dark overcast condition, all the three DF types in Sky 1 are independent of χ. For non-overcast skies (Skies 8 and 13), the DFs change with χ. The peak DFs appear at low χ of 36.5° and the DFs drop with the increase of χ. When χ is between 120°and 130°, the DFs are at their minimum values and then increase slightly as the χ goes further.

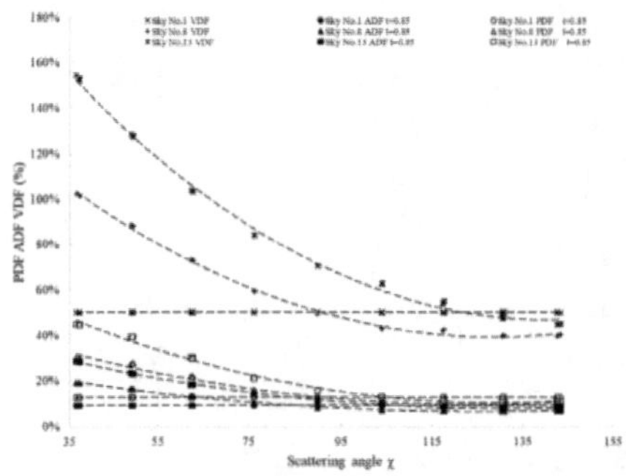

Figure 3. The χ against the ADF, PDF and VDF for 15 CIE unobstructed skies.

Figure 4 displays the correlation between the VDF and ADF for all cases. Three linear trends can be observed and Eqs. 6 to 8 present the regression equations for Skies 1-10, Skies 11, 13-15 and Sky 12, respectively. The R^2 values close to 1 indicating quite perfect relation. It shows that once the C and D values are obtained, the required ADF can be computed using appropriate room parameters.

- ADF = 0.186VDF - 0.008 $R^2 = 0.993$ (Sky No.1-10) (6)
- ADF = 0.194VDF - 0.03 $R^2 = 0.998$ (Sky No.11,13-15) (7)
- ADF = 0.193VDF – 0.057 $R^2 = 1$ (Sky No.12) (8)

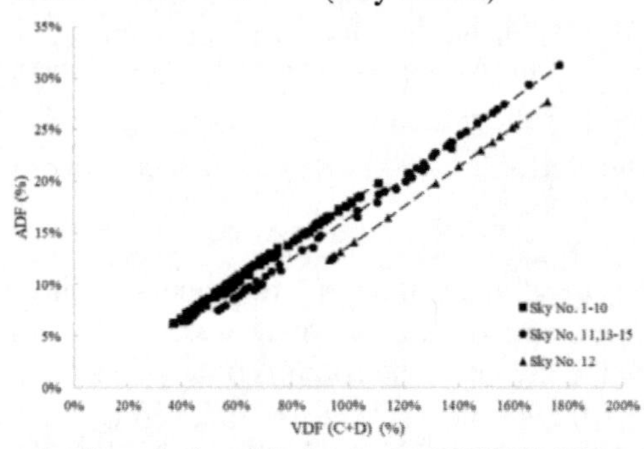

Figure 4. The relationship between VDF(C+D) and ADF

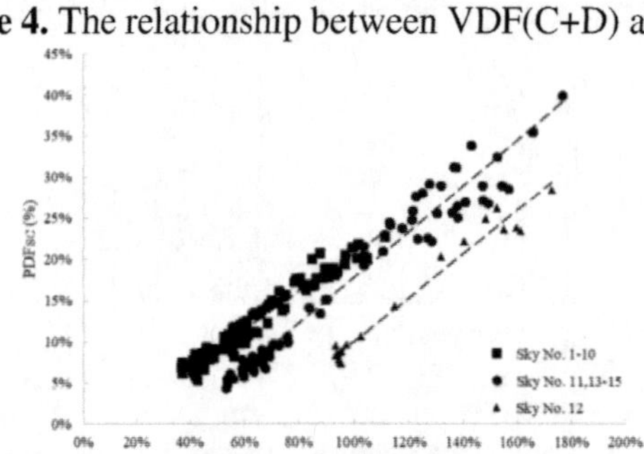

Figure 5. The relationship between VDF and PDF_{SC}

Similarly, the simulated data of VDF and the SC of PDF (PDF_{SC}) are plotted in Fig. 5 and.Eqs.9-11 give the mathematical functions for these two components. The R^2 ranges from 0.96 to 0.97 indicating that 96-97% of the PDF_{SC} can be explained by the VDF. It means that the PDF_{SC} and PDF can be computed by the corresponding VDF [10]. Such issue can help using open loop daylight-linked lighting controls.

- PDFsc = 0.23VDF + 0.03 $R^2 = 0.955$ (Skies.1-10) (9)
- PDFsc = 0.29VDF + 0.1 $R^2 = 0.968$ (Skies.11,13-15) (10)
- PDFsc = 0.26VDF + 0.16 $R^2 = 0.972$ (Sky 12) (11)

Likewise, the PDF and ADF are depicted in Fig. 6 and the linear regression is shown in Eq. 12. The strength of correlation is considered very good as R^2 is 0.97. The mathematical model can estimate the point that PDF is equal to ADF.

- ADF = 0.59 PDF + 0.014 $R^2 = 0.97$ (All 15 CIE Skies) (12)

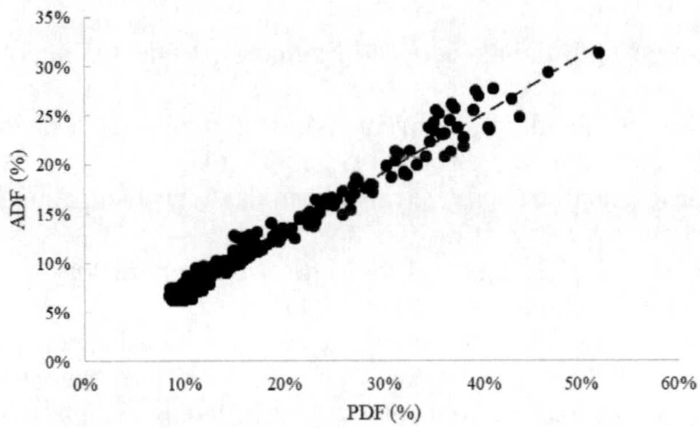

Figure 6. The relationship between PDF and ADF

The daylight glare index (DGI) and the corresponding subjective perception levels are : '<16 imperceptible', '16-20 perceptible', '20-24 acceptible', '24-28 uncomfortable' and '>28 intolerable', respectively. The DGIs were calculated when the vertical window faces south on 21 December solar noon (χ=44.3°) and Table 2 shows the results. For overcast skies, the DGIs are almost under 22. For non-overcast skies, the indoor daylight environment is only acceptible under Sky 11. The DGIs of other clear skies are in the range of 24-28 which cause the uncomfortable daylight environment. The results show that discomfort glare is caused by high luminance between the glare source (window) and its surroundings. Glare from window usually arises when large amount of direct sunlight enters the room. A high availability of daylight in an indoor environment could be a disadvantage for achieving optimal visual conditions, not only for psychological reactions, but also for the variability of the characteristics of daylight during the time.

Table 2. Vertical daylight factor (VDF) and daylight glare index (DGI) under 15 CIE skies

Sky No.	VDF	DGI	Sky No.	VDF	DGI	Sky No.	VDF	DGI
1	0.5	18.08	6	0.72	23.61	11	1.28	23.98
2	0.58	21.06	7	0.85	24.63	12	1.61	22.73
3	0.56	19.31	8	0.94	25.87	13	1.38	25.71
4	0.66	21.92	9	0.93	24.81	14	1.49	25.45
5	0.6	20.62	10	1.05	24.57	15	1.57	27.29

5. Conclusions

Calculating procedures for indoor daylight determination under the 15 CIE Standard Skies were elaborated. A number of equations, figures and charts were presented for estimating the VDF, ADF and PDF. The methods for determining VSC and ground reflected components were adopted to estimate the configuration factors C and D such that the IRC can be computed. With the required SC, VSC and IRC, the PDF, VDF and ADF can then be obtained for subsequent analysis. The calculation procedures explicitly illustrate the interdependency of various daylight parameters. The DGI has also been calculated to show the visual discomfort resulting from high window luminance. The analysis and calculation of DFs and DGIs are necessary in order to achieve a good indoor daylight environment and avoid visual discomfort. As the 15 CIE skies represent the actual skies for many places and cover the whole probable spectrum of skies found in nature, the findings could be globally adopted and would be useful to practitioners engaged in architectural and daylighting designs and evaluations. Further research analyses for validating more complex cases using parametric study will be conducted in near future.

References

[1] Li DHW. A review of daylight illuminance determinations and energy implications. Appl. Energy 2010;87(7):2109-2118

[2] Chartered Institution of Building Services Engineers. Code for interior lighting. London, UK: CIBSE, 1997.

[3] Yu X, Su Y. Daylight availability assessment and its potential energy saving estimation – A literature review. Renew. Sustain. Energy Rev. 2015;52:494-503.

[4] Lynes J, Littlefair P. Lighting energy savings from daylight: Estimation at the sketch design stage. Light Res Technol 1990;22(3):129-137.

[5] Cheung HD, Chung TM. Calculation of the vertical daylight factor on window facades in a dense urban environment. Archit Sci Rev 2005; 48(1):81-92.

[6] Hraska J. Criteria of daylighting and sunlight access in sustainable construction evaluation systems. In: Proceedings of the 5th International Conference on Solar Radiation and Daylighting, 10-11 Aug. 2011, Brno University of Technology, Czech Republic, pp.98-103.

[7] Littlefair PJ. Predicting hourly internal daylight illuminances for dynamic building energy modeling. Garston, UK: Building Research Establishment: 1989.

[8] International Standardisation Organisation. CIE S 011/E:2003. Spatial Distribution Daylight-CIE Standard General Sky, ISO Standard 15469: 2004, Geneva: 2004.

[9] Li DHW, Lau CCS, Lam JC. A study of 15 sky luminance patterns against Hong Kong data. Archit Sci Rev 2003;46(1):61-68.

[10] Li DHW, Lou S, Ghaffarianhoseini A, Alshaibani KA, Lam JC. A review of calculating procedures on daylight factor based metrics under various CIE Standard Skies and obstructed environments. Building and Environment 2017; 112:29-44.

[11] Li DHW, Cheung GHW, Lau CCS. A simplified procedure for determining indoor daylight illuminance using daylight coefficient concept. Build Environ 2006;41(5):578-589.

[12] Li DHW, Li C, Lou SW, Tsang EKW, Lam JC. Analysis of vertical sky components under various CIE Standard General Skies. Indoor Built Environ 2016;25(4):703-711.

[13] Kittler R, Darula S, Perez R. A set of standard skies, Polygrafia, Bratislave. 1998.

[14] Li DHW. Daylight and energy implication for the CIE standard skies. Energy Convers Manage 2007;48(3):745-755.

[15] Li DHW, Cheung GHW. Average daylight factor for the 15 CIE standard skies. Light Res Technol 2006;38(2):137-152.

[16] Longmore J. Daylighting: a current view. Light and Lighting 1975;68:113-119.

Acknowledgments

Work described was fully supported by a Strategic Research Grant from the City University of Hong Kong (Project no. 7004629). Wenqiang Chen was supported by a City University of Hong Kong Postgraduate Studentship.

Influencing factor Analysis of collecting capability over a year by the panel typed solar collector for seven representative cities in Yunnan, China

Ming yang[1] Long enshen[1*] Xu kuo[2] Li yanru[1] Zhang yin[1] Jin zhenghao[2]
[1]College of Architecture and environment, Sichuan University, Chengdu, China
[2] Institute for Disaster Management and Reconstruction, Sichuan University, Chengdu, China
E-mail: 502463684@qq.com

Abstract. Solar energy is abundant in Yunnan province, China, and buildings in this area have a surging demand for heat supply over winter season. Targeting on seven representative cities in Yunnan, this paper analyzed the optimum tilt and azimuth angles of the panel typed solar collector within these cities over a year by applying the orthogonal method, and an improved mathematical model in calculating the collected solar radiation was also developed, and the influence of the varying tilt and azimuth angles on the solar heat collection was also analyzed. Results showed that, the collected solar radiation varied in a parabola curve and was slowly decreasing after the azimuth was reaching 0° and the dip angle was approaching the location latitude, when the azimuth angle was varying from -65° to 65° and the dip angle was increasing from 0° to 90°, respectively. And among these cities, the optimum dip angles were ranging from 29° to 33° and the optimum azimuth angles were ranging from -18° to 5°, respectively, which were within the recommended ranges of the local standard. However, the maximum decrease rates of the captured solar radiation can amount to 12.15% and 2.8%, respectively, if choosing the tilt and azimuth angles of the panel typed solar collector randomly.

Keywords: solar energy; dip angle; azimuth angle; orthogonal method

1. Introduction

Most cities in the Yunnan province, China, have a typical plateau climate and are pretty cold in winter, and the heat supply demand is surging with the improving living standard and social development. Whereas most public and residential buildings are not equipped with central heating facilities, since this area is excluded from the regulated concentrated heating area, and the building energy consumption is high and is deteriorating the local environment. In addition, the annual sunlight hours in Yunnan province are about 2200hrs and the solar energy application is of high practical significance, thus it is imperative to utilize abundant solar energy resources locally.

Because the azimuth and elevating angle of sunshine are various in every hour, the solar radiation projected onto the fixed collector varies from time to time. Therefore, it is very important to optimize the dip angle and azimuth angle to ensure the maximum solar capture capability of the collector in the long run state. This is not only related to the initial investment of the system, but also has an important impact on improving the efficiency of solar energy utilization.

At present, many scholars have studied the optimal installation Angle of solar collector. Sachin Muralee Krishna et al. [6] have taken the city of Bangkok as an example to determine the optimum Angle of inclination in order to improve the efficiency of solar energy utilization by establishing mathematical model; Kamal Skeiker et al. [7] have used a mathematical model to calculate the solar radiation from the tilted surface and determined the optimal dip angle and azimuth angle of the solar collector in the main areas of Syria in a certain period of time; Li rongling et al. [8], based on the solar radiation data of Beijing 54511 meteorological observation station and aiming at the maximum cumulative solar radiation on inclined plane, have obtained the optimal installation angle of Beijing at different time periods; Zhu dandan et al. [9] based on the analysis of the influence factors of the solar energy capture capability of the collector plate in the whole year, have obtained the optimal dip Angle of some parts of China through the calculation model of the solar radiation of the inclined plane.

On the basis of a large number of research results at home and abroad for reference, our country has enacted a norm[10] in 2009, in which the scope of solar collector's azimuth angle is 0±30° and the range of dip angle is local latitude +20° or -10°. Although the solar energy application engineering has a certain guiding role, solar radiation characteristics are different in different cities, thus their best angles of solar collector are unlike, and the designers will easily get lost when they choose installation angle from a wide range which will cause more investment or less efficiency of the system.

Under the background of the solar energy heating applications in Yunnan province, this paper has analyzed influence rule of the connected heat amount effected by azimuth and dip angle and got the optimization angle of 7 cities. What's more, the author has contrasted the optimized results with the scope in the norm. The outcome can provide references for the design of solar heating engineering in this area.

2. Objects and method of study

2.1 Objects of study

The basic information of 7 representative cities in Yunnan province studied in this paper is given by table 1 in which the meteorological data is from literature [11]. From this table, the range of 7 cities' longitude is from 98° to 103°, and the range of latitude is from 22.7° to 28.5°. Besides, 7 cities' elevation is between 1000 m to 3400 m and their winter temperature is between -5~11℃ which shows that it has cold climate and a rigid demand for heating. The annual total solar radiation energy of cities is 1733.9~ 2112.8 kWh /m^2, with a maximum of 2112.8 kWh /m^2. Thus it can be seen that the solar energy resources in Yunnan have great application potential.

Table 1. The basic information of 7 representative cities in Yunnan province

cities	longitude (°)	latitude (°)	elevation (m)	Winter temperatures (℃)	Annual total direct radiation (kWh/m^2)	Annual total direct radiation (kWh/m^2)
Kunming	102.73	25.04	1930	3.0	1138.2	1951.8
Deqin	98.92	28.48	3350	-5.0	964.5	1733.9
Lijiang	100.25	26.86	2384	3.0	1479.0	2112.8
Lancang	99.90	22.70	1024	11.1	1209.4	2017.9
Chuxiong	101.54	25.01	1777	5.8	1050.8	1826.5
Lincang	100.08	23.88	1493	9.2	1199.8	1973.9
Simao	101.03	23.07	1330	9.0	1045.2	1897.6

2.2 Method

Any installation angle of collector can be expressed by dip angle and azimuth angle. The relationship between the position of the sun and the surface of the earth can be expressed by the height angle and azimuth angle of the sun. On this basis, we can get collected solar radiation at different times by calculating the amount of collected direct radiation, sky scattered radiation and ground reflected radiation.

The following equation can be used to calculate the collected direct solar radiation per unit area H_{bT}:

$$H_{bT} = H_{DN} \cdot (\cos\varepsilon \cdot \sin\beta + \sin\varepsilon \cdot \cos\beta \cdot \cos(A-\alpha)) \quad (1)$$

Where H_{DN} is normal radiation intensity, W/m^2; ε is the dip angle of solar collector, °; A is the azimuth angle of the solar collector, °, in south to 0 °, by east is positive; β is height angle of the sun, °; α is azimuth angle of the sun, °.

To calculate the sky scattered radiation H_{dT}, the following formula can be used:

$$H_{dT} = H_d \cdot (1+\cos\varepsilon)/2 \quad (2)$$

Where H_d is solar scattering radiation, W/m^2. It can be seen that when the plate is horizontal, the amount of sky scattered radiation is the largest.

The amount of ground reflection radiation capture H_{rT} is related to different ground materials, and the formula is given by:

$$H_{rT} = \rho \cdot H \cdot (1-\cos^2\varepsilon)/2 \quad (3)$$

Where ρ is ground reflectance, the ρ of grass is 0.17~0.22, the concrete floor is 0.3~0.37 and the snowfield is 0.7. In this paper, we have set concrete as the ground, and its ρ is 0.3; H is total horizontal radiation intensity, W/m^2.

Finally, we can get the amount of total collected solar radiation per unit area plane at any installation angle H_T:

$$H_T = H_{bT} + H_{dT} + H_{rT} \quad (4)$$

The range of azimuth we chose, apart of 0°, is -65° ~ 65° to 10 ° for interval, a total of 15 bearing point for calculation. And the issue in 0 ~ 90 ° dip angle (from horizontal to vertical), every 10 ° for a total of 10 Angle between points are combined to calculate. At last, the influence rule of solar radiation change under different dip angle and azimuth combinations is obtained by orthogonal method.

3. Results analysis

Taking Lijiang city as an example, this article will first respectively analysis the influence of tilt and azimuth angle of collected solar radiation, and then will find the optimization by the orthogonal method. At last, the results of seven representative cities will be compared with the existing norms.

3.1 Influence of azimuth on collected solar radiation

The effect of different azimuth angles on solar radiation capture under three dip angles is given by Fig 1. It can be seen that when the azimuth was varying from -65° to 65°, the collected solar radiation first increase and then decrease while the dip angle is invariant, and the maximum of the curve is 1750 kWh/m^2 when azimuth is 0°. That is because the sun rises from the east and sets in west and usually has max radiation at noon when the sun is just in south. When the dip angle is different, the influence degree of azimuth change on the collected solar radiation amount is also different. The larger the dip angle is, the steeper the curve is, and the maximum values of the curves are also different.

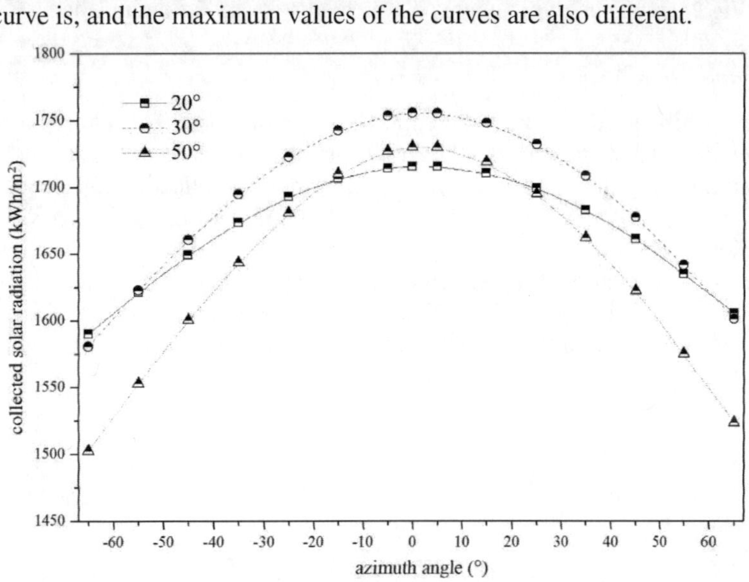

Fig 1. Influence of azimuth angle on collected solar radiation in Lijiang

IOP Conf. Series: Materials Science and Engineering **556** (2019) 012045 doi:10.1088/1757-899X/556/1/012045

3.2 Influence of dip angle on collected solar radiation

The effect of different dip angles on collected solar radiation under three azimuths has been showed by fig 2 in which the influence rule of collected solar radiation is lower open parabola with the increasing dip angle from 0° to 90° when the azimuth is fixed. Taking 0° azimuth angle for example, the maximum value of the curve is 1750 kWh/m^2 when the dip angle is 35° and the minimum value is 1150 kWh/m^2 when it is 90°. This is because the collected solar radiation is mainly influenced by the height angle of the sun, so it is usually best when the dip angle close to the city's latitude. What's more, when the azimuth changes, the influence rule of the curves are also different. According to the three curves showed in the picture, they have absolute difference when dip angle closed to Lijiang's latitude, and they are nearly invariant when dip angle is 0° or 90°. That's because the collector is horizontal when dip angle is 0°, there will be no effect if we change the azimuth angle. Then, when dip angle is 90°, the collector will catch lower direct radiation which means the proportion of scattered radiation which is not affected by azimuth will increases.

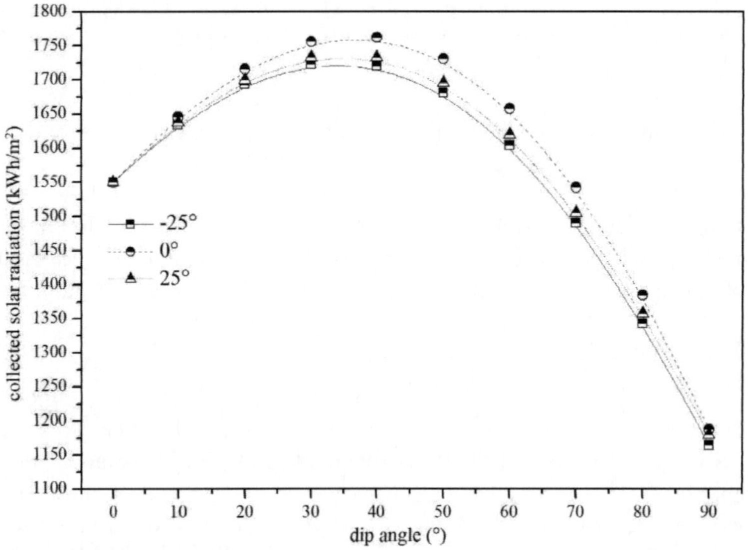

Fig 2. Influence of dip angle on collected solar radiation in Lijiang

3.3 Orthogonal optimization results

Assuming that the installation azimuth of the collector is fixed, and the dip angle of the collector is changed at some interval, the change curve of several heat collection capacity with installation dip angle can be obtained. Similarly, by the same way, several curves of heat collection capacity with the azimuth can be obtained. Finally the optimum dip angle and azimuth angle of the collector will be obtained by coupling optimization.

Taking Lijiang's result for example (Fig 3), we can get the amount of collected solar radiation under different azimuth and dip angles by orthogonal method and the maximum point of the change surface formed by each point. It is showed that the collected solar radiation is max when the dip angle is 35° and the azimuth is 0°.

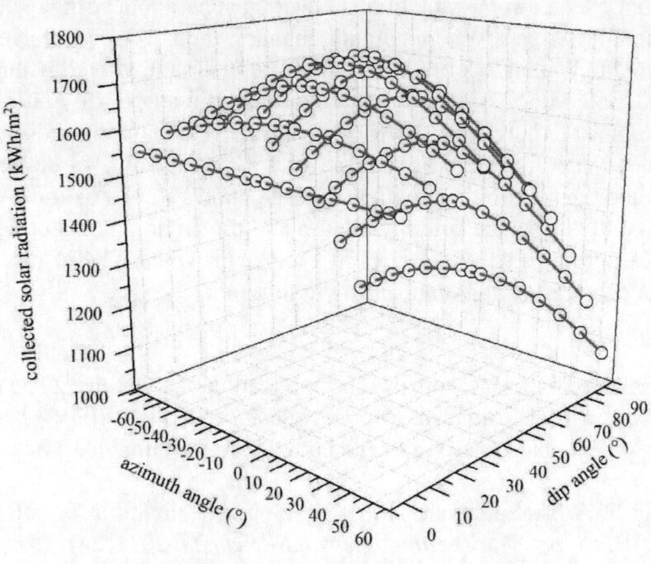

Fig 3. Result from orthogonal method in Lijiang

By using the same method, we can get the optimal angle of solar collector in 7 cities (Fig 4). It is showed that the optimal dip angles of 7 cities are ranging from 29° to 33°, and the azimuth angles are ranging from -18° to 5° with difference among those cities. 7 cities' optimization results are within the recommended range of the local standard, but the maximum attenuation of solar energy capture capability can be 12.15% if the value of dip angle is taken blindly from the specification, and the maximum attenuation of the blind value of azimuth can also amount to 2.8%.

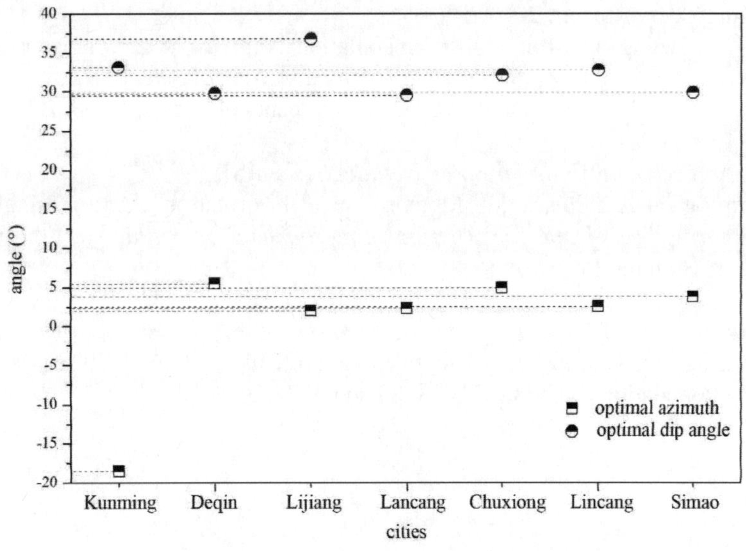

Fig 4. The best installation angle of 7 cities

4. Conclusions

Making solar collectors' optimal installation angle of 7 cities in Yunnan province as a goal, this paper has set up solar radiation collecting math model which was based on hourly meteorological parameters and has analyzed the influence caused by azimuth and dip angle. Finally got the optimization results by orthogonal method. The main results are as follows.

(1) With azimuth angle ranging from -65° to 65°, the collected solar radiation decreased after first increased while the dip angle is invariant, the greater the dip angle, the more drastic the change;

(2) The influence rule of collected solar radiation is lower open parabola with dip angle increasing from 0° to 90° when the azimuth is fixed. When azimuth changes, the curves will have absolute difference when the dip angle closed to Lijiang's latitude, and they are invariant when dip angle is 0° or 90°.
(3) Through orthogonal method, it is showed that the optimal dip angles of 7 cities are between 29° to 33°, and the azimuth angles are between -18° to 5° with difference among those cities. the maximum attenuation of solar energy capture capability can be 12.15% if the value of dip angle is taken blindly from the specification, and the maximum attenuation of the blind value of azimuth can also reach 2.8%. The research has showed the influence rule of azimuth and dip angle on heat collection effect of solar collector and has got the optimal angle of 7 cities in Yunnan province. The results can provide a better reference for the design of solar heating engineering in this area.

References

[1] Wang qianwei, Chen hongbing, Hao youzhi. The application of solar energy heating system in some areas of Yunnan province[J]. *Building Energy Efficiency*, 2011, 39(**03**): 42-44.
[2] Xu jinming. Development status of solar energy utilization in Yunnan[J]. *Solar Energy*. 2000(**01**): 2-3.
[3] Wang guohai, Long li. A brief analysis on the development direction of solar energy industry in yunnan province[J]. *China New Technologies and Products*. 2011(**24**): 252.
[4] Lishan. Analysis and countermeasure research on the development of yunnan solar energy industry. *Agriculture Engineering Technology(New Energy Industry)*, 2013.
[5] Amaya V Novo, Joseba R Bayon, Daniel Castro-Fresno, et al. Review of seasonal heat storage in large basins: Water tanks andgravel-water pits [J]. *Applied Energy*, 2010, 87: 390-397.
[6] Sachin Muralee Krishna,Nimal Madhu M,Vivek Mohan,Reshma Suresh M P,Jai Govind Singh. A Generalized Approach for Enhanced Solar Energy Harvesting Using Stochastic Estimation of Optimum Dip angles: A Case Study of Bangkok City[J]. *Green*, 2015, 5(1-6).
[7] Kamal Skeiker. Optimum dip angle and orientation for solar collectors in Syria[J]. *Energy Conversion and Management*, 2009, **50**(9).
[8] Li rongling, Cao jiazong. Optimization of installation Angle of solar planar receiver based on radiation prediction[J]. *Journal of Donghua University(Natural Science)*. 2009, 35(**06**): 721-725.
[9] zhu dandan, Yan da. Study on optimal dip angle of solar panel placement[J]. *Building Science*. 2012, 28(**S2**): 277-281.
[10] GB 50495-2009. *Technical Code for Solar Heating System*[S].
[11] China Meteorological Administration Meteorological Information Center, Tsinghua University. *Special Meteorological Data for Thermal Environment of Buildings in China*[M]. China Architecture & Building Press,

Acknowledgments

This project is funded by the National Key R&D Program of China (2016YFC0700400), and the Natio nal Natural Science Foundation of China (51478280 and 51778382).

Analysis on the influence of solar radiation on heating load of existing residential buildings in Western Sichuan

Yi Huang[1] **and Yong Liu**[2]
[1] Chongqing University, Chongqing, China
[2] Chongqing University, Chongqing, China
E-mail: hy571756417@sina.cn

Abstract. The calculation of the solar radiation heat is important to the accuracy of heating load prediction, especially in area with sufficient solar radiation, e.g. western Sichuan in China. However, only a few researches are focused on the specific effects of solar radiation heat in this area. Therefore, based on a practical residential building in Western Sichuan and the survey data, the heating load during the day and night in a heating season under different working conditions are compared in this paper. The influence of solar radiation on the heating load of residential buildings in Western Sichuan is analyzed, and the correctness of the simulation software is verified by experiments. The results show that the solar radiation will greatly increase room base temperature in the daytime. In a heating season， the heating load of the building increases by 26.12% without considering solar radiation heat. During the day (9:00 to 18:00) and night (18:00 to 9:00), the heating load increases by 35.69% and 19.80%, respectively. Besides, solar radiation mainly affects the south exterior wall in a north-south residential building. The heating load of south rooms is 27.52% less than north rooms, with 35.92% less during the day and only 22.08% less during the night.

Keywords: residential building, solar radiation heat, heating season, heating load

1. Introduction

The Western Sichuan Plateau is part of the southeastern margin of the Qinghai Tibet Plateau and the Hengduan Mountains, about 2500~4500 meters above sea level. It is located in cold climate areas. The climate characteristics can be summarized as: cold climate, low humidity, low pressure, high wind speed and strong solar radiation. [1] In the previous experiments, it was found that the room base temperature of some non-heating rooms could reach more than 15 ℃ in the daytime in winter. This demonstrates that the solar radiation has a great influence on the heating load in this area. So the effects of solar radiation on heat load have a high research value.

For a long time, many scholars have studied the influence of solar radiation and the application of solar heating system. A S El-Asfouri[2], R Levinson[3], G Oliveti[4], Chen B[5] analyzed the influence of solar radiation on thermal performance of building envelope. Govind NK[6], Xingxing Zhang[7], Qi Qi[8] made in-depth researches on different forms of solar heating system. In the study of solar radiation in high latitude and cold regions, there are many researches on solar energy applications. Hongmao Wang [9] and Yongtao Ma[10] studied the application of single solar system in this areas. Besides, Runliang Shi[11], Lei Wang[12], Zonghe Zheng[13] and Jianguo Cheng[14] studied the application of solar energy combined with other systems in high latitude and cold areas. And Wei Xiao[15], Xianhui Li et al[16], Hancheng Yu[17] studied the application of solar house in these regions. However, there are few studies on the specific impact of solar radiation on heating load in high latitude and cold region.

Numerous researchers have studied the solar radiation and its application of the buildings, however, few studies have been done for the influence on solar radiation of buildings, especially in high altitude and cold regions. Therefore, in this paper, DeST software is used to simulate the hourly heating load of solar radiation in heating season. And the effect of solar radiation on residential buildings in Western Sichuan is researched. It lays the foundation for the research on the application of the solar energy.

2. Theoretical calculation model

2.1. Room load calculation model

Room load includes sensible heat load and latent heat load. Because the room has no wet source and air conditioning fresh air, the heat load only consider sensible heat load in this paper. The calculation of sensible heat load is based on the theory of room air heat balance. According to Reference [18], the mathematical expression of the heat balance equation of room air is in Equation (1):

$$\sum_{k=1}^{N} f_k \hbar_{n,k}\left[t_{n,k}(n) - t_r(n)\right] + \left[q_1(n) - q_2(n)\right] + \frac{L_a(n)c_{p,a}\rho_a[t_a(n)-t_r(n)]}{3.6} + q_3(n) = Vc_{p,r}\rho_r\frac{t_r(n)-t_r(n-1)}{3.6\times\Delta\tau} \quad (1)$$

where N is the number of enclosure structures. f_k is the inside surface area of the enclosure structure k, m^2. $\hbar_{n,k}$ is surface convective heat transfer coefficient between inside surface k and indoor air, W/(m^2 • ℃). $t_{n,k}(n)$ is the inside surface temperature of the enclosure structure k at the moment n, ℃. $t_r(n)$ is the indoor air temperature at the moment n, ℃. $q_1(n)$ is the convection heat dissipation from lighting, sensible heat of the human body and explicit heat of equipment at the moment n, W. $q_2(n)$ is the room sensible heat consumed by water evaporation, W. $L_a(n)$ is the air infiltration at the moment n, m^3/h. $c_{p,a}$、$c_{p,r}$ are the indoor and outdoor air specific heat capacity, respectively, kJ/(kg • ℃). ρ_r、ρ_a are the indoor and outdoor air density, respectively, kg/m^3. $t_a(n)$ is the outdoor air temperature at the moment n, ℃. $q_3(n)$ is air conditioning system sensible heating energy, W. V is the room volume, m^3. $\Delta\tau$ is the time interval, s.

2.2. Solar radiation calculation model

According to Equation (1), $t_{n,k}(n)$ is the related item after considering the factors such as solar radiation. Moreover, the intensity of solar radiation in Western Sichuan is large, its influence on heating load in winter cannot be ignored. Therefore, the calculation of solar radiation heat is critical. There are many calculation models of solar radiation at present, and this paper uses the solar radiation calculation model in DeST to calculate.

For direct solar radiation, DeST use projection transformation method to calculate the shadow of each surface at first. And then the direct solar radiation of each surface is calculated by combining the direct radiation and the incidence angle at the present time.

For diffuse solar radiation, diffuse radiation from the sky, the ground and the surrounding building surface reflection radiation are considered in DeST. The total diffuse radiation is the sum of three radiations, and the calculation is in Equation (2).

$$I_{j,d} = I_{j,s} + I_{j,g} + I_{j,b} = \varphi_s I_{s,s} + \varphi_g \rho_g I_{sh} + \varphi_b \rho_b I_b \quad (2)$$

where $I_{j,d}$ is the total diffuse radiation, W/m^2; $I_{j,s}$ is diffuse radiation from the sky, W/m^2. $I_{j,g}$ is ground reflection radiation, W/m^2. $I_{j,b}$ is the surrounding building surface reflection radiation, W/m^2. φ_s、φ_g、φ_b are corresponding angular coefficients. $I_{s,s}$ is the diffuse radiation from the sky of unsheltered horizontal plane, W/m^2; $I_{s,s}$ is the total solar radiation of horizontal plane, W/m^2; I_b is the average total reflection radiation of surrounding buildings, W/m^2。

For solar transmission radiation, DeST uses a universal recursive algorithm [19] based on the energy balance equation to calculate the absorption rate (shown in Equation (3))of the radiation energy input from the outermost layer to the dielectric layer, so that the solar radiation absorbed by each layer can be calculated.

$$\xi(K, K-1) = \frac{Q_{K\to K+1}}{G_1} = \frac{Q_{K\to K+1}}{G_K} \times \frac{G_K}{G_1} = \xi_{K\to K+1} \times \prod_{i=1}^{K-1}\left(\alpha_i \tau_{a,i\to i+1}\right) \quad (3)$$

3. Modeling and analysis

3.1. Survey data in Western Sichuan

IOP Conf. Series: Materials Science and Engineering **556** (2019) 012046 doi:10.1088/1757-899X/556/1/012046

According to the actual situation, Kangding (low altitude) and Litang (high altitude) are selected as the research objects for field investigation. The survey results of the local enclosure structure are shown in Table 1.

Table 1. Survey data of residential building envelope in Western Sichuan.

Locat ion	Building Type	Building Construction	Exterior wall		Exterior Window Type	Floor Material	Insulati on measur es
			Building Materials	Thickness (mm)			
Kang ding	Old residential building	Brick concrete structure	Reinforced concrete (50%)/ brick wall (50%)	260-300	Double-layer plastic steel window (over 70%)/Single-layer plastic steel window l(under 30%)	Reinforce d concrete	None
	New residential building	Concrete shear wall	Reinforced concrete	260-400 (Around 300)	Double-layer plastic steel window	Reinforce d concrete	None
Litan g	Old residential building	Concrete frame wall (over 80%)/ Brick concrete structure(under 20%)	Reinforced concrete (over 80%)/ brick wall(under 20%)	280-320	plastic steel windows(over 70%)/ Aluminum alloy windows(under 30%)/ Singel-layer window(50%) /Double-layer window(50%)	Reinforce d concrete	None
	New residential building	Concrete frame wall	Reinforced concrete	200-380 (Around 280-300)	Double-layer plastic steel window	Reinforce d concrete	None

Note: the completion time of old residential building is 2000 ~2010, and the completion time of new residential building is 2010.

The survey results show that there is little regional difference in the enclosure structure of residential buildings in West Sichuan. The exterior wall materials of the new and old residential buildings are usually concrete blocks with a thickness of about 300mm and without heat preservation. The exterior windows of the building are usually double plastic steel windows and a few are aluminum alloy window frames. As the results of the survey show that the new style is similar to the old style residential building envelope, the new and old buildings in the West Sichuan area after 2000 are not analyzed in the latter simulation.

3.2. Model establishment

A residential building in Kangding is selected as the research object. The area of the building is about 181 m^2, about 3000m above sea level. There are 3 floors on the ground, 1 storey is restaurant, 2nd and 3rd floors are residential rooms. It has no basement, and the floor heights are 3m. Because this paper studies residential buildings, it only analyzes 2 and 3 storey. The standard floor plan is shown in Figure 1.

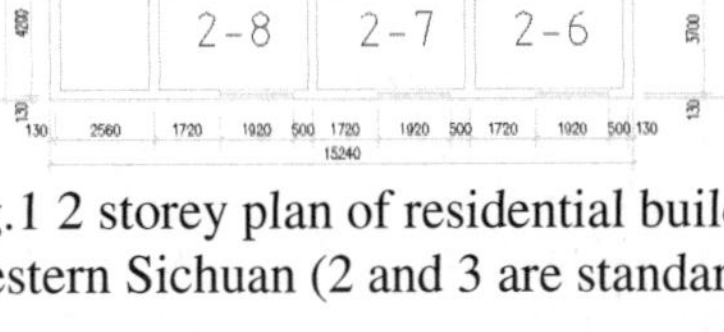

Fig.1 2 storey plan of residential building in Western Sichuan (2 and 3 are standard floor)

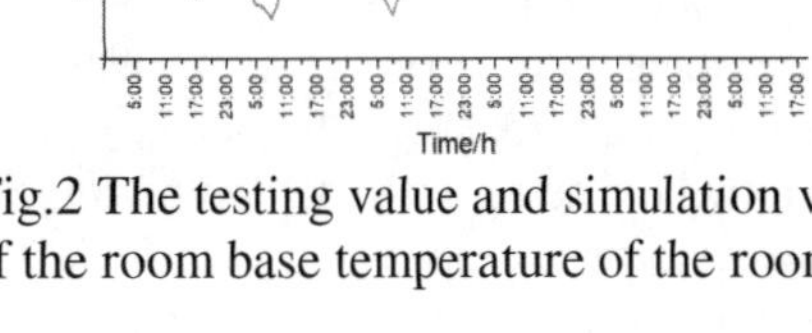

Fig.2 The testing value and simulation value of the room base temperature of the room 2-5

3.3. Boundary conditions of simulated building

The outdoor meteorological parameters are taken from the standard or DeST's databases, and the enclosure structure is selected according to the survey data in Table 1. The thermal parameters are shown in Table 2. In high altitude area, the physical parameters of the air change greatly due to the low atmospheric pressure. According to the heat transfer theory, the surface convection heat transfer coefficient of the enclosure structure changes greatly. Therefore, the surface heat transfer coefficient is determined according to the Equations (4) ~ (7) [15]. Other parameters are set according to the standards.

Table 2. Enclosed structure thermal parameters of simulation building.

Enclosure type	Material and thickness	Heat transfer coefficient /(W/(m² • K))
Exterior wall	Cement mortar 20mm, reinforced concrete 250mm, lime mortar 20mm	2.773
Interior wall	Cement mortar 20mm, porous concrete 200mm, cement mortar 20mm	1.308
Floor slab	Cement mortar 25mm, reinforced concrete 100mm, cement mortar 20mm	2.944
roof	Cement mortar 20mm, reinforced concrete 150mm, lime mortar 20mm	3.413
Outside window	Plastic steel window frames, double-layer, 3mm thick ordinary glass, air layer 9mm	2.449
Door	Monolayer solid wood door	2.4

$$\begin{cases} h_{in} = 0.282\left(\frac{\rho^2 \Delta t}{l}\right)^{\frac{1}{6}} + 1.131(\rho^2 \Delta t)^{\frac{1}{3}}, \ 10^9 < GrPr < 10^{12}, \ \text{Vertical wall} & (4) \\ h_{in} = 1.614(\rho^2 \Delta t)^{\frac{1}{3}}, \ 8 \times 10^6 < GrPr < 1.5 \times 10^9, \ \text{Horizontal wall with hot surface upward} & (5) \\ h_{in} = 0.643\left(\frac{\rho^2 \Delta t}{l}\right)^{\frac{1}{4}}, \ 10^5 < GrPr < 10^{10}, \ \text{Horizontal wall with cold surface upward} & (6) \end{cases}$$

$$\overline{h_{out}} = \frac{\int_0^L 0.023\lambda\left(\frac{V}{\nu}\right)^{0.891} l^{-0.109} dl}{L} = 0.0258\lambda\left(\frac{\overline{V}}{\nu}\right)^{0.891} L^{-0.109} \quad (7)$$

Note: the indoor air temperature for heating room is 18 ℃, and for non-heating room is 3 ℃. (very small items are neglected)

3.4. Correctness verification

3.4.1 Boundary conditions of Actual building

The outdoor meteorological parameters are input according to the measured values, and the residential plan is shown in Figure 1. The actual enclosure structure of the residence is shown in Table 3, with a building facing 25° west to south.

Table 3. Enclosed structure thermal parameters of actual building.

Enclosure type		Material and thickness	Heat transfer coefficient /(W/(m² • K))
Exterior wall	North wall	Heavy mortar clay 100mm, lime mortar 20mm	3.276
	The others	Heavy mortar clay 240mm, lime mortar 20mm	2.095
Interior wall		Cement mortar 20mm, heavy mortar clay 240mm, cement mortar 20mm	1.761
Floor slab		Cement mortar 25mm, heavy mortar clay 100mm, cement mortar 20mm	2.527
roof		Cement mortar 20mm, heavy mortar clay 120mm, lime mortar 20mm	2.871
Outside window		Plastic steel window frames, double-layer, 6mm thick ordinary glass	4.503
Door		Monolayer solid wood door	2.4

3.4.2 Experimental test

This experiment takes room 2-5 and 3-2 as the test object. During the test, the indoor disturbance of the test was accurately recorded. Indoor disturbance mainly includes three categories: personnel, lighting and equipment. The tested room temperature, adjacent room temperature, outdoor air temperature and humidity were measured by the self recording instrument of temperature and humidity, and the air velocity are measured by temperature and humidity self-recording instrument. The inside surface temperature of enclosure structure and infiltration air velocity are measured by the infrared thermometer and hot ball anemometer, respectively.

3.4.3 Comparison between simulation and testing values

Input the relevant information to DeST. The testing value and simulation value of the room base temperature of the room 2-5 are shown in Figure 2. The calculation value (according to Equation (1)) and simulation value of the heating load of the room 3-2 are shown in Figure 3.

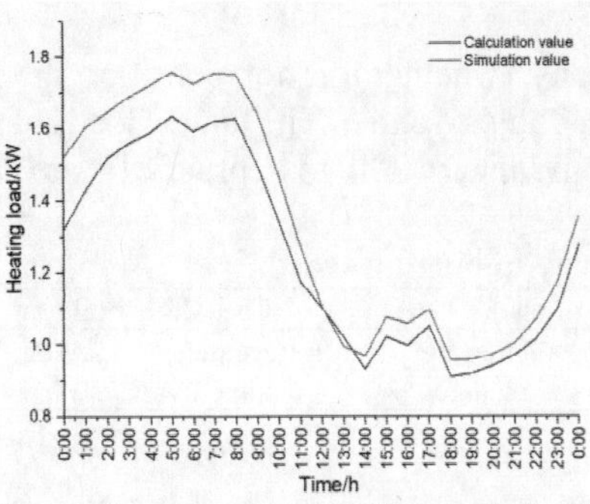

Fig.3 The calculation value and simulation value of the heating load of the room 3-2

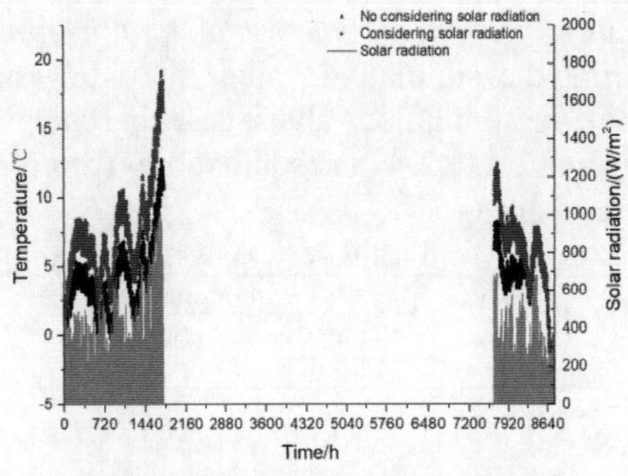

Fig.4 Room base temperature of the room 2-2 in heating season

Fig. 2 shows that the simulation temperature is gradually consistent with the actual natural temperature from the third day. because of the different wall heat storage before the experiment, the testing temperature is larger than the simulation temperature at the beginning. As the influence of the heat storage weakened with time, the gap gradually becomes smaller. According to the calculation of relative error, from 12 pm in the fourth day, the largest relative error appears at 19 pm in the fourth day (11.9%). And for most of the time, the relative error is less than 10%, which can meet the requirements of engineering application. It means that the parameter setting is correct. Fig.3 shows that the simulation value and calculation value are quite different at first because of the different wall heat storage, too. But the change law of the two is almost the same. And the gap gradually becomes smaller with time, the final relative error is about 6%. Therefore, the DeST software calculation model can be applied to the load calculation in Western Sichuan.

4. Analysis of simulation results

By changing the orientation of buildings, solar radiation and other parameters, the influence of solar radiation on existing residential buildings in West Sichuan is analyzed. According to the standard, combined with the actual situation, the heating season is unified from November 15th to March 15th for a total of 120 days.

4.1. Influence of solar radiation on room base temperature

Take the south room 2-2 as the research object. The room base temperature during heating season in two working conditions (whether or not consider the solar radiation) is shown in Figure 4.

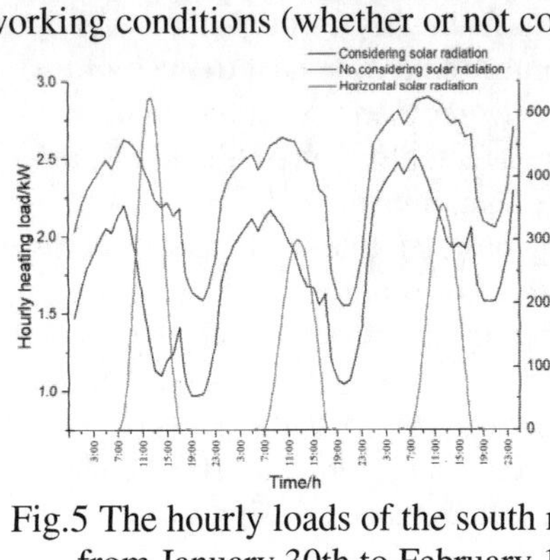

Fig.5 The hourly loads of the south rooms from January 30th to February 1st

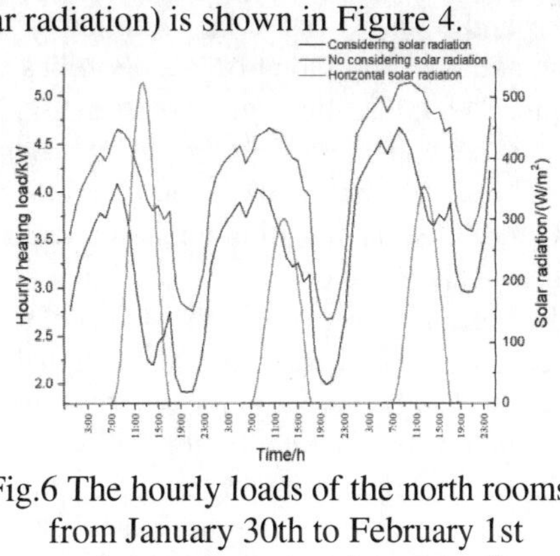

Fig.6 The hourly loads of the north rooms from January 30th to February 1st

Figure 4 indicates that considering the solar radiation, the room base temperature is greater than that of the other condition. Moreover, the solar radiation is positively correlated with the difference of room base temperature between the two working conditions. Besides, when the solar radiation is greater than 600W/m^2, the room base temperature can be kept above 8 ℃. And when the solar radiation is greater than 800W/m^2, the room base temperature can be kept above 10℃ .

4.2. Influence of solar radiation on building heating load

IOP Conf. Series: Materials Science and Engineering **556** (2019) 012046 doi:10.1088/1757-899X/556/1/012046

4.2.1 Influence of solar radiation on total heating load of building

Comparing the simulation values in two working conditions (whether or not consider the solar radiation), the simulation values of heating energy are shown in Tables 4 and 5. The hourly loads of north rooms (room 2-1 to 2-4) and south rooms (room 2-6 to 2-8) from January 30th to February 1st are shown in Figures 5 and 6.

Table 4. Rooms and building heating energy in heating season.

	North rooms		South rooms		The whole building	
	Before the change	After the change	Before the change	After the change	Before the change	After the change
Heating energy / (kW • h)	8391.81	10856.97	4431.03	6202.44	40097.98	50577.16
More than before the change /%		29.38		39.98		26.12

Table 5. Building heating energy in the daytime and nighttime in heating season.

	Daytime heating energy		Nighttime heating energy	
	Before the change	After the change	Before the change	After the change
Heating energy / (kW • h)	15992.25	21699.14	24105.73	28878.02
More than before the change /%		35.69		19.80

Note: Before the change means considering the solar radiation and after the change means no considering the solar radiation.

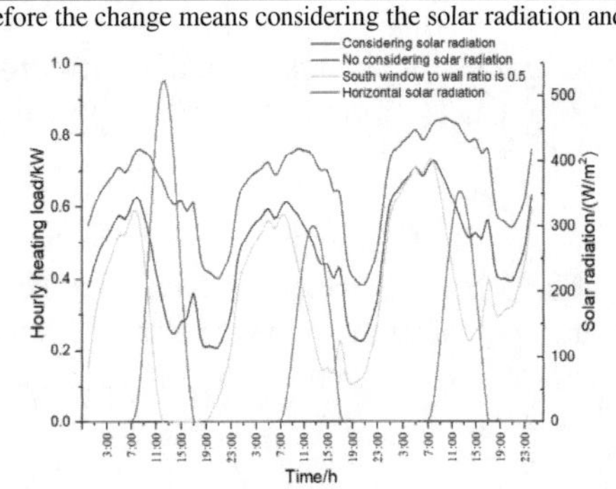

Fig.7 The hourly loads of room 2-7 under 3 conditions from January 30th to February 1st

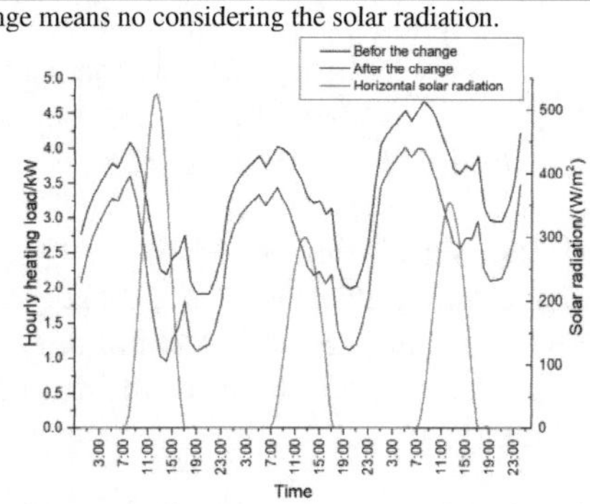

Fig.8 The hourly loads from January 30th to February 1st

Tables 4 and 5 show that solar radiation is converted into heat directly through the window in the daytime, and the solar radiation mainly relies on wall heat storage to provide heat to the building in the nighttime. So the heating load increase ratio in the daytime is greater than that in the nighttime if solar radiation is not considered. Figures 5 and 6 demonstrate that the greater the solar radiation is, the greater the increase in daytime hourly heating load is when the solar radiation is not considered. Among them, the influence of solar radiation on the south rooms is more obvious than that on the north rooms.

To further analyze the effect of solar radiation on the heating load of south rooms, take room 2-7 as an example. Figure 7 shows the hourly loads of room 2-7 under 3 conditions from January 30th to February 1st. According to the calculation, the original room window to wall ratio is about 0.25. And according to the standard, take the Kangding energy-saving building south window to wall ratio as 0.5. The figure indicates that when the window to wall ratio as is 0.5, the hourly heating load during the day is much lower. Therefore, increasing the window to wall ratio within a certain range can significantly reduce the heating load of residential buildings in Western Sichuan.

4.2.2 Influence of solar radiation on heating load of different number of external walls

Room 2-2 (a room with one exterior wall) and room 2-1 (a room with two exterior walls) are compared in this paper. The area of the two room is the same by adjusting the parameters, and the area of south exterior wall of the room 2-1 is the same as that of the room 2-2. The simulation values of heating energy per unit exterior wall of the two rooms in the heating season are shown in Tables 6 and 7.

Table 6. Heating energy per unit exterior wall of the two rooms

	Room 2-1	Room 2-2
Heating energy per unit exterior wall /(kW • h/ m)	361.40	382.89
Larger than room 2-1/%		5.95

Table 7. Heating energy per unit exterior wall of the two rooms in the daytime and nighttime in heating season

	Daytime heating energy		Nighttime heating energy	
	Room 2-1	Room 2-2	Room 2-1	Room 2-2
Heating energy per unit exterior wall /(kW • h/ m)	146.92	150.87	214.48	230.02
Larger than room 2-1/%		2.69		7.25

Note:: heating energy per unit exterior wall = heating energy /length of exterior wall

Tables 6 and 7 show that the heating energy per unit exterior wall of room 2-2 is 5.95% larger than that of room 2-1, with 2.69% larger in the daytime and 7.25% larger in the nighttime. This means the non-south exterior wall is affected little by the solar radiation. Thus, the solar radiation has no direct influence on the heating load of different number of exterior walls, and the outdoor air temperature is also one of the main factors.

4.3. Influence of solar radiation on heating load of different orientation rooms

Change the orientation of the building, the room facing the north (room 2-1 to 2-4) is turned south. The simulation values of the rooms in the heating season are shown in Table 8. The hourly loads under two conditions from January 30th to February 1st is shown in Figure 8.

Table 8. Room heating energy in heating season.

	Daytime heating energy		Nighttime heating energy		Total heating energy	
	Before the change	After the change	Before the change	After the change	Before the change	After the change
Heating energy / (kW • h)	3301.86	2115.92	3966.21	3966.21	8391.81	6082.13
Less than before the change /%		35.92		22.08		27.52

Note: Before the change the room is toward the north and after the change means the room is toward the south.

Table 8 shows that the heating load in the heating season decreased by 27.52% after changing the room orientation, and the heating load in the heating season decreased by 35.92% in the daytime, which's rate is more than that in the nighttime (22.08%). Figure 8 indicates that from 8:00 am to14:00 pm, the reduction ratio of the heating load gradually increased after the change. From 14:00 pm to 8:00 am the next day, the reduction ratio of the heating load gradually decreased after the change. Besides, the change reduction ratio of the heating load in the nighttime is more slow than that in the day after changing the orientation.

5. Conclusion

The heating load during the day and night in a heating season under different working conditions are simulated with DeST in this paper, and the simulation values are verified by experiments. The impact of solar radiation on heating load in different conditions is analyzed in this paper, and the conclusions arising from this study are summarized as follows:

(1) Solar radiation will greatly increase the room base temperature. When the solar radiation is greater than 800W/m^2, the room base temperature can even be kept above 10 ℃. Therefore, for some areas with high solar radiation in Western Sichuan, residential buildings may even have no heating load during the day.

(2) In Western Sichuan, solar radiation is an important parameter affecting the thermal load of the enclosure structure. The heating load of the building increases by 26.12% without considering solar radiation. In the daytime and nighttime, the heating load increases by 35.69% and 19.80%, respectively.

Therefore, residential buildings in Western Sichuan should consider increasing window to wall ratio in a certain range to increase solar radiation heat and reduce heating load.
(3) The influence of solar radiation on the heating load of is mainly related to the area of the south exterior wall, but has little to do with the number of external walls.
(4) For typical residential buildings in Western Sichuan, the heating load of the south rooms is 27.52% less than that of the north rooms. Among them, the heating load of south rooms is 35.92% less in the daytime and only 22.08% less in the nighttime than that of north rooms. Therefore, the heating rooms of residential buildings in West Sichuan should be set to the south if possible.

References

[1] China Architecture southwest Design Research Institute Co., Ltd. *Research on building energy planning and heating technology program in Western Sichuan Plateau* [R]. Sichuan:.2014 of China Construction southwest design and Research Institute Ltd.
[2] A S El-Asfouri, M F El-Refaie, M M Karawya. *Effect of variousfactors on the shading coefficient of different types of glazing* [J]. Building and Environment, 1988, 23 (1): 45-55.
[3] R Levinson, H Akbari, P Berdahl. *Measuring solar reflectance-Part I: Defining a metric that accurately predicts solar heat gain* [J]. Solar Energy, 2010 (84): 1717-1744.
[4] G Oliveti, N Arcuri, R Bruno, M D Simone. *An accurate calculation model of solar heat gain through glazed surfaces* [J]. Energy & Buildings, 2011 (43): 269-274.
[5] B Chen, H J Chen, S R Meng, X Chen, P Sun, Y H Ding. *The effect of Trombe wall on indoor humid climate in Dalian, China* [J]. Renewable Energy, 2006 (31): 333-343.
[6] G N Kulkarni, SB Kedare, S Bandyopadhyay. *Determination of design space and optimization of solar water heating systems* [J]. Solar Energy, 2007 (81): 958-968.
[7] Xingxing Zhang, Xudong Zhao, Jihuan Xu, Xiaotong Yu. *Characterization of a solar photovoltaic/loop-heat-pipe heat pump water heating system* [J]. Applied Energy, 2013 (102):1229-1245.
[8] Qi Qi, Shiming Deng, Yiqiang Jiang. *A simulation study on a solar heat pump heating system with seasonal latent heat storage* [J]. Solar Energy, 2008 (82): 669-675.
[9] Hongmao Wang, Zonghe Zheng, Jinshui Gao, *Yongtao Ma. Design of direct heating system with solar energy in Tibet region* [J]. Gas and heat, 2007, 27 (1): 67-68.
[10] Yongtao Ma, Zonghe Zheng, Xi Ge, Hongmao Wang. *Performance analysis of heating system with solar energy heat pump in Tibet region* [J]. Gas and heat, 2007, 27 (1) :69-70.
[11] Runliang Shi. *Study on the Application of Pluripotent Complementary Heating Supply System in Shigatse* [D]. Chongqing: Chongqing University, 2016
[12] Lei Wang. *Optimization of solar heating system combined with water source heat pumps in Tibet* [J]. Journal of HV&AC, 2007, 37 (11): 90-94.
[13] Zonghe Zhong, Jinshui Gao, Xi Ge. *Application of heat-supply system with solar energy in Tibet region* [J]. Gas and heat, 2006 , 26 (12) :52-54.
[14] Jianguo Cheng. *Application of solar energy and water source heat pump combined heating in Tibet* [D]. Sichuan: Southwest Jiao Tong University, 2005.
[15] Wei Xiao. *Study of the Direct-gain Solar Heating in Remote Southwest Tibet* [D]. Beijing: Tsinghua University, 2010.
[16] Xianhui Li, Bing Song, Haisong Xu. *Simulation analysis and evaluation of passive solar house of peasants and herders in Qinghai Tibetan area* [J]. Journal of HV&AC, 2011, 41 (8) :74-76.
[17] Hancheng Yu, Wensheng Ma, Zhijian Liu, Guangya Jin. *Field Measurement of the Indoor Thermal Environment of Passive Solar House in Alpine Pastoral Areas in Qinghai* [J]. Architectural Science, 2017, 33 (4): 39-42.
[18] QiSen Yan, Qingzhu Zhao. *Building thermal process* [M]. Beijing: China Construction Industry Press, 1996.
[19] Yi Jiang, Yuanzhe Li, Hongfa Di. *A research of the method for calculation of the transmittance of partially transparent mediums* [J]. Acta Energiae Solaris Sinica, 1980 (2): 54-63.

Experimental and theoretical research on bending behaviour of photovoltaic panels with a special boundary condition

T Y Zhang[1], L Z Xie[1,2], Y X Li[2], T K Mallick[3], Q Z Wei[4], X Hao[5], B He[1]
[1]Institute of New Energy and Low-Carbon Technology, Sichuan University, Chengdu, 610207, P. R. China
[2]College of Architecture and Environment, Sichuan University, Chengdu, 610065, P. R. China
[3]Environment and Sustainability Institute, University of Exeter, Penryn Campus, Penryn, TR10 9FE, UK
[4]Suzhou Talesun Solar Technologies Co., Ltd., Chang Shu, 215542, P. R. China
[5]College of Materials Science and Engineering, Sichuan University, Chengdu, 610065, P. R. China

E-mail: yongxuel@scu.edu.cn

Abstract. Currently, the photovoltaic panels widely manufactured on market are composed of stiff front and back layers and the solar cells embedded in a soft polymeric interlayer. The wind and snow pressure are the usual loads to which working photovoltaic panels need to face, and it needs the panels keep undamaged under those pressure when they generate electricity. Therefore, an accurate and systematic research on bending behaviour of photovoltaic panels is important and necessary. In this paper classical lamination theory (CLT) considering soft interlayer is applied to build governing equations of the panel. A Rayleigh-Rita method is modified to solve the governing equations and calculate the static deformation and inter force of the photovoltaic panel. Different from many previous researches only analysing simply supported boundary condition for four edges, a special boundary condition which consists of two opposite edges simply supported and the others two free is studied in this paper. A closed form solution is derived out and used to do the numerical calculation. The corresponding bending experiments of photovoltaic panels are completed. Comparing the numerical results with experiment results, the accuracy of the analytical solutions are verified. The influence of boundary condition is shown by comparing the results with previous researches, and a guide for the photovoltaic panel installation is finally proposed based on those conclusions.

Keywords: photovoltaic panel; bending behaviour; classical lamination theory; Rayleigh-Rita method; special boundary condition; bending experiment

1. Introduction

According to the report from Ministry of Housing and Urban-Rural Department of P.R. China, building consumes 40% to 50% of the total energy in P.R. China each year, which is one of the main reasons for the aggravation of energy consumption. In order to solve it, building integrated photovoltaic (BIPV) has gained a great of attention and great advances in recent years. In BIPV, the photovoltaic (PV) module is designed and constructed with buildings at the same time, which makes it as one part of the building structure. Due to the requirements of lighting inside the building, the double glass PV module with better photopermeability are widely applied in the real structures. The photovoltaic panels studied in the present paper are just double glass PV module which consists of two glasses and an interlayer in where the cells are sealed by ethylene vinyl acetate (EVA) or polyvinyl butyral (PVB). Since the bending deformation under distributed force are both important to structural components and PV module [1], the bending behaviour of PV panels is studied carefully in the present paper to make sure that it has qualified stiffness and can work safety in BIPV. From 1990s, PV module designed to be integrated into building became commercially available and it promoted some relative research works [2]. One decade later, U.S. National Renewable Energy Laboratory (NREL) promulgated a brief overview, in which the history of BIPV was summarized and some challenges about the BIPV future were stated [3]. Since it must satisfy the requirements of building component when the PV panel is installed as roof, wall or window in the BIPV building, the mechanical behaviour of it should be studied systematically. Naumenko and Eremeyev [4] believed that PV panel is the layered composite with relatively stiff skin layer and relatively soft core, and the layer-wise theory is suitable for it. They derived the differential equations based on layer-wise theory, but the equations are only solved for plate strip with simply supported boundary condition. Eisentrager etc. [5] adopted first order shear deformation theories (FSDT) to study the PV panel and laminated glass with weak shear stiffness. As to the theories and mechanic models for laminate composite, Vedrtnam and Pawar [6] made a review work on them and their introductions are mainly about the laminate glass plate which is extremely similar to PV panel. However, in many of works mentioned in that paper, the boundary conditions of the plates are all simply-supported for four edges.

The bending behaviour of PV panel is studied carefully in the present paper by both experimental and theoretical work. Different from many previous researches, a special boundary condition which is two opposite edges free and the other two edges simply-supported is considered. Hoff model which is one of the classical lamination theory (CLT) is adopted in this research. By the modified Rayleigh-Rita method, a closed form solution is derived out. The experiment is designed using water pressure to produce uniformly distributed force, so it is much more accurate than other research works using sand or brick. A frame is manufactured to simulate the special boundary conditions. Comparing the theoretical results with experimental results, the accuracy of the analytical solutions are verified. The theoretical model and solutions can be the fundamental works for the optimal design work in future.

2. Structural analysis

2.1. Mechanical model and basic hypothesis

The basic components of PV panel are shown in figure 1, including cover glass, ethylene-vinylacetate (EVA), silicon solar cells and back glass. To simplify the problem and emphasize the main characters of the PV panel, a laminate plate model is applied and several hypothesises are made as follows。

(1) The cover and backboard glasses are treated as top and bottom surfaces of the laminate plate, respectively. And both of them are simulated as isotropic plates with constant flexural rigidity.
(2) The silicon solar cells layer is ignored and two EVA layers are merged as one layer which is defined as the interlayer of EVA.
(3) The PV module is a soft core laminate plate and the stress of interlayer in x-y plan is ignored.
(4) Only the anti-symmetrical deformation is studied in present paper, so $\sigma_z=0$, $\varepsilon_z=0$.

Based on those hypothesises, the mechanical model of PV panel under uniformly distributed force and the corresponding coordinate system are shown in figure 2.

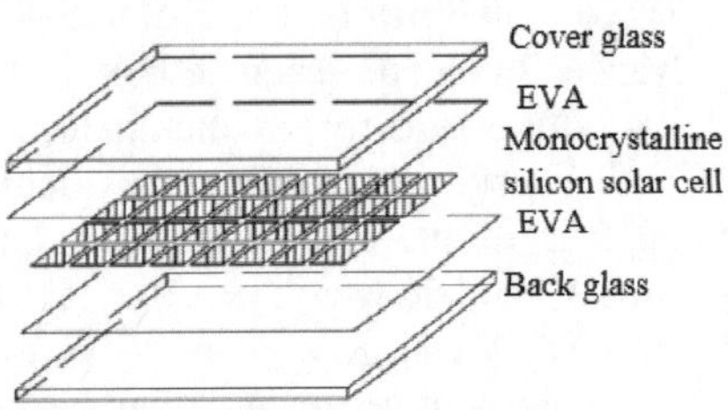

Figure 1. Structural diagram ofmonocrystalline silicon double glass photovoltaic panel.

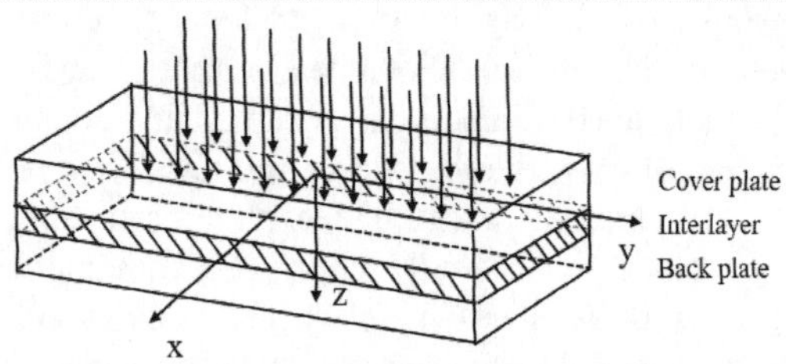

Figure 2. Mechanical model of PV panel and corresponding coordinate system.

2.2. Hoff model and governing equations

Hoff [7] modified Reissner theory and developed a Hoff model for laminated plate. In Hoff model, the flexural rigidities of surface plates are calculated but the interlayer is a relative soft layer. According to Hoff model, we could obtain the governing equations of PV panel under uniformly distributed force as

$$w=\omega-\frac{D}{C}\nabla^2\omega \tag{1}$$

$$\left(D+2D_f\right)\nabla^2\nabla^2\omega-\frac{2DD_f}{C}\nabla^2\nabla^2\nabla^2\omega=q \tag{2}$$

$$\frac{1}{2}D\left(1-\nu_f\right)\nabla^2 f-Cf=0 \tag{3}$$

with $D=\dfrac{E_f\left(h+t\right)^2 t}{2\left(1-\nu_f{}^2\right)}$, $D_f=\dfrac{E_f t^3}{12\left(1-\nu_f{}^2\right)}$, $C=G_C\dfrac{\left(h+t\right)^2}{h}$

where E_f is the elastic modulus of the cover and the back glass plate, ν_f is the Poisson's ratio of the cover and the back glass plate, G_C is the shear modulus of EVA, t and h are the thickness of surface plate and EVA interlayer, respectively.

2.3. Special boundary condition

In previous reseachs, in order to simplify the problem, all four edges simply supported boudnary condition is mostly studied. In present paper, a special boudnary condition which is two opposite edges simply supported and the other two free is studied. The PV panel is simply supported at the edges x=0 and a, and free-free at the edges $\mathrm{y}=\pm 0.5b$.

The boundary condition studied in present paper should satisfy the formulas as follows.

$$\left(M_{\mathrm{x}}^{'}\right)_{x=0,a}=0,\left(w\right)_{x=0,a}=0,\left(\varphi_{\mathrm{y}}\right)_{x=0,a}=0,\left(M_{\mathrm{x}}^{''}\right)_{x=0,a}=0 \tag{4}$$

$$\left(M_{\mathrm{y}}^{'}\right)_{\mathrm{y}=\pm\frac{b}{2}}=0,\left(M_{\mathrm{xy}}^{'}\right)_{\mathrm{y}=\pm\frac{b}{2}}=0\left(M_{\mathrm{xy}}^{''}\right)_{\mathrm{y}=\pm\frac{b}{2}}=0,\left(Q_{\mathrm{y}}\right)_{\mathrm{y}=\pm\frac{b}{2}}=0,\left(M_{\mathrm{y}}^{''}\right)_{\mathrm{y}=\pm\frac{b}{2}}=0 \tag{5}$$

2.4. Modified Rayleigh-Rita method and closed-form solutions

A modified Rayleigh-Rita method is applied in present paper due to the special boundary condition. The deflection of PV panel under uniformly distributed force could be derived out as follows.

$$\begin{aligned} w&=\omega-\frac{D}{C}\nabla^2\omega \\ &=\sum_{n=1}^{\infty}\left\{\sum_{r=1}^{6}e^{\lambda_{nr}y}\omega_{nr}+\frac{2q}{n\pi}\left[1-\cos\left(n\pi\right)\right]\frac{1}{\mathrm{A}_1k_n^4+\mathrm{A}_2k_n^6}\right\}\sin\left(\mathrm{k}_n x\right) \\ &-\frac{D}{C}\sum_{n=1}^{\infty}\left\{\sum_{r=1}^{6}\left[\left(-k_n^2\right)e^{\lambda_{nr}y}\omega_{nr}+\lambda_{nr}^2 e^{\lambda_{nr}y}\omega_{nr}\right]+\left(-k_n^2\right)\frac{2q}{n\pi}\left[1-\cos\left(n\pi\right)\right]\frac{1}{\mathrm{A}_1k_n^4+\mathrm{A}_2k_n^6}\right\}\sin\left(\mathrm{k}_n x\right) \end{aligned} \tag{6}$$

2.5. Finite element analysis

To verify the results of Hoff model, a finite element analysis is performed by the use of ANSYS code. SHELL181 composite shell element is used for modeling. The rectangular shell structure is divided into five layers and they are cover glass, EVA, silicon battery sheet, EVA and back glass and the material of each layer is simulated as isotropic material. The mechanical properties of the model materials are shown in table 1.

IOP Conf. Series: Materials Science and Engineering **556** (2019) 012047 doi:10.1088/1757-899X/556/1/012047

Table 1. Material parameter values.

Material	Parameter values		
	Modulus of elasticity /MPa	Poisson ratio	Thickness /mm
Reinforced glass	7.2E+4	0.2	2
Crystalline silicon battery	1.44E+5	0.28	0.2
EVA	3.5E+1	0.3	0.8

3. Experimental analysis

Bending testing is performed for 8 specimens at room temperature and all of them are double glass photovoltaic modules (as shown in figure 3) which are provided by Suzhou Tenghui Photovoltaic Technology Co., Ltd. Five specimens are with size 1658×995×5 (unit: mm) and three specimens are with size 1658×995×7.4 (unit: mm). The cover and back glasses are 2mm for first group and 3.2mm for the second group, but the thickness of interlayer is same as 1mm.

Figure 3. Monocrystalline silicon double glass photovoltaic module.

Figure 4. The test frame for installing photovoltaic module.

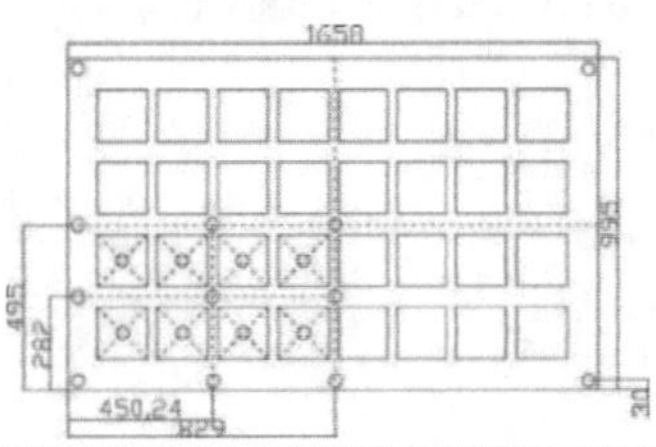

Figure 5. Arrangement diagram of strain measurement point (unit: mm).

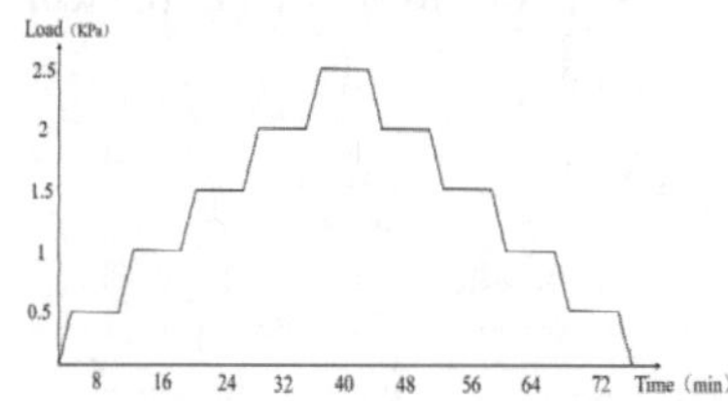

Figure 6. Test loading scheme for PV panels with 2mm glass.

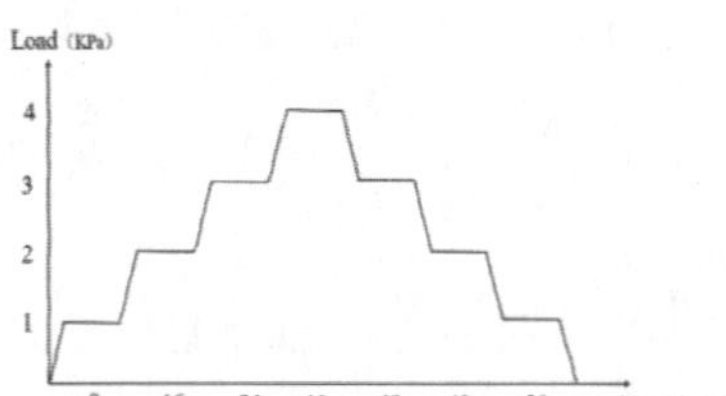

Figure 7. Test loading scheme for PV panels with 3.2mm glass.

Figure 8. Test site.

The bending test was completed in National Photovoltaic Product Quality Supervision and Inspection Center at Chengdu, referring to the current quality inspection certification, IEC 61215[1]. The test frame (as shown in figure 4) can simulate the discussed special boundary condition. The strain measurement points are only set on a quarter part which is shown as figure 5. The PV panel strains are collected by DH3816 static strain gauge, and the deflection at panel center is measured by a laser displacement meter installed under the panel. Different from previous experiments, water pressure is used to simulate the uniformly distributed force. The loading plan is different for two groups of PV panels since the bending behavior of them is different (as shown in figure 6 and figure 7).

4. Verifications and discussions

4.1. PV panel deflection

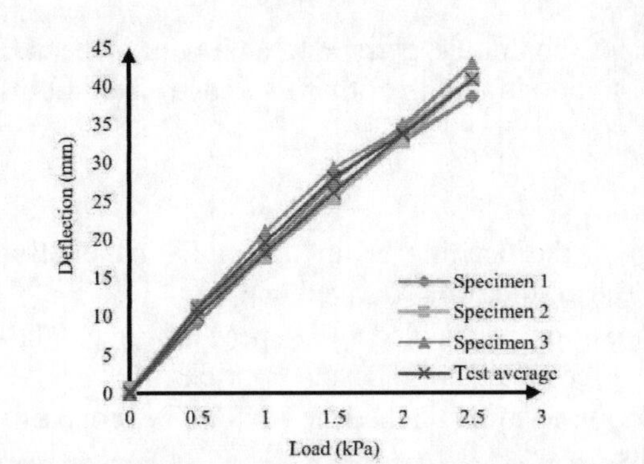

Figure 9. Central deflection of PV panels with 2mm glass.

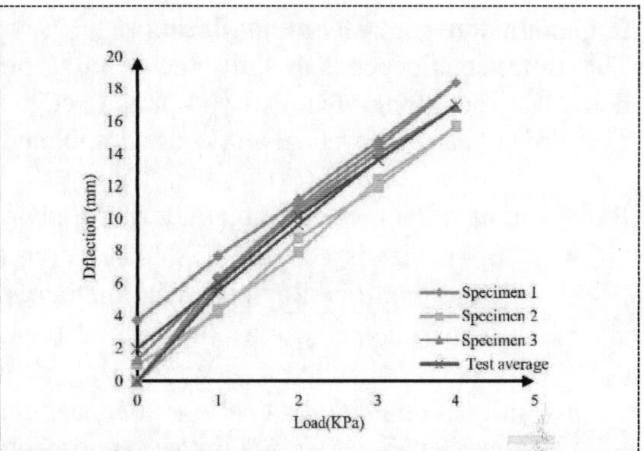

Figure 10. Central deflection of PV panels with 3.2mm glass.

From figure 9, the deflections calculated by both ANSYS and equation (6) are very close to the data measured in the test, so the accuracy of the proposed equations is proved. The deflections from ANSYS and equation (6) behave a linear elastic relationship, which is same as the test data. But equation (6) is better than ANSYS to predict the maximum deflection of PV panel. The figure 10 states clearly the linear elastic deformation of PV panel with 3.2mm glass too. And the proposed equations are more accurate. After the test, all the specimens were checked carefully and there was not nay cracks or breakages on the surface glass. It proves the safety of the PV panels, which just satisfy the requirements from the PV module certification, IEC 61215[1].

4.2. PV panel stress

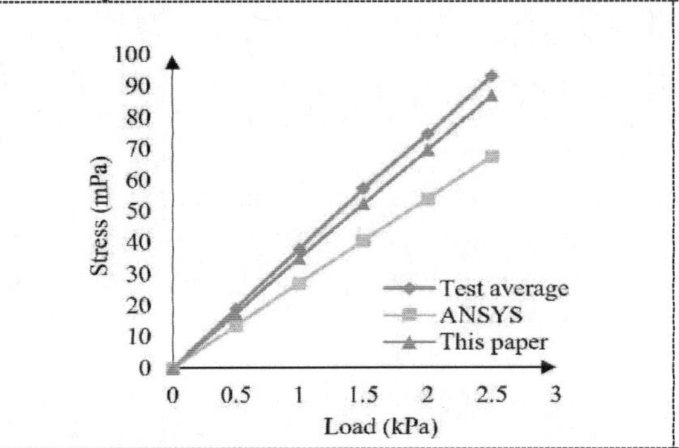

Figure 11. Maximum principal stress of PV panels with 2mm glass.

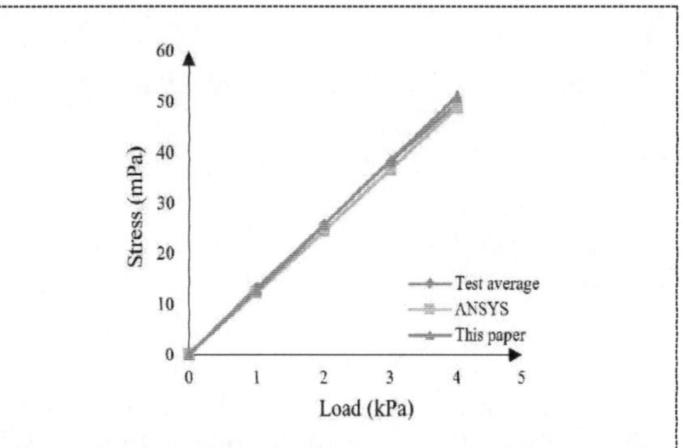

Figure 12. Maximum principal stress of PV panels with 3.2mm glass.

In figure 11, the central stress calculated by proposed equations is much better than the ones from ANSYS. It is the same in figure 12 to the PV panels with 3.2mm glass. The accuracy of the proposed equations on the stress calculation is verified by those comparisons. From both figure 11 and figure 12, the central stresses calculated by equation (6) or ANSYS have a linear relationship to the water pressure, which is same as the deflection behaviour concluded in section 4.1. Comparing figure 11 with figure 12, with the increase of surface glass thickness, the central stress decreases clearly. The maximum stresses on the surface glass of PV panels are smaller than the limit stress of reinforced glass, so it is safe when they are used under those loads.

5. Conclusions and recommendations

The aim of this paper is to study the bending behaviour of the double glass PV panel with a special boundary condition, which is two opposite edge simply supported and the other two edges free. Both experimental and theoretical works are completed.

Based on the results we may conclude as follows:

- Hoff model is adopted in this research to describe the bending behaviour of PV panel. By using a modified Rayleigh-Rita method, a closed form solution is derived out.
- In experimental works, the special boundary condition is realized by a specific frame. The water is applied to provide uniformly distributed force instead of the sand or bricks.
- In both deflection and stress analysis, the calculation results obtained by ANSYS or proposed equation are very close to the experimental results.
- The deflection and stress calculated by proposed equation are more accurate than the ones simulated by ANSYS, so it can be used in the optimal design work of BIPV component.
- When the load is in a range of 2.4Kpa which is required by current certification of PV module, the PV panels behave a linear elastic deformation and there is not any damage.
- Both maximum deflection and maximum stress are located at the middle of PV panel, and it should be chosen as the key position and key point in future design work.
- With the increase of surface glass, the stress of PV panel decreases clearly. It is safer to choose thicker glass in the design work but the economy will be a problem.

Acknowledgments

The authors are grateful for the financial support from the National Key Research and Development Program of China: Newton Fund - China-UK Research and Innovations Bridges (No. 2016YFE0124500).

References

[1] IEC 61215-2005, *Crystalline Silicon Terrestrial Photovoltaic (PV) Modules*[S]. International Electrotechnical Commission, 2005

[2] Patrina Eiffert, *An Economic Assessment of Building Integrated Photovoltaics*, Oxford Brookes School of Architecture, 1998

[3] A. Kylili, and A. Paris, Investigation of building integrated photovoltaics potential in achieving the zero energy building target. *Indoor and Built Environment*, Retrieved October 30, 2014

[4] Konstantin Naumenko and Victor A. Eremeyev, A layer-wise theory for laminated glass and photovoltaic panels, *Composite Structures*, 112(1), 2014, pp 283-291

[5] J. Eisentrager, K. Naumenko, H. Altenbach and H. Koppe, Application of the first-order shear deformation theory to the analysis of laminated glasses and photovoltaic panels, *International Journal of Mechanical Sciences*, 96-97, 2015, pp 163-171

[6] Ajitanshu Vedrtnam and S. J. Pawar, Laminated plate theories and fracture of laminated glass plate-A review, *Engineering Fracture Mechanics*, 186(1), 2017, pp 316-330

[7] N. J. Hoff, Bending and buckling of rectangular sandwich plates, *NACA*, TN2225, 1950

Mechanical analysis of photovoltaic panels with various boundary condition

Y X Li[1], L Z Xie[1, 2], T Y Zhang[2], Y P Wu[3], Y Y Sun[3], Z C Ni[4], J Q Zhang[5], B He[2]
[1]College of Architecture and Environment, Sichuan University, Chengdu, 610065, P.R. China
[2]Institute of New Energy and Low-Carbon Technology, Sichuan University, Chengdu, 610207, P.R. China
[3]Department of Architecture and Built Environment, Faculty of Engineering, University of Nottingham, University Park, Nottingham NG7 2RD, UK
[4]Suzhou Talesun Solar Technologies Co., Ltd., Chang Shu, 215542, P. R. China
[5]College of Materials Science and Engineering, Sichuan University, Chengdu, 610065, P.R. China

E-mail: xielingzhi@scu.edu.cn

Abstract. The photovoltaic (PV) panels currently existed on market are a kind of laminated plate structure, which is composed of two stiff glass skins and a soft interlayer. Some of those panels are installed on the buildings and integrated as the components of the structures, such as wall and roof. In different locations, the installations of the panels are different and the boundary conditions are not simply supported any more. The generating electricity requirement of the panels needs the mechanic behaviour of them is stable with any boundary condition. In this paper the bending behaviour of PV panels with various boundary conditions is analysed and the influence of boundary condition is studied carefully. The Kirchhoff theory which is one of the classical lamination theory (CLT) is adopted to build governing equations of photovoltaic panels under static force. A Rayleigh-Rita method is modified to solve the governing equations and calculate the static deformation and stress. Different boundary condition requires different assumptions of the deflection function and a modified general function is developed to solve that problem. A theoretical solution is derived out and used to do the numerical calculation. A bending experiment of PV panel with two opposite edges simply supported and the other two free is used to verify the correctness and accuracy of the proposed solution. Finally, the influence of different boundary condition is stated by comparing the numerical results and some guides for the PV panel installation are proposed.

Keywords: photovoltaic panel; various boundary condition; classical lamination theory; Rayleigh-Rita method; modified deflection function; bending experiment

1. Introduction

U.S. Department of Energy stated that building consumes more than 40% of the electricity produced in U.S. every year. To reduce building energy consumption, building integrated photovoltaic (BIPV) has attracted much attention. It requires that the photovoltaic (PV) component must generate electricity for the building, and bear external loads to keep the safety of the building at the same time. Currently, only in the standards of PV module, such as IEC 61215 (2005) [1], there are several codes about its mechanical properties and the corresponding test methods. It is lack of specific codes to the mechanical characters of PV modules applied in BIPV. In present paper, it focuses on the bending behaviour of the PV panels under wind load or snow load. In BIPV, the double glass PV module with better photopermeability are widely applied. Therefore, the PV panels studied in here are double glass PV module which consists of two glasses and an interlayer. In buildings, different installation ways mean different boundary conditions for the PV module, and the mechanical behaviours of them are different. It is necessary to study the influences of different boundary conditions in here. From 1990s,

PV module started to be integrated into building, and some relative research works were promoted [2]. Naumenko and Eremeyev [3] used the layer-wise theory to analyze PV panel and they treated the PV panel as a layered composite with relatively stiff skin layer and relatively soft core. Vedrtnam and Pawar [4] made a review work on laminate composite, and laminate glass plate which is very like PV panel is mainly introduced. First order shear deformation theory (FSDT), Zig-zag theory and layer-wise theory (LWT) were applied in many previous research works. However, in many of those works, the boundary conditions of the plates are only four edges simply-supported. And there are very few works discussing the influence of different boundary conditions.

In present paper, the mechanical properties of double glass PV panels with two different boundary conditions are analysed by both experimental and theoretical research. A classical lamination theory, Hoff model, is applied to build the constitutive equations of whole panel. The specific boundary equations are given based on two boundary conditions. The Rayleigh-Riata method is modified to derive the closed form solution. By using water pressure, the bending experiment of PV panel was completed. And changing the test frame, two boundary conditions were realized in the experiments. Comparing the theoretical results with experimental results, the accuracy of the analytical solutions are verified. The influences of boundary condition are also concluded. The theoretical model and solutions obtained in this paper can be the foundations for the optimal design work in future.

2. Theoretical analysis

2.1. Mechanical model and basic hypothesis

In figure 1, it shows the basic components of PV panel. A laminate plate model (as shown in figure 2) is applied and several hypothesises are made at the beginning of the theoretical analysis.

(1) The cover and backboard glasses are treated as top and bottom surfaces of the laminate plate, respectively. And both of them are simulated as isotropic plates with constant flexural rigidity.
(2) The silicon solar cells layer is ignored and two EVA layers are merged as one layer.
(3) The PV module is a soft core laminate plate and the stress of interlayer in x-y plan is ignored.
(4) Only the anti-symmetrical deformation is studied in present paper, so $\sigma_z=0$, $\varepsilon_z=0$.

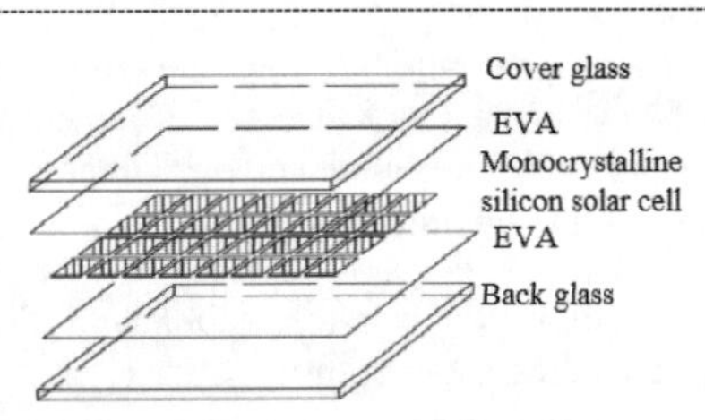

Figure 1. Structural diagram of monocrystalline silicon double glass PV panel.

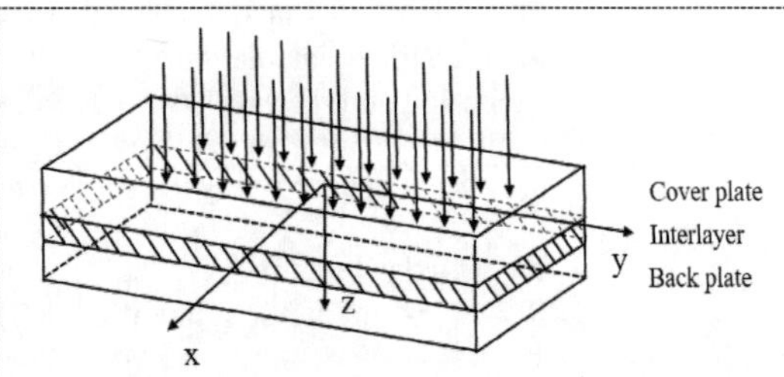

Figure 2. Mechanical model of PV panel and corresponding coordinate system.

2.2. Hoff model and governing equations

Reissner theory is modified by Hoff [5], and a Hoff model is developed for the laminated plate.We could obtain the governing equations of PV panel under uniformly distributed force as follows.

$$w=\omega-\frac{D}{C}\nabla^2\omega \tag{1}$$

$$\left(D+2D_f\right)\nabla^2\nabla^2\omega-\frac{2DD_f}{C}\nabla^2\nabla^2\nabla^2\omega=q \tag{2}$$

$$\frac{1}{2}D\left(1-\nu_f\right)\nabla^2 f-Cf=0 \tag{3}$$

with $D=\frac{E_f\left(h+t\right)^2 t}{2\left(1-v_f^{\ 2}\right)}$, $D_f=\frac{E_f t^3}{12\left(1-v_f^{\ 2}\right)}$, $C=G_C\frac{\left(h+t\right)^2}{h}$.

where E_f is the elastic modulus of the cover and the back glass plate, v_f is the Poisson's ratio of the cover and the back glass plate, G_C is the shear modulus of EVA, t and h are the thickness of surface plate and EVA interlayer, respectively.

2.3. Boundary conditions and boundary equations

Two boundary conditions are studied in present paper to find the influece of bundary condition: one is four edges simply supported; one is two opposite edges simply supported and the other two edges free.

1. Boundary equations of four edges simply supported

$$\left(M_x^{'}\right)_{x=0,a}=0,\left(w\right)_{x=0,a}=0,\left(\varphi_y\right)_{x=0,a}=0,\left(M_x^{"}\right)_{x=0,a}=0 \tag{4}$$

$$\left(M_y^{'}\right)_{y=\pm\frac{b}{2}}=0,\left(w\right)_{y=\pm\frac{b}{2}}=0,\left(\varphi_x\right)_{y=\pm\frac{b}{2}}=0,\left(M_y^{"}\right)_{y=\pm\frac{b}{2}}=0 \tag{5}$$

2. Boundary equations of two opposite edges simply supported and the other two edges free

$$\left(M_x^{'}\right)_{x=0,a}=0,\left(w\right)_{x=0,a}=0,\left(\varphi_y\right)_{x=0,a}=0,\left(M_x^{"}\right)_{x=0,a}=0 \tag{6}$$

$$\left(M_y^{'}\right)_{y=\pm\frac{b}{2}}=0,\left(M_{xy}^{'}\right)_{y=\pm\frac{b}{2}}=0\left(M_{xy}^{"}\right)_{y=\pm\frac{b}{2}}-0,\left(Q_y\right)_{y=\pm\frac{b}{2}}-0,\left(M_y^{"}\right)_{y=\pm\frac{b}{2}}=0 \tag{7}$$

2.4..Modified Rayleigh-Rita method and closed-form solutions

1. Closed-form solutions of four edges simply supported

In present paper, a modified Rayleigh-Rita method is applied to solve the governing equations and a modified general assumption is developed for the solutions. The deflection of PV panel under uniformly distributed force could be calculated based on equation (8).

$$\begin{aligned} w=\omega-\frac{D}{C}\nabla^2\omega=\sum_{n=1}^{\infty}\left\{\sum_{r=1}^{6}e^{\lambda_{nr}y}\omega_{nr}+\frac{2q}{n\pi}\left[1-\cos\left(n\pi\right)\right]\frac{1}{A_1k_n^4+A_2k_n^6}\right\}\sin\left(k_nx\right) \\ -\frac{D}{C}\sum_{n=1}^{\infty}\left\{\sum_{r=1}^{6}\left[\left(-k_n^2\right)e^{\lambda_{nr}y}\omega_{nr}+\lambda_{nr}^2e^{\lambda_{nr}y}\omega_{nr}\right]+\left(-k_n^2\right)\frac{2q}{n\pi}\left[1-\cos\left(n\pi\right)\right]\frac{1}{A_1k_n^4+A_2k_n^6}\right\}\sin\left(k_nx\right)\end{aligned} \tag{8}$$

2. Closed-form solutions of two opposite edges simply supported and the other two edges free

The modified Rayleigh-Rita method is also applied but eight unknown variables should satisfy the new boundary condition as equation (9) to equation (12).

$$\sum_{r=1}^{6}\left[\lambda_{nr}^2e^{\lambda_{nr}y}+v_f\left(-k_n^2\right)e^{\lambda_{nr}y}\right]\omega_{nr}+\sum_{r=1}^{2}\left[v_f\eta_{nr}\left(-k_n\right)e^{\eta_{nr}y}-\eta_{nr}\left(-k_n\right)e^{\eta_{nr}y}\right]f_{nr}=v_f\bar{\omega}^*k_n^2 \tag{9}$$

$$\sum_{r=1}^{6}\begin{bmatrix}\lambda_{nr}^2e^{\lambda_{nr}y}-\frac{D}{C}\lambda_{nr}^2\left(-k_n^2\right)e^{\lambda_{nr}y}-\frac{D}{C}\lambda_{nr}^4e^{\lambda_{nr}y} \\ +v_f\left(-k_n^2\right)e^{\lambda_{nr}y}-v_f\frac{D}{C}k_n^4e^{\lambda_{nr}y}-v_f\frac{D}{C}\lambda_{nr}^2\left(-k_n^2\right)e^{\lambda_{nr}y}\end{bmatrix}\omega_{nr}=v_f\bar{\omega}^*k_n^2+v_f\frac{D}{C}\bar{\omega}^*k_n^4 \tag{10}$$

$$\sum_{r=1}^{6}\left[k_n\lambda_{nr}e^{\lambda_{nr}y}+\frac{D}{C}k_n^3\lambda_{nr}e^{\lambda_{nr}y}-\frac{D}{C}k_n\lambda_{nr}^3e^{\lambda_{nr}y}\right]\omega_{nr}=0 \tag{11}$$

$$\sum_{r=1}^{6}\begin{bmatrix}D\lambda_{nr}k_n^2e^{\lambda_{nr}y}-D\lambda_{nr}^3e^{\lambda_{nr}y}+2D_f\lambda_{nr}k_n^2e^{\lambda_{nr}y}-2D_f\lambda_{nr}^3e^{\lambda_{nr}y} \\ +\frac{2D_fD}{C}\lambda_{nr}k_n^4e^{\lambda_{nr}y}+\frac{4D_fD}{C}\lambda_{nr}^3\left(-k_n^2\right)e^{\lambda_{nr}y}+\frac{2D_fD}{C}\lambda_{nr}^5e^{\lambda_{nr}y}\end{bmatrix}\omega_{nr}+\sum_{r=1}^{2}C\left(-k_n\right)e^{\eta_{nr}y}f_{nr}=0 \tag{12}$$

The new eight unknown variables are solved, the deflection of PV panel is calculated by equation (8).

3. Finite element analysis

A finite element analysis is performed by the use of ANSYS. The material of each layer is simulated as isotropic material and the mechanical properties of them are shown in table 1. Because it could be layered, SHELL181 composite shell element is used for modeling.

Table 1. Material parameter values.

Material	Parameter values		
	Modulus of elasticity /MPa	Poisson ratio	Thickness /mm
Reinforced glass	7.2E+4	0.2	2
Crystalline silicon battery	1.44E+5	0.28	0.2
EVA	3.5E+1	0.3	0.8

4. Experimental analysis

The 6 specimens for bending test are all the double glass photovoltaic modules (as shown in figure 3) which are provided by Suzhou Tenghui Photovoltaic Technology Co., Ltd. The size is 1658×995×7.4 (unit: mm), in which the cover and back glasses are 3.2mm and the interlayer thickness is 1mm. The two boundary conditions are realized as figure 4 and figure 5, respectively. Water pressure is adopted in the tests. The test site with frame, PV panel and equipment are just shown as figure 6.

Figure 3. Monocrystalline silicon double glass photovoltaic module.

Figure 4. Four edges simply supported.

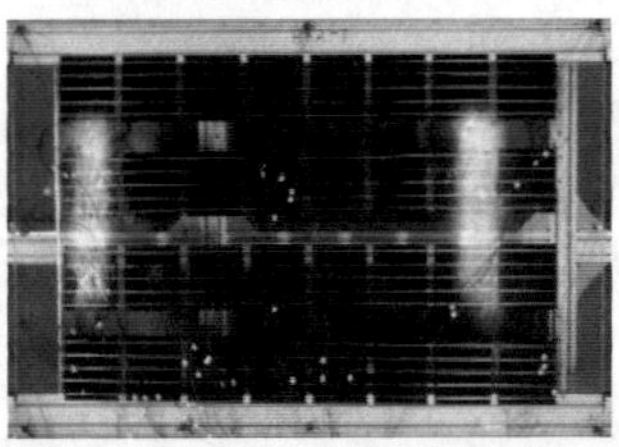

Figure 5. Two opposite edges simply supported and other two edges free.

Figure 6. Test site.

5. Verifications and discussions

5.1. PV panel deflection

In figure 7, the deflection calculated by both ANSYS and equation (8) is changed linearly with water pressure but the test results are changed nonlinearly. It is because the Hoff model is based on linear elastic deformation hypothesis, but the real deformation of PV panel with four edges simply supported is a nonlinear elastic deformation. A nonlinear theory should be applied in future study. The figure 8 states clearly the linear elastic deformation of PV panel with two opposite edges simply supported and the other two edges free. The results from equation (8) are closer to test results than ANSYS, and the accuracy of the proposed equations is verified. All the specimens were checked carefully after the text, and there was not nay cracks on the surface glass. The deformation of PV panels is in safe range.

IOP Conf. Series: Materials Science and Engineering **556** (2019) 012048 doi:10.1088/1757-899X/556/1/012048

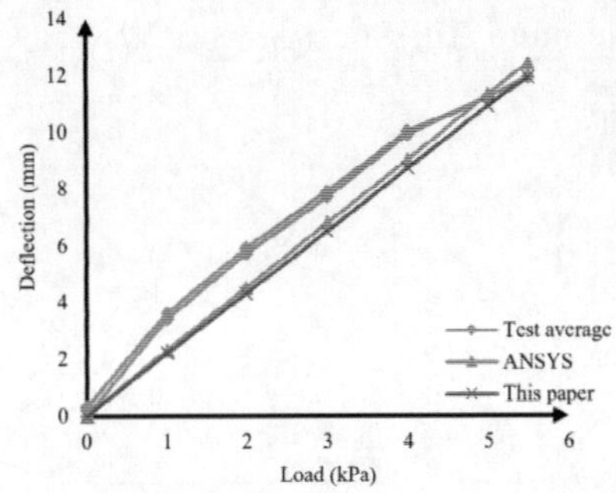

Figure 7. Central deflection of PV panels with four edges simply supported.

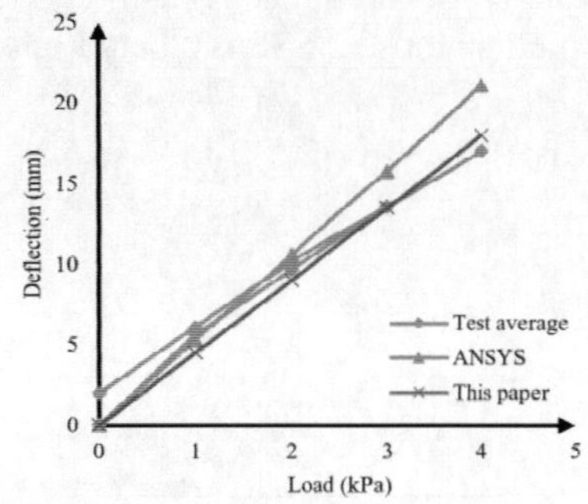

Figure 8. Central deflection of PV panels with two edges simply supported and two edges free.

5.2. PV panel stress

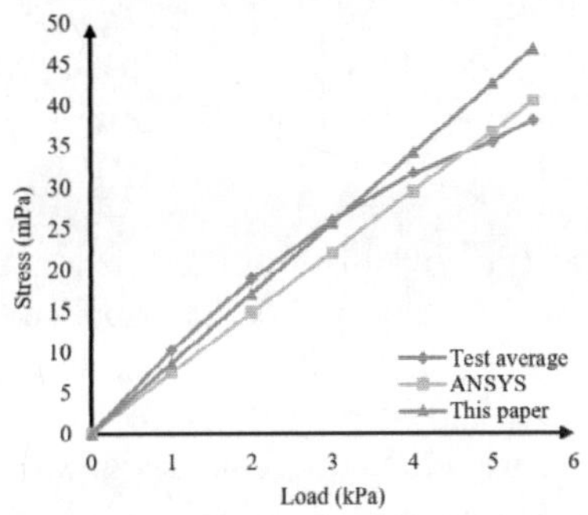

Figure 9. Central stress of PV panels with four edges simply supported.

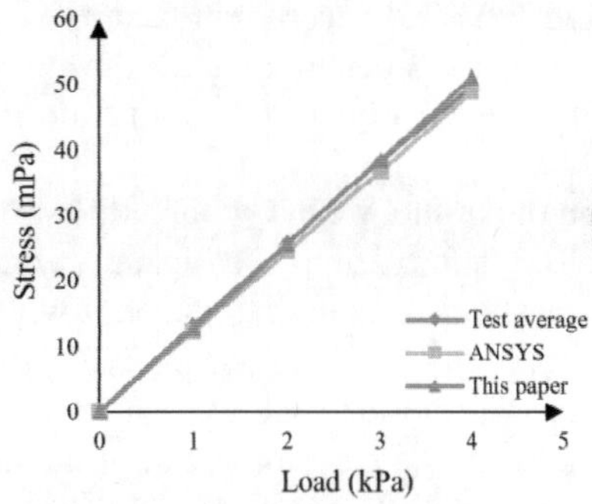

Figure 10. Central stress of PV panels with two edges simply supported and two edges free.

As shown in figure 9, the central stress of PV panel behaves same as the central deflection discussed in figure 7 and it is indeed a nonlinear elastic deformation for the PV panels with four edges simply supported. The data from proposed equations are better when the water pressure is small, but the ones from ANSYS are better under the large water pressure. It is still different in the second group as shown in figure 10, all the stress data has a linear relationship to the water pressure. The deformation of those PV panels is indeed a linear elastic deformation which is also concluded by the deflection data in section 5.1. Although both proposed equation data and ANSYS date match the experimental data very well, the errors are smaller in proposed equation than ANSYS. Moreover, as shown in both figure 9 and figure 10, the maximum stresses on the surface glass of PV panels are all smaller than the limit stress of reinforced glass, so it is safe for the PV panels when they are utilized under those loads.

5.3. Influence of boundary condition

Some discussions about the influence of boundary condition are made. One boundary condition is four edges simply supported, which is marked as SSSS in the figures. The other boundary condition is two opposite edges simply supported and the other two edges free, which is marked as SSFF in figures.

In figure 11, the deflections of PV panels with SSSS are much better. It shows the SSSS can bear a larger uniformly distributed force with a smaller deflection. The central stresses are shown in figure 12 and it can be seen clearly that the maximum central stresses of two cases are both close to 50mPa, which is the limit strength of the reinforced glass. It means that the final damage of the PV panel is because the central stress of panel exceeds the limit stress of surface glass. Under the limit stress of surface glass, SSSS has a larger ultimate load so its mechanical behaviour is proved better than SSFF again. Based on those, SSSS is a better choice for engineers in future BIPV design work.

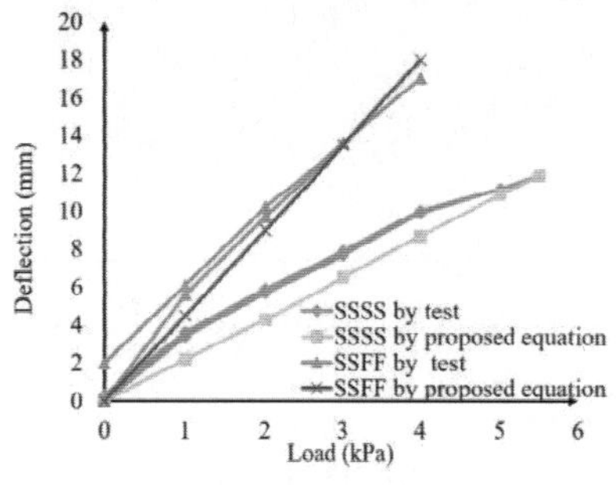

Figure 11. Central deflection of PV panels with two boundary conditions.

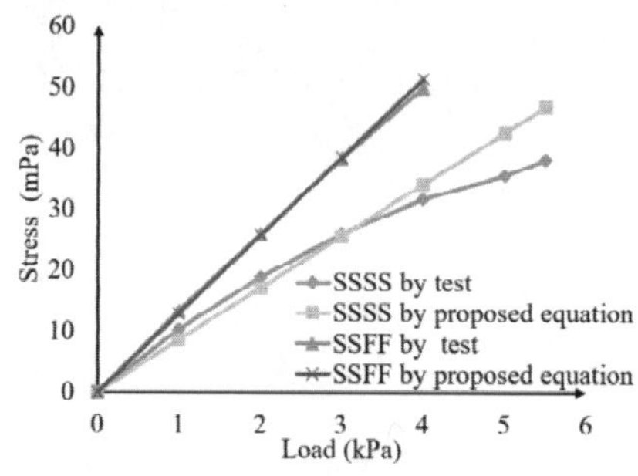

Figure 12. Central stress of PV panels with two boundary conditions.

6. Conclusions and recommendations

The aim of this paper is to study the mechanical properties of the double glass PV panel with two different boundary conditions. Both experimental and theoretical works are completed.

Based on the results we may conclude as follows:

- The deformation of PV panels with SSFF is a linear elastic deformation. But it is a nonlinear elastic deformation to the ones with SSSS. A nonlinear elastic theory should be applied later.
- As to the PV panels with SSFF, the calculation results from ANSYS and proposed equations are all close to the experimental results. The deflection and stress calculated by proposed equations are even more accurate than ANSYS. It can be used in the optimal design work.
- As to the PV panel with SSSS, the ANSYS and proposed method are based on linear elastic deformation theory but the real deformation of PV panels is nonlinear. In central deflection, the proposed equations are more accurate for the maximum deflection. In central stress, it is suitable to calculate the stress when the PV panel is under small deformation.
- Comparing the central deflection and central stress from different boundary conditions, the PV panels with SSSS have better effects. The PV panels with SSSS can bear a larger load with a smaller deformation. It should be considered as the primary choice in the BIPV.

Acknowledgments

The authors are grateful for the financial support from the National Key Research and Development Program of China: Newton Fund - China-UK Research and Innovations Bridges (No. 2016YFE0124500).

References

[1] IEC 61215-2005, *Crystalline Silicon Terrestrial Photovoltaic (PV) Modules*[S]. International Electrotechnical Commission, 2005
[2] Patrina Eiffert, *An Economic Assessment of Building Integrated Photovoltaics*, Oxford Brookes School of Architecture, 1998
[3] Konstantin Naumenko and Victor A. Eremeyev, A layer-wise theory for laminated glass and photovoltaic panels, *Composite Structures*, 112(1), 2014, pp 283-291
[4] Ajitanshu Vedrtnam and S. J. Pawar, Laminated plate theories and fracture of laminated glass plate-A review, *Engineering Fracture Mechanics*, 186(1), 2017, pp 316-330
[5] N. J. Hoff, Bending and buckling of rectangular sandwich plates, *NACA*, TN2225, 1950

Study on the hysteresis characteristic of solar thermal storage system

N Mao[1], **E S Long** [1], **J Li** [1] **and D T Xiao** [1]
[1]Sichuan university, Chengdu, China
E-mail: longes@163.com.

Abstract. The thermal storage efficiency is one of the important indicators for evaluating the operation of solar thermal storage system. Affected by the thermal inertia of the system, the analysis of the relationship between the thermal storage efficiency and solar radiation, which is inadequate at the smaller time scale. This paper measured the operation of a flat plate solar collector system, studied the hysteresis characteristics of the system's collector efficiency changes with solar radiation, took a numerical solution method for the actual lag time of the system and finally obtained the actual influence rule of solar radiation on the collection efficiency per minute. The results show that the solar collector's efficiency change lags behind the solar radiation by 27 min; combined the lag time, the relationship between the modified thermal storage efficiency and solar radiation also were analysed; initially the modified storage efficiency per minute increases with the solar radiation, then gradually decrease after reach a certain value, due to the overall system temperature is higher and the system temperature rise is smaller when the solar radiation is large enough; the varying curve conforms to quadratic function. It provides theoretical support and practical reference for contrasting and evaluating the thermal storage efficiency changes with solar radiation in a smaller time scale.

Keywords: thermal storage efficiency; solar radiation; hysteresis characteristic; flat plate collector

1. Introduction

With the energy crisis become serious, the heat-collecting system that solar energy and other renewable energy sources as heat sources will take the main body for development and utilization [1] under the urgent need of social operation. The effective increase and accurate determination of the actual thermal storage efficiency of the solar system is a key factor to ensure the maximization of solar thermal conversion.

In recent years, various scholars have conducted extensive in-depth research on the collection efficiency by optimizing various components within the heat collection system. Such as Cheng W P [2] and other scholars enhanced the collection efficiency of the system by change the heat transfer area and the distribution of the heat pipe; Zhao J [3] of Tianjin University analysed the effects of the collector area-tank volume ratio (A/V) variation on the thermal storage efficiency; Ji Y M, Duan M L [4] studied a solar-assisted buried pipe compound system in the transition season, the two operating modes with heat storage and the thermal storage efficiency of the system is evaluated. Li X G [5] and other scholars designed a solar thermal system with a ground-source heat pump, stored abundant solar energy in underground soil and improved the system thermal storage efficiency and realized the transfer and utilization of solar energy finally. Lin Y [6], Han Z W [7] added a phase change thermal storage device based on the original solar-soil source heat pump, which increased the system's thermal storage efficiency up to 60% and the average heating COP reached to 6.5.

Influenced by the thermal inertia of the heat-collecting system, the whole temperature rise of the system will change lags behind the solar radiation for a certain period of time, all the above researches

IOP Conf. Series: Materials Science and Engineering **556** (2019) 012049 doi:10.1088/1757-899X/556/1/012049

use the thermal storage efficiency of the large time scale to evaluate the solar heat collection system, which cannot accurately reflect the actual influence of the solar radiation on the thermal storage efficiency.

In order to research the quantitative effect of the solar radiation on the thermal storage efficiency at a small time scale, this paper established a planar solar collector-thermal storage system, measured the solar radiation, outdoor temperature and system water temperature during the operation of the system, took advantage of the numerical solution to accurately calculate the lag time of the system's heat gain and combined with the lag time, obtained the actual rule of the modified collector efficiency (min) change with the solar radiation finally.

2. Experimental system and methods

2.1. Experimental system

In combination with the meteorological conditions, a flat-plate solar collector-thermal storage system has been designed and installed. The system consists of 2 plate-type heat collector (FPC-1-CTL), 1 booster pump, 1 heat storage tank, some pipeline, 1 flow meter, 1 intelligent solar radiation recorder (JR-08) and 1 multi-channel temperature data acquisition system (JTDL-80).

During the operation of the system, clean water is adopted as the system heat storage medium for those large specific heat capacity relatively, heated up rapidly in the plate heat collector by the solar radiation and transported to the hot water storage tank through the pipe network, then send into the inlet of a plate collector after uniformly mixed in the tank. The experimental system and temperature measurement point layout are shown in Figure 1.

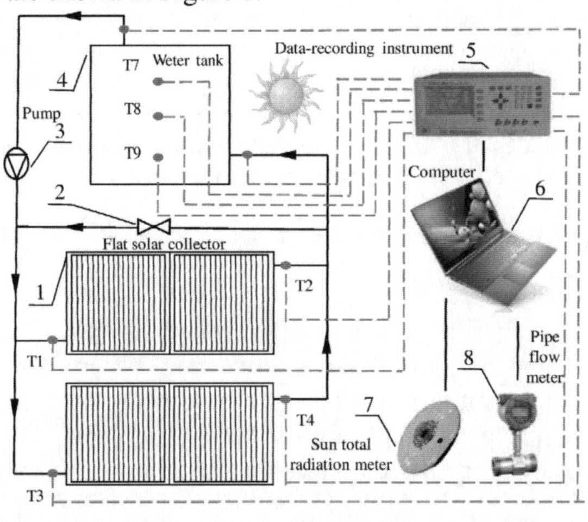

Figure 1. The experimental system and the arrangement of temperature testing points

The heat collector and the heat storage tank are connected through the distribution networks of the PVC pipe of 20 mm diameter and all the transmission pipes are covered with a 35mm-thick rubber sponge to prevent excessive loss of heat. The temperature measurement data is recorded by multi-channel temperature data acquisition system (JTDL-80) matched with the thermocouple, to record the water temperature and ambient temperature of the system meanwhile. Intelligent solar radiation recorder (JR-08) is applied to the solar radiation record, which installed at an inclination of 40° to the south as same as that of the solar radiation panel. The specific parameters of the system components are shown in Table 1.

Table 1. System equipment parameter table.

System equipment	model	size（mm）	amount	power(w)	range
Flat Solar Collector	FPC-1-CTL	2000×1000×70	4		>0℃
Pump	PB-H090	300×150×200	1	60	

IOP Conf. Series: Materials Science and Engineering **556** (2019) 012049 doi:10.1088/1757-899X/556/1/012049

Heat storage tank		600×600×1750	1		0~414.27L
Solar-recording instrument	JR-08	50×2000×30	1	20	
Temperature-recording instrument	JTDL—80	400×300×200	1	35	0~800℃
Thermocouple	WRNT		9		0~800℃
Pipe flow meter	LDG03	20	1	15	0.1~11.3m3/h

2.2. Experimental contents
In order to investigate the actual heat storage characteristics of the flat plate solar collector-thermal storage system, the experimental measured the solar radiation, outdoor temperature and system water temperature during 4.20 to 4.23. The specific experimental process and the starting and ending conditions are shown in Table 2.

Table 2. Specific conditions of the experiment

Date	Test time	Average outdoor temperature(℃)	Total solar radiation （w）	System temperature （℃）
4.20	10：35-17：50	18.9	23751.2	32.9-55.9
4.21	10：50-16：30	20.1	31577.4	26.1-41.3
4.22	10：50-17：35	24.9	79453.2	47.5-64.8
4.23	10：41-18：29	25.1	77748	47.1-65.7

3. Calculation method

3.1. Calculation of heat collecting efficiency
According to the US ASHRAE Standard: Testing and analysis on thermal property has been widely used in solar collectors evaluating and the instantaneous efficiency of collectors is a decisive factor for the thermal characteristics of collectors [8].

The following setup steps assume that the system is fully filled with clean water; the specific heat capacity of water is constant; and the quality of water storage in the system remains stable, which could be converted to the total volume of the system directly. In this experiment, the useful energy gain of the collector can be expressed as the system heat storage, which is calculated by formula (3.1). The calculation formula of the system average thermal storage efficiency per unit time [9] is shown in formula (3.2).

$$Q_A = C_w \rho_w V_s (t_{i+1} - t_i) \quad (3.1) \qquad \omega_A = \frac{Q_A}{I_T} = \frac{C_w \rho_w V_s (t_{i+1} - t_i)}{I_T} \times 100\% \quad (3.2)$$

Q_A: heat storage in the system per unit time, J; C_w: the specific heat capacity of water, 4.2×10^3 J/（kg.℃); ρ_w: the density of water, 1kg /m^3; V_s: the volume of the system, m^3; t_{i+1}、t_i: the average water temperature of the system before and after the unit time, ℃; ω_A: collecting efficiency in the system per unit time, %; I_T: in the unit time, whole solar radiation energy received on the collector theoretically, J。

3.2. Calculation of lag time
Affected the system's thermal inertia and other reasons, the system water temperature fluctuations would change lag behind the solar radiation during testing, causing that the thermal storage efficiency of a small time scale inaccurately reflect the relationship between solar radiation and the system heat gain.

In the experimental operation, it is assumed that a mass point moves with the heat storage medium in the system. the time required for a complete circulation of the mass point and the overall of the system evenly heating up is the essence of the lag time.
The time needed for a complete cycle of the mass point can be computed by formula (3.3), which is the ratio of the total volume of the system to the flow rate of the pipeline per unit time [10].

$$T_1 = \frac{V_A}{v_Q} = \frac{V_s+V_b+V_t}{s_v \times A_s} \qquad (3.3)$$

T_1: the time needed for a complete cycle of the mass point in the system, min; V_s、V_b、V_t: the volume of the Plate collector、storage tank and pipeline, m^3; v_Q: the volume of flow in the pipe per unit time, m^3/h; s_v: the distance of flow in the pipe per unit time, m/s; A_s: Section area of the pipe, m^2。

4. Discussion

In this paper, the typical experimental results during one day (April 21st) are taken as an example to deeply study the relationship between the thermal storage efficiency and the solar radiation.
The varying curve of solar radiation and the water temperature of each measuring point as shown in the Figure 2.a): Analysed the water temperature at each measuring point, the order is flat water outlet, water tank upper, water tank inlet, flat water inlet, water tank middle, the water tank outlet and water tank lower, were 41.7°C, 36.77°C, 36.75°C, 36.72°C, 35.8°C, 35.3°C and 35.1°C respectively.
The temperature at the outlet of the flat plate is highest, partly because the water flows through the flat plate and converts the heat of solar radiation into internal temperature rise. The temperature in the lower part of the water tank is lowest due to the delamination phenomenon of the water tank, Hot water floats on the top of the tank and the lower water is not heated sufficiently.
Considering the influence of system thermal inertia, the overall temperature of the system will change lag behind that of solar radiation and the calculation of the collection efficiency per minute is inaccurate. As a result, the time scale of 0.25h and 0.5h is often used as a unit time to calculated the system thermal storage efficiency, as shown in Figure 2.b). When 0.5h is selected as the unit time, the range of efficiency varies from 18% to 81% and the average heat storage efficiency is 47.4%. When 0.25h is selected as the unit time, the range of efficiency varies from 11% to 89% and the average heat storage efficiency is 49.2%.

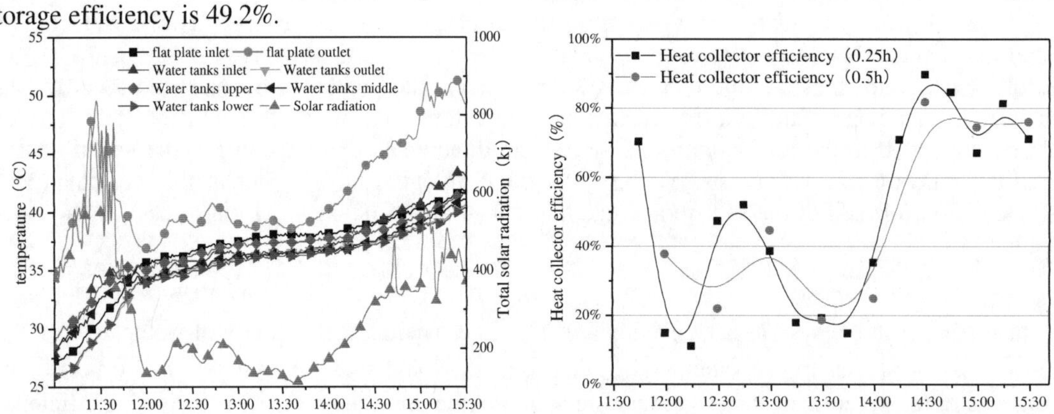

Figure 2. a) Solar radiation and water temperature b) The system thermal storage efficiency

From the above analysis, it can be seen that the variation of heat storage efficiency and solar radiation with time is more consistent at smaller time scales.
Taking 1min as a unit time, the calculated results of total heat gain, solar radiation and thermal storage efficiency of the system are shown in Figure 3.a). The total heat of the system come to 7487.9 KJ, the solar radiation reach 15788 KJ whole day and the average daily thermal storage efficiency is 47.4%, what's more, the solar radiation reaches the maximum value at 11:55 and 14:07 respectively and the variation range is 15-226KJ. While the collector efficiency (min) of the system reaches the extreme value at 12:29 and 14:34 respectively, with a range of 2% to 109%. This is because the bulk

temperature of the system is lag behind the solar radiation due to the overall thermal inertia, which also has a corresponding effect on the thermal storage efficiency.

The relationship between the thermal storage efficiency (min) and solar radiation can be used to evaluate and compare the thermal storage performance of the system. As shown in Fig. 3.b), the total solar radiation per minute is mainly distributed in the range of 20 to 90 KJ during testing. With the solar radiation gradually increases, the variation of thermal storage efficiency (min) is unpredictable. This is because the effect of the system thermal inertia, the real-time thermal gain of the system does not match the real time solar radiation, the calculated collector efficiency (min) can't accurately reflect the thermal storage performance of the system.

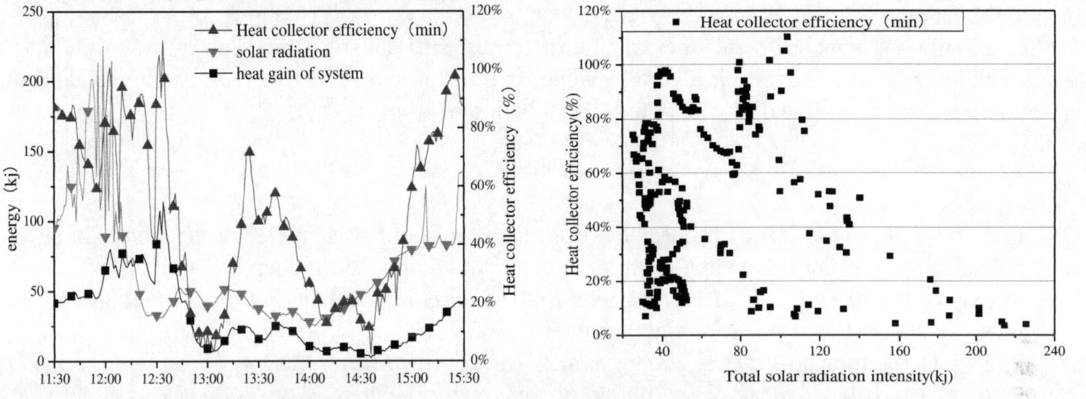

Figure 3. a) Solar radiation, system heat gain and thermal storage efficiency

b) The relationship between the thermal storage efficiency (min) and solar radiation

According to the calculation of formula (3.3), the overall temperature changes of the system lags behind the variation of the solar radiation about 27 min.

Applying this lag time to the correction calculation of the thermal storage efficiency (min), the varying curve of solar radiation, modified system heat gain and modified thermal storage efficiency are shown in Figure 4.a). The modified thermal storage efficiency (min) is consistent with the fluctuation of solar radiation, and concentrated at 10% to 60%.

The relationship between the modified storage efficiency (min) and solar radiation is shown in Figure 4.b). Firstly, with the gradual increase of solar radiation, the system thermal storage efficiency increases rapidly, whose rate of increase is nearly 0.57%/KJ, then gradually decreases after reach to 55%. The varying curve conforms to quadratic function, and the fitting formula is: $y = -0.004x^2 + 0.0113x - 0.0535$, $R^2 = 0.7566$. When the solar radiation reaches a certain intensity, the overall temperature of the system is higher accordingly, and It is more difficult to increase the temperature of the system, causing gradually reduce in the thermal storage efficiency.

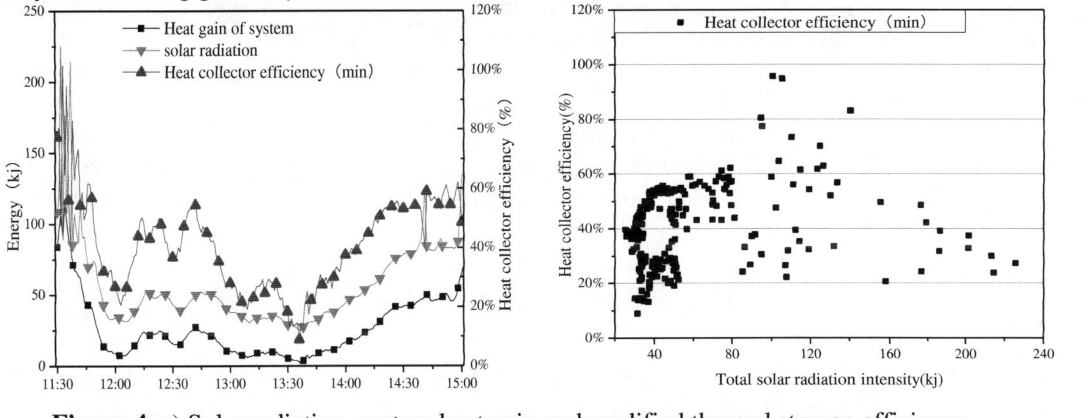

Figure 4. a) Solar radiation, system heat gain and modified thermal storage efficiency

b) The relationship between the modified storage efficiency (min) and solar radiation

5. Conclusion

(1) During the experiment, the highest and the lowest temperatures of the system appear at the flat outlet and the water tank lower respectively. Selecting 0.25h and 0.5h as the unit time, the variation of heat storage efficiency and solar radiation with time is more consistent at smaller time scales.

(2) Affected by the thermal inertia of the system, the variation of thermal storage efficiency (min) with solar radiation is unpredictable. The time required for a complete circulation of the mass point is the essence of the lag time, which is about 27 minutes.

(3) Applying this lag time to the correction calculation of the thermal storage efficiency (min), the varying curve of the modified thermal storage efficiency (min) and solar radiation conforms to quadratic function.

(4) The modified storage efficiency per minute increases with the solar radiation In the initial stage, then gradually decrease after reach a certain value, due to the overall temperature is higher and the system temperature rise is smaller when the solar radiation is large enough

6. References

[1] Liu X Y and Jing H 2014 Comparative test and analysis of solar collector efficiency in severe cold area *(*Scientific and technological innovation and Application*)* chapter 2 pp 12–18

[2] Cheng W P and Hu X W 2017 *Research and analysis of solar energy phase-change thermal storage system* (Electronic Materials) chapter 3 p 65

[3] Zhao J, Qu H and Cui J k et al 2008 *Simulation analysis of solar energy central heating system with cross seasonal heat storage* (Journal of solar energy) chapter 2 pp 214-219

[4] J Y M and Duan M L 2017 *solar assisted ground source heat pump composite system transition season operation mode* (HVAC) chapter 10 pp 127-131

[5] Li X G, Z J and Wang Y P et al 2009 *Application and experiment of solar energy, heat storage and ground source heat pump combined system* (Journal of solar energy) chapter 12 pp 1658-1661

[6] Li Y 2009 *experimental study of solar heat pump heat storage system* residential *technology)* chapter 3 pp 1–5

[7] Han Z w, Zheng M Y and Liu W et al 2006 *Solar energy ground source heat pump phase change heat storage heating system in severe cold area* (Journal of solar energy) chapter 12 pp 1214-1218

[8] Han Z w, Zheng M Y and Liu W et al 2007 *Experimental research on thermal performance of glazing cover of high-efficiency solar* (Energy Conservation)

[9] A. Hobbi and Siddiqui K Liu W et a 2009 *Solar energy ground source heat pump phase change heat storage heating system in severe cold area* (International Journal of Heat & Mass Transfer) chapter 52 pp 4650-4658

[10] Mao N and Long E S 2017 *Comparative study on indoor thermal comfort between capillary floor and fan coil heating* (heating and cooling) chapter 12 pp 20-22

Acknowledgments

This work is supported by grant no.51778382 from the National Natural Science Foundation and grant no.2016YFC0700400 form the National Key Research and Development Program of China. We do thank Dr. Li Y R and Ren J Y for valuable discussion and assistance with the theoretic and experiment part.

A smart hybrid pole with street lighting and EV charging based on DC micro-grid and renewable energy

Yongming Zhang[1], **Li Li**[1], **Zhe Yan**[1] and **Jiawei Yao**[1*]
[1]Tongji University, Shanghai, China
*E-mail: jiawei.yao@tongji.edu.cn

Abstract. In order to efficiently utilize renewable energy and solve roadside fast charging of Electric Vehicle (EV) problem, a smart hybrid pole system is proposed in this paper, which integrated with renewable energy Street Lighting Pole (SLP) and EV charging. Besides, smart hybrid pole can also serve as the backup power for traffic lights. In this system, several renewable energy SLPs are coupled together by Direct Current (DC) micro-grid with storage battery, so it is suis for both DC fast charging and slow charging for EV. More importantly, the energy efficiency in this system can become much higher than alternating current (AC) EV charging system, because renewable energy power such as photovoltaic (PV), load and energy storage are all DC subsystem. Then for studying the suitable number in a group of smart hybrid pole, the group model of smart hybrid pole is presented. In addition, the hybrid pole energy storage system can be controlled by a cloud management platform, so it is convenient to make the decision to charge or discharge according to the different power prices. The smart hybrid pole solution is conducive to EV charging and the efficient application of renewable.

Keywords: smart pole; street lighting; EV charge; micro-grid; renewable energy; energy storage; cloud management.

1. Introduction

Nowadays, energy crisis and environment pollution have become the global problem, and the increasing use of energy caused climate change. However, the renewable energy, clean energy, other distributed generation (DG) such as solar-based photovoltaic (PV) and wind generation, some changes of energy consumption such as the use of light-emitting diode (LED) lighting [1-3] and new transport EV [4-6], positively contribute to the reduction of the carbon footprint of electricity generation.

The new-energy vehicles, especially EV, has been quickly developed in recent years [4,5]. In fact, automakers, including Germany-based BMW and Audi, Japan-based Toyota and Nissan, the US-based General Motors and Tesla, and China-based BYD and Roewe, launched a series of new-energy vehicle models, which marks the initial popularity of new-energy vehicles. As expected, China has a rapidly growing market of new-energy vehicles. In 2016, 517,000 new-energy vehicles were produced and 507,000 new-energy vehicles were sold in China, with the growth of 52% and 53% respectively compared to the last year.

However, EV charging points and service do not match the growing trend of EV in China. EV charging points increased with the growth of 12% compared to 2015. Until June 2016, China's total public charging poles are 81,000 and the private charging piles are 55,000. The ratio of car to pole is about 8:1, so it is far below the theoretical ratio of 1:1~1.2. Currently, it needs to build more charging poles [6]. In 2020, 435 thousand charging points will be built in Beijing to satisfy the requirement for 600 thousand EV.

The construction of charging points needs to lay plenty of cable in every corner of the city, which means large cost. To reduce expenses, street-lighting pole with EV charging system is presented. The distribution characteristic of street-lighting is similar with the distribution characteristic of EV charging, so charging points can be constructed based on the system of street-lighting. However, it is only suitable for EV slow charging because the capacity of cable for street lighting pole is not enough for EV fast charging. In this paper, a smart hybrid pole with street-lighting and EV-charging systems, which can be used to both fast charging and slow charging for EV, is proposed based on DC micro-grid, renewable energy and energy storage.

2. Architecture of lighting system

It is well known that the architecture of traditional street lighting system is presented in Figure 1. In this system, traditional alternating current (AC) street light can be supported by AC power directly, and DC light, such as LED light is required to be linked by an AC/DC converter [7,8].

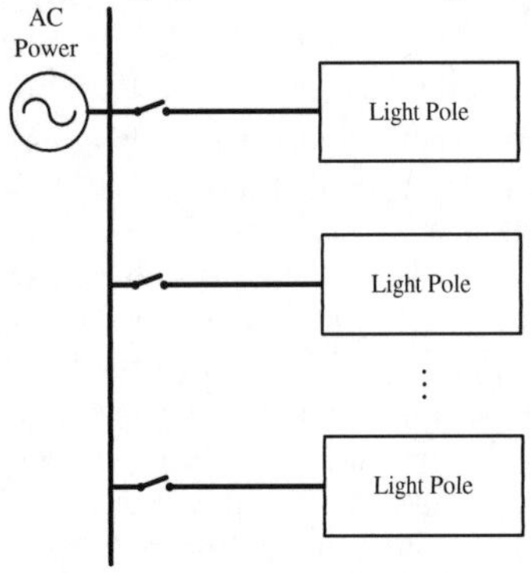

Figure 1. The traditional street lighting system

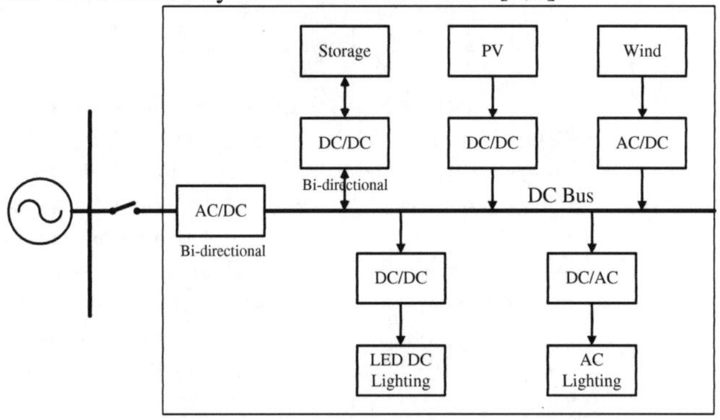

Figure 2. The architecture of renewable energy street lighting pole

When the renewable energy DG, such as PV and wind generation, are introduced into street lighting system, DC bus and storage devices such as battery are necessary in order to reduce the times of energy conversion to get a highly efficient system. Meanwhile, the energy storage can ensure the voltage stability and operate in island mode in long time. The architecture of renewable energy street lighting pole is presented in Figure 2. In this system, both AC and DC light can be considered.

3. Architecture of smart hybrid pole

In recent years the hybrid poles with street lighting and EV changing have been developed by different companies respectively. In fact, this kind of hybrid pole is only suitable for slow charging for EV.

In order to fast charging for EV, a smart hybrid pole with street-lighting is proposed in this section. EV charging system is added to the renewable energy street lighting pole as shown in Figure 3, then the hybrid poles are integrated by use of a DC micro-grid [9] as shown in Figure 4. In this micro-grid system, smart hybrid poles are connected by DC bus. In this case, the part of EV charging can get power from AC grid, DC micro-grid and storage unit directly, so theoretically it is suitable for EV fast charging if the power controlled reasonable.

Compared to island hybrid pole, besides it can provide fast charging service, the integrated smart hybrid poles provide following benefits:

- The micro-grid system is fairly flexible. The smart hybrid poles can operate in grid-connected mode or island mode;
- The system has more safety and reliability. With the connection of DC micro-grid, the power can be dispatched between hybrid poles, so the DC bus voltage is in stable condition;
- It has more economical efficiency for consumer. Every smart hybrid pole with storage can be management by DC micro-grid, so it can reduce its operational expenses in accordance with peak-valley price;
- The system can provide UPS services for emergency.

Additionally, smart hybrid pole can also serve as the backup power for traffic lights, as shown in Figure 5. When the power grid malfunctions, the storage battery in smart hybrid pole can provide emergency electric power for traffic signal, which avoids traffic disturbance.

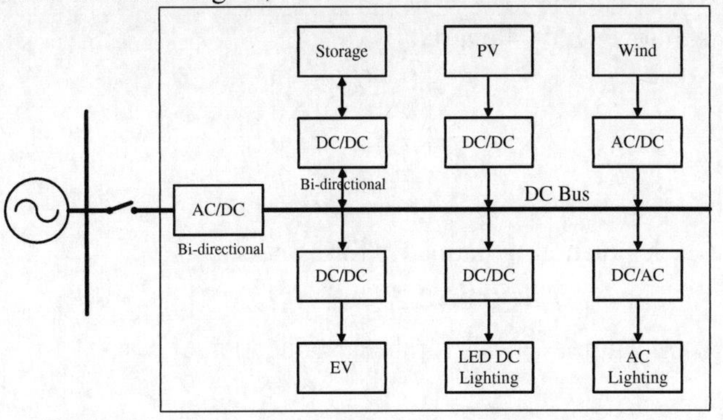

Figure 3. The smart hybrid pole with street lighting and EV charging

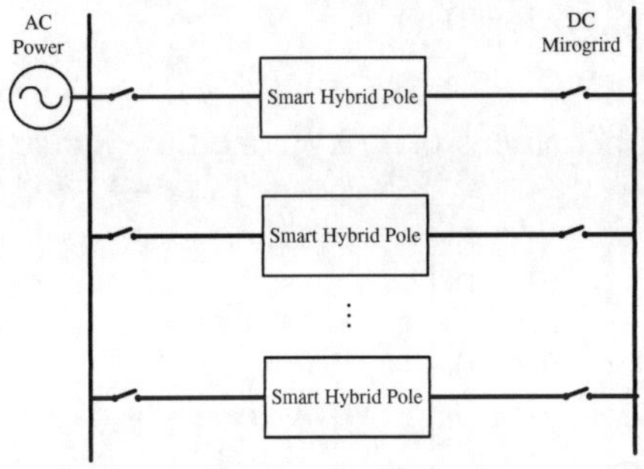

Figure 4. The integrated hybrid poles by a DC micro-grid

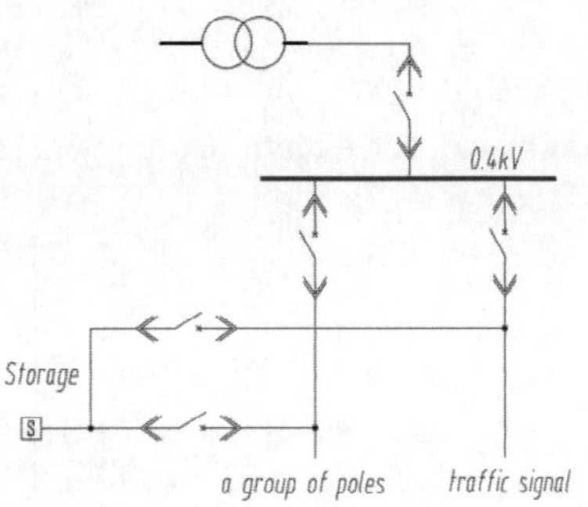

Figure 5. The backup power for traffic signal

4. The group model of smart hybrid pole

For studying the suitable number in a group of smart hybrid pole, the group model of smart hybrid pole is proposed in this section.

The power of street lighting is usually only 0.2 kW, and the power of slow charging for EV is 7 kW (220 V×32 A). Therefore, the power of the EV charging poles should be primary part of the capacity of the distribution box of municipal road.

Supposing that the rated slow charging power is a kW, and the rated fast charging power is $n \times a$ kW, which is n times more than the rated slow charging power ($n \in Q^{+}$, $n > 1$); the number of poles is t ($t \in Q^{+}$, $t \geq n$). Traditional poles with street light and EV charging only support slow charging. Taking $t = 4$ as an example, there should be 5 cases as shown in Table 1. The possibility of each case is hard to be calculated due to different time and position. For simplifying, supposing that the probability of each situation is equal, the average power per case is $2a$ kW, and the average power per pole is $0.5a$ kW.

Table 1. Average power in the case of 4 poles with only slow charging

Case	Pole 1	Pole 2	Pole 3	Pole 4	Total power (kW)
1	0	0	0	0	0
2	a	0	0	0	a
3	a	a	0	0	$2a$
4	a	a	a	0	$3a$

5	*a*	*a*	*a*	*a*	4*a*
		Average power per case			2*a*
		Average power per pole			0.5*a*

Table 1 is changed to the following 5×4 matrix:

$$A_{5\times4}=\begin{bmatrix}0&0&0&0\\a&0&0&0\\a&a&0&0\\a&a&a&0\\a&a&a&a\end{bmatrix} \quad (1)$$

The equal probability of each situation f_t in matrix form under $t=4$ is:

$$f_4=\begin{bmatrix}0.2 & 0.2 & 0.2 & 0.2 & 0.2\end{bmatrix} \quad (2)$$

Supposing that the matrix M_t under $t=4$ is for the accumulation of cases:

$$M_4=\begin{bmatrix}1 & 1 & 1 & 1 & 1\end{bmatrix}^T \quad (3)$$

The average power per pole $P_{avg,t}$ under $t=4$ is:

$$P_{avg,4}=0.25\times f_4\times A_{5\times4}\times M_4=0.5a \quad (4)$$

Extending to each *t*, there should be $(t+1)$ cases as shown in the following $(t+1)\times t$ matrix; with the same probability of each situation, the average power per case is 0.5*ta* kW, and the average power per pole is also 0.5*a* kW.

$$A_{(t+1)\times t}=\begin{bmatrix}0&\cdots&\cdots&0&0\\a&0&\cdots&0&0\\a&a&\ddots&\vdots&\vdots\\\vdots&\vdots&\ddots&0&0\\a&a&\cdots&a&0\\a&a&\cdots&a&a\end{bmatrix} \quad (5)$$

$$f_t=\begin{bmatrix}(t+1)^{-1} & \cdots & (t+1)^{-1}\end{bmatrix} \quad (6)$$

$$M_t=\begin{bmatrix}1 & \cdots & 1\end{bmatrix}^T \quad (7)$$

$$P_{avg,t}=t^{-1}\times f_t\times A_{(t+1)\times t}\times M_t=0.5a \quad (8)$$

If the poles can provide the serves of slow and fast charging, *t* smart hybrid poles will maximally at the same time support $\lfloor t/n \rfloor$ poles with fast charging with total *t***a* kW power limit, ($\lfloor t/n \rfloor$ means that t/n is rounded down). In the case of *t* = 4 and *n* = 4 as shown in Table 2, the average power per case is 2.6*a* kW, and the average power per hybrid pole is 0.65*a* kW. So, the expectation value of each hybrid pole power with slow and fast charging is more than with only slow charging, which means efficiency promotion.

Table 2. Average power in the case of 4 hybrid poles with slow and fast charging

Case	Pole 1	Pole 2	Pole 3	Pole 4	Total power (kW)
1	0	0	0	0	0
2	*a*	0	0	0	*a*
3	*a*	*a*	0	0	2*a*
4	*a*	*a*	*a*	0	3*a*
5	*a*	*a*	*a*	*a*	4*a*
		Average power per case			2.6*a*
		Average power per hybrid pole			0.65*a*

Table 2 is changed to a 5×4 matrix as following:

$$B_{5\times4}=\begin{bmatrix}0&0&0&0\\4a&0&0&0\\a&a&0&0\\a&a&a&0\\a&a&a&a\end{bmatrix}=A_{5\times4}+\begin{bmatrix}0&0&0&0\\3a&0&0&0\\0&0&0&0\\0&0&0&0\\0&0&0&0\end{bmatrix} \tag{9}$$

The average power per pole $P'_{avg,t}$ under $t = 4$ is:

$$P'_{avg,4}=0.25\times f_4\times B_{5\times4}\times M_4=0.65a \tag{10}$$

Extending to each *t* and *n*, the cases is shown in the following $(t+1)\times t$ matrix:

$$B_{(t+1)\times t}=\begin{bmatrix}0&\cdots&\cdots&0&0\\na&0&\cdots&0&0\\na&na&\ddots&\vdots&\vdots\\\vdots&\vdots&\ddots&0&0\\a&a&\cdots&a&0\\a&a&\cdots&a&a\end{bmatrix}=A_{(t+1)\times t}+\begin{bmatrix}0&\cdots&\cdots&0&0\\(n-1)a&0&\cdots&0&0\\(n-1)a&(n-1)a&\ddots&\vdots&\vdots\\\vdots&\vdots&\ddots&0&0\\0&0&\cdots&0&0\\0&0&\cdots&0&0\end{bmatrix} \tag{11}$$

The average power per pole $P'_{avg,t}$ is:

$$P'_{avg,t}=t^{-1}\times f_t\times B_{(t+1)\times t}\times M_t \tag{12}$$

Supposing that the Average power per pole is $f(t,n,a)$, which can be calculated as formula:

$$f(t,n,a)=P'_{avg,t}=[0.5+\frac{2\left\lfloor\frac{t}{n}\right\rfloor(t+1)-\left\lfloor\frac{t}{n}\right\rfloor(\left\lfloor\frac{t}{n}\right\rfloor+1)n}{2(t+1)t}\cdot(n-1)]a \tag{13}$$

In the actual project, hybrid poles have rated slow and fast charging power with stationary, so n is constant. Take 7 kW slow charging and 28 kW fast charging as an example ($n = 4$, $a = 7$), $f(t,4,7)$ is shown in Figure 6 by MATLAB. $f(t,4,7)$ increases rapidly when t is from 4 to 20, and it rises slowly when t is more than 20. Therefore, a set of 20 smart hybrid poles is more efficient in ideal conditions. In actual situation, the optimum number of smart hybrid poles also depend on the cost of electric system and actual road light planning and design.

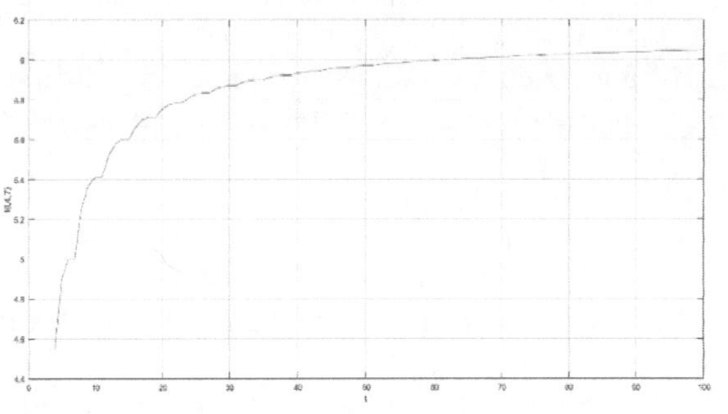

Figure 6. The functional image of average power per hybrid pole with power limit

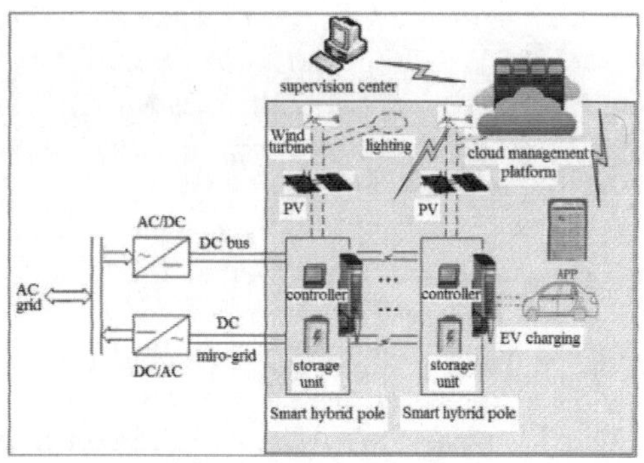

Figure 7. The management system of smart hybrid poles

5. Management of Smart hybrid pole

In order to control system operation and maintain stability, a powerful management system is extremely essential. A management system is proposed for DC micro-grid integrated hybrid poles. In the management system, as shown in Figure 7, the charging terminal control and metering system, and the client application are combined to cloud intelligence management platform to collect big data, integrate scheduling, optimize storage, remote monitoring and other functions. Besides, management system can control the charge and discharge of storage unit judging by time-of-use electricity price. This management system has economic benefit.

6. Conclusions

In this paper, a smart hybrid pole with street lighting and EV charging was proposed, and the integrated hybrid poles with management system were studied. The conclusions and suggestion are summarized as following:

- A structure of smart hybrid pole system was proposed. The proposed hybrid pole can operate in grid-connected mode or island mode, and the stability of DC bus voltage can be supported by the storage.
- The smart hybrid poles system integrated by DC micro-grid is more safe and reliable. With the cloud management platform, the power can be dispatched between AC grid and DC micro-grid or between hybrid poles.
- With the use of cloud management platform, it is easy to make the decision on the basis of the peak-valley price to charge or discharge, so it has more economical efficiency for consumer.

In addition, the next step we will focus on further research on specific calculations and experiments and analyse the parameters and energy efficiency of different poles.

Acknowledgements

This paper is supported by Jiangsu Province Science and Technology Support Plan Project (Grant No. BE2014830), Fundamental Research Funds for the Central Universities (Grant No. 102472015118), China Postdoctoral Science Foundation funded project (Grant No.2018M632162) and Open Projects Fund of Key Laboratory of Shanghai Urban Renewal and Spatial Optimization Technology (Grant No.201810101).

References

[1] Liu J, Zhang W, et al. 2016 Fuzzy logic controller for energy savings in a smart LED lighting system considering lighting comfort and daylight. *Energy & Buildings* **127** 95-104

[2] Chew I, Kalavallya V, et al 2016 Design of an energy-saving controller for an intelligent LED lighting system *Energy & Buildings* **120** 1-9

[3] Kim S, Kim I, et al 2014 Evaluation of optimized PV power generation and electrical lighting energy savings from the PV blind-integrated daylight responsive dimming system using LED lighting. *Solar Energy* **107** 746-757

[4] Honarmand M, Zakariazadeh A, Jadid S 2014 Integrated scheduling of renewable generation and electric vehicles parking lot in a smart microgrid. *Energy Conversion & Management* **86** 745-755

[5] Kam, Mart Van Der, and W. V. Sark 2015 Smart charging of electric vehicles with photovoltaic power and vehicle-to-grid technology in a microgrid; a case study. *Applied Energy* **152** 20-30

[6] Wang Y, Wang B, et al 2016 Energy management for a commercial building microgrid with stationary and mobile battery storage. *Energy & Buildings* **116** 141-150

[7] Jagadeesh Y, Akilesh S, Karthik S, et al 2015 Intelligent Street Lights. *Procedia Technology* **21** 547-551

[8] Sperber AN, Elmore AC, Crow ML, et al 2012 Performance evaluation of energy efficient lighting associated with renewable energy applications. *Renewable Energy* **44** 423-430

[9] Zhang YM, Ding B, Fu WD, et al 2015 Electrical Energy Conservation Based on DC Distribution and DC microgrid. *Transactions of China Electrotechnical Society*. **30** 389-397

Skylight Sizing based on balancing Daylighting Performance and Visual Comfort in Atrium Buildings

Jie Li[1], Xueming Chen[1], Qichao Ban[1,2] and Jiawei Yao[3]
[1] Xi'an Jiaotong-Liverpool University, Suzhou, China
[2] Qingdao University of Technology, Qingdao, China
[3] Tongji University, Shanghai, China
E-mail: jie.li@xjtlu.edu.cn

Abstract. As one of the most traditional and ubiquitous elements in architecture, atrium design has remained an imperative topic. With the growing concerns of energy conversation, carbon emission, comfort, and well-being recently, atrium design optimization to match the new crises and demands becomes a big challenge. This study investigated the impact of atrium shapes and building heights to daylighting performance. It was found that round atrium performs the best, next is square, and the worst one is the rectangle. And the higher the buildings are, the worse the daylight performance is. To further guide design work, generic atrium design criteria from a view of balancing daylight performance and visual comfort was proposed. The more reliable and valid annual dynamic simulation methods and climate-based daylighting modeling metrics are adopted.

Keywords: skylight sizing; atrium; daylighting; visual comfort; simulation

1. Introduction

Daylighting is a fundamental, primary and historic element in the architectural design. With the worldwide efforts in energy conservation and well-being nowadays, daylighting is brought to a new level. The benefits and advantages of daylighting have been examined and illustrated by a body of studies. Generally summarized is energy saving and health. The energy saving here is a broad-sense concept, could be expanded to economics, carbon emissions and global warming etc.[1-6]. For instance, Energy Center of Wisconsin indicated that a 22% annual lighting and HVAC cost savings were achieved by using daylighting in an experiment. A survey produced by Southern California Edison reported that the saving range of daylight was between 6% and 80%[1]. In the aspect of health, by hundreds of literature review, Edwards and Torcellini demonstrated that daylight could enhance learning efficiency, improve learning outputs, reduce patients' recovery time and increase turnover[5]. Evidence presented in studies that daylight plays a crucial role in vitamin D producing, which has a close relation to diseases like rickets, osteomalacia, depression and possible cancers[1, 6]. Besides, daylighting is also good to sleep quality and productivity.

Due to the prominent ability to draw daylight deep into the core of the building, atrium has become a ubiquitous architectural component in many building types [3, 7-9]. A number of investigations on atrium daylighting have been carried out in previous research, but most are focused on the office building, therefore, there is a strong demand on extending the study area to other types of building [10-12]. In other hands, along with the maximum of the glazing portion, unexpected discomfort glare is growing at the same time, so how to balancing these two ambivalent factors becomes a challenge. As a result, this study chooses commercial building type as object, to exam the relationship of atrium shape and glazing proportion to daylight performance, thereby further propose a skylight sizing criteria from a view of balancing daylighting illuminance level and discomfort glare for atrium design.

1.1. Daylighting assessment

Various daylighting performance investigating methods are developed in the past, which could be generalized into four types: Scale models, Mathematical models, Full-scale models, and Computer simulations [13-17]. Among all these methods, full-scale models are the most expensive, difficult and time-consuming one to implement because of full-size building construction while the results are always the most reliable and practical. By contrast, building of scale models is much easier in cost, but more technical issues raised up in a daylighting model along with the size scaling, like the geometry, material choice, furniture, etc. Previous research has shown that the data simulated from reliable software is highly in accord with the site measurement[15, 18], plus the economic, time effective and realistic visualization, etc. characteristics, computer daylight simulation has become a trend and popular tool currently [13]. Survey indicates that the total number of daylighting simulation program has reached to about 42 [4]， which can be classified into two categories generally, static simulation and annual dynamic simulation. Ecotect and Radiance are two typical and frequently used static modeling tool, annual dynamic simulation tools are developed in recent decades, DAYSIM and DIVA for Rhino are good examples.

1.2. Daylighting Metrics

1.2.1. Illuminance. Along with the exploration on daylighting, two main kinds of daylighting evaluating metrics are formed, Static Daylight Modeling Metrics and Dynamic or Climate-Based Daylight Modeling Metrics. Daylight factor (DF) was first developed in the early of the 20th century by Moon and Spencer in 1942, is the most traditional and prevalent static metric to predict daylight in building regulations. It is defined as the ratio between the daylight illuminance level at a target indoor point and the outdoor illuminance on the ground in the International Commission on Lighting (CIE) overcast sky mode[1]. However, this simple and easy method exists many limitations, for instance, avoidance of seasonal variations, time variations, and different sky conditions etc.[3, 19]. An example is that clear sky usually rarely appears in many places, like Boston[19]. In last decades, a series of time and climate based metrics emerged, which remedy the limitations of static daylight modeling metrics appropriately, is called Dynamic or Climate-based Modeling Metrics. The calculation span over the entire year and take the solar radiation data of the building located in. A couple of these metrics are as below:

- Daylight autonomy (DA): the percentage of a given time frame of a year when a minimum illuminance on a workplace is met by daylight[3, 18].
- Spatial daylight autonomy (sDA): is based on the floor area percentage instead of time percentage, represents the ratio of the floor area that reach the certain illuminance level for a specified amount hours. One frequently used indicator is $sDA_{(300,50\%)}$, which means the space percentage with over 300lux illuminance level for more than 50% time in the schedule[3, 20, 21]. This method usually is used to evaluate the daylight availability of a daylight space.
- Maximum Daylight Autonomy (DA_{max}): presents the area with illuminance is ten times than the target illuminance level, usually 5% of time is used as the line, which means when the sensor receiving over ten times illuminance is greater than 5% of the annual schedule time, it counted as maximum daylight autonomy[21, 22]. This indicator could be a kind of benchmark for uncomfortable high illuminance or glare.
- Useful daylight illuminance (UDI): calculates the time percentage of the interior daylight illuminance which within a specific range, 100 lux to 2000 lux[3, 18, 21, 23, 24]. As the name states, this approach suggests the illuminance both lower than 100 lux and over 2000 lux is not merited to consideration due to the poor visibility and visual discomfort, which was also the first metric to take visual comfort into daylighting modeling considerations[1].

1.2.2. Glare. Glare is the concept to measure the physical discomfort for light. To quantify the glare, a number of indices are developed during the past years, such as Daylight Glare Index (DGI), CIE Glare Index (CGI), Visual Comfort Probability (VCP), Unified Glare Rating (UGR) and Daylight Glare

IOP Conf. Series: Materials Science and Engineering **556** (2019) 012051 doi:10.1088/1757-899X/556/1/012051

Probability (DGP). As these metrics is invented from diverse experimental conditions, in different methods, and for various intentions, select an appropriate metric is still difficult. According to Jakubiec and Reinhart (2012), DGP is the most recommended one among all the above indices after a detail theoretical analysis and a computer-based software simulation[19], however, one nonnegligible limitation is that the glare calculation should base on a certain point and one specific view direction, so it is actually hard to interpret that how about if the occupant changing a position or facing to another direction.

2. Methodology

Annual dynamic and climate-based computer simulation was utilized as the major method of this study. Program DIVA for Rhino was entirely used, which is a highly advanced annual daylighting modeling plug-in. Its daylight performance calculation is based on Radiance backward ray tracing techniques, while discomfort glare evaluation utilizes Evalglare to calculate glare indices. It was initially developed at Harvard University and has been trusted and validated by the industry[21, 25].

2.1. Model description

Generic shape, dimensional proportions, and fenestration system are three main determinants to an atrium. According to the taxonomy of rectilinear form developed in 1980s, atrium can be identified by four types: four-sided atrium, three-sided atrium, and two-sided or linear atrium. Four-sided enclosed atrium is chosen as the study object, three most common forms, Rectangle, Square and Round, are taken in account, as illustrated in Figure 1.

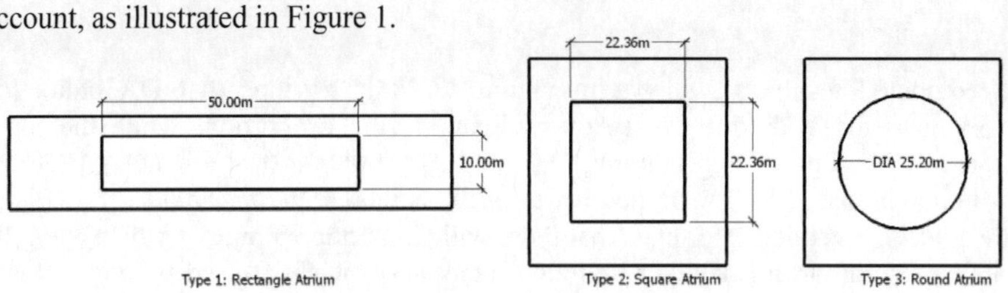

Figure 1. Plan of different type of atrium.

The variable is the percentage of skylight glazing area to atrium roof area. Meanwhile, five variations of building heights from 24m to 56m were set for each atrium type. Table 1 shows all the detail model information.

Table 1. Model variation of Rectangle, Square and Round atrium type.

Atrium Type	Height (*m*)	Atrium Size (Length/Width, *m*)	Skylight Size (Length/Width, *m*)				
			20%	40%	60%	80%	100%
Rectangle	24	10/50	4.47/22.36	6.32/31.62	7.75/38.73	8.94/44.72	10/50
	32	10/50	4.47/22.36	6.32/31.62	7.75/38.73	8.94/44.72	10/50
	40	10/50	4.47/22.36	6.32/31.62	7.75/38.73	8.94/44.72	10/50
	48	10/50	4.47/22.36	6.32/31.62	7.75/38.73	8.94/44.72	10/50
	56	10/50	4.47/22.36	6.32/31.62	7.75/38.73	8.94/44.72	10/50
Square	24	22.36/22.36	10/10	14.14/14.14	17.32/17.32	20/20	22.36/22.36
	32	22.36/22.36	10/10	14.14/14.14	17.32/17.32	20/20	22.36/22.36
	40	22.36/22.36	10/10	14.14/14.14	17.32/17.32	20/20	22.36/22.36
	48	22.36/22.36	10/10	14.14/14.14	17.32/17.32	20/20	22.36/22.36
	56	22.36/22.36	10/10	14.14/14.14	17.32/17.32	20/20	22.36/22.36
Round	24	12.6	5.6	8	9.8	11.3	12.6
	32	12.6	5.6	8	9.8	11.3	12.6
	40	12.6	5.6	8	9.8	11.3	12.6
	48	12.6	5.6	8	9.8	11.3	12.6
	56	12.6	5.6	8	9.8	11.3	12.6

To keep the comparability among three types, assume atrium floor area of the three models the same, adjoining space area is in a fixed ratio with the atrium floor area, and the aperture type of skylight is horizontal. As the pedestrian corridor in the atrium is an essential element of a commercial building, a 4m wide corridor is set in the model as the import analysis area.

2.2. Metric

Three annual dynamic daylight metrics are chosen to analysis illuminance and glare. sDA and UDI are used to shown the daylight results, some comparison is carried to get a mutual validation. Since the above-discussed limitation of DGP and comparison demands in the subsequent analysis, DA_{max} is chosen to indicate the glare level.

2.3. Simulation settings

Essential assumptions are set for the model. As is well known, surface reflectance is the key factor to affect the daylight distribution in spaces, generic materials surface reflectance is adopted in the simulation, with floor at 20%, outdoor wall at 35%, ceiling at 80% and skylight at 65% transmittance double glazing pane low-E. The atrium partition between atrium and adjoining is typically transparent glazing in commercial, so it is assumed as 88% transmittance single glazing pane. The schedule is set from 10 am to 8 pm every day, totally 3650 hours are calculated for each case of a year. Shanghai is utilized as the weather location, weather data is downloaded from EnergyPlus weather source in TMY3 format[26].

3. Results

The overall computed results are shown in Figure 2. It is obvious that DA300lux [50%] and DA_MAX>5% is rising with the growth of skylight glazing proportion, while the increase rate decreases from low building to high building. And both DA300lux [50%] and DA_MAX>5% are the highest in 24m height and reduce with the growth of the height. *UDI100-2000lux* rises distinct at the beginning then become gentle, it portends that there will be a critical point, at which even the glazing area becomes larger, the useful daylight will stop increase, which is caused by the exclusion of the spaces with illuminance lower than 100lux and higher than 2000lux.

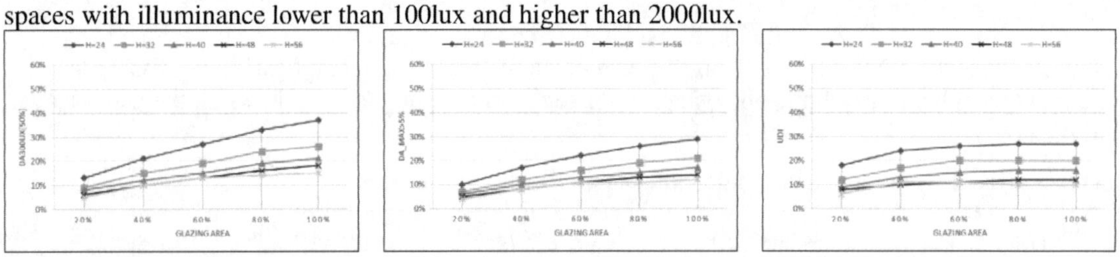

(a) Rectangle atrium

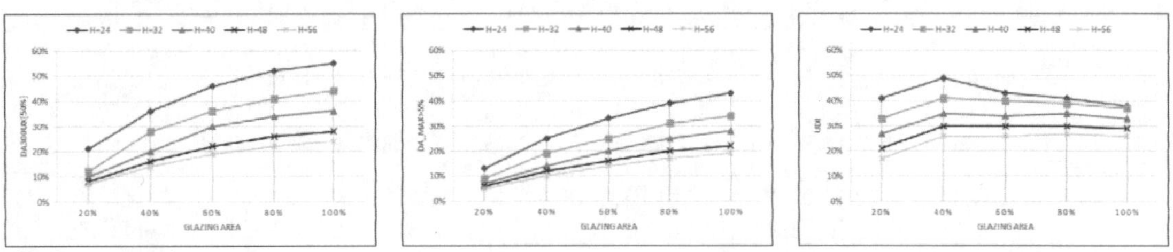

(b) Square atrium

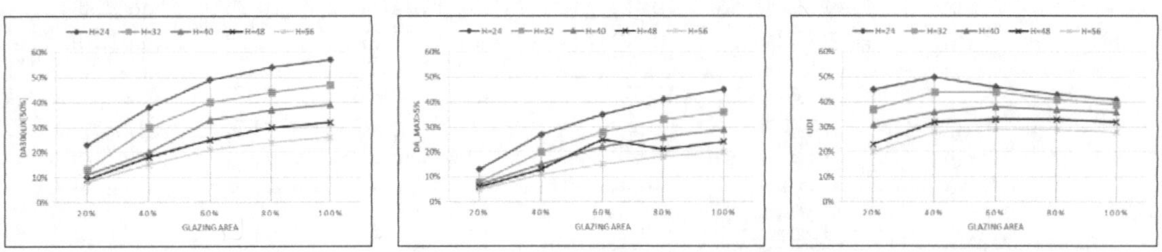

(c) Round atrium

Figure 2. Daylight metrics of various building heights.

To show the impact atrium shapes to daylighting, averages of five building heights are calculated, as offered in Figure 3. It can be seen seemingly that round atrium performs the best, square is a little weaker than round but quite close, rectangle is the worst one.

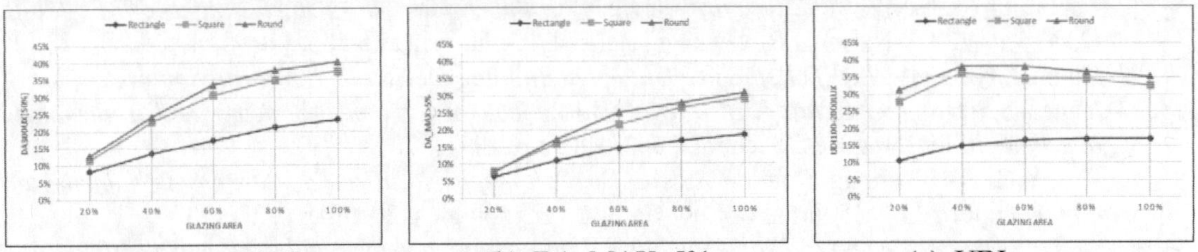

(a) DA_{300lux} [50%] (b) DA_MAX>5% (a) $UDI_{100\text{-}2000lux}$

Figure 3. Daylight performance comparison of atrium types.

To find out the relationship between the two increases, the growth rate is introduced. The principles are when the growth rate of illuminance is higher than glare, we deem that the increase is positive and advantageous, and when it is contrary, the growth is negative and not suggested, thus the intersection point is the critical point of the design glazing proportion. By calculating and analyzing the growth rates of the two selected metrics, design criteria of skylight is summarized in table 2. One significant point should be highlighted is that the design criteria doesn't mean the glare will be completely avoided, thus shading or other effective strategies should still be taken into consideration in actual design process.

Table 2. Skylight design criteria.

	H=24m	H=32m	H=40m	H=48m	H=56m
Rectangle	100%	100%	100%	100%	100%
Square	80%	75%	78%	80%	80%
Round	78%	75%	80%	98%	80%

4. Discussion and Conclusion

Overall, this paper reveals the relationship between skylight glazing proportion and daylight performance, the influence of building height to daylight level and the contribution of atrium geometry to daylight illuminance, and proposes a skylight design criteria from a view of both maximizing daylight illuminance and minimizing discomfort glare of indoor environment, intend to provide designers a guideline on skylight area sizing. In conclusion, the findings of this study are as following:

- Atrium geometry is a significant effect factor to the daylight performance, by contrast, round atrium shows best, square atrium comes second, and rectangle takes the last.
- With the increase of building height, both illuminance and glare are reducing, and the impact of glazing area to daylight performance becomes smaller and smaller at the same time.
- The skylight design criteria based on the purpose both maximizing daylight illuminance and minimizing discomfort glare in design stage is proposed for three types of atrium geometry.

The annual and climate-based method makes the conclusions more reliable and valid, but the deficiencies of glazing area interval problem, limited geometry types, and integrating energy use etc. are existing as well, therefore, there are further and extensional work could be done in the future, we would welcome replication and extension.

5. References

[1] B, M., *Daylighting Design: Planning Strategies and Best Practice Solution*. 2014, Basel: Birkhauser Verlag GmbH.

[2] W, I.L., *A review of daylighting design and implementation in buildings.* Renewable and Sustainable Energy Reviews, 2017(74): p. 959-968.
[3] H, M.M.a.J., *Assessing daylight performance in atrium buildings by using Climate Based Daylight Modeling.* Solar Energy, 2015(119): p. 553-560.
[4] W, W., *The Key Research Points of International Daylighting.* Zhaoming Gongcheng Xuebao, 2016. **27**(4): p. 95-104.
[5] Torcellini, L.E.a.P., *A Literature Review of the Effects of Natural Light on Building Occupants*. 2002, National Revewable Energy Laboratory: Colorado.
[6] Council, U.S.G.B., *LEED Reference Guide for Building Design and Construction v4*. 2013.
[7] Yang, S.S.a.F., *daylighting in atria: The effect of atrium geometry and reflectance distribution.* Lighting Res. Technol, 2007. **39**(2): p. 147-157.
[8] Ran Y, L.S., Yuehong S and Saffa R, *Daylighting performance of atriums in subtropical climate.* International Journal of Low-Carbon Technologies, 2009(4): p. 230-237.
[9] S, S., *A critical review of articles published on atrium geometry and surface reflectances on daylighting in an atrium and its adjoining spaces.* Architectural Science Review, 2010(53): p. 145-156.
[10] R, B.P. *Lighting quality: the unanswered questions* in *First CIE Symposium on Lighting Quality CIE*. 1998. Vienna: CIE Central Bureau.
[11] A, G.A.D.a.V.J., *Occupant preferences and satisfaction with the luminous environment and control systems in daylit offices: a literature review.* Energy and Buildings, 2006. **7**(38): p. 728-742.
[12] P, B., *Lighting quality for all*, in *Session 3 Keynote Speaking, CIBSE & SLL International Lighting Conference Dublin 2013*. 2013: Ireland.
[13] CIBSE, *Lighting guide 10: daylighting - a guide for designers*. 2014: London.
[14] Aghemo C, P.A.L.V., *The approach to daylighting by scale models and sun and sky simulators: a case study for different shading systems.* Build Environ, 2008(43): p. 917-27.
[15] Chen YY, L.J., Pei JJ, Cao XD, Chen QY and Jiang Y., *Experimental and simulation study on the performance of daylighting in an industrial building and its energy saving potential.* Energy Build, 2014(73): p. 184-91.
[16] Li DHW, C.A., Chow SKH and Lee EWM. , *Study of daylight data and lighting energy savings for atrium corridors with lighting dimming controls.* Energy Build, (72): p. 457-64.
[17] T, K., *Fuzzy logic model to classify effectiveness of daylighting in an office with a movable blind system.* Build Environ, 2013(69): p. 22-23.
[18] A, R.C.a.F., *Findings from a survery on the current use of daylight simulations in building dsign.* Energy Build, 2006(38): p. 824-35.
[19] Reinhart, J.J.a.C., *The 'adaptive zone' - A concept for assessing discomfort glare throughout daylit spaces.* Lighting Res. Technol, 2012(44): p. 149-170.
[20] Society, I.E., *IES LM-83-12 - Approved Method: IES Spatial Daylight Autonomy (sDA) and Annual Sunlight Exposure (ASE)*. 2012.
[21] *Rhino Training*. [cited 2018 5.23]; Available from: http://solemma.net/TrainingRhino.html.
[22] Y, P., *Lighting Environment Simulation of Building*. 2010, Beijing: China Building Industry Presss.
[23] Nabil, A., J. Mardaljevic, *Useful Daylight Illuminance: A Replacement for Daylight Factors.* Energy and Buildings, 2006. **38**(7): p. 905-913.
[24] Dubois, F.C.a.M.-C., *Daylighting metrics based on illuminance, distribution, glare and directivity.* Lighting Res. Technol, 2011(43): p. 291-307.
[25] *DIVA User Guide*. [cited 2018 5.23]; Available from: http://diva4rhino.com/user-guide.
[26] *weather data*. [cited 2018 0525]; Available from: https://energyplus.net/weather.

Movable insulation in building integrated semi-transparent photovoltaic thermal (BiSPVT) system

G K Mishra[1*], **G N Tiwari**[2] **and T S Bhatti**[1]
[1]Center for Energy Studies, Indian Institute of Technology Delhi, New Delhi, India
[2]Bag Energy Research Society (BERS), Ghaziabad, India

E-mail: mishra.gourav@gmail.com (G K Mishra)

Abstract. The BiSPVT system involves an integration of semi-transparent photovoltaic (SPV) modules with buildings along with the provision of harvesting thermal energy. The system generates electrical and thermal energy for the building as well as provides day-lighting. However, during summer season the production of thermal energy is higher and thermal demand is lesser as compared to winter season. Also, the higher thermal energy production is unfavourable for the generation of electrical energy by SPV modules. The use of movable insulation (MI) on glazed wall can be one of the method to reduce the thermal energy generation. In this paper, a thermal model has been developed for BiSPVT system and the effect of MI on thermal performance of the system is studied. The energy balance equations are obtained for the inclined SPV modules (top), reinforced cement concrete (RCC) floor (bottom), air of room (SPV modules, floor and four brick/glazed side walls). The room temperature and SPV cell temperature are derived as a function of a) climatic parameters (solar irradiance and ambient temperature), b) design parameters and c) heat transfer coefficients by using the energy balance equations. It is found that by using MI with less or no air cavity thickness, the temperature of room has reduced by 1.93°C. However, the effect of MI on SPV cell temperature is marginal. The proposed model can be utilized to evaluate cell and room temperatures for BiSPVT system installed at different places of the world provided solar irradiance and ambient temperature of that place are known.

Keywords: Building integrated semi-transparent photovoltaic thermal (BiSPVT) system, Thermal modelling, Movable insulation, Photovoltaic system.

1. Introduction

The energy demand of building sector accounts for approximately 30% and due to urbanisation this percentage is increasing annually by 8% [1]. The significant energy demand results in higher contribution of greenhouse gas emissions. However, the energy saving potential of the building sector will result in impactful reduction of energy consumption. The incorporation of the passive techniques like cross ventilation, trombe wall, orientation, earth shelter, wind tower, shading etc. will curtail the energy demand. In view of this, the energy generation within the building by using photovoltaic (PV) modules is in the trend. But again, the creation of thermal energy during the conversion of low grade solar energy into electrical energy limits the efficiency of the PV systems [2]. Therefore, to make the thermal energy into usable form photovoltaic thermal (PVT) systems were developed, which increases the overall efficiency of PV systems [3]. At the same time, it also reduces the temperature of PV cells thereby increasing its conversion efficiency [4]. Many researchers have integrated PVT systems with building by using opaque PV modules to bring the generating unit near electrical and thermal load [5,6]. Further the researchers have found that for identical packing factor and dimension of semi-transparent PV (SPV) modules and opaque PV (OPV) modules, the former is more efficient. In addition to higher efficiency, SPV modules also provide daylighting. The integration of semi-transparent PV modules with buildings along with the provision to harvest the generated thermal energy is termed as building integrated photovoltaic thermal (BiSPVT) system. In consideration of energy savings, the responsible authority is watchful on reducing the transfer of heat by increasing the insulation levels [7]. However, in the region of temperate climate where summers are too hot, the use of BiSPVT system with high level of insulations reduces the occupant thermal comfort [8]. Yang and Tian [9] had found that solar shading is a productive measure of increasing the comfort level since the glazed walls are usually the lowest performing part of building in controlling excess heat gain and energy loss. Many studies have suggested that by using optimal overhangs depth blocks solar beam irradiance of altitude sun trajectories in summer season and also allow the solar irradiance of low altitude sun trajectories in winter season. Thereby curtails the cooling load of summers and heating load demand of winters [10]. However, the solar diffuse irradiance are not blocked in summers due to which an occupant thermal comfort is compromised. On the other hand, movable insulation can be adjusted to block global solar irradiance in summer season and allow the global solar irradiance in winter season. Stazi et. al [11]found the potential of glazed walls in heating energy saving and thermal comfort in winter and intermediate seasons. Moreover, they suggested the use of double glazing to improve the performances. Further, the drawback of the glazing regarding the risk of overheating during summer season is also mentioned. Chen and Zheng [12] have analysed the feasibility of a microporous ceramic board based movable insulation in boilers to maintain a high constant temperature near the burner zone. Arinze et. al. [13] stated that in average climatic conditions, by using the movable thermal insulation between the double glazed greenhouses the heating requirement is reduced by as much as 60 to 80% in extremely cold regions. Yao [14] investigated the effect of movable insulation installed at south wall of a residential building on indoor thermal and visual comfort.

At present, there are only a few literatures on movable insulation installed in buildings which covers the vital thermal comfort. Furthermore, the performance of movable insulation is very much dependent on climatic condition. Hence the present work is done to bridge the gap between the use of MI in BiSPVT system and thermal comfort. A validated theoretical model has been developed for the BiSPVT system installed at Sodha BERS Complex (SBC), Varanasi, India.

Table 1. Nomenclature

U_{tca}	Overall heat transfer coefficient from top of SPV cell to ambient ($Wm^{-2}K^{-1}$)	U_{sba}	Overall heat transfer coefficient from bricked side walls to ambient ($Wm^{-2}K^{-1}$)
U_{bcr1}	Overall heat transfer coefficient from bottom of SPV cell to room ($Wm^{-2}K^{-1}$)	U_{sga}	Overall heat transfer coefficient from glazed side walls to ambient ($Wm^{-2}K^{-1}$)
h_{cfr1}	Convective heat transfer coefficient from floor to room ($Wm^{-2}K^{-1}$)	$U_{sga,MI}$	Overall heat transfer coefficient from glazed side walls to ambient with MI ($Wm^{-2}K^{-1}$)
U_{bcr2}	Overall heat transfer coefficient for floor to room 2 ($Wm^{-2}K^{-1}$)	C(ac)	Air conductance at various depth of cavity ($Wm^{-2}K^{-1}$)
α	Absorptivity	V_1	Volume of room (m^3)
τ	Transmissivity	N	Number of air change of room
β_m	Packing factor of SPV module	η_m	Efficiency of SPV module
β_k	Ratio of glass to total wall area	A	Area (m^2)
M_{a1}	Mass of air in room (kg)	K	Thermal conductivity of materials ($Wm^{-1}K^{-1}$)
C_{a1}	Specific heat of air in room ($Jkg^{-1}K^{-1}$)	L	Thickness of materials (m)
c	solar cell	gc	glass used in solar cell module
f	floor of room	gw	glass used in glazed side wall
g	Glass	bw	plastered brick used in wall
k	East, west , north, south	r	Roof formed by SPV modules
RCC	Reinforced cement concrete floor	2	Room-2
cur	Curtain (Movable insulation)	1	Room
s	south	w	wall

2. System description:

The schematic diagram of BiSPVT installed on the third floor (terrace) of Sodha BERS Complex (SBC), Varanasi, India is shown in Figure 1. The figure also outlines the dimensions, number of semi - transparent PV modules, glazing portion of side walls etc. In the present theoretical analysis, the temperature of room2 (2nd floor) underneath the room shown is taken constant because it is air conditioned. The reinforced cement concrete (RCC) beam structure is made to support the closed structure and withstand the heavy thrust of wind. The SPV modules are tightened with the RCC beams with the help of aluminium frame. The peak power rating of the system is 7.2 kW_p ($75W_p \times 96$), which has three power conditioning units (2 kVA, 48 V rating each), three set of batteries (48 V, 150 Ah rating each). The total area and ratio of glazed to bricked area of side walls facing east, west, north, south direction is given in Table 2. To remove hot air of the room, two exhaust fans (100 W, 220 V rating each) are installed.

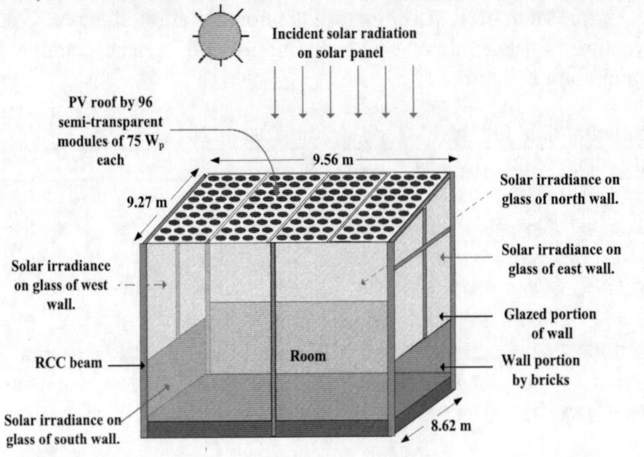

Figure 1. Schematic diagram of the BiSPVT system installed at Sodha BERS Complex, Varanasi.

Table 2. Design parameters and heat transfer coefficients

Parameters	Values	Parameters	Values	Parameters	Values	Parameters	Values	Parameters	Values
β_r	0.05	L_{cur}	0.003	K_{cur}	0.06	A_r	88.66	$U_{sga,MI}$(0 ac)	12.30
β_m	0.85	L_{bw}	0.1	τ_g	0.85	α_c	0.85	$U_{sga,MI}$(1cm ac)	4.03
β_e	0.64	L_{gc}	0.003	η_c	0.09	α_f	0.6	$U_{sga,MI}$(3cm ac)	3.55
β_w	0.44	L_{gw}	0.004	A_f	82.41	U_{tca}	19.77	C(1 cm ac)	6
β_n	0.12	K_{gc}	1.1	A_e	41.06	U_{bcr1}	2.77	C(3 cm ac)	5
β_s	1	K_{gw}	1.1	A_w	41.06	U_{sba}	18.93		
T_{r20}	20	K_{bw}	3.98	A_n	64.13	U_{bcr2}	4.7		
L_{RCC}	0.2	K_{RCC}	1.28	A_s	27.05	U_{sga}	31.91		

3. Thermal Modelling

The assumptions made for the development of theoretical thermal model are as follows:

- The heat is transfer from various nodes of the system namely SPV roof, side walls and floor is one dimensional and the system is in quasi-steady state.
- The thermal properties of the room air is constant with no stratification.
- The radiative heat transfer between walls and room is negligible due to small temperature difference.

The energy balance equations used in the formulation of theoretical model are as follows:

At cell of SPV module:

$$\alpha_c\tau_g\beta_m G_{inc} = U_{tca}(T_c - T_a) + U_{bcr1}(T_c - T_r) + \eta_m \tau_g\beta_m G_{inc} \quad (1)$$

At floor of the room:

$$\alpha_f\tau_g^2(1-\beta_m)(1-\beta_r)A_r G_{inc} + \alpha_f\tau_g\Sigma A_k\beta_k G_k = h_{cfr1}(T_f - T_r)A_f + U_{bcr2}A_f(T_f - T_{r20}) \quad (2)$$

At air of room:

$$h_{cfr1}(T_f - T_r)A_f + U_{bcr1}(T_c - T_r)\beta_m(1-\beta_r)A_r + (1-\alpha_f)\tau_g^2(1-\beta_m)(1-\beta_r)A_r G_{inc}$$
$$+(1-\alpha_f)\tau_g\Sigma A_k\beta_k G_k = M_{a1}C_{a1}\frac{dT_r}{dt} + \Sigma A_k\beta_k U_k(T_r - T_a) + 0.33NV_1(T_r - T_a) \quad (3)$$

The Eq. (1)-(3) are solved to get the expression of T_{rl} and T_c as a function of input climatic parameters i.e solar irradiance and ambient air temperature. Hence,

$$\frac{dT_{r1}}{dt} + P_1 T_r = f(t) \quad (4)$$

$$\text{where, } f(t) = P_2 G_{inc} + X_3 \sum A_k \beta_k G_k + P_3 T_a + X_{40}$$

$$T_c = Z_1 T_a + Z_2 G_{inc} + Z_3 T_r \quad (5)$$

Eq. 4 is solved in MATLAB to get the expression of T_r. After getting the non-differential expression of T_r, the expression of SPV cell temperature (T_c) is evaluated with the help of Eq. 5. The details of expressions are defined in Appendix A and nomenclature.

Table 3. The obtained hourly experimental data for BiSPVT system on 19/Jan/2018.

Time	1	2	3	4	5	6	7	8	9	10	11	12	13	14	15	16	17	18	19	20	21	22	23	24
G_{inc}	0	0	0	0	0	0	20	80	217.5	427.5	620	745	795	747.5	582.5	382.5	157.5	10	0	0	0	0	0	0
G_e	0	0	0	0	0	0	110	320	517.5	622.5	562.5	367.5	200	150	125	95	55	15	0	0	0	0	0	0
G_w	0	0	0	0	0	0	10	50	90	110	125	190	200	150	172.5	202.5	125	20	0	0	0	0	0	0
G_n	0	0	0	0	0	0	5	15	32.5	37.5	42.5	47.5	50	50	47.5	42.5	27.5	7.5	0	0	0	0	0	0
G_s	0	0	0	0	0	0	15	62.5	147.5	280	422.5	532.5	595	615	380	117.5	62.5	15	0	0	0	0	0	0
T_a	16	15.6	15.1	14.5	13.5	12.5	12.3	13.7	15.8	18.3	21.3	23.8	25	25.5	25.9	25.6	24.1	22.5	21.5	20.3	19	18	17.24	16.5
$T_{r,exp}$	17	16.5	16	15.5	15	14	14.2	17	19.3	22.1	24.9	26.5	27.9	28.9	28	27.2	23.2	22.8	22	21.4	20.4	19.8	18.8	17.8
$T_{c,exp}$	-	-	-	-	-	-	-	17.2	25.2	35.3	41.1	42.9	45.4	43.8	37.2	32.4	23.9	-	-	-	-	-	-	-

4. Methodology:

(i) The hourly experimental data of solar irradiance (at inclined SPV roof, vertical side walls) and ambient air temperature were taken on Jan 19th 2018 by using solarimeter (Central electronic limited, India) and two laboratory thermometers (ZEAL, United Kingdom). The experiments were performed on Jan 19th 2018 at the site and the mentioned day was a clear day during the end of winter season in India. The experimental data is shown in Table 3.

(ii) To evaluate theoretical temperature of room (T_r) and SPV cell (T_c), computational codes were developed in the MATLAB 2015 software by using the thermal model described above.

(iii) The theoretical T_r and T_c were validated by the experimental data of the same which are also mentioned in Table 3. The hourly experimental data of T_r and T_c were taken by four laboratory thermometers (ZEAL, United Kingdom) and infrared thermometer (Fluke, U.S.A) respectively.

(iv) The good correlation between the experimental and theoretical temperatures validate the thermal model. Further, the effect of movable insulation on room and cell temperature were evaluated by using the thermal model.

(v) Moreover, to analyse the effect of MI in hot climatic condition of summer (Apr 03rd 2017, clear day) also, the computation has been done.

5. Result and discussion:

Figure 2 shows the hourly room air temperature on 19/Jan/2018 obtained through the theoretical model and experimental observations. The values of percentage root mean square deviation (e) and correlation coefficient (r) between the theoretical and experimental data are 6.43% and 0.99 respectively.

Figure 3 shows the hourly variation on 19/Jan/2018 of SPV cell temperature obtained theoretically

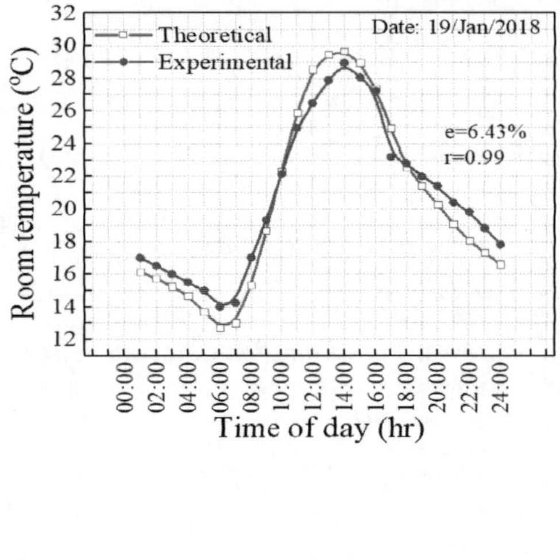

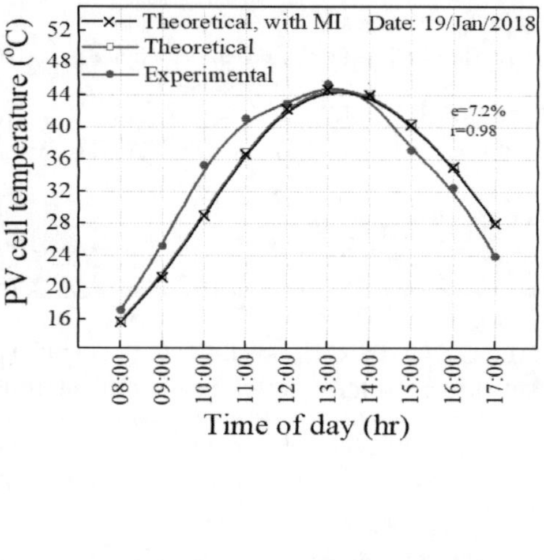

IOP Conf. Series: Materials Science and Engineering **556** (2019) 012052 doi:10.1088/1757-899X/556/1/012052

and experimentally. The variation of SPV cell temperature by applying movable insulation (MI) for intermittent duration i.e. 0700 to 1800 hour is also shown. By applying MI, the decrease in SPV cell temperature is negligible, thereby the increase in SPV electricity generation is marginal. The value of 'e' and 'r' are 7.2% and 0.98 respectively. The values of 'e' and 'r' shown in Figures 2 and 3 proofs the authenticity of the theoretical thermal model.

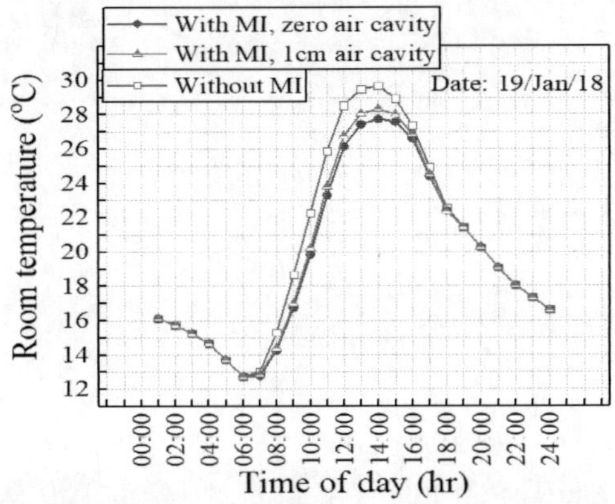

Figure 4. Effect of MI air cavity on the room temperature.

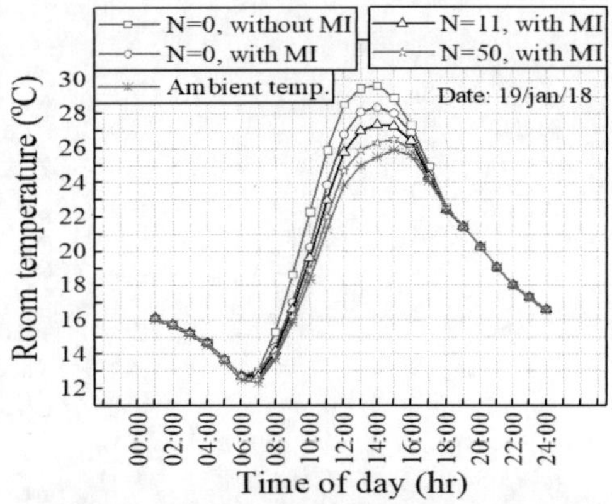

Figure 5. Effect of MI and number of air change on room temperature.

Through the validated thermal model, the effect of movable insulation on room temperature has been evaluated. The movable insulation (MI) is applied on glazed portion of side walls for intermittent duration (0700 to 1800 hours). Figure 4 depicts the plots of hourly room temperature of the BiSPVT system under various condition namely i) without MI, ii) with MI having zero air cavity and iii) with MI having 1 cm air cavity. The maximum temperature under the conditions are 29.66°C, 27.73°C and 28.32°C respectively. It has been found that MI with zero air cavity offers highest decrement in room temperature. This is because the glazed wall with zero air cavity has higher heat transfer coefficient ($U_{sga,MI}(0\ ac)$), hence the thermal conductance from room to ambient is higher in comparison to MI having 1cm ($U_{sga,MI}(1cm\ ac)$) air cavity.

Further, to decrease the room temperature the help of exhaust fan has been taken. Figure 5 shows the plots of hourly room temperature of the BiSPVT system under various condition namely i) number of air change (N)=0 and without MI, ii) N=0 and with MI, iii). N=11 and with MI and iv) N=50 and with MI. The MI used in this plots are with 3 cm air cavity ($U_{sga,MI}(3cm\ ac)$) and both N, MI are applied for intermittent duration as mentioned above. The maximum temperature under the conditions are 29.66°C, 28.37°C, 27.37°C and 26.30°C respectively. The decrement is observed with number of air change but it has also a limit that it cannot decrease the room temperature more than ambient temperature.

Further, the potential of MI combined with N to decrease the room temperature for a day in summer (03/Apr/2017) has been analysed as shown in Figure 6. The various conditions such as, MI material, depth of air cavity and applied duration of N/MI are the same as discussed in figure 5. The maximum temperature under the conditions are 43.68°C, 42.61°C, 41.32°C and 39.99°C respectively. By comparing the figures, it has been found that the decrement in room temperature by applying MI on 03/Apr/2017 is lesser in comparison to 19/Jan/2018 whereas the effect of number of air change is higher. Hence, the decrement offered by applying MI and N are very much dependent on climatic conditions.

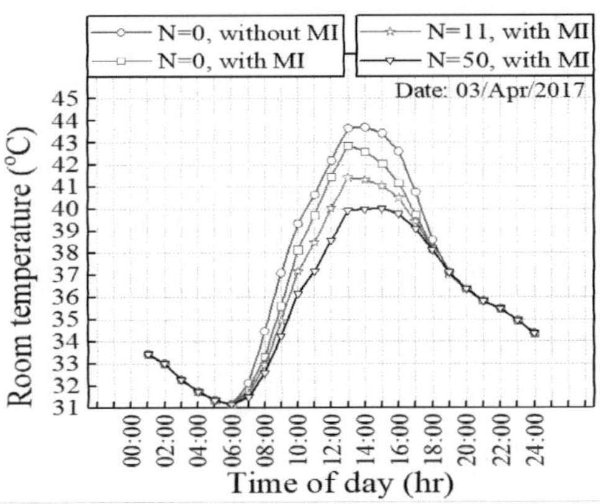

Figure 6. Effect of MI and number of air change on room temperature on 03/Apr/2017.

6. Appendix

$S_2 = \sum_{k=e,w,n,s}[A_k U_{sga}\beta_k + A_k U_{sba}(1-\beta_k)]$ A1

$Q_2 = h_{cfr1}A_f + U_{bcr2}A_f$ A2

$Q_5 = h_{cfr1}A_f/Q_2$ A3

$R_2 = 0.33NV_1 + S_2$ A4

$A_c = A_r(1-\beta_r)\beta_m$ A5

$X_1 = R_2 - h_{cfr1}A_fQ_5 + h_{cfr1}A_f + U_{bcr1}A_c$ A6

$U_{tc} = U_{tca} + U_{br1}$ A7

$Z_3 = U_{bcr1}/U_{tc}$ A8

$X_5 = U_{bcr1}A_c/(M_{a1}C_{a1})$ A9

$P_1 = X_1/(M_{a1}C_{a1}) - Z_3X_5$ A10

$R_1 = (1-\alpha_f)\tau_g^2(1-\beta_m)(1-\beta_r)A_r$ A11

$Q_1 = \alpha_f\tau_g^2(1-\beta_m)(1-\beta_r)A_r$ A12

$Q_3 = Q_1/Q_2$ A13

$X_2 = (h_{cfr1}A_fQ_3 + R_1)/(M_{a1}C_{a1})$ A14

$Z_2 = (\alpha_c\tau_g\beta_m - \eta_c\tau_g\beta_m)/U_{tc}$ A15

$P_2 = X_2 + Z_2X_5$ and $Z_1 = U_{tca}/U_{tc}$ A16

$X_6 = R_2/(M_{a1}C_{a1})$ and $P_3 = Z_1X_5 + X_6$ A17

$Q_4 = \alpha_f\tau_g/Q_2$ A18

$X_3 = (h_{cfr1}A_fQ_4 + (1-\alpha_f)\tau_g)/(M_{a1}C_{a1})$ A19

$Q_0 = U_{bcr2}A_fT_{r20}$ A20

$Q_{60} = Q_0/Q_2$ A21

$X_{40} = (h_{cfr1}A_fQ_{60})/(M_{a1}C_{a1})$ A22

$U_{tca} = [1/h_{0t} + L_{gc}/K_{gc}]^{-1}$ A23

$h_{0t} = 5.7 + 3.8\,v_{at}$, $v_{at} = 4$ m/s A24

$U_{bcr1} = [1/h_{cfr1} + L_{gc}/K_{gc}]^{-1}$ A25

$h_{cfr1} = 2.8 + 3v_{r1}$, $v_{r1} = 0$ m/s A26

$U_{bcr2} = [1/h_{cfr2} + L_{RCC}/K_{RCC}]^{-1}$ A27

$h_{cfr2} = 2.8 + 3v_{r2}$, $v_{r2} = 5$ m/s A28

$U_{sga} = [1/h_{0s} + L_{gw}/K_{gw}]^{-1}$ A29

$U_{sba} = [1/h_{0s} + L_{bw}/K_{bw}]^{-1}$ A30

$U_{sga,MI} = [L_{cur}/K_{cur} + 1/C(dep) + 1/h_{0s} + L_{gw}/K_{gw}]^{-1}$ A31

where, $h_{0s} = 5.7 + 3.8\,v_{as}$, $v_{as} = 8$ m/s A32

7. References

[1] Babu K R and Vyjayanthi C 2017 Implementation of net zero energy building (NZEB) prototype with renewable energy integration *IEEE Region 10 Symposium (TENSYMP) 2017* pp. 1-5.

[2] Notton G, Cristofari C, Mattei M and Poggi P 2005 Modelling of a double-glass photovoltaic module using finite differences *Applied Thermal Engineering* Vol 25 pp. 2854 - 77.

[3] Kim J H and Kim J T 2012 A simulation study of air-type building-integrated photovoltaic-thermal system *Energy Procedia* Vol 30 pp. 1016 – 24.

[4] Chen F and Yin H 2016 Fabrication and laboratory-based performance testing of a building-integrated photovoltaic-thermal roofing panel *Applied Energy* Vol 177 pp. 271–84.

[5] Chow T T 2010 A review on photovoltaic/thermal hybrid solar technology *Appl. Energy* Vol 87 pp. 365–79.

[6] Agrawal B and Tiwari G N 2010 Optimizing the energy and exergy of building integrated photovoltaic thermal (BiPVT) systems under cold climatic conditions *Appl. Energy* 87 pp. 417–26.

[7] IEA. Technology roadmap - energy efficient building envelopes. International Energy Agency; 2013.

[8] Chvatal S K M and Corvacho H 2009 The impact of increasing the building envelope insulation upon the risk of overheating in summer and an increased energy consumption *Journal of Building Performance Simulation* Vol 2 pp. 267-82.

[9] Yu J, Yang C and Tian L 2008 Low-energy envelope design of residential building in hot summer and cold winter zone in China *Energy Buildings* Vol 40 pp. 1536-46.

[10] Feuermann D and Novoplansky 1998 A Reversible low solar heat gain windows for energy savings *Solar Energy* Vol 62 169-75.

[11] Stazi F, Mastrucci A and Perna C D 2012 The behaviour of solar walls in residential buildings with different insulation levels: An experimental and numerical study *Energy and Buildings* Vol 47 pp. 217–29.

[12] Chen D and Zheng C 2002 Experimental investigation on the feasibility of a movable heat-insulation device *Applied Thermal Engineering* Vol 22 pp. 1905–18.

[13] Arinze E A, Schoenau G J and Besant R W 1986 Experimental and computer performance evaluation of a movable thermal insulation for energy conservation in greenhouses, *Jour of agricultural Engineering* Vol 34 pp. 97-113.
[14] Yao J 2014 An investigation into the impact of movable solar shades on energy, indoor thermal and visual comfort improvements *Building and Environment* Vol 71 pp. 24-32.

IOP Conf. Series: Materials Science and Engineering **556** (2019) 012053 doi:10.1088/1757-899X/556/1/012053

Challenges and research needs in life cycle analysis of building-integrated photovoltaic

Xiaoyu Yan[1,2]
[1]Environment and Sustainability Institute, University of Exeter, Penryn, UK
[2]College of Engineering, Mathematics and Physical Sciences, University of Exeter, Penryn, UK

E-mail: xiaoyu.yan@exeter.ac.uk

Abstract. Building-integrated photovoltaic (BIPV) is a promising solar energy technology that looks set to grow in popularity in the pursuit of a sustainable future. It has the potential to mitigate some of the main concerns over ground-mounted solar energy systems such as land use. However, there is an apparent gap in our understanding of its life cycle environmental impacts. Very few life cycle analysis (LCA) studies have evaluated BIPV comprehensively in comparison with standalone PV systems and other energy technologies. In this paper, we review the limited existing LCA studies on BIPV and identify the challenges and future research needs. The findings will help researchers, industries and policy makers better understand the environmental sustainability of BIPV to facilitate its development.

Keywords: solar energy; building-integrated photovoltaic; life cycle analysis; environmental sustainability

1. Introduction

Building-integrated photovoltaics (BIPV) is an emerging renewable energy technology that has great potential for meeting a significant portion of the electricity needs of cities globally [1,2]. It also has the potential to mitigate some of the main concerns over ground-mounted solar energy systems such as land use, which may result in competition with food production or ecological impacts. However, the overall environmental performance of BIPV is less well understood compared with standalone solar farms or building-applied photovoltaics (BAPV) such as roof-mounted systems. Life cycle analysis (LCA) is a key tool to evaluate the environmental sustainability of any products or technologies and has been applied extensively in energy [3–5]. Here, we review the limited existing LCA studies on BIPV and identify the challenges and further research needs in this area.

2. Review of the literature

The literature search using keywords including building integrated, solar energy, PV, environmental impacts and LCA resulted in more than 30 publications. There are studies investigating BI solar thermal energy systems (e.g., [6]) and solar PV and thermal (PVT) systems (e.g., [7]). These are excluded from the present review as PV only systems were analysed. Some of the studies investigated BAPV instead of BIPV and therefore are excluded as well.

Li et al. (2018) presented an LCA of a novel high optical performance low-concentration concentrator PV module for building south wall integration in China [8]. However, it is unclear whether the PV system evaluated was BIPV or BAPV as no details of the integration were given in the paper. In addition, there was no mentioning of the PV system replacing conventional building materials or influencing building energy performance.

Jayathissa et al. (2016) assessed a novel Adaptive Solar Façade based dynamic PV system in comparison with existing static PV systems [9]. Although the authors used BIPV to describe this technology in the paper, it is very difficult to justify that as the system was added to existing buildings.

Wang et al. (2016) performed a life cycle energy and GHG analysis of a 10 kW monocrystalline silicon based BIPV system as roof for a test building in Shanghai [10]. They found that the EPBT, GPBT and GHG footprint of the BIPV system are 3.1 years, 0.4 years and 60 gCO_2e/kWh, respectively. However, many methodological details are lacking in this study. For example, the goal and scope of the LCA were not clearly defined and the inventory data was not presented at all.

Lamnatou et al. (2016) reported an environmental assessment of a linear dielectric-based building-integrated concentrating PV (BICPV) system using several different LCIA methods including ReCiPe Endpoint, Eco-indicator 99, USEtox and Ecological footprint [11]. They found that reflective film considerably improves the eco-profile of the BICPV system and that material/module manufacturing is the stage with the highest impact for all the studied categories. However, the results were only presented using highly aggregated indicators. This made them difficult to be compared with those from most other studies.

Lamnatou et al. (2017) conducted a comprehensive LCA to evaluated the environmental performance of a dielectric-based 3D BICPV device [12]. Several scenarios and life-cycle impact assessment (LCIA) methods including Cumulative Energy Demand (CED), IPCC 2013 GWP, ReCiPe Endpoint, Ecological footprint, USEtox and Eco-indicator are adopted to calculate environmental performance for six different cities: Barcelona, Seville, Paris, Marseille, London and Aberdeen based on simulated power generation. Most of the results were presented using highly aggregated indicators, making direct comparisons with other studies difficult. The GHG footprint was found to range from 105 (Barcelona) to 171 (Aberdeen) gCO_2e/kWh. The energy return on investment (EROI) varies from 5 to 18 for the different cities.

Kristjansdottir et al. (2016) presented a comparison of the life cycle GHG emissions from the PV systems in 3 different Zero Emission Buildings (ZEBs) in Norway [13]. The 3 ZEBs had different types and configuration of PV systems: 1) Poly-Si BAPV mounted on top of the roof; 2) Mono-Si BIPV fully integrated as roof; and 3) Poly-Si PV semi-integrated into the roof with the mounting structure replacing normal roofing and the modules mounted on top of a solid board and able to be removed without any impact on the building physics. They found that under the baseline scenario, life cycle GHG emissions of electricity generated was 45, 80 and 85 gCO_2e/kWh for the BAPV, BIPV and semi-BIPV systems, respectively. The results also show that benefits from the reduction in demand for traditional roofing material are small (13 and 3 $kgCO_2e/m^2$ for BIPV and semi-BIPV systems, respectively) relative to the total emissions (350 and 280 $kgCO_2e/m^2$ for BIPV and semi-BIPV systems, respectively). Under the baseline scenario, the greenhouse gas payback time (GPBT) was found to be 3, 7 and 8 years for the BAPV, BIPV and semi-BIPV systems, respectively. However, the calculation of the GPBT numbers involved some projections of the emissions intensity of European grid electricity and is not very easy to interpret.

Belussi et al. (2015) assessed a BIPV ceramic tile prototype developed in a research project [14]. The system consists of a PV layer formed by a thin film of amorphous silicon deposited on a ceramic substrate and is intended for installation on ventilated facades. A entirely attributional LCA approach was adopted with a goal to evaluate the environmental impact of BIPV module production and identify the stages with the greatest impact. The system boundary was "cradle-to-gate" and the LCIA method used was CML2001 which included 12 impact categories. Primary data for the LCI was claimed to be collected through field interviews and questionnaires but key data such as the inventory of the materials used in the BIPV system was not presented. The results for the 3 cities in Italy simulated (Milan, Rome and Agrigento) show that the environmental impacts of the BIPV system are comparable to the conventional PV systems. In particular, the GWP and embodied energy of electricity produced from the BIPV system were found to be 27.8-38.4 gCO_2e/kWh and 0.49-0.68 MJ/kWh, respectively.

Ng and Mithraratne (2014) evaluated life cycle environmental performance of 6 different commercially available thin-film modules as semi-transparent BIPV windows under the tropical conditions of Singapore [15]. Previously performed energy simulations were adopted in the LCA. It was found that the CED of the 6 systems varies from 240 to 2754 MJ/kWh, with the corresponding EPBT and EROI being 0.7-2 years and 12-35, respectively. The modules with the worst performances were mainly due to their low visible transmittances, energy efficiencies and higher thermal conductivities.

Menoufi et al. (2013) assessed a BICPV scheme assembled and tested on a small building model at the University of Lleida (Spain) in comparison to a hypothetical BIPV scheme. The scheme is to represent the installation of the reflectors as windows blinds. Two CPV modules of 250 Wp each are used as the receiver units. A hypothetical BIPV scheme mainly consisting of two transparent PV modules achieving the same power as the CPV ones is used as a comparison. The LCIA methods used include Eco-Indicator99 and EPS 2000. The results show that the environmental impact of BIPV scheme is 10%-14% higher than that of the BICPV scheme.

Hammond et al. (2012) performed energy analysis and LCA on modern BIPV roof tiles (mono-crystalline) connected in a modular arrangement. The study was completed with the assistance of a UK manufacturer and was used as a proxy for (domestic) BIPV products within the UK. The LCIA method chosen was Eco-indicator 99 and the LCA software SimaPro 7.1 was used to perform the calculations. The energy analysis revealed a short (displaced) energy payback period of 4.5 years for the system studied. The embodied energy and GHG emissions were found to be 83 GJ and 4500 $kgCO_2e$ for a 2.1 kWp mono-crystalline BIPV roof tile system (the functional unit chosen). This may be offset against the avoided impact of roof tiles, 217 $kgCO_2e$ and the avoided impact of 25 years of UK grid electricity, 26,700 $kgCO_2e$, resulting in a net saving of 22,400 $kgCO_2e$ over its lifetime.

3. Challenges and further research needs

During the review, many issues with existing LCAs on BIPV systems became apparent. For example, many studies actually evaluated BAPV even though the term BIPV was used [16–19]. There needs to be a clearer definition for these terminologies to reduce the ambiguity.

While some studies (e.g., [13]) follows the ISO guidelines for LCA and clearly presented the majority of the key information needed for readers to really understand the details of the study and findings, most studies fail to do so. Some studies only focused on limited energy or carbon/GHG indicators [15] whereas some studies look at many different impacts using various impact assessment methods [12].

In some studies there are many scenarios to explore the effects of many parameters such as location, system lifetime, with or without material replacement effects on the life cycle impacts [12]. While this is important (as it can offer insights into which parameters are key to the impacts), it makes comparisons across studies difficult. It would be beneficial to always use a baseline case and then vary different parameters to see the effects. This way the key results from the baseline case would be much clearer and easier to be compared with other studies. Different studies also use different functional units, system boundaries, data sources for key system parameters and inventory. More consistent approaches in these aspects and clearer presentation of these key information will help not only the readers of the individual papers but also researchers doing reviews and meta-analysis.

Another issue is that many studies mainly report results in figures without presenting all numerical values in the text (e.g., see [12]). This also makes it difficult to compare the results with other studies. One way around this issue is to show numerical values in the figures. Some studies included an LCA analysis but only cited results from other LCA studies rather than actually performing the LCA within their studies (e.g., [20]). In these cases it should be clearly stated in the methodology of the papers.

In terms of the coverage of the entire life cycle of BIPV, most studies focused on the materials and manufacturing of the PV systems, some included transport. In general, studies tend to exclude or use very simplified assumptions for the end-of-life stage. Future LCA studies need to give more emphasis on the end-of-life stage, even if the uncertainties might be high or assumptions need to be made. BIPV systems can also influence building energy performance [21]. This needs to be taken into account in LCA.

As an emerging technology, BIPV suffer from barriers such as data availability and uncertainties induced by rapid technology change. Therefore, there is a need for more "Anticipatory"

IOP Conf. Series: Materials Science and Engineering **556** (2019) 012053 doi:10.1088/1757-899X/556/1/012053

LCA studies that incorporates technology forecasting, risk analysis and stakeholder engagement in order to synthesise the available social, environmental, and technical knowledge [22]. In addition, there appears to be no study comparing BIPV with standalone PV systems and other types of renewable energy systems. Therefore, more research needs to be directed towards this area in order to understand the advantages and disadvantages of BIPV in the context of sustainable energy development.

4. Conclusions

BIPV is an important new technology with potentially huge environmental benefits. This review of the existing literature on LCA of BIPV suggests that it is still premature to determine the comprehensive environmental performance in relation to other energy technologies. Differences in scope and methods of existing LCA studies make it difficult to compare different studies and draw useful synthesis. Further research should focus on developing clear and consistent terminology, using LCA approaches, collecting data on the effects of BIPV on building energy performance as well as comparisons with other energy technologies covering a wide range of environmental impact categories.

5. References

[1] Ballif C, Perret-Aebi L-E, Lufkin S and Rey E 2018 Integrated thinking for photovoltaics in buildings *Nature Energy* **3** 438–42

[2] Petter Jelle B, Breivik C and Drolsum Røkenes H 2012 Building integrated photovoltaic products: A state-of-the-art review and future research opportunities *Solar Energy Materials and Solar Cells* **100** 69–96

[3] Wang Z, Li Z, Lei T, Yang M, Qi T, Lin L, Xin X, Ajayebi A, Yang Y, He X and Yan X 2016 Life cycle assessment of energy consumption and environmental emissions for cornstalk-based ethyl levulinate *Applied Energy* **183** 170–81

[4] Yan X, Tan D K Y, Inderwildi O R, Smith J a. C and King D A 2011 Life cycle energy and greenhouse gas analysis for agave-derived bioethanol *Energy Environ. Sci.* **4** 3110–21

[5] Ou X, Yan X and Zhang X 2011 Life-cycle energy consumption and greenhouse gas emissions for electricity generation and supply in China *Applied Energy* **88** 289–97

[6] Lamnatou C, Notton G, Chemisana D and Cristofari C 2015 The environmental performance of a building-integrated solar thermal collector, based on multiple approaches and life-cycle impact assessment methodologies *Building and Environment* **87** 45–58

[7] Tripathy M, Joshi H and Panda S K 2017 Energy payback time and life-cycle cost analysis of building integrated photovoltaic thermal system influenced by adverse effect of shadow *Applied Energy* **208** 376–89

[8] Li G, Xuan Q, Pei G, Su Y, Lu Y and Ji J 2018 Life-cycle assessment of a low-concentration PV module for building south wall integration in China *Applied Energy* **215** 174–85

[9] Jayathissa P, Jansen M, Heeren N, Nagy Z and Schlueter A 2016 Life cycle assessment of dynamic building integrated photovoltaics *Solar Energy Materials and Solar Cells* **156** 75–82

[10] Wang W, Liu Y, Wu X, Xu Y, Yu W, Zhao C and Zhong Y 2016 Environmental assessments and economic performance of BAPV and BIPV systems in Shanghai *Energy and Buildings* **130** 98–106

[11] Lamnatou C, Baig H, Chemisana D and Mallick T K 2016 Environmental assessment of a building-integrated linear dielectric-based concentrating photovoltaic according to multiple life-cycle indicators *Journal of Cleaner Production* **131** 773–84

[12] Lamnatou C, Baig H, Chemisana D and Mallick T K 2017 Dielectric-based 3D building-integrated concentrating photovoltaic modules: An environmental life-cycle assessment *Energy and Buildings* **138** 514–25

[13] Kristjansdottir T F, Good C S, Inman M R, Schlanbusch R D and Andresen I 2016 Embodied greenhouse gas emissions from PV systems in Norwegian residential Zero Emission Pilot Buildings *Solar Energy* **133** 155–71

[14] Belussi L, Mariotto M, Meroni I, Zevi C and Svaldi S D 2015 LCA study and testing of a photovoltaic ceramic tile prototype *Renewable Energy* **74** 263–70

[15] Ng P K and Mithraratne N 2014 Lifetime performance of semi-transparent building-integrated photovoltaic (BIPV) glazing systems in the tropics *Renewable and Sustainable Energy Reviews* **31** 736–45

[16] Cucchiella F and D'Adamo I 2012 Estimation of the energetic and environmental impacts of a roof-mounted building-integrated photovoltaic systems *Renewable and Sustainable Energy Reviews* **16** 5245–59

[17] Lu L and Yang H X 2010 Environmental payback time analysis of a roof-mounted building-integrated photovoltaic (BIPV) system in Hong Kong *Applied Energy* **87** 3625–31

[18] Radhi H 2010 Energy analysis of façade-integrated photovoltaic systems applied to UAE commercial buildings *Solar Energy* **84** 2009–21

[19] Seng L Y, Lalchand G and Sow Lin G M 2008 Economical, environmental and technical analysis of building integrated photovoltaic systems in Malaysia *Energy Policy* **36** 2130–42

[20] Wang Y, Li M, Hassanien R H E, Ma X and Li G 2018 Grid-Connected Semitransparent Building-Integrated Photovoltaic System: The Comprehensive Case Study of the 120 kWp Plant in Kunming, China *International Journal of Photoenergy*

[21] Chae Y T, Kim J, Park H and Shin B 2014 Building energy performance evaluation of building integrated photovoltaic (BIPV) window with semi-transparent solar cells *Applied Energy* **129** 217–27

[22] Wender B A, Foley R W, Prado-Lopez V, Ravikumar D, Eisenberg D A, Hottle T A, Sadowski J, Flanagan W P, Fisher A, Laurin L, Bates M E, Linkov I, Seager T P, Fraser M P and Guston D H 2014 Illustrating Anticipatory Life Cycle Assessment for Emerging Photovoltaic Technologies *Environ. Sci. Technol.* **48** 10531–8

Acknowledgments
Financial support by Innovate UK (grant 102880) is greatly acknowledged.

The Design of Sustainable Retrofitting Strategies and Energy-efficiency Optimization for Residential Buildings

Guopeng Li, Yingqing Xu, and Yue Fan
School of Architecture and Fine Arts, Dalian University of Technology, Dalian, China

E-mail: liguopeng@dlut.edu.cn

Abstract. There is a huge stock of existing residential buildings in China, with high energy consumption but low comfort, so it is of great significance to retrofit the residential buildings in order to enhance energy-efficiency of the buildings. In this paper, residential buildings, built between 1980 and 2000 in typical cities of the Northeast China, are defined as the research objects. This project tries to systematically put forward an applicable design method and an overall retrofitting plan of residential buildings. Dealing with the issues of the existing research on practical retrofitting of residential buildings in China, such as short of basic data and statistics, neglect of preliminary design planning and lack of guidance evaluation criteria, this project, based on the classification of residential building envelope system, builds a façade system information modelling database of residential buildings in typical cities of the Northeast China and abstracts a design prototype. Relying on the design prototype, Matrix method is used to organize, discover and edit the retrofit design strategies. BIM - EnergyPlus integrated technique is also introduced to evaluate and optimize the design strategies suitable for reference methods for realizing the sustainable retrofitting and energy efficiency optimization. The findings of this research will extend and improve the role and scientificity of architectural design in the retrofitting of residential buildings.

Keywords: Residential Buildings; Sustainable Retrofitting; Energy-efficiency Optimizations ; Design Strategies; Facade Refurbishment

1. Research Background and the Issue

Since the reform and opening up, China's urban housing construction has developed rapidly and achieved tremendous success especially the construction of residential buildings. However, cities in china have a huge housing stock which accounts for more than half of China's existing building area. However, its building performance lags behind the needs. On the one hand, it is of great significance to improve the energy efficiency of existing homes because of the low design standards for energy efficiency in buildings especially in the Northeast. On the other hand, the designed life is far from the useful life when the functional have expired. So, it is imperative to renovate existing housing, which is the focus of China's construction industry in the new era.

Building energy efficiency is an important part of the national energy conservation and emission reduction work. And the energy-saving renovation of existing buildings is an important part of building energy conservation. However, there are still problems in the retrofitting of existing residential buildings. In the scope of architectural design, the following are mainly reflected in: lack of basic information data; ignoring the previous design planning; and lack of effective design evaluation.

Above problems have led to a single retrofitting method, extensive operation processes, and unclear energy-saving effects. It is difficult to achieve the best comprehensive optimization.

2. Research Definition

There is a lack of research on the collection and arrangement of basic data and the specific application guidance of the design strategy in the actual retrofitting projects of existing residences in China. Therefore, this study proposes that it is necessary to establish a design strategy system which is applicable for complex existing residential situations. And it should be the focus of exploration. Based on the idea of hierarchical and partial reform, this study proposes should focus on the weak links of thermal performance in the various components of the existing residential enclosure structure. It is also important to advance the quantitative evaluation of the existing residential renovation performance to the design phase. This requires the integration of optimization technology methods to optimize the energy efficiency of the building renovation design.

This study takes the retrofitting performance of the existing residential buildings in the Northeast China as an example, selecting the typical urban residences in the northeastern region which constructed between 1980 and 2000 as the research object to explore the selection and optimization methods of green residential design strategies for existing residences. Green retrofitting is the source of design strategies in this topic and is also the goal of energy efficiency optimization.

The ultimate goal of this project is to select green building retrofit design strategies and energy efficiency optimization methods for existing dwellings to provide effective design guidance and reference for designers and related practitioners.

3. Research content

3.1 Construct an information model database for the facade system of old cities in typical cities in Northeast China and establish a design prototype:

- Sort out the constituent elements (walls, windows and balconies) and facade system information (architectural form, shape factor, orientation of each facade, and ratio of window to wall) of the old residential facade system. The corresponding materials, structural forms, structural characteristics (heat preservation measures and thermal performance, presence or absence of special structural nodes, etc.) and degradation degree of the elements of the old residential facade system should be recorded and arranged. The BIM information data is constructed as Carrier of northeastern region typical old house facade system information model database;
- Based on the facade system information model database, some typical old residential design prototypes in Northeast China were established through type selection.

3.2 Reorganization, editing and integration of existing housing renovation design strategies:

- Collect and sort out existing home renovation design strategies.
- Based on the design prototypes of typical old residential buildings in the northeastern region, it carried out exploration and integration of retrofitting design strategies, including: solar direct and thermal collectors (winter), natural lighting (winter, summer), natural ventilation and passive cooling (summer), and other low-cost passive design integration methods (ventilation form module design, double Layers of windows and walls integrated technology applications, etc.).

The retrofitting and design strategy of collation and integration applies BIM information data as a form of expression in the editing process, establishing the corresponding relationship between the retrofitting design strategy and the components of the facade system of the research object.

3.3 The selection, evaluation and optimization of green retrofitting design strategies for old houses in typical cities in Northeast China:

- According to the corresponding relationship between the various components of the facade system and editing design strategies, taking a comprehensive analysis of different factors are required to select green retrofitting design strategy;
- According to the current normative standards or set up energy efficiency promotion goals, use the BIM-EnergyPlus integrated technology optimization platform to form green retrofitting design strategies which are targeted and operational.

4. Research method
To achieve research goals, specific research methods include:

4.1 Field survey and data collection methods based on element sorting and type processing:
The important foundations of this topic are field research and data collection work. The field survey of the existing residential facade system form and status sorts current existing residences (each component of the facade system and the architectural form, shape coefficient, the orientation of each facade and the ratio of window to wall) and type of treatment (materials, structural forms, thermal insulation properties, structural characteristics, and degradation levels) and data collection work. The results of field research and data will be edited by Autodesk Revit Architecture based on BIM information data to form the existing housing renovation information model library material. This provides the basis for further deepening integration and research.

Façade System Information Survey : No. 117-2,Taishan Street

Location: Taishan Street, Shahekou District, Dalian
Construction year: 1985-1990
2800 households, with a total area of 163,300 m2
Construction company: Dalian City Construction Development Corporation
Peripheral conditions: The supporting facilities are relatively complete, with secondary schools and elementary schools evenly distributed around the area, and large areas of natural greening parks on the west and southeast sides.

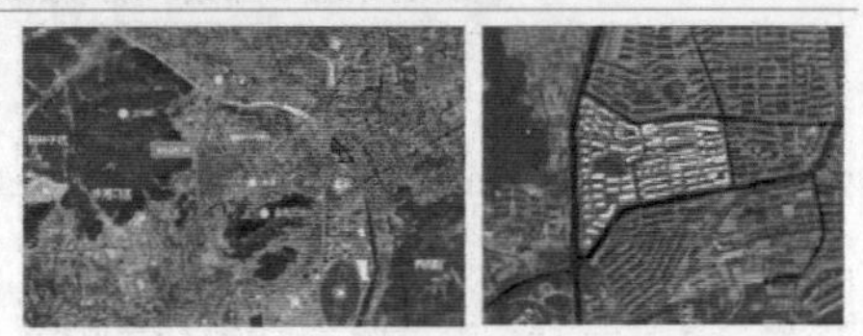

Standard floor plan : East façade / South façade / West façade / North façade

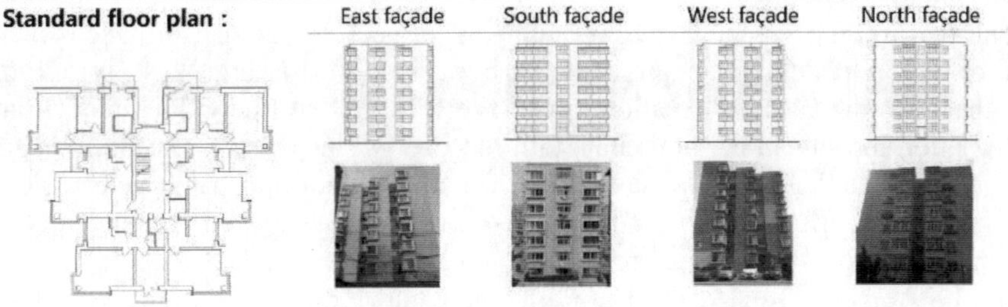

shape coefficient : 0.31			Window to wall ratio: 18%	Window to wall ratio : 50%	Window to wall ratio : 18%	Window to wall ratio : 30%
Façade	**Wall**	Form	Lightweight concrete/ Non-insulated hollow brick	Lightweight concrete/ Non-insulated hollow brick	Lightweight concrete/ Non-insulated hollow brick	Lightweight concrete/ Non-insulated hollow brick
		Performance	0.8 W/m²K	0.8 W/m²K	0.8 W/m²K	0.8 W/m²K
System	**Window**	Form	L-shaped window , double-layered glass	Early stage double glass	L-shaped window , double-layered glass	Early stage double glass
Design		Performance	3.3 W/m²K	3.3 W/m²K	3.3 W/m²K	3.3 W/m²K
Elements	**Balcony**	Form		Continuous plate without insulation		Continuous plate without insulation
		Performance		6.0 W/m²K		封闭，6.0W/m²K

Figure 1. Façade system information survey (taking No.117-2, Taishan Street, Dalian as an example)

IOP Conf. Series: Materials Science and Engineering **556** (2019) 012054 doi:10.1088/1757-899X/556/1/012054

4.2 The hierarchical matrix editing method applied to the consolidation and integration of green retrofitting design strategies:

The research focuses of this project are selecting and designing a suitable retrofitting strategy for the retrofitting site of the facade system and the design integration method between the various components of the facade system. Therefore, this project uses the hierarchical matrix editing methods to organize design strategies, emphasis the progressiveness of retrofitting design, and formulate a hierarchical retrofitting method that transforms the individual elements of the facade system elements into facades.

Measure	Description	Strategy	Material	Functions,	Limitations	Variations
Upgrade window	Double glazing applied on existing window frame	Add-in	Glazing	Heat protection Solar gains Increase thermal resistance of window pane	Additional weight and width of glazing Air tightness not improved	Different types of glass and frames
Secondary single glazing	Secondary single glazing window in the inside	Add-in	Glazing, window frames, sealant	Heat protection buffer zone. External appearance unchanged Acoustic performance improves	Adequate ventilation needed in the cavity. Secondary glazing draught-stripped, while the existing widows left without seals.	Different types of glazing and frame
Secondary double glazing	Secondary insulated window pane in the inside	Add-in				
Replaced windows with double glazing	Replace the existing glazing and frame with double-insulated glazing windows	replace		Heat protection Solar gains Increase thermal resistance and air-tightness. Solves thermal bridging in window frame	Thermal break in fame needed. Correct site installation includes sealing round window and door frames. Improves infiltration, ventilation consideration needed	
Replaced windows with triple glazing	Replace the existing glazing and frame with triple glazed windows	replace				
Shading fixed	Shading devices placed outside the window pane, in form of overhang, fixed louvers, etc.	Wrap-it	Shading devices, Aluminum, plastic, timber, etc.	Solar protection	Consider orientation for the type of shading	Different types of shading devices
Shade adjustable	Shading devices placed in or outside the window pane, in form of movable louvers, venetian blinds etc.	Wrap-it				
Enlarged windows	Existing window and parapet removed and replaced with enlarged, improved window pane	replace	Glass, window frames, sealing material	Insulation, lighting, absorption of higher solar heat, greater transparency, creating new indoor and outdoor relationships	Thermal break in fame needed. Correct site installation includes sealing round window and door frames. Improves infiltration, ventilation consideration needed	Different types of glazing and frames

Façade system retrofitting design			
	Wall	**Window**	**Balcony**
Original structure	Brick-concrete structure, hollow wall without insulation	single glazing	Non-insulated continuous floor
	Lightweight concrete structure, hollow brick without heat insulation	Early double glazing	No or very few insulation boards
	Poor thermal insulation material		
Retrofit measures	Increase air space as insulation buffer	Upgrade window	Balcony floor insulation
	Insulation	Secondary single glazing	Demolition of the balcony
	External insulation and veneer integration	Secondary double glazing	Balcony closed - single glazing
	Wooden frame thermal insulation wall	Replaced windows with double glazing	Balcony Closed - double glazing
	Ventilated facade	Replaced windows with triple glazing	
	Second facade, single glass	Shading fixed	
	Second facade, double glass	Shade adjustable	
Space intervention	Increase the space and integrate the second façade	Enlarged windows	Balcony integration
Green building design strategy	Direct Solar Heating and Passive Heat Harvesting Strategies (Winter)		
	Natural lighting related design strategies (winter, summer)		
	Natural Ventilation and Passive Cooling Related Design Strategies (Summer)		

Figure 2. Matrix method (taking window retrofitting as an example-left, and its position in the façade renovation system-right)

4.3. Building energy efficiency design optimization method based on BIM-EnergyPlus integrated optimization technology platform:

The important part of the research purpose of this project is the performance evaluation and optimization of the remodeling design strategy by using BIM-EnergyPlus. Based on the data convertibility between the BIM information model file IFC and the EnergyPlus model data file IDF, the BIM information model can be integrated with the EnergyPlus energy consumption simulation to form an integrated technology. At the same time, a genetic algorithm batch generation file can be introduced, which can efficiently promote design interaction between architectural design and performance quantification, the specific process is:

- Adopt the Autodesk Revit Architecture to integrate retrofitting design strategies of existing home prototypes to form a BIM information database, and use the IFC conversion function of Revit to generate an IFC data exchange file of the BIM information model;
- Create IDF files with geometric information through IFC and IDF files. It is necessary to embed a material thermal parameter database and establish the correspondence between the material name and the parameter. This can search the corresponding parameter according to the material name in the conversion process, and integrate energy-related information into the IDF file with geometric information for EnergyPlus energy consumption simulation;
- Track specific retrofitting sites through genetic algorithms, change geometry and material design parameters in batches within a certain range, generate a variety of design options, and conduct energy evaluation by EnergyPlus to perform verification and verification. Plans that meet the design criteria will go to the next step;
- Transfer the IDF file that meets the requirements of the design standard to the IFCViewer graphical interface through the VRML/DXF format, and then fed back to the Revit Architecture in the IFC/CAD format to be reflected in the building information model to complete the energy efficiency optimization process.

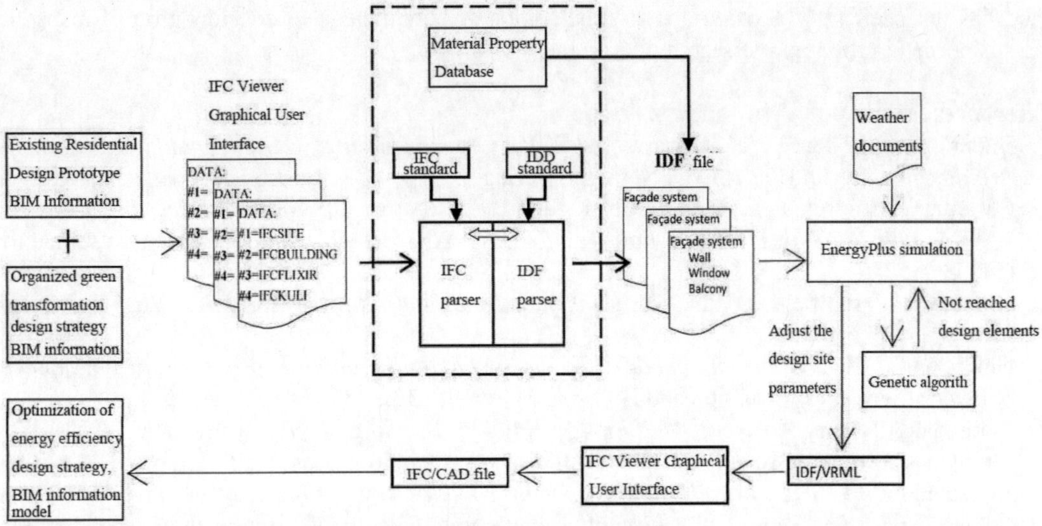

Figure 3. energy efficiency design optimization process based on BIM-EnergyPlus integrated optimization technology

5. The Technical Framework

Combining the research contents and research methods, the technical framework of this project is shown as follows:

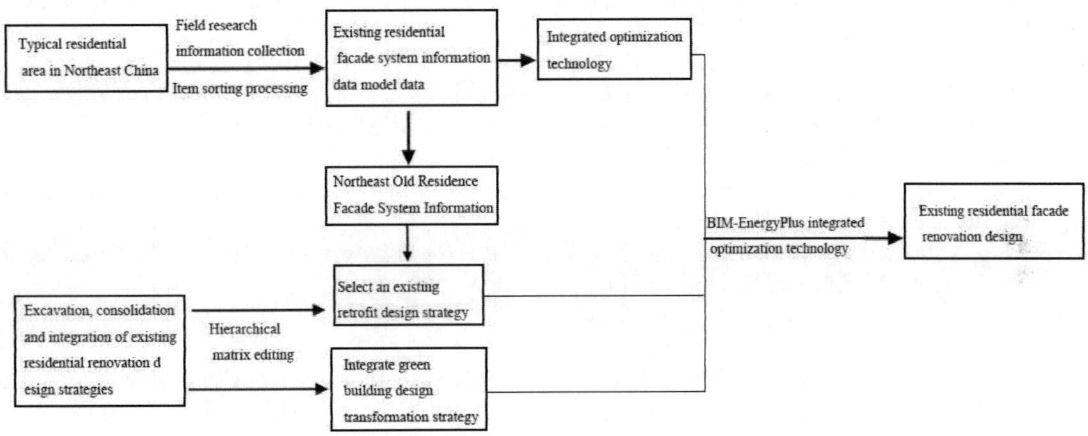

Figure 4. the technical framework showing the research procedure

6. Summary and Discussions

This research study attempts to explore a design approach to improve energy efficiency in existing residential buildings by utilizing passive design strategies. The study takes existing residential buildings built between 1980 and 2000 in Northern China heating areas as research objects. This study can provide integrated passive design strategies that quantify the impact of residential building facade components refurbishment, as well as decision-making information towards energy-efficiency up-grade for designers and related groups, such as contractors and home owners.

The research methods involve case studies and energy performance calculation. Existing residential buildings in different cities of Northern China heating areas are investigated. And based on the case studies, the early stage of this study has laid a solid foundation for energy performance calculation. The key methods of energy performance calculation rely on computer simulations and physical model testing. These approaches are relatively reliable and repeatable. However, the feasibility of this method has not been fully verified due to time limits. A comprehensive consideration of inspection process is needed in the future research plan.

7. References

[1] Bauer, M., Mösle, P. & Schwarz, M. 2010 *Green Building – Guidebook for Sustainable Architecture*. Berlin Heidelberg : Springer-Verlag.

[2] Carletti, C. Sciurpi, F. & Pierangioli, L. 2014 *The Energy Upgrading of Existing Buildings: Window and Shading Device Typologies for Energy Efficiency Refurbishment*, Sustainability (6): 5354-5377.

[3] Gorse, C. & Highfield, D. 2009 *Refurbishment and Upgrading of Buildings.* New York: Spon Press.

[4] Häkkinen, T. 2012 *Systematic method for the sustainability analysis of refurbishment concepts of exterior walls*, Construction and Building Materials (37): 783–790.

[5] Industrial Advisory Ad-Hoc. 2012 *Energy-Efficient Buildings PPP beyond 2013: Research & Innovation Roadmap*; Energy Efficient Buildings Association: Brussels, Belgium.

[6] Konstantinou, T. 2014 *Façade Refurbishment Toolbox: Supporting the Design of Residential Energy Upgrades.* Delft University of Technology, Faculty of Architecture and The Build Environment.

[7] Kwok, A. & Grondzik, W. 2007 *The Green Studio Handbook - Environmental strategies for schematic design.* Oxford: Architectural Press.

[8] Poel, B., van Cruchten, G. & Balaras, C. A. 2007 *Energy performance assessment of existing dwellings,* Energy and Buildings (39): 393-403.

[9] Ruparathna, R., Hewage, K. & Sadiq, R. 2016 *Improving the energy efficiency of the existing building stock: A critical review of commercial and institutional buildings*, Renewable and Sustainable Energy Reviews (53): 1032-1045.

[10] Ruut, P., Sverre, H., Stig, G., Jonas, H., Christopher, T., & Kruti, G. 2011 *Guidelines for the use of building physical modelling methods and tools in the development of sustainable refurbishment technologies for external walls.* SUSREF Deliverable 2.2; 2011. <http://cic.vtt.fi/susref/node/20> [22.12.2015].

[11] Tommerup, H. and Svendsen, S. 2006 *Energy savings in Danish residential building stock.* Energy and Buildings (38): 618-626.

[12] Xing, Y., Hewitt, N. & Griffiths, P. 2011 *Zero carbon buildings refurbishment—A Hierarchical pathway*, Renewable and Sustainable Energy Reviews (15): 3229-3236.

Acknowledgments

The project is funded by the National Natural Science Foundation of China (no.51808093, no.51638003) and the Fundamental Research Funds for Central Universities (DUT17RC(4)21).

The relative change of drying rate as a function of the cumulative ventilation air drying potential in a thin-layer solar drying facility

P Krawczyk[1]
[1] Division of Power Machines and Devices, Institute of Heat Engineering, Warsaw University of Technology, Nowowiejska st.25, 00-665 Warsaw, Poland
E-mail: piotr.krawczyk@itc.pw.edu.pl

Abstract. Controlling low-temperature drying facilities which utilise non-prepared air is quite difficult, due to very large variability of ventilation air parameters – both in daily and seasonal cycles. The results of experimental studies of sewage sludge solar drying process in thin-layer solar dryer have been presented. It was found that regardless of weather conditions, the rate of sewage sludge drying is significantly changed during drying process. The results showed that the rate of drying decreases with increasing so called cumulative ventilation air drying potential.Knowledge on current dryin rate provides new possibilities for controlling such systems. Based on the results, operating recommendations for analysed type of dryer have been formulated.
Experimental data analysed in the paper was collected in 2012 - 2014 during operation of a test solar drying facility in a sewage treatment plant in Błonie near Warsaw, Poland.

Keywords: solar drying, sewage sludge, drying rate calculation, thin-layer solar dryer

1. Introduction

An inseparable element of wastewater treatment is sewage sludge created in this process. This waste is highly hydrated and it is a mixture of various living and dead microorganisms as well as organic and mineral components [1]. Its management causes major problems for the operators of sewage treatment plants. Rules of law in this range prefer thermal methods of waste transformation relative to other methods of their management, i.e. storage or reuse for natural purposes [2], [3], [4], [5], [6]. Dryers are elements that allow to transform sewage sludge from a hydrated form, which contain about 80% of water to a form enabling using it as a fuel (about 20% of water).

A reasonable solution for small and medium-sized sewage treatment plants seems to be solar dryers [7], [8], [9]. Despite their increasing popularity, there are only few publications describing the thermal-flow processes taking place in this type of facilities. There is also a lack of design and operational recommendations concerning the construction of sludge solar dryers. In the article, there are presented the results of research regarding a new type of thin-layer solar dryer.

2. Pilot plant

Research on the process of sewage sludge solar drying was carried out in an experimental thin-layer solar dryer built in the sewage treatment plant in Blonie near Warsaw, Poland. The basic design assumptions of the dryer were presented in [10], [11]. In the test dryer, a bed of dried sludge was laid on a horizontal band conveyor placed in a tunnel roofed by a transparent material for solar radiation (Fig.1).

Figure 1. The experimental thin-layer solar drying facility of sewage sludge in Blonie (Poland).

The band conveyor is used to transport dried sludge inside the tunnel (drying chamber). The air flow inside the drying chamber was forced by mechanical ventilation (4 exhaust fans with rate of delivery each about 6'000 m^3/h). The sludge during the feeding was subjected to the forming process by a spreader. The diameter of formed bed elements during the tests was 12-15 mm (Fig.2). The drying facility has a length of 30 m and a width of 3 m.

Figure 2. The bed of dried sewage sludge in experimental thin-layer solar drying facility

The research program included continuous acquisition:

- temperature and relative humidity of the supply air to the drying facility and leaving the drying facility;
- the mass of sediments remaining on the separated area of the belt by a weighbridge installed under the drying belt;
- content of dry matter and humidity of dried sludge - samples analyzed in the treatment plant laboratory.

The temperature and relative humidity of the air were recorded with a time base of 5 minutes. The results of the measurements were used to analyze the nature of changes in the drying rate of sludge.

3. Method of measurement parameters' analysis

In the articles [12], [13] the concept of the drying potential of the ventilation air was implemented. It is defined as the theoretical value of the driving force of the drying process (difference between the moisture content in the air having contact with moist material and in the free flow), which occurs in the first drying period with a ventilation air of known temperature and relative humidity.
The rate of moisture evaporation from the dried material can be expressed using the equation [14], [15]:

$$W = A \cdot k_y \cdot (Y_s - Y) \tag{1}$$

The moisture content Y_s corresponds to the moisture content in the saturation state at the temperature of the liquid film covering the dried material t_s and it depends on the value of this temperature. The driving force of the moisture evaporation process is the difference between the moisture content in the air having contact with the dried material Y_s and the moisture content in the main flow Y. Based on equation (1), this value can be expressed as:

$$\frac{W}{A \cdot k_y} = (Y_s - Y) \tag{2}$$

The mass exchange coefficient k_y depends mainly on the hydrodynamic conditions of the process [16]. The thickness of the boundary layer for mass transfer, which determines the k_y coefficient, depends on the velocity of the gas above the surface of the dried body according to the dependence: $k_y \approx U^{0,8}$ [17]. Considering a drying facility with a known active surface of a bed of dried material and invariable settings of the ventilation, during the process, it can be assumed that:

$$A \cdot k_y = C = const. \tag{3}$$

In this case, the drying rate depends only on the driving force of the drying process:

$$W = C \cdot (Y_s - Y) \tag{4}$$

According to the theory of drying in the area of constant drying rate (first drying period), the temperature of the material surface in contact with air is constant and is equal to the temperature of the wet thermometer determined for the parameters of the ventilation air [18]. Being in accordance with the theory of drying, it is feasible to determine the surface temperature of the dried material $t_s = t_m(t_z, \varphi_z)$, thus the moisture content in the air in contact with the dried material Y_s, only on the basis of knowledge of the temperature and humidity of the ventilation air. Therefore, it is possible to determine the theoretical value of the driving force of the drying process, which occurs under the conditions of the first drying period using ventilation air of known temperature and relative humidity. This quantity for the purpose of this work has been defined as the drying potential of the ventilation air P_z:

$$P_z = Y_m(t_m, \varphi = 100\%) - Y_z \tag{5}$$

where: P_z – drying potential of the ventilation air, Y_m – moisture content in the temperature of the wet thermometer (t_m) and in the saturation state (φ=100%), Y_z – moisture content in the ventilation air (external).
Having the temperature t_z and the relative humidity φ_z of the ventilation air, it is possible to determine the value of the moisture absorption potential P_z for this air.
It follows from considerations that the instantaneous drying rate should be proportional to the instantaneous value of moisture removal potential of ventilation air.

$$W = C \cdot (Y_m - Y_z) = C \cdot P_z \tag{6}$$

The influence of energy supplied to the drying facility by direct solar radiation on the rate of drying of sludge, except the short-term periods of the largest solar activity, is low (isenthalpic process) [11].
During the drying, the parameters of the ventilation air change continuously. Integral over time of drying potential of the ventilation air was called the cumulative potential of the ventilation air S [12] - equation 7.

$$S = \int_0^t P_z(t) \cdot dt \tag{7}$$

It is a quantity directly proportional to the amount of evaporated water in a given drying cycle from its beginning until the time of t.
The collected experimental data (over 20 drying cycles) from the research installation was analyzed for the drying rate as a function of ventilation air potential for different values of cumulative potential. Consideration of the drying rate as a function of drying potential of the ventilation air allows the analysis results to be independent from the continuously changing atmospheric conditions.

4. **Results and discussion**

For each of the analyzed drying cycles, the values of the drying potential and the cumulative ventilation air drying potential were determined. The drying rate can be expressed as the product of water gain content in the ventilation air and the flow rate of this air. Because during the research, the air flow was constant in further analyzes only the values related to the air mass unit were used. The results of the drying process are shown in Figures 3 and 4. The collected measurement data were divided into four groups depending on the value of the cumulative ventilation air drying potential. For each group, the average proportionality coefficient was determined between the ventilation air drying potential and the drying rate expressed by increasing the moisture content in the ventilation air. It turned out that the mentioned dependence changes substantially with the drying process, i.e. the gain the cumulative potential value. The higher value of cumulative potential causes the lower drying rate under the same atmospheric conditions, i.e. with the same ventilation air drying potential (table 1).

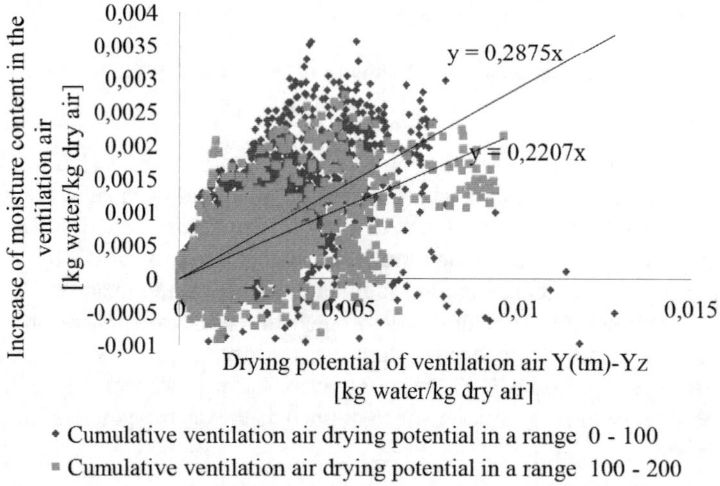

Figure 3. Increase of moisture content in the ventilation air as a function of drying potential for that air – experimental data (♦ – cumulative ventilation air drying potential in a range 0 - 100, ■ – cumulative ventilation air drying potential in a range 100 - 200).

Table.1. The average coefficient of proportionality between the drying potential of ventilation air and the rate of drying for different values of cumulative potential S

Cumulative potential S, (kg_{steam}•s/kg_{air})	The average coefficient of proportionality between the drying potential of ventilation air and the rate of drying
0-100	0,2875
100-200	0,2207
200-300	0,1753
300-400	0,0931

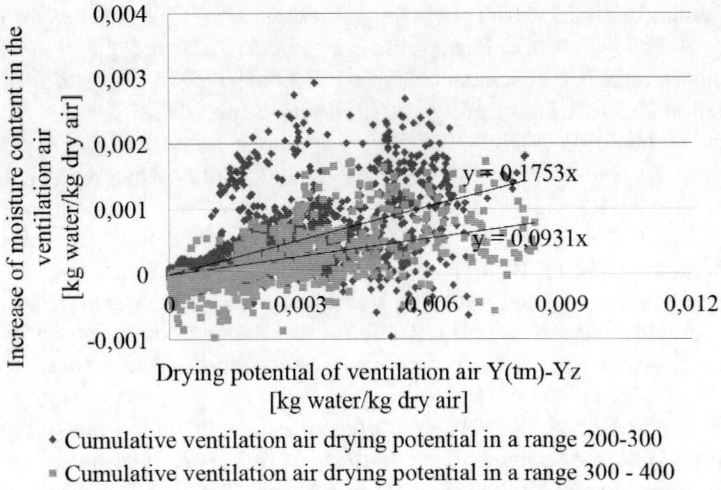

Figure 4. Increase of moisture content in the ventilation air as a function of drying potential for that air – experimental data (♦ – cumulative ventilation air drying potential in a range 200 - 300, ■ – cumulative ventilation air drying potential in a range 300 - 400)

At the same time, it is clear that for the analyzed drying cycles, the average sludge humidity remained higher than the critical humidity at which occurs transition from the not bound water area to the bound water area and was minimum about 2 $kg_{water}/kg_{dry\ matter}$ [12]. As research show [16], [19], [20], the critical humidity for sewage sludge is approx. 0.3 $kg_{water}/kg_{dry\ matter}$.
According to the author, it is caused by the lack of moving the bed of dried material during the drying process and the drying of its top layer. This, in turn, hinders the diffusion of water from the interior of the dried material to the surface of contact with air. Consequently, it causes a reduction of the average drying rate in relation to realisable under the given conditions.

5. Conclusions and summary

In a large thickness of the dried material bed (about 20-30 cm) solar dryers, mixer supports the transport of moisture from the interior of the sludge to the contact surface with the drying air. In the presented solution of thin-layer drying facility, the similar component is missing. In the design assumptions, the bed was supposed to be such a small thickness that the moisture could effectively diffuse from the interior of the dried material to the contact surface with the air. As the presented results show, with the progress of drying expressed by the cumulative ventilation air drying potential, the drying rate drops significantly despite the average humidity of the material remaining above the limit value for its transition to the second drying period. Therefore, it seems, that to better utilization the ventilation air drying potential, and hence to increase the efficiency of the drying facility, this kind of component, which forces the move of sludge, should be introduced into the analyzed type of installation. This effect can be achieved, for example, by dividing the drying belt and transferring the dried material from one belt to another.

REFERENCES

[1] Bień J 2002 *Sewage sludge. Theory and practice.* Częstochowa University of Technology Częstochowa.
[2] Bień J and Bień B 2015 *Disposal of municipal sewage sludge - thermal methods* Inżynieria Ekologiczna vol. 45, pp 36 - 43 (in polish)
[3] Kelessidisa A and Stasinakis A 2012 *Comparative study of the methods used for treatment and final disposal of sewage sludge in European countries* Waste Management vol. 32 Issue 6 pp 1186-1195

[4] Sebastian W and Wilk R 2010 *A review of methods for the thermal utilization of sewage sludge: The Polish perspective* Renewable Energy vol. 35 Issue 9 pp 1914-1919
[5] Stasta P and Borana J and Bebara L Stehlika P Oral J 2006 *Thermal processing of sewage sludge* Applied Thermal Engineering vol 26, Issue 13, pp 1420-1426
[6] Fytili D and Zabaniotou A 2008 *Utilization of sewage sludge in EU application of old and new methods—A review* Renewable and Sustainable Energy Reviews vol. 12 Issue 1 pp 116-140
[7] Bennamoun L 2012 *Solar drying of wastewater sludge: A review* Renewable and Sustainable Energy Reviews vol. 16 pp 1061–1073
[8] Shannon R and Nathan S and Luboschik V 2004 *Solar Sludge Drying* Water Works pp 38–43
[9] Krawczyk P and Badyda K 2011 *Modeling of thermal and flow processes in a solar wastewater sludge dryer with supplementary heat supply from external sources* Journal of Power Technologies vol. 91 pp 37–40
[10] Krawczyk P and Badyda K 2011 *Modeling of heat and flow processes in thin layer solar sludge dryer. Impact of using mixing baffles on the basic parameters of the dryer* Polish Journal of Environmental Studies vol. 20 Issue 4A pp 185 - 188
[11] Krawczyk P and Badyda K 2012 *Investigation of thermo-flow processes in a solar thin-layer wastewater sludge drying facility. Ambient conditions impact on drying rates* Polish Journal of Environmental Studies vol.21 Issue 5A pp 246-250
[12] Krawczyk P 2013 *Cumulative ventilation air drying potential as an indication of dry mass content in wastewater sludge in a thin-layer solar drying facility* Archives of Thermodynamics vol. 34 Issue 4 pp 23–34
[13] Krawczyk P 2016 *Control strategy for ventilation system of sewage sludge solar dryer* Journal of Power Technologies vol. 96 Issue 2 pp 145–148
[14] Korczak-Niedzielska M and Gromiec M 1998 *Drying of sewage sludge* Materials of Scientific-Technical Conference Problems of sludge in wastewater treatment plants Czestochowa 1998 (in Polish).
[15] Krawczyk P 2008 *Research of the thermal-hydraulic processes in sewage sludge solar dryer PhD thesis* Division of Power Machines and Devices Institute of Heat Engineering Warsaw University of Technology Warsaw (in Polish).
[16] Palipane K and Driscoll R 1992 *Moisture sorption characteristics of – Stell macadamia nuts* Journal of Food Engineering vol. 18 pp 63–76.
[17] Kneule F 1970 *Drying* Arkady Warsaw (in Polish).
[18] Strumiłło Cz 1983 *Fundamentals of Theory and Techniques of Drying* WNT Warsaw (in Polish)
[19] Katsiris N and Koszeli-Katsiris A 1987 *Bound water content of biological sludges in relation to filtration and dewatering* Water Research vol. 21 pp 1319 - 1327
[20] Vaxelaire J 2001 *Moisture sorption characteristics of waste activated sludge* Journal of Chemical Technology and Biotechnology vol. 76 p 377
[21] Luboschik U 1999 *Solar sludge drying — Based on the IST process* Renewable Energy vol. 16 Issues 1 pp 785-788
[22] Nezih K and Salihoglu V and Salihoglu G 2007 *Solar drying in sludge management in Turkey* Renewable Energy vol. 32 Issue 10 pp 1661-1675
[23] Slim R Zoughaib A Clodic D 2008 *Modeling of a solar and heat pump sludge drying system* International Journal of Refrigeration vol. 31 Issue 7 pp 1156-1168
[24] Mathioudakis V and Kapagiannidis A and Athanasoulia E and Paltzoglou A and Melidis P and Aivasidis A 2013 *Sewage Sludge Solar Drying: Experiences from the First Pilot-Scale Application in Greece* Journal Drying Technology An International Journal vol. 31 Issue 5 pp 519-526

IOP Conf. Series: Materials Science and Engineering **556** (2019) 012056 doi:10.1088/1757-899X/556/1/012056

It's not just power generation: The positive effect of building integrated photovoltaic design on architecture

A Ju Xiaolei[1], B Lu Yongfei [1],and C Guo Junnan[2]
[1]China Arch Design and Research Group, Beijing, China
[2] ShanDong Jianzhu University, Jinan, China
E-mail: juxl@cadg.cn

Abstract. From the perspective of solar photovoltaic power generation system and the building integration, studied the practical application and functionality of the PV tile, Aluminium Honeycomb panel PV Components, Vacuum Glass PV Components and FRP Board PV Components. Five positive effects of photovoltaic system on buildings are summarized: beautifying building facades, replacing original building components, upgrading or expanding the building use function, improving the building use comfort and improving the building energy saving effect.

Keywords: PV; PV Building; Building energy conservation

1. The positive effect of PV on architecture

Figure 1. Turpan BIPV Comprehensive Building.

Figure 2. PV Tile.

1.1. Landscaping Building Facade

The aesthetic characteristics of PV components, such as colour, shape and texture, can affect the overall appearance of the building. Such as Turpan BIPV Comprehensive Building, using the colour and size of PV components, arranging the texture of the local grape cool house. (Figure 1)

1.2. Replace the original building elements

Building materials photovoltaic components through the solar cell and metal substrates, glass substrates, organic materials, such as the integration of different types of substrate, so that it can have the corresponding function with the building components. It can replace the original building components installed in the corresponding parts of the building.

1.2.1. PV Tile. Photovoltaic tiles are the integration of solar panels and roof tiles,

IOP Conf. Series: Materials Science and Engineering **556** (2019) 012056 doi:10.1088/1757-899X/556/1/012056

many solar tiles can form a solar roof (Figure 2). When applied to building roofing, it has the same power generation function as the ordinary battery board, and maintains the harmony of the architectural style.

Figure 3. Application of Aluminium Honeycomb panel PV Component Engineering.

1.2.2. Aluminum Honeycomb panel PV Components. Aluminium Honeycomb panel PV components is the solar cell integrated with the common building decoration material Aluminium Honeycomb board. (Figure 3)Compared with the ordinary PV modules, the aluminium Honeycomb panel PV members can greatly enhance the insulation, temperature rise and noise reduction performance of the components, and can be used directly instead of building elements for building walls, facades, roofing and other parts. (Table 1)

Table 1. Comparison of aluminium Honeycomb panel PV components and general PV modules.

Component Type	Heat transfer coefficient (W/m^2•K)	Noise reduction (db)	Back plate temperature (℃)
Aluminium Honeycomb Panel PV components	1.0 ~3.0	$\geqslant$35	48.6
General PV components	6.0-6.8	$\leqslant$25	65

Figure 4. Application of Vacuum glass PV component.

1.2.3. Vacuum Glass PV Components. Vacuum Glass PV components is a new type of flat panel PV component made of vacuum glass as the substrate, which has two advantages of solar cell and vacuum glass. (Figure 4)The heat transfer coefficient of the vacuum glass PV component is less than 1 W/m^2•K and the thermal insulation performance reaches the level 10. Compared with the hollow glass PV components, it can effectively reduce the energy consumption of building more than 50%. (Table 2) Often used in building walls, lighting roof and other parts of the installation.

Table 2. Comparison of vacuum glass PV components with the middle-controlled glass PV components.

Component type	Heat transfer coefficient (W/m^2•K)	Surface temperature (℃)
Vacuum Glass PV Components	$\leqslant$1.0	61.2
Central control Glass PV component	$\geqslant$2.4	78.1

IOP Conf. Series: Materials Science and Engineering **556** (2019) 012056 doi:10.1088/1757-899X/556/1/012056

Figure 5. FRP panel PV component engineering.

1.2.4. FRP Board PV Components.
FRP Panel PV Components is a flat panel PV component made of FRP plate as substrate. (Figure 5)The organic material has the characteristics of light weight, strong structural safety and convenient transportation and installation. FRP daylighting Board has certain heat insulation performance, has good daylighting performance, and does not exist the expansion crack and so on. Compared with insulating glass, it can reduce the cost and reduce the surface temperature of the component. (Table 3) It can be used for building curtain wall, lighting roof, building outdoor light facilities and other parts.

At present, the heat transfer coefficient K value of single-layer plate is 2.7 w/㎡ K, and multilayer plate is 1.0 w/㎡ K, which meets the requirement of building performance in the FRP plate photovoltaic members have been developed.

Table 3. Comparison of FRP panel PV components with hollow glass PV components.

Component type	Heat transfer coefficient (W/㎡•K)	Surface temperature (℃)
FRP Board PV Components	1.5～3.2	64.5
Central control Glass PV component	⩾2.4	78.1

1.3. Promote or extend the building's use function
Using part of the physical properties of PV components, through the means of architectural design to upgrade, improve the use of the original building, or the existing building function to expand, so that it has a new use function to create more benefits.

Such as an office building in Shanghai, on the roof of the podium layout of the steel structure of the rain shed, combined with the steel structure of the photovoltaic roof. (Figure 6)Solar cells can absorb more solar energy; reduce the direct radiation to the roof of the solar energy, with thermal insulation effect. In addition, photovoltaic roof can absorb the direct sunlight and partially reflected light, which converts most of the solar radiant energy into electrical energy. The photovoltaic roofs on both sides of the podium can get a good shade effect, and the commercial facilities such as coffee, cold drink and so on are expanded in commercial space. (Figure 7)

Figure 6. Design of steel structure canopy.

Figure 7. PV roof shading effect.

1.4. Improve the comfort of building use
Research and Development building of Trina Solar Co., Ltd. to improve indoor daylighting comfort through PV components. (Figure 8) The PV components and the coated glasses are arranged, and the low transmittance of

Figure 8. Tianhe light Energy Limited research and development building facade daylighting analysis.

Figure 9. Map Central Control Building.

the PV components is used to prevent excessive sunlight from entering the room. At the same time, using the coating glass between PV components for lighting and ventilation, improve the indoor lighting comfort.

The central control building combines the local climatic conditions, and sets up the solar daylighting atrium. Not only solve the atrium of the relevant room lighting, but also to prevent too much sunlight into the room to avoid indoor temperature rise according the use of photovoltaic components. (Figure 9)

1.5. Improve the efficiency of building energy saving

PV components in the construction of a variety of installation forms, generally based on the construction project itself, the basic conditions of the comprehensive consideration and reasonable design. Additional functions are derived from the installation of different PV components (table 4).

Table 4. Energy-saving effect of buildings with different installation forms of PV components.

Installation form	Additional Features
Parallel roof Installation	Thermal insulation, ventilated roof
Wall Combination Installation	Effectively reduce the temperature of building wall and reduce the cold load of air conditioner
PV curtain Wall	It has the function of shading and saves the building area
PV sunshade Panel	Has the function of shading
Lighting Roof	Reduce lighting Load

1.5.1. PV Roof Energy Saving. Building Roof is the best layout of PV panels, it has incomparable advantages: good sunshine conditions, not easy to occlusion, to maximize the reception of solar radiation, and PV modules can be close to the building roof structure installation, reduce the adverse effects of wind. At the same time, solar PV modules can also be used as insulation layer shielding roofing.

Ministry of Housing and Urban and rur al construction of the building will be tiled on the roof and with the roof 150~200mm distance, photovoltaic components to prevent direct sunlight roof, reduce the roof temperature. (Figure 10) The heating cavity air is heated by the back plate of the component, which accelerates the air flow, thus taking away the heat in the cavity, reducing the component and the roof temperature, also reducing the loss of the power generation efficiency.

Figure 10. Housing and urban and rural construction Office PV Roofing.

1.5.2. Energy Saving of photovoltaic wall surface. The PV system is arranged on the exterior wall of the building, and the solar energy is collected rationally. Not only can the solar energy photovoltaic modules be used to convert the solar energy into electricity, but also can effectively reduce the temperature of the building wall and reduce the cooling load of indoor air-conditioning.

Figure 11. PV Wall of Shanghai solar energy engineering research center.

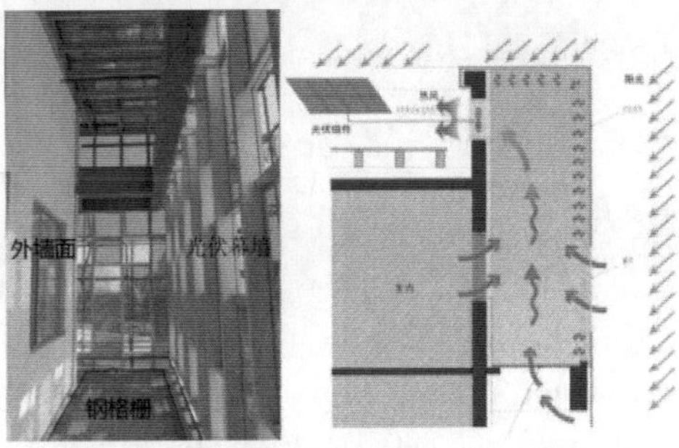

Figure 12. On the hollow cavity of the central control building PV curtain wall.

Shanghai Solar Energy Engineering Technology Research Centre arranges the photovoltaic curtain wall in the outside 100~150mm of the building. (Figure 11) After heating the back plate of the component, the air is heated and the air flow is accelerated, thus taking away the heat in the cavity, reducing the temperature of the component and the wall.

1.5.3. PV Curtain Wall Energy Saving. Photovoltaic curtain wall and the main building can take off a distance, effectively blocking the sun on the wall of direct, while avoiding the back of the photovoltaic wall heat directly into the room.

In the two floor of the central Control building, the east and south facade is provided with a thin film photovoltaic curtain wall. Photoelectric curtain wall has shading effect, reduce the heat generated by indoor radiation. The photovoltaic curtain wall and the outer surface of the building form cavities, photovoltaic curtain wall absorbs solar radiation after heating the air in the cavity, through the top of the daughter wall set ventilation openings, combined with the bottom of the ventilation steel grille, so that the photovoltaic wall and building outside the wall and indoor heat from the bottom of the effective discharge, can effectively reduce the energy consumption of building refrigeration. (Figure 12)

Test data show that the building through the shielding of photovoltaic curtain wall in the summer can effectively block direct sunlight, save a lot of air-conditioning energy consumption, energy-saving effect is very obvious. (Table 5)

Table 5. PV curtain wall test data.

Test points	Temperature (℃)
No blocking outside wall on the south Side	53.5
Inner surface temperature of curtain wall	49.5
The South Side exterior wall (exterior has the photovoltaic curtain wall)	38.2

1.5.4. PV Shading Energy saving. PV shading System uses PV modules as exterior sunshade components, and it is a typical application form of PV architecture, which has the functions of shading and power generation. The form includes plate shading, vertical shading, louver shading, etc.

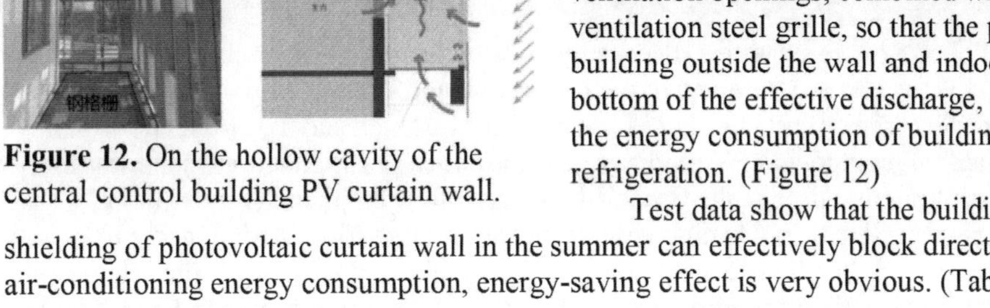

Figure 13. Map of the central control building PV shading system.

On the atrium roof of the central control building, set the light transmittance of 30%-40% glass-clamped crystal silicon photovoltaic ceiling. In the construction of the east, south, west facade set Sunshade, the second floor of the east and south to set the light transmittance 20% film photovoltaic curtain wall, to meet the needs of building lighting. (Figure 13) At the same time, it can effectively block the strong light direct to the exterior wall of the building, block excessive light into the room, so as to improve the indoor environment effect.

1.5.5. Energy saving of PV lighting roof. Although the photovoltaic roof can save the atrium illumination Energy, it may cause the indoor temperature to rise and increase the cooling load because of the heating of the PV component. The central control building is provided with a double glass clamp crystal silicon photovoltaic roof at the top of the atrium, and an open ventilation window is arranged around the roof of the skylight. (Figure 14)

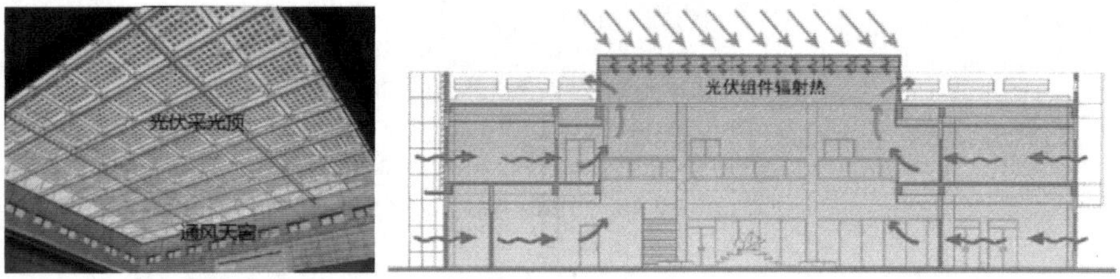

Figure 14. Map of the central control building lighting top signal.

2. Conclusion

At present, the design of photovoltaic system is divorced from architectural design which restricts the application of PV in architecture. In order to realize the integration and synergy of the integrated design process of PV building. First of all, we should speed up the research and development of more types of building materials and component PV modules, and form the industrialization capability to promote the development of photovoltaic scale. Secondly, we need to strengthen the training of building PV-related talents and provide a full range of talents for building PV. Third, in order to improve investors ' confidence in the photovoltaic industry, We should further improve the relevant standards, standards and related Atlas of building PV , and establish the testing centre and certification organization of the products. As well as the completion of construction acceptance and maintenance of technical procedures. Finally, the government agencies should subsidize the building integrated PV System project, and introduce relevant policies to encourage the development of new photovoltaic components, and gradually expand the scope of PV component applications.

Acknowledgments

State Grid Corporation of China 《Research on comprehensive development and utilization technology of renewable energy in multi format ecological development zone》 science and technology projects.

IOP Conf. Series: Materials Science and Engineering **556** (2019) 012057 doi:10.1088/1757-899X/556/1/012057

A novel Fluid-Structure Interaction modelling and optimisation of roofing designs of buildings for typhoon resilience

Conrad Allan Jay Pantua[1,2], **John Kaiser Calautit**[1] **and Yupeng Wu**[1]
[1]Department of Architecture and Built Environment, University of Nottingham, Nottingham, UK
[2]Mechanical Engineering Department, De La Salle University, Manila, Philippines
E-mail: conrad.pantua@nottingham.ac.uk

Abstract. Stronger typhoons have been more frequent in the Western Pacific region. Typhoon Haiyan caused widespread loss of life and destruction to properties when in made landfall in the Philippines in 2013. An estimated 1.1 million homes were damaged or destroyed in the aftermath. Damage surveys show extensive roofing damage evident in most detached structures attributed to strong winds. Clearly, there is a need to evaluate the current roofing designs and its structural integrity for it to properly respond to extreme environmental events in the future. Using a novel Fluid Structure Interaction (FSI) approach, this study evaluated a single detached gabled building made of timber which is common in the Philippines. The building was subjected to typhoon strength winds in an urban environment using Computational Fluid Dynamics (CFD) analysis. Atmospheric Boundary Layer (ABL) flow simulation was conducted to predict the pressure distribution around the structure. A structural model of the roofing support was then developed, and the structural analysis performed using FSI to predict failure in the sheathing and the supports. The results of the study show the structural weaknesses in the current design considering the wind angle, structural frame and materials.

Keywords: Typhoon; Computational fluid dynamics; Finite element analysis; Fluid structure interaction; Urban environment, Buildings

1. Introduction

The western pacific region has been experiencing a lot of relatively stronger typhoons (tropical cyclones) in the past years. The Philippines a nation with a 100 million people [1] is visited by an average of 19 to 20 tropical cyclones every year [2]. The 2015 World Risk report ranked the country as the third most risk prone country in terms of natural disasters in the world [3]. Last 2013, the Philippines made headlines as it was hit by Typhoon Haiyan which was the strongest storm ever recorded at landfall. The super typhoon had sustained wind speeds up to 315 kilometres per hour (195 mph) and gustiness up to 380 kilometres per hour (235 mph) [4]. The damage to infrastructure and casualties were staggering. A total of 6,300 people died in the aftermath with 28,000 injuries [5]. Around 1.1 million homes were damaged or destroyed with an economic impact of 12.9 billion USD [6]. Visual evaluation and survey of the damage along Typhoon Haiyan's path showed that majority of low rise detached buildings suffered roof damage caused by strong winds [7]. Immediately after the disaster there was interest and clamour for better typhoon resilient building in the country. Unfortunately, very few researches were conducted on typhoon resilient infrastructure in the Philippines. Most of the studies were concentrated on the effect of hurricanes to low rise structures in the United States. There is a clear need to evaluate current roofing designs and infrastructure support to minimise the damage in future climactic events.

Numerous studies on wood frame low rise building performance under hurricane wind loads have been conducted in the past. The studies are usually categorised into hurricane hazard modelling, wind structure interactions, building representation and performance assessment [8]. Majority of wind structure interactions were conducted using wind tunnel tests and have been the basis of the design codes. The problem with wind tunnel tests is that it is highly dependent on the availability of high quality experimental facilities which is almost non-existent in a country like the Philippines. Costs are also prohibitive especially with the use of full scale models combined with the actual cost of wind tunnel facilities. The use of Computational Fluid dynamics (CFD) in the built environment to accurately predict aerodynamic behaviour in different Atmospheric boundary layer conditions (ABL) has been emerging lately [8-11]. The preference for CFD over wind tunnel tests is due to its practicality, shorter development time and lower costs compared to a full-scale prototype. Despite of its many advantages compared to wind tunnel tests [12-15], CFD cannot predict the structural response of the model which is only possible through actual wind tunnel tests or the use of Finite element analysis (FEA). The FEA method has been utilised in the assessment of light-framed wood roof structures to wind loads as seen in numerous studies [16-18]. However, the pressure loads required as boundary conditions are still obtained from wind tunnel tests. An alternative is to combine CFD and FEA into a fluid-structure interaction (FSI) analysis. This method was already performed successfully by Lee et al. [19] in the analysis of a phase VI wind turbine.

In the FSI analysis, the surface pressure results computed from the CFD analysis is transferred to the structural surface pressure loading and then a linear or non-linear finite element analysis is performed using the FEA module. The structural response of the model such as forces, stresses and displacements are calculated from the structural analysis. This paper utilises the FSI method to analyse the structural response of a low rise gabled timber building under typhoon strength wind conditions. The wind speed conditions and geometry of the building were patterned after He et al. [16] for experimental validation. The CFD and linear static analysis for the gabled building was performed using a commercial program ANSYS.

2. Methodology

2.1 FSI framework

FSI generally uses 1-way or 2- way modelling approach. In the 1-way approach, a converged solution is obtained for one field then used as a boundary condition for the second field. On the other hand, the 2-way approach solves fluid and solid equations separately and iterates within each time step to obtain an implicit solution. Implicit means the dependencies between the fluid and solid fields are converged within a time step. Unsteady transient turbulence models are more suitable for 2-way FSI analysis while steady Reynolds-averaged Navier-stokes equations (RANS) models use a 1-way FSI approach. 1-way FSI might be easier to implement compared to 2-way FSI due to uncertainties of obtaining convergence and the potential of having numerical singularities during the simulation in the interface and fluid domain [19]. Applications with weak physical coupling wherein the primary objective is to obtain maximum structural stresses and with strains not significant enough to affect the CFD results are most suitable for a 1-way FSI. Tominaga et al. [20] compared and validated the different RANS turbulence models and concluded that the RNG κ-ε exhibited the best performance compared to the standard κ-ε and κ-ω SST. The modelling approach used in this paper is a 1-way FSI analysis utilising an RNG κ-ε turbulence model. The 1-way FSI methodology flowchart used in this study is described in Fig 1.

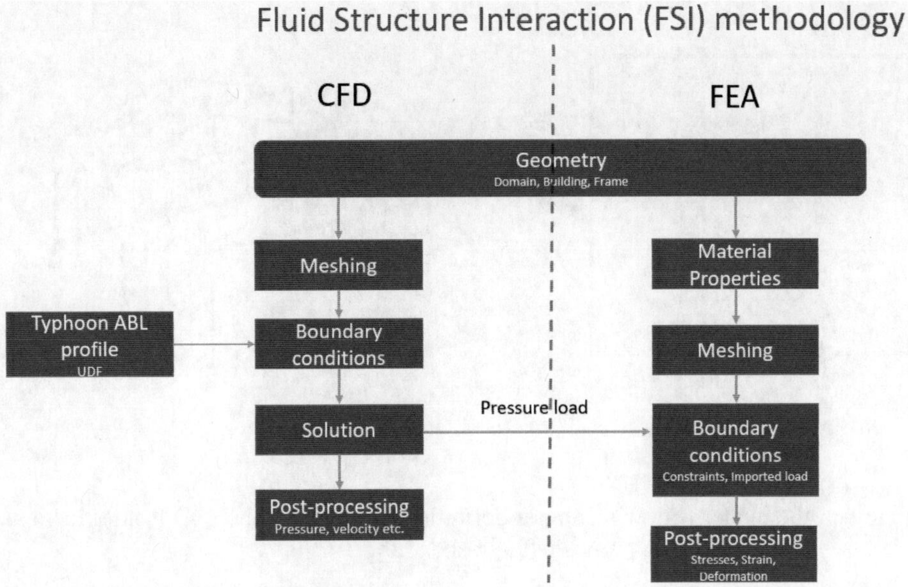

Figure 1. Framework of the study.

2.2 Flow field

The building model used in the simulation is a gabled-roof building with a roof pitch of 16.7 °. This gabled roof design is typical in the Philippines [21]. The size of the computational domain which is 84 x 46 x 32.76 m as described in Figure 2 complies with the guidelines recommended by Blocken [22] for directional blockage ratio. The size of the building 3.57 x 2.29 x 0.91 m is the same one used by He et al. [16] in their study. The velocity used is 40.68 m/s (147 kph) with mean wind speed profile of $\alpha = 0.15$ (power law). This wind speed is categorised to a Typhoon category 3 in the Philippines four cyclone warning categories.

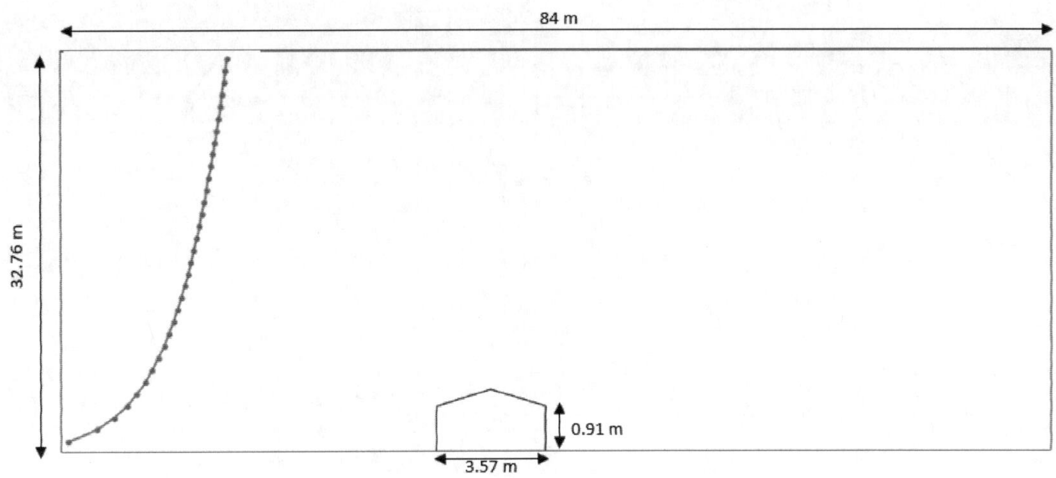

Figure 2. Computational domain

2.3 Structural Members

The structure is composed of frames and sheathing panels made from 7/16 in Oriented Strand Board (OSB) which the material properties derived from literature are incorporated in the FEA model. The frame design and wind angle definitions are shown in Figure 3.

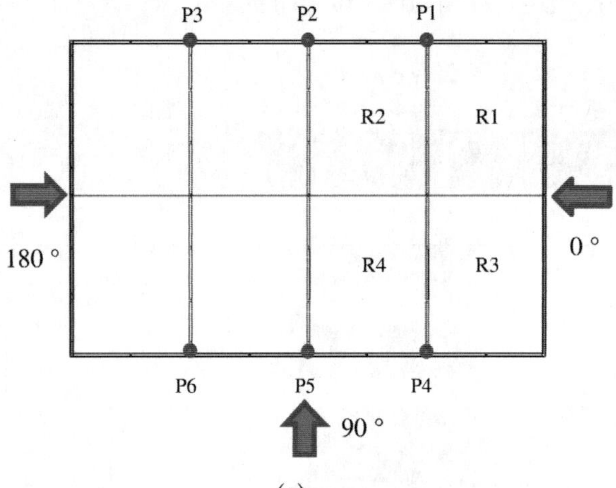

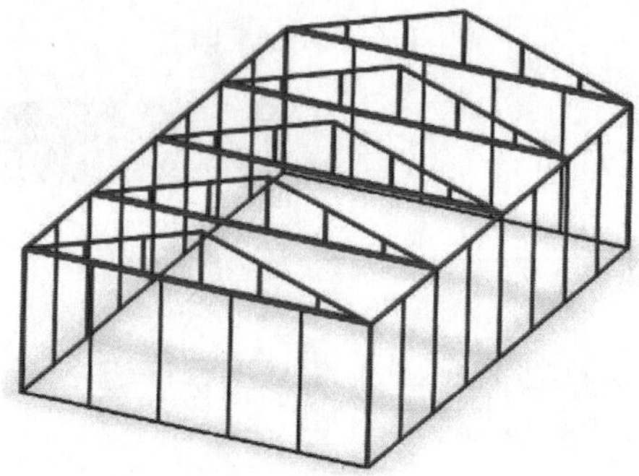

Figure 3. a.) Top view of the model with wind angles definitions and frame labels b.) Isometric view showing frames without sheathing

3. Results

3.1 Model Validation

The model was subjected to a constant wind speed at different angles from 0 to 180° at a 15° interval. The displacement at various locations were compared with the wind tunnel displacement results of the study conducted by He et al. [16]. The displacement results taken at two different sides in the model for the simulation and experimental runs are shown in Figure 4.

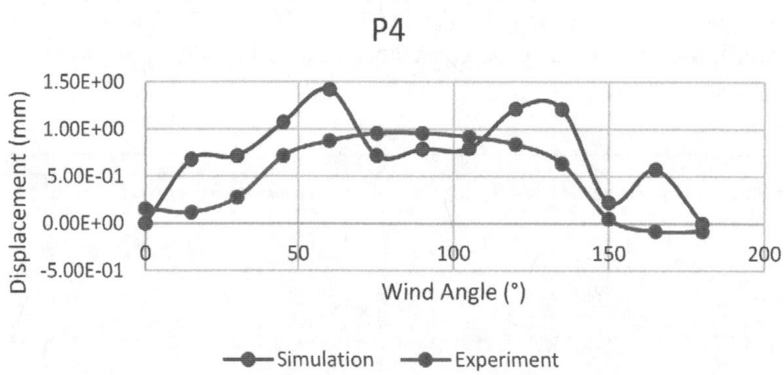

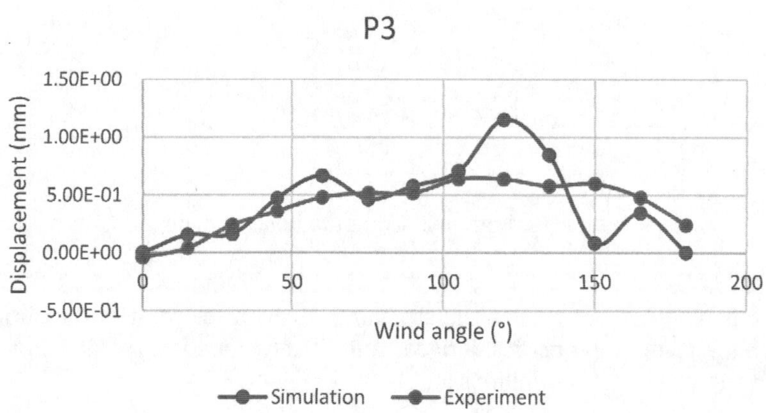

Figure 4. Deflection of the structure at positions P3 and P4

IOP Conf. Series: Materials Science and Engineering **556** (2019) 012057 doi:10.1088/1757-899X/556/1/012057

Both simulation charts followed the experimental trend with a little bit of overestimation for P4. The trend in the leeward side at P3 is much more better than at P4 except for some noticeable differences at the 120° and 150° wind angles. This may be attributed to the imported pressure data which is derived from the CFD results. The FSI analysis is highly sensitive to the CFD results which may result into some errors in the structural simulation. Another possible cause of error is that the structural model is assumed to be linear wherein the actual scenario follows a non-linear behaviour. Nevertheless, the differences in the displacement magnitudes of the simulation and experimental data are less than a mm.

3.2 FSI structural analysis

The equivalent von mises stresses and total displacements were obtained for wind angles cases. The maximum stress and displacements in the roof occurred in the 90° wind angle which was 38.288 MPa and 122 mm respectively. The imported pressure, stress and displacement plots are shown in Figure 5.

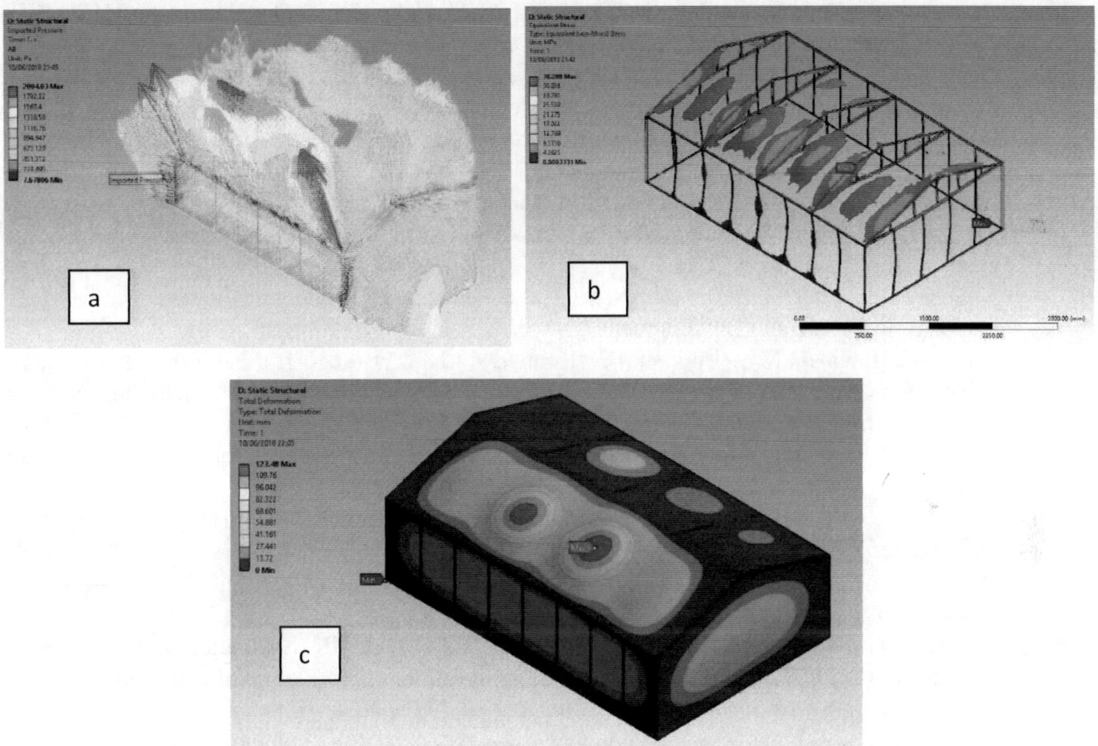

Figure 5. a.) Imported pressure results b.) Maximum and minimum equivalent stresses in the frame c.) Maximum and minimum displacements in the sheathing

Upon further observation, the maximum stress occurred in the frame of the roof near P5 and the maximum deflection occurred in the roof sheathing at area R4. Although the highest stresses are experienced at the structural frame, the highest deformation occurs at the roof section which can cause ripping. The pressure along the roof surface is highest at the windward (front) and lowest at the leeward side (back). The minimum deformation scenario in the roof occurred at the 0° angle which was around 0.65 mm maximum at the windward side. A relatively high stress of 44.65 MPa was experienced in the frame wall but it resulted into a small deformation of about 0.95 mm.

4. Conclusions

The study has successfully demonstrated the capability of FSI analysis in low rise building performance under typhoon winds. The model was validated and matched the response of the wind tunnel results from a previous study. Based from the results, the area of the highest roof uplift can be properly estimated based on a worse case wind angle scenario. The highest stress concentration on the roofing frame can be properly determined as well. Both occurrences (highest stress and deflection) happened in the windward side. The importance of the wind angle can play the importance role in the orientation of houses. Urban planners can properly plan the orientation of houses in the known path of the typhoon to minimise destruction. Future directions of this study can include the evaluation of different roofing designs such as gabled with eaves overhang, mono-sloped, domed and hip. Optimisation of the pitch, height and length together with the structural parameters such as frame reinforcement and proper material selection are also essential for this FSI framework to be applicable in the real setting.

References

[1] World Bank. *World Development Indicators 2017*. Washington, DC: World Bank; 2017.

[2] Craig A, Cinco T, Monteverde C, Hilario F, Celebre C, Tudao A, et al. Tropical Cyclone Severe Wind Risk Modelling in GMMA. *2013 APEC Typhoon Symposium*; NTU GIS convention center, Taipei: APEC Research Center for Typhoon and Society (ACTS) 2013.

[3] Garschagen, M, Hagenlocher M, Kloos J, Pardoe J, Lanzendörfer M, et al. *World Risk Report 2015*. Berlin, Germany: Alliance Development Works; 2015.

[4] Evans A. *Annual Tropical Cyclone Report 2013*. Pearl Harbor, Hawaii: Joint Typhoon Warning Center; 2014.

[5] National Disaster Risk Reduction and Management Council. *Final Report Effects of Typhoon "Yolanda" (Haiyan)*. National Disaster Risk Reduction and Management Council. Quezon City, Philippines 2014.

[6] National Economic and Development Authority. *Reconstruction Assistance for Yolanda*. National Economic and Development Authority. Pasig City, Philippines 2013.

[7] Chen SE, Leeman ME, English BJ, Kennedy AB, Masters FJ, Pinelli JP, et al. Basic Structure System Rating of Post-Super Typhoon Haiyan Structures in Tacloban and East Guiuan, Philippines. *Journal of Performance of Constructed Facilities*. 2016;**30**(5):11.

[8] He J, Pan F, Cai CS. A review of wood-frame low-rise building performance study under hurricane winds. *Engineering Structures*. 2017;**141**:512-29.

[9] Enteria N. CFD Evaluation of Philippine Detached Structure with Different Roofing Designs. *Infrastructures*. 2016;**1**(1):3.

[10] Hughes BR, Calautit JK and Ghani SA. The Development of Commercial Wind Towers for Natural Ventilation: a review. *Applied Energy*. 2012;**92**:606-27

[11] Hosseini SH, Shokry E, Hosseini AJ, Ahmadi G and Calautit JK. Evaluation of airflow and thermal comfort in buildings ventilated with wind catchers: Simulation of conditions in Yazd City, Iran, *Energy for Sustainable Development*. 2016;**35**:7-24

[12] Calautit JK and Hughes BR. Measurement and prediction of the indoor airflow in a room ventilated with a commercial wind tower. *Energy and Buildings*. 2014;**84**:367-377

[13] Calautit JK and Hughes BR. Wind tunnel and CFD study of the natural ventilation performance of a commercial multi-directional wind tower. *Building and Environment*. 2014;**80**:71-83

[14] O'Connor D, Calautit JK and Hughes BR. A study of passive ventilation integrated with heat recovery. *Energy and Buildings*. 2014;**82**:799-811

[15] Calautit JK, Chaudhry HN, Hughes BR and Sim LF. 2014, A validated design methodology for a closed-loop subsonic wind tunnel. *Journal of Wind Engineering and Industrial Aerodynamics;***125**:180-94

[16] He J, Pan F, Cai CS, Habte F, Chowdhury A. Finite-element modeling framework for predicting realistic responses of light-frame low-rise buildings under wind loads. *Engineering Structures*. 2018;**164**:53-69.

[17] Jacklin RB, El Damatty AA, Dessouki AA. Finite-element modeling of a light-framed wood roof structure. *Wind and Structures*. 2014;**19**(6):603-21.

[18] Satheeskumar N, Henderson DJ, Ginger JD, Wang CH. Three-Dimensional Finite-Element Modeling and Validation of a Timber-Framed House to Wind Loading. *Journal of Structural Engineering*. 2017;**143**(9):11.

[19] Lee K, Huque Z, Kommalapati R, Han SE. Fluid-structure interaction analysis of NREL phase VI wind turbine: Aerodynamic force evaluation and structural analysis using FSI analysis. *Renewable Energy*. 2017;**113**:512-31.

[20] Tominaga Y, Akabayashi S, Kitahara T, Arinami Y. Air flow around isolated gable-roof buildings with different roof pitches: Wind tunnel experiments and CFD simulations. *Building and Environment*. 2015;**84**:204-13.

[21] Enteria N, Awbi H, Yoshino H. Application of renewable energy sources and new building technologies for the Philippine single family detached house. *International Journal of Energy and Environmental Engineering*. 2015;**6**(3):267-94.

[22] Blocken B. Computational Fluid Dynamics for urban physics: Importance, scales, possibilities, limitations and ten tips and tricks towards accurate and reliable simulations. *Building and Environment*. 2015;**91**:219-45.

The potential of phase change energy storage for office cooling load shifting

J P Cunha[1] and P C Eames[2]
[1,2]Centre for Renewable Energy and Systems Technologies, Loughborough University, LE11 3TU, UK
[1]E-mail: j.m.p.cunha @lboro.ac.uk

Abstract. An analysis of the performance of four cooling approaches to meet daytime office cooling requirements for an open plan office was undertaken using a finite-volume model implemented in Matlab. The four cooling approaches simulated were: a conventional HVAC system (A), using extra ventilation on demand (B), using cool night time air to charge a store with RT15 (C) and charging a store with RT8HC using night time off peak electricity (D). Charging the thermal store with cool night time air reduced the yearly cooling demand by 39% compared to the conventional HVAC cooling system. Mechanical ventilation on demand achieved a reduction of 34% compared to the same conventional system. Using off peak electricity to charge the PCM store achieved a reduction of 15%, and successfully shifted daily cooling demands into night time operation during spring and autumns months. Due to the high capital cost of the thermal stores, the lowest cost of energy was obtained using extra ventilation on demand, with a cost of 84.45£/MWh for an assumed 20-years of operation.

Keywords:

Phase change materials, Thermal energy storage, Lumped capacitance numerical model, Night time cooling, Encapsulated latent heat storage, Office load shifting, HVAC.

1. Introduction
The demand for space cooling is rapidly growing worldwide and will be exacerbated by global warming [1]. Currently, it is estimated that 40 to 45% of the EU's total energy demand [1] is used in buildings. According to the European Union approved directive 91 in 2002 [2] and ratified in 2010, all new buildings in the EU are expected to reduce or consume "nearly zero" energy after 2020 [3]. Although currently marginal, the energy used for cooling and humidification is expected to grow and efforts should be made to reduce its associated emissions [1].

2. The developed simulation model
To assess the potential reduction in CO_2 emissions achievable by the integration of a compact latent heat storage container into a building space cooling system, a typical open plan office was chosen, based on that used in a study by Korolija et al [4]. The office building is a three-storey building schematically shown in Figure 1A with a 32 by 16 meter footprint, and floor-to-ceiling height of 3.5 meters. The proposed heating and cooling solution is the installation of a centralized air distribution system with a compact thermal store, heat pump and the centrifugal blower installed on the building rooftop, as seen in **Error! Reference source not found.**. For the conventional approach, the thermal store and centrifugal fan are removed, retaining the heat pump to supply the required heating and cooling needed.

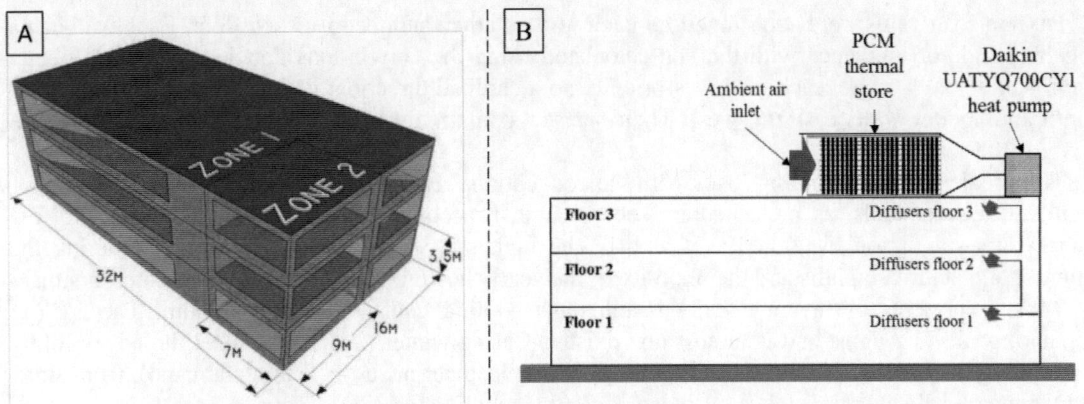

Figure 1 - Schematic drawing of the simulated 3-storey building (A), from [4] and a representation of the thermal store installed on the rooftop (B).

The indoor temperature of the building was considered uniform with no thermal stratification, heat exchange with the environment was calculated based on the recommended building fabric U-values from the CIBSE guide A [5] (0.18W/m^2.K for Cavity wall, 0.15 W/m^2.K for flat roof and ground floor and 1.98 W/m^2.K for the glazing). The predicted building yearly energy consumption was then compared with that from an EnergyPlus [6] simulation of the same building presented by Korolija et al. [4]. The heat balance calculation for each floor was based on equation 1. A fresh air ventilation rate of 10 l/s per person is required to meet the building ventilation standards. The auxiliary heat input (Q_{aux}) will be positive or negative according the required auxiliary energy to meet the set temperature.

$$\dot{Q}_{indoor} = \rho_{air} Cp_{air}(F_{air} + V_{person} \cdot N_{person}) \cdot (T_{amb} - T_{indoor}) + \dot{Q}_{solar} + \dot{Q}_{gain} + +\dot{Q}_{LHS} + \dot{Q}_{aux} + \dot{Q}_{fabric} \qquad 1$$

Adopting the recommended values from the CIBSE guides [5], the building indoor temperature was either heated to 20°C or cooled to 24°C during the occupied hours (from 07.00 to 19.00) [4], with the indoor temperature maintained between 12 and 28°C during unoccupied hours. Casual gains from office appliances and lighting were assumed to be 27 W/m^2, for the occupants casual gains were assumed to be 108W/person and the number of occupants per floor were based on an occupant density of 9 m^2/person, in line with the CIBSE guide recommendations [5]. The building fabric heat transfer with the indoor air (Q_{fabric}) was calculated using equation 2.

$$\dot{Q}_{fabric} = \Big[3 * Area_1 * \big(T_{wall\,N}(k) + T_{wallS}(k) + T_{glassN}(k) + T_{glassS}(k)\big) + 3 * Area_2 * \big(T_{wallE}(k) + T_{wallW}(k) + T_{glassE}(k) + T_{glassW}(k)\big) + A_{space} * \big(T_{roof}(k) + T_{floor}(k) + 4 * T_{deck}(k) + 3 * k_{solid} * T_{solid}(k)\big) - T_{indoor}(k) * \big(12 * (Area_1 + Area_2) + (6 + k_{solid}) * A_{space}\big)\Big] * \alpha_{indoor} \qquad 2$$

All the exterior building surface temperatures were calculated assuming an instantaneous heat balance between the absorbed shortwave solar radiation, long wave radiation and convective heat transfer to the external environment. The external convective heat transfer correlation used was that from the EnergyPlus reference handbook [7], presented in equation 3. Coefficients D, E and F varied according to the external surface characteristics ([7]) and V_z is the local wind speed.

$$h_{cv} = \mathrm{D} + \mathrm{E}V_z + FV_z^2 \qquad 3$$

To calculate the respective thermal mass of each building component exchanging heat with the indoor zone temperature (as expressed in equation 2), the materials used were aerated concrete for the inner wall leaf, concrete slabs for the floor and roof and a single layer of glass for the double glazing. The internal convective heat transfer was assumed to be 4 W/m^2.K for all vertical and 2 W/m^2.K for all horizontal surfaces, based on the EnergyPlus engineering reference [7]. To account for the extra thermal mass summed by office appliances (tables, desks, chairs, etc) a solid mass was included on each floor, occupying 2% of each floor air volume. The total building thermal mass (the thermal mass of the building fabric exchanging heat by convection with the indoor air) was 420675kJ/K, the cavity walls capacitance was 25128kJ/K, the glass capacitance was 19044kJ/K, the floors and roof combined capacitance was 290487kJ/K and the extra thermal mass 28672kJ/K per floor.

Passive solar gains were determined for each wall and then summed for each floor. Each wall had a specific angle of incidence with the sun calculated using the correlations detailed in [8]. To obtain each wall azimuth angle, each wall was oriented to a cardinal direction; the longer walls to north and south and shorter walls east and west. Their surface azimuth angle was their relative azimuth angle towards the south.

The predicted yearly cooling loads with the conventional HVAC system were considerably higher than the heating loads, as presented in Table 1 and in accordance to the values obtained in the EnergyPlus simulation by Korolija et al [4]. The highest cooling loads were obtained during the summer and autumn months and the majority of the yearly heating demand during the winter months.

Table 1 presents the average daily heating and cooling demand for each month. The cooling demand, peaks in August and is almost non-existent during winter months. To meet the base cooling load during the cooling season (when the daily cooling load surpasses its heating demand), from April to October, a daily average of 349 kWh/day cooling is required.

Table 1 – Predicted office daily heating and cooling loads for each month.

Month	Jan	Feb	Mar	Apr	May	Jun	Jul	Aug	Sep	Oct	Nov	Dec
Cooling load [kWh/day]	6	0	11	91	309	403	577	594	340	110	13	0
Heating load [kWh/day]	223	249	140	62	10	2	0	1	4	35	106	236

The average daily cooling required to maintain comfort values suggested by CIBSE [6], varies drastically between spring and autumn and summer months. The rooftop heat pump chosen provided 72.6kW of cooling and 69.8kW of heating [9]. To charge a thermal store in economy 7 tariff times, between 23.00 and 06.00, the total charging time available is 7 hours, consequently limiting the maximum storage capacity to 508kWh with the selected heat pump.

Two thermal energy storage system capacities were designed and simulated with the aim of reducing the cooling load of the building by: 1) using cold night time air temperatures (with a thermal storage capacity of 349kWh between 8 and 24°C using PCM RT15), and 2) using cold night time air temperatures with off-peak electrical cooling (with a thermal storage capacity of 508kWh between 1 and 24°C using PCM RT8HC) charged during economy 7 electricity tariff [10] times. Both thermal stores were designed to be fully charged in 7h and discharged in 10h, leading to charging/discharging cooling rates of 50/35kW for the store using RT15 and 73/51kW for the store using RT8HC.

2.1. Thermal storage model developed

A 2D finite volume model was developed in Matlab to simulate the performance of a store comprised of parallel slabs containing PCM in an office cooling system, similar to the geometry developed by Osterman et al. [11]. It assumes simple thermal diffusion, with no volume changes during the melting process and isotropic heat propagation within the PCM. Although the duct contained many PCM slabs it was assumed that flow between each pair of slabs would be similar allowing the simulation domain to be simplified. The model thus simulated air flow through a rectangular duct containing PCM slabs on its outer walls. The ducts side walls contained half of a rectangular PCM slab, the duct it was insulated on the top and bottom walls.

The air flow Reynolds number was calculated using the hydraulic diameter of the air gap between the slabs. The laminar Nusselt number was calculated using the expression for constant heat flux detailed in the VDI heat atlas [12]. For turbulent flow regimes, the Gnielinski's expression for turbulent pipe flow was used, also detailed in [12], using the hydraulic diameter of the flow passage .

2.2. Storage model validation

To determine the accuracy of the numerical models developed, an open-loop air flow experimental rig was assembled to test a developed thermal store. The temperature of the air supply was controlled using a shell and tube heat exchanger with a silicon oil circulating on the shell side, with temperature controlled by a Huber Tango temperature control unit. A centrifugal fan mounted on the air inlet circulated air through the shell and tube heat exchanger, delivering the required air flow to the thermal store, as shown in Figure 2B. A PCM thermal store was then constructed to verify the accuracy of the algorithm developed. It was constructed from 1mm thick HDPE sheet and wooden battens. Figure 2A presents a photograph of the parallel slab store constructed. It consisted of 5 PCM slabs 9 mm thick of 1000x1000 mm dimensions with 12 mm air spaces between them. It contained 40 L of PCM. Wooden battens were inserted between the PCM slabs to ensure a 12mm air gap was maintained and enabled the 1m long air flow path to be extended to 5m as seen in Figure 2B. Using a Huber Tango the store inlet air temperature could be set to temperatures between 5 and 30°C to fully charge and discharge the PCM store.

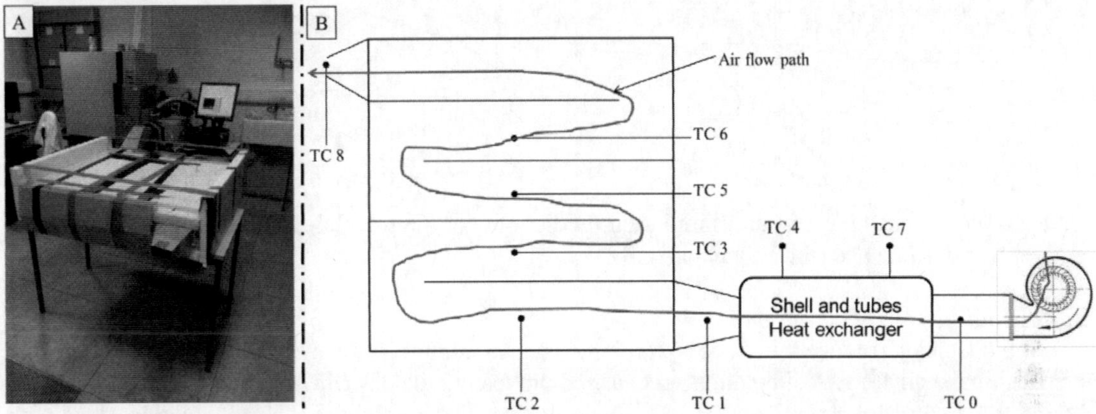

Figure 2 - General view of the assembled parallel slab store (A) and a schematic diagram showing the thermocouples location in the test rig (B).

2.2.1. Shape stabilized PCM used

A shape stabilized PCM composite was prepared in situ by blending Croda 21, an organic ester mixture developed by Crodatherm [13] with a 18°C melting point, with an enhanced polymeric adsorbent and additional materials in collaboration with Phase Energy Ltd [14]. The composite had a higher thermal conductivity and an extra 10 J/g enthalpy compared to the base mixture of Croda21 with the block copolymer only as seen in Table 2.

Table 2 - Material properties of the selected PCMs used in the experiment.

Material (%wt)	T_{min} - T_{max}	H_{stored}	λ_{solid}	λ_{liquid}	Price	
	°C	kWh/m^3	W/m.K		£/kg	£/kWh
Croda 21		53	0.180	0.150	4.00	69.7
Croda 21/adsorbent (80/20)	12-24	40	0.200	0.180	4.02	91.9
Croda 21/composite adsorbent (77/23)		43	0.470	0.350	3.50	81.1

The shape stabilized PCM produced could store up to 173.2J/g in the temperature range between 6 and 24°C, corresponding to a volumetric storage capacity of 45kWh/m^3. The thermal store developed had a storage capacity of 1.89kWh over the same temperature range.

Figure 3 presents the experimental results obtained and the simulation predictions. The obtained simulation results are in good agreement with the experimental results obtained (both charging and discharging enthalpies have a relative error of less than 5%), with a slight variation in the heat transfer rate during the charging process (freezing of the PCM). This might be due to dimensional changes not considered during the freezing of the shape stabilized PCM slab (contraction of the storage material allowing the formation of air gaps within the PCM slabs).

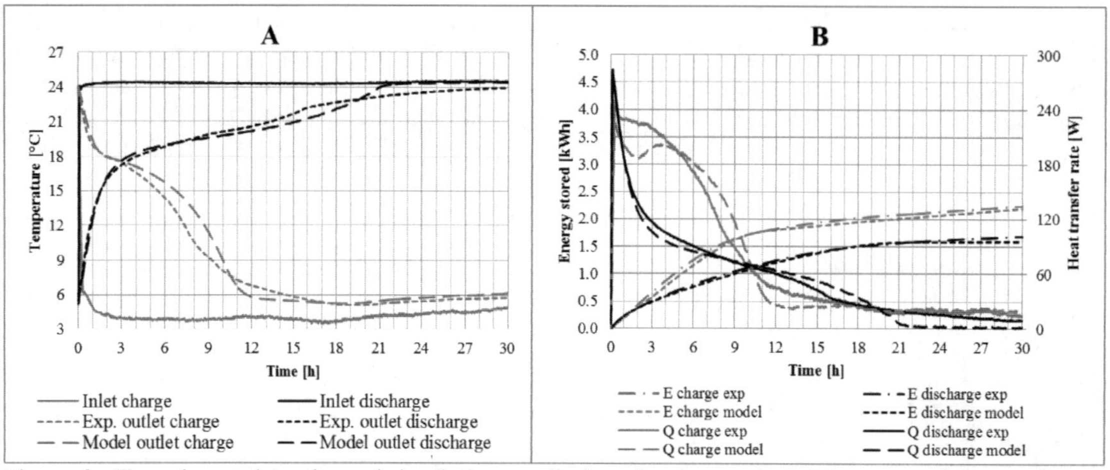

Figure 3 - Experimental results and simulation predictions for the parallel slab store outlet temperature (A), heat transfer rate and storage capacity (B).

2.3. *Thermal storage optimization*

To cool the simulated office building, larger flow rates are required, achieved with thermal stores with no baffles in the air flow. A simulation was carried out testing the day time cooling reduction potential of using free night cooling with the RT15 [15] composite. The possibility of shifting cooling load with night time charging was simulated with RT8HC [15]. The models were used to simulate a scenario in which each latent heat store was fully utilised. In regards to the operational temperature range, the store filled with RT8HC was charged at 1°C and the store filled with RT15 charged at 7°C; both were discharged at 24°C.

Figure 4 presents the simulated results for the parallel slab optimized stores with and without fins. The model simulated a full utilization scenario for each latent heat store. The store using RT8HC was charged at 1°C and at 7°C using RT15; both were discharged at 24°C. Both stores were designed to provide 10 hours of cooling at a constant rate of 50.8kW for the store filled with RT8HC and 34.9kW for the store filled with RT15. Both stores were charged in 7 hours, the predicted charging rates were 72.6kW for the store filled with RT8HC and 49.86kW for the store filled with RT15. The centrifugal fan flow rate was controlled between 50% and 150% of the nominal flow rate, according to the heat demand needed.

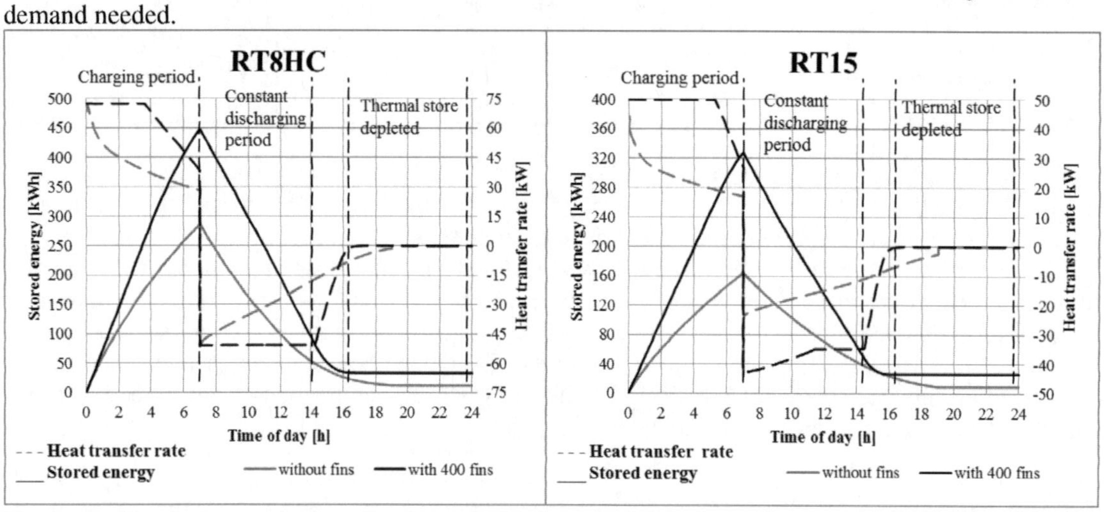

Figure 4 - Simulation predictions for the optimized parallel slab thermal store with RT8HC and for the thermal store with RT15 with and without fins.

IOP Conf. Series: Materials Science and Engineering **556** (2019) 012058 doi:10.1088/1757-899X/556/1/012058

3. Yearly simulation results

Using London Gatwick design weather data and the building design parameters, simulations were performed with HVAC systems and thermal stores for a whole year to determine potential enhancements in the daily performance and consequently reductions in yearly energy requirements with each approach. Four cases with extra fresh air provision were compared to the conventional HVAC system (A): on demand (B), using a store filled with RT15 (C) and using a store filled with RT8HC and with electrical night cooling (D). The integration of the thermal store into the space cooling network is schematically illustrated in Figure 5A.

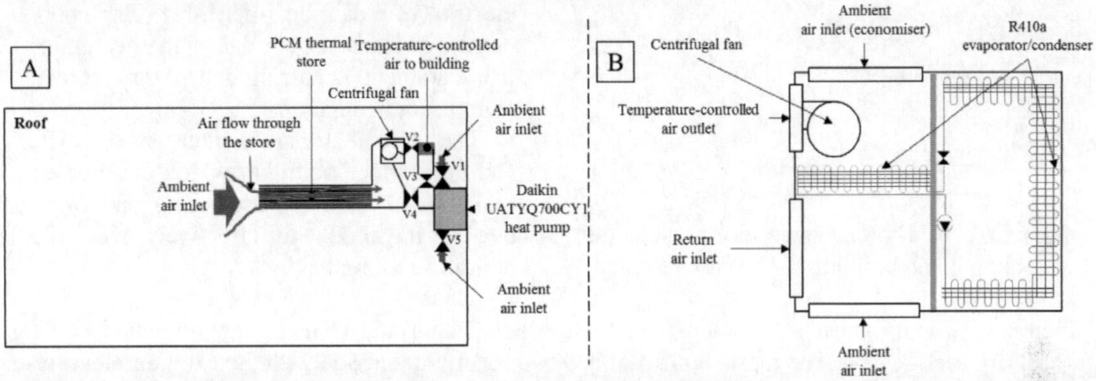

Figure 5 - Schematic representation of the parallel slab latent heat storage integration into the building HVAC system (A) and detailed view of the rooftop heat pump (B)

All cooling systems modelled were activated during daytime office hours when the indoor air temperatures were above 22.5°C. The heat pump was activated only if the indoor air temperature increased above 24°C. The coefficient of performance (COP), used for the heat pump during office working hours was 3 [9]. For the simulation using extra ventilation on demand, valve V2 would open (shown in Figure 5) and the centrifugal fan activated when indoor temperature increased above 22.5°C, allowing colder outdoor air to be introduced reducing the building cooling load during office working hours.

For the simulation with a thermal store filled with RT15, during the night valve V1 and V4 would open, directing the outside ambient air flow to cool/charge the thermal store. During office working hours (from 07.00 to 18.00), if the indoor temperature increased above 22.5°C valve V4 closes, V2 and V3 open and the centrifugal fan is activated, reversing the air flow through the store and consequently discharging it to provide space cooling to the office building. The air flow from the centrifugal fan was allowed to vary between 50% and 150% of its nominal flowrate.

For the simulations using the heat store filled with RT8HC, the heat pump was assumed to charge the latent heat store with a COP of 4.5, based on the Daikin UATYQ-CY1 rooftop air conditioners service manual [9]. During store charging, valve V5 and V4 are open, allowing the ambient air flow to be cooled by the heat pump if the ambient air temperature is above the required cooling temperature of 3°C. To effectively work and improve energy efficiency, all of the specified cooling systems are only operated if their COP calculated according to equation 4, is above 3.5. For the extra ambient fresh air provision on demand system, the ambient temperature is used instead of the thermal store outlet temperature. Using the centrifugal fan specifications from SODECA [16], the 3kW electrical input to the fan had an estimated 65% efficiency reported at 15 000m^3/s (varying its flowrate and electrical consumption according to the required cooling demand), meaning the designed cooling systems were only activated for cooling loads above 16kW.

$$COP(t) = \frac{\rho \times \dot{V}_{fan}(t) \times Cp \times (Tstore\ out(t) - Tindoor(t)) * \eta_{fan}(t)}{\Delta P_{fan}(t)} ; \quad 4$$

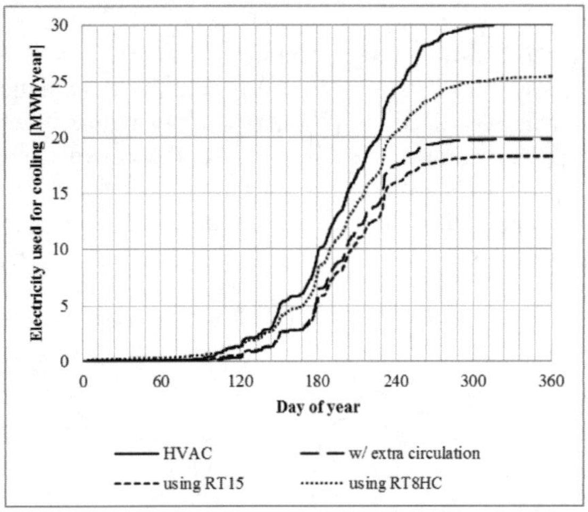

Figure 6 – Yearly total electricity consumed for cooling in the building for all cases studied.

Figure 6 presents the aggregated yearly results for each studied case. It can be seen that the obtained reduction in electricity use for cooling by using only extra air circulation can reach up to 34%. Using RT15 with ambient night time cooling used 39% less electricity for cooling. Shifting daily cooling loads with RT8HC obtained a yearly electricity reduction of 15%. The poor performance values are due to the heat gains from ambient occurring when the thermal store is not being used.

The thermal storage system with RT8HC and night time cooling provided the greatest reduction in electricity use during peak times due to its ability to effectively shift the building cooling load.

Figure 7 presents the daily cooling load for the heat pump and COP of the simulated cooling approach for each season. The most significant levels of energy reduction with the extra ambient fresh air provision on demand are predicted mainly during the spring and autumn months and the reductions are minimal in summer months due to the higher ambient temperature.

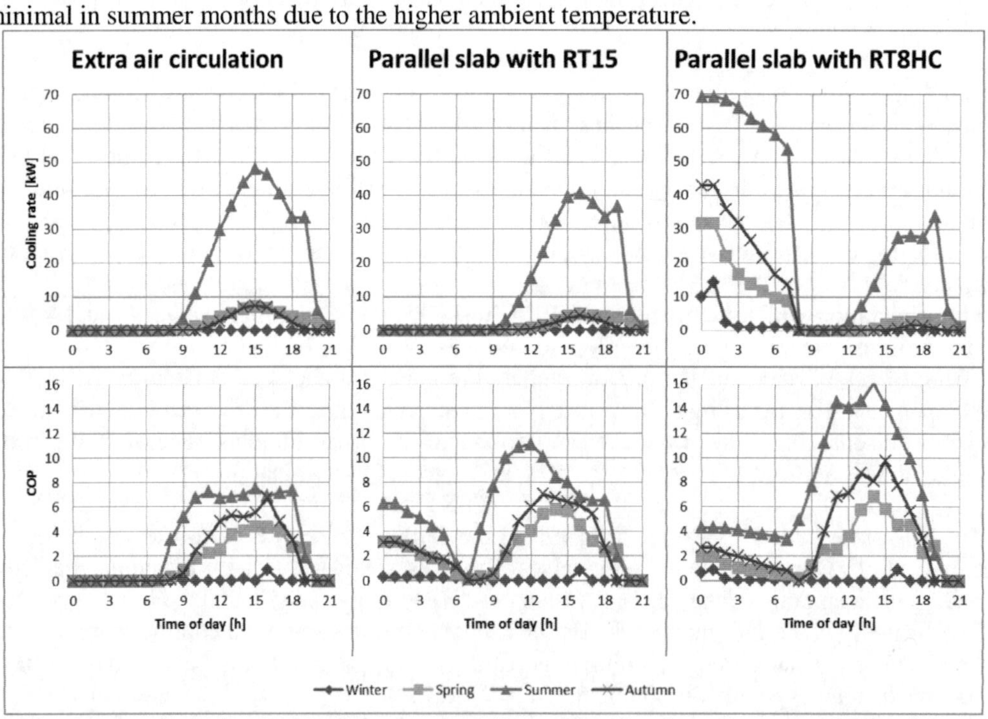

Figure 7 - Seasonal results for the heat pump daily cooling loads and system COP with extra air circulation, for the parallel slab thermal stores with RT15 and with RT8HC

The reduction in the cooling load during office working hours in summer is larger using a store filled with RT15 with free night cooling; however, the overall gain is reduced by the additional energy required for the fan to charge the store during the night. Using RT8HC with electrical night cooling successfully enabled daily cooling loads to be shifted into off peak times in both spring and autumn months and significantly reduced the cooling load during the summer months, shifting the summer peak cooling load to 19.00. Full load shifting during summer months was prevented by the capacity of the selected heat pump (due to restricted charging period, the maximum energy able to be charged was 7h*16.13kW*4.5=508kWh, less than the maximum required in summer months, 594kWh).

The increase in predicted COP for the thermal storage system using RT15 compared to the extra ambient fresh air provision on demand system was more pronounced during summer months and only slightly higher during spring and autumn months. The predicted daily COP profile for the storage system with RT8HC was very similar to that of the system with RT15 during spring and autumn months, but considerably higher during summer months.

3.1. Economic comparison of all modelled systems

A general financial assessment was made to assess the three approaches described and compared to the conventional HVAC system. Table 3 presents the initial capital expense (CAPEX) and operational expense (OPEX) costs for the 4 cases studied. For the heat pump system, Daikin UATYQ700MCY1 [9] was selected due to meeting the heating and cooling required. An industrial-scale centrifugal fan [17] with a nominal flow rate of 15 000m^3/h was selected from SODECA [16]. For the parallel slab thermal stores, assuming that the shape-stabilized PCM slabs are enclosed in steel sheets and that compressed corrugated sheets are included as fins in the air gaps to increase heat transfer area, a cost of 650£/ton was used [18] as a cost for the metal.

Due to the large size of the designed systems, the whole system installation costs were assumed to be 5% of the initial CAPEX [19]. In the thermal stores, the composite PCM was the largest capital component, representing 53% of the initial CAPEX. Operation and maintenance costs were assumed to be 0.9£/m^2 of office space, in accordance with [20]. The waxes used were assumed to have an indicative price of 4£/kg [15] and the composite adsorbent material developed by [14] an indicative price of 2.65£/kg, when mixed with a 80/20 mass ratio providing a composite storage material cost of 3.73£/kg. For the standard electricity tariff, Npower provided a price of 0.1664£/kWh with a standing charge of 55.1£/year. Using economy 7 electrical tariffs, Npower provided a peak tariff of 0.2131£/kWh and an off-peak tariff of 0.0765£/kWh, with the same standing charge. The levelized cost of energy (LCOE) for each system was calculated using equation 5, from [21].

$$LCOE = \frac{\left(\frac{CAPEX \times r}{1-(1+r)^{-n}} + OPEX + E_{used} * E_{tariff}\right) \times n}{W_{produced} \times n} \left[\frac{£}{kWh}\right]; \qquad 5$$

Table 3 – CAPEX and OPEX costs comparison for the systems studied.

Component		HVAC (heat pump only)	+extra circulation	Parallel Slab With RT15	Parallel Slab With RT8HC	Ref.
Thermal Store cost [£]	Insulation+ structure			1 057	1 089	[22]
	PCM enclosure			6 317	6 527	[18]
	PCM + composite			32 474	33 556	[15]
	Centrifugal fan		3 142			[23]
Heat pump cost [£]		17 305				[24]
Installation costs [£]		865	1 022	3015	3081	[19]
CAPEX [£]		**18 170**	**21 469**	**63 310**	**64 700**	
Yearly system maintenance [£/year]		1 382				[20]
Yearly electricity used [kWh]		7 206	5 504	5 245	6 087 (econ7)	
OPEX [£/year]		**8 588**	**6 886**	**6 627**	**7 469**	-
LCOE [£/MWh]		**97.17**	**84.45**	**120.77**	**129.89**	

For a 20-year life cycle with an annual interest rate of 8%, the LCOE calculated for each system is presented in Table 3. The system with extra air circulation presents the lowest LCOE due to the reduction in the energy required to meet the cooling load achieved with ambient air cooling. Due to the high CAPEX costs of the thermal stores, their yearly reduction in electricity costs compared to the conventional system were not enough to provide a more economical cost of energy.

4. Conclusions

Medium to highly glazed buildings with lower U-value building fabric generally consume more energy for cooling than for heating due to the high solar gains transmitted through the glazing, high occupancy and electronic devices emitting heat. The use of extra mechanical ventilation gave the lowest price for cooling of 84.45£/MWh. This is a consequence of the typical UK external ambient air temperature during office working hours being mostly below 24°C, (97% for the London Gatwick IWEC weather data).

Using a thermal store with RT8HC with off peak electrical cooling successfully shifted office daily cooling loads during spring and autumn months and in summer months was capped by the capacity of the heat pump used. Night time ambient cooling with RT15 obtained the highest reduction for cooling, but the high capital cost makes it less feasible than using extra ventilation only.

5. References

[1] D. Connolly, B. V. Mathiesen, P. A. Østergaard, S. Nielsen, U. Persson, and S. Werner, "Heat Roadmap Europe 2050: Second Pre-study for the EU27," *Energy Eng.*, 2013.
[2] K. Sipilä, *Advanced District Heating and Cooling (DHC) Systems*, 1st ed. Elsevier, 2016.
[3] M. Chaudry, M. Abeysekera, S. H. R. Hosseini, N. Jenkins, and J. Wu, "Uncertainties in decarbonising heat in the UK," *Energy Policy*, 2015.
[4] I. Korolija, Y. Zhang, L. Marjanovic-Halburd, and V. Hanby, "Selecting HVAC Systems for Typical UK Office Buildings," Jun. 2009.
[5] CIBSE, *CIBSE Guide A: Environmental Design*. 2015.
[6] D. B. Crawley, C. O. Pedersen, and L. K. Lawrie, "Energy plus: Energy simulation program," *ASHRAE J.*, vol. 42, no. 4, pp. 49–56, 2000.
[7] EnergyPlus, "Engineering Reference," *US Dep. Energy*, 2010.
[8] J. A. Duffie and W. A. Beckman, *Solar Engineering of Thermal Processes: Fourth Edition*. John Wiley & Sons, 2013.
[9] Daikin, "ESIE05-04 Service Manual R-410A Sky-Air Indoor," *Service manual*, 2005. .
[10] F. Murray and S. Stern, "Innovation Policy and the Economy 7," *When ideas are not free: The impact of patents on scientific research*. 2006.
[11] E. Osterman, V. Butala, and U. Stritih, "PCM thermal storage system for 'free' heating and cooling of buildings," *Energy Build.*, vol. 106, pp. 125–133, Nov. 2015.
[12] W. Wagner, *VDI Heat Atlas*, vol. 2010. Springer Berlin Heidelberg, 2010.
[13] CRODA;, "Crodatherm international Plc." [Online]. Available: http://www.crodatherm.com/.
[14] I. Biggin, "Phase Energy Ltd," *Holly Bush House, Sheriff Highway, Hedon, Hull, HU12 8HA, United Kingdom*, 2017. [Online]. Available: http://phase-energy.com/.
[15] Rubitherm Technologies GmbH, "Rubitherm GmbH," *Rubitherm GmbH*, 2016. [Online]. Available: http://www.rubitherm.eu/.
[16] SODECA, "Sodeca, efficient work fans," 2017. [Online]. Available: http://www.sodeca.com/en.
[17] W. Cory, *Fans and Ventilation*. Elsevier Ltd, 2005.
[18] L. Wuxi Raymond Steel Co., "Hot Dipped Galvanized Steel Sheet," *Jiangsu, China (Mainland)*, 2017. [Online]. Available: https://www.alibaba.com/product-detail/Hot-Dipped-Galvanized-Steel-Sheet-Sheets_60630874458.html?spm=a2700.7724857.main07.269.42adbc51WA5teT.
[19] A. Teke and O. Timur, "Assessing the energy efficiency improvement potentials of HVAC systems considering economic and environmental aspects at the hospitals," *Renew. Sustain. Energy Rev.*, vol. 33, pp. 224–235, May 2014.
[20] K. Nissinen, P. Takkunen, and H. Ito, "Maintenance and repair cost of office buildings in Japan," in *Durability of Building Materials and Components 7*, 2014.
[21] A. Smallbone, V. Jülch, R. Wardle, and A. P. Roskilly, "Levelised Cost of Storage for Pumped Heat Energy Storage in comparison with other energy storage technologies," *Energy Convers. Manag.*, vol. 152, no. Supplement C, pp. 221–228, 2017.
[22] 203801000, "EPS 70 Polystyrene Insulation Board," *Insulation Express*, 2017. [Online]. Available: http://www.insulationexpress.co.uk.
[23] SODECA, "Sodeca, efficient work fans," 2017.
[24] D. technical team, "Daikin Air Conditioning Rooftop Packaged Unit UATYQ700CY1," *Orion air*, 2017. [Online]. Available: www.orionairsales.co.uk.

Acknowledgments

The research presented in this paper is funded by the EPSRC through grant [EP/K011847/1], the Interdisciplinary centre for Storage, Transformation and Upgrading of Thermal Energy (i-STUTE) and a Loughborough University PhD studentship.

IOP Conf. Series: Materials Science and Engineering **556** (2019) 012059 doi:10.1088/1757-899X/556/1/012059

Steady-Flow-Type Particle Receiver for High-Temperature Solar Thermal Storage

Koji Matsubara[1,4], Yudai Suzuki[2,4], Atsushi Sakurai[1,4], Selvan Bellan[3,4] and Tatsuya Kodama[1,4]
[1]Faculty of Engineering, Niigata University, Niigata, Japan
[2]Graduate School of Science and Technology, Niigata University, Niigata, Japan
[3]Center for Transdisciplinary Research, Niigata University, Niigata, Japan
[4]Pacific Rim Solar Fuels Research Center, Niigata University, Niigata, Japan

E-mail: matsu@eng.niigata-u.ac.jp

Abstract. The concentrated solar power is utilized for industrial thermal usage, electricity generation and fuel production. To increase the solar thermal temperature is essential for enhancing the thermal efficiency of the energy conversion and for widening application. The synthetic oil, widely used for solar thermal storage, pylorizes at 400℃, and this cannot be used for very high temperatures. The authors used quartz sand as thermal storage medium in the fluidized bed receiver, and achieved the thermal receiver temperatures beyond 1000℃ in the previous report. The research work was further made on using the particle steady flow for high-temperature solar thermal storage. The steady-flow-type particle receiver was designed through visualization of cold model and numerical simulation. The experimental apparatus was fabricated and tested using beam-down sun simulator of 1.0 kWth. The apparatus comprises of the internal-circulating fluidized bed receiver placed between a loop seal with an inlet port and another loop seal with a weir and an exit port of the particles. The tested particles are quartz sand of 0.1 – 0.2 mm and fused brown alumina of 0.09 – 0.1 mm. The experimental set up produces hot particles of quartz sand at 428℃ and those of fused brown alumina at 491℃. The highly efficient steady-flow-type particle receiver created steady particle flow smoothly. This apparatus takes in the concentrated light through the quartz window to continuously discharge the particles at high temperatures beyond the criterion in synthesis oil (400℃) for solar thermal storage.

Keywords: high-temperature solar thermal storage; sun simulator; experiment; visualization; simulation

1. Introduction

The concentrated solar power (CSP) is commercialized in the sunbelt of Europe, United States and Asia. The condensed light collected from the solar field using the heliostats is used for heat source of desalinization, power generation and fuel production [1-3]. The solar thermal is advantageous against other renewable energy in capability of the massive storage at low cost. The solar thermal storage enabled 24 hour power generation in Europe and United States. Recent progress on solar fuel production is opening the way to recover carbon dioxide into synthetic gas. The newest CPS plants use the molten nitrate as heat transfer and storage medium. The solar receiver receives the concentrated irradiation to create the sensible enthalpy at as high temperatures as 560 ºC [4]. To increase the solar thermal temperatures is very important for enhancing the thermal efficiency of energy conversion and for widening the industrial application. However, the nitrate salt cannot be used for high temperatures beyond 600 ℃ since the thermal medium is thermally broken up at such high temperatures. New

IOP Conf. Series: Materials Science and Engineering **556** (2019) 012059 doi:10.1088/1757-899X/556/1/012059

technologies of heat transfer media are studied by national institutes and universities for high-temperature solar absorption and thermal storage.
This study aims to create new solar particle receiver which utilizes the steady stream of the particles for solar thermal acquisition. The present authors develop the fluidized bed solar receiver for high temperature solar absorption, and achieved very high temperatures above 1000 ℃ [5]. The previous apparatus applied the cylindrical fluidized bed with quartz window to take in the concentrated light which directly irradiated onto the bed surface. The present study describes the newer apparatus extended to utilize the steady stream in the fluidized bed configuration. Visualization was made for cold model of the apparatus to investigate the smoothness of the polystyrene particle flow. Numerical simulation was then conducted for analyse the flow field of the ceramic particles. The experimental model was designed based on these analysis, and this was fabricated and tested using sun simulator of 1.0 kWth.

2. Visualization of cold model

Prior to the design of the hot model of steady-flow-type particle receiver, visualization was conducted for cold model of it to confirm the validity of the configuration.

2.1. Experimental method

Figure 1 shows experimental apparatus for visualization of the cold model of steady-flow-type particle receiver. The transparent channel comprises of the inner-circulating fluidized bed placed between two loop seals. One seal on the left is equipped with inlet port for particle feeding and the other seal on the right has the weir and the outlet port for particle discharging. The feeder feeds the polystyrene particles with 0.7 - 1.4 mm diameter into the transparent channel for visualization. The air was provided by the compressor into the channel through the bottom distributor plate. Four lines of air stream fluidized the particles and exit from outlet port on the top.

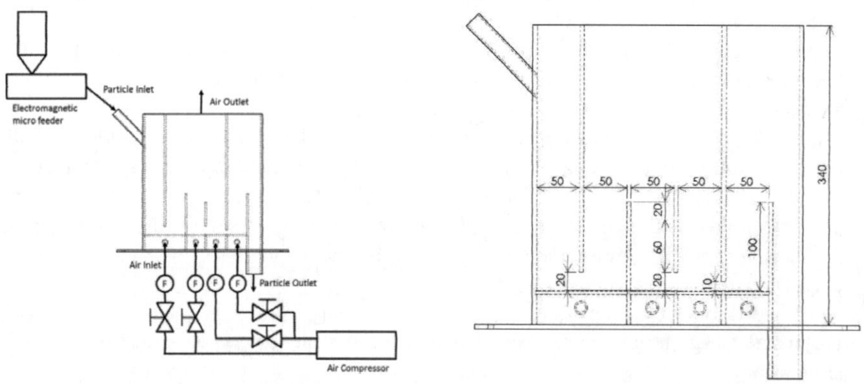

(a) Experimental configuration (b) Transparent channel of receiver

Figure 1. Experimental apparatus for visualization of cold model of steady-flow-type particle receiver

2.2. Experimental results

The experimental visualization was conducted changing the flow rate of air and particles. Figure 2 exemplifies the photograph of particles, where particle flow rate was 52.7 g/min and air flow rate was 40 NL/min, 60 NL/min, 20 L/min and 20 NL/min for each part of the air distributors. The superficial velocities are the same at the inlet side and outlet side of loop seals. However, this velocities differ between two towers, which creates the internal circulation of particles. At the initial state, blue and white particles fill separately each sub-channels as shown in figure 1 (a). These two colour particles enabled to observe the stream of the particles. As presented in figure 2 (b), the particles drop from the feeder onto the inlet side of the loop seal, stream inside the channel and internally circulates in the central two-tower channel and reaches the exit side of the loop seal and overflows from the weir. The experimental visualization was made for systematically changing the condition and the flow pattern was summarized by the flow map which separates the no-stream and stream area.

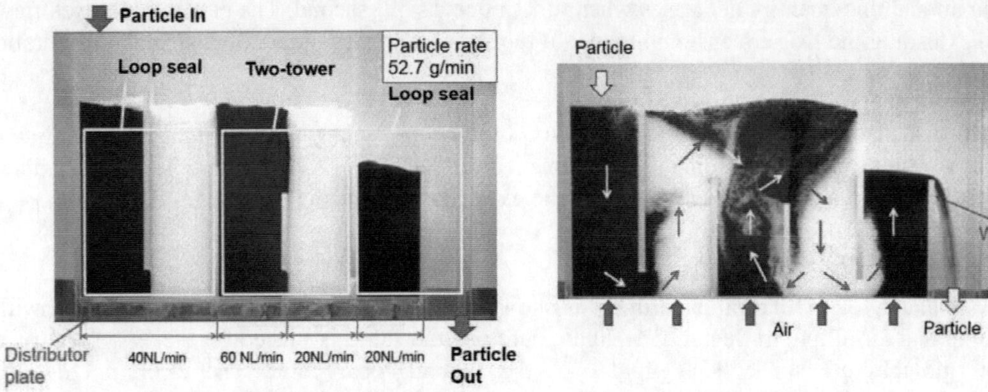

(a) Initial state (b) t = 3 s

Figure 2. Photograph of particle stream

3. Numerical simulation of cold model

Two-dimensional numerical simulation was made for design of the irradiation-experiment model. The particle considered in the simulation was quartz sand for high-temperature experiment.

3.1. Numerical method

The simulation was performed using ANSYS Fluent Ver. 16.0. Eulerian-Eulerian approach based on granular method was applied for two-dimensional model of steady-flow-type receiver. The particles simulated are quartz sand with 0.1 mm diameter. This particle was selected since they are used for irradiation experiment. The computational domain resembles the cold model for visualization. Some minor changes were made in preparation for irradiation experiment; inlet port was shaped in triangle and the whole height and width were shrunken to half of the visualization apparatus. The down-sizing was performed considering capacity of the sun simulator and easy handling. Numerical conditions are air flow rate = 52.7 g/s, particle flow rate = 29.3 g/s assuming 30mm receiver depth.

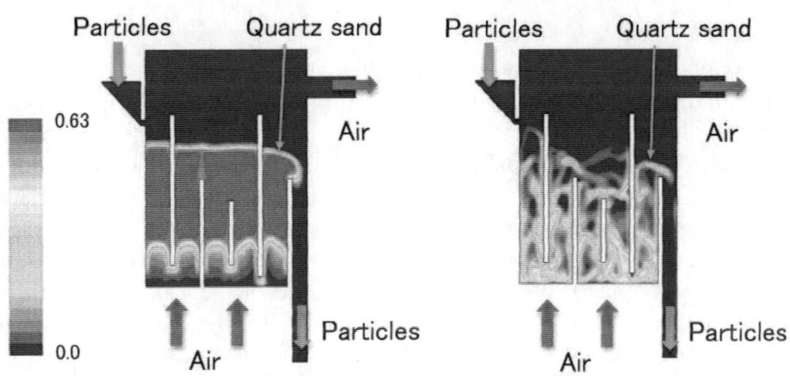

(a) t = 0.1 s (b) t = 5.0 s

Figure 3. Volume fraction of particles in cold model

3.2. Numerical results

The volume fraction of particles is indicated in figure 3. The particles are continuously fed from the inlet port. These particles are observed to stream inside and exit from the channel. The particle stream was smooth in this down-sized model. The volume fraction is large at 0.1 second from the start of the providing air and this quantity decreases when time proceeds 5.0 second. The experimental set up was designed extending the two-dimensional model of the simulation into three-dimensional configuration.

4. Experiment of irradiating receiver

Experiment is made for steady-flow-type particle receiver using sun simulator. Experimental conditions are scrutinized by preliminary experiment with electrical heating wire before the irradiation experiment. Two kinds of particles are used for the experiment with condensed light irradiation.

4.1. Experimental method

The sun simulator for irradiation experiment is shown in figure 4. This apparatus is equipped with a xenon lump and an elliptic reflector. The light source of the lump is placed on the first focal of the reflector, and the reflector collects light onto the second focal point. The figure includes the distribution of heat flux at the focal point. The Gordon gouge was traversed at 5.0 mm interval to measure the heat flux distribution point-by-point. The heat flux distribution was peaky as shown in figure 4. The peak value is almost 500 kWth/m^2. Thermal power enclosed by 50 mm x 50 mm square is 0.75 kWth.

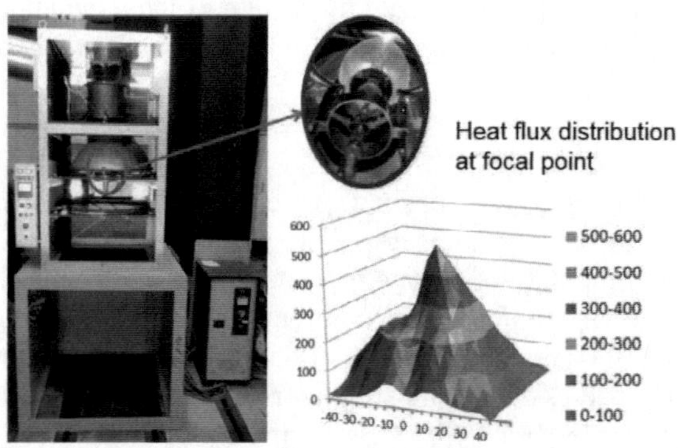

Figure 4. Sun simulator

The experimental model of steady-flow-type receiver has a frustum-shape head with a quartz window (120 mm x 120mm) on the top to take in the concentrated light as shown in figure 5. The main parts of the receiver are made of stainless steel, SUS304. The apparatus is equipped with a stainless steel distributor with mesh size 40μm. The visualized and simulated model separated the air room below the distributor in order to differentiate the superficial velocity to each channel of the receiver. However, the experimental model was simplified so that the room is not separated and the same superficial velocity is provided to each channel. The apparatus is set inside the cabin of sun simulator. The top surface of the particle bed is placed on the second focal point of the elliptic reflector for efficiently absorbing the concentrated light. The receiver is surrounded by the adiabatic sheet for insulation. Temperatures are measured at several positions of the apparatus using thermocouples. Air installation and irradiation started at the same time, and they continue during 90 minutes. Experiments are made changing the particle supply rate at different levels of flow rate of air. Two kinds of the particles are used as heat transfer medium; quartz sand with 0.1 - 0.2 mm diameter and brown fused alumina with 0.09 - 0.1 mm diameter.

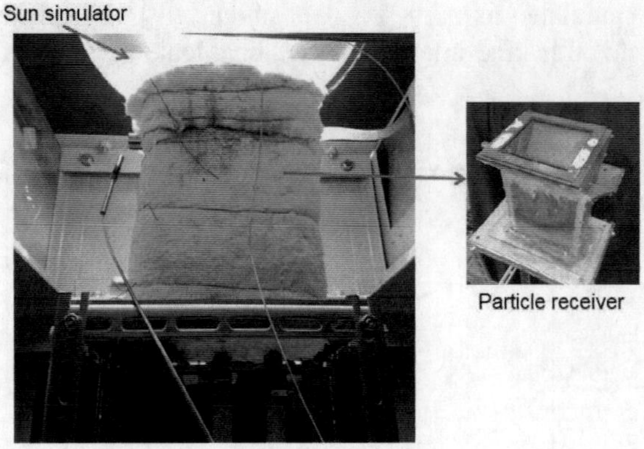

Figure 5 Experimental set up for irradiation experiment.

4.2. Experimental result

Figure 6 shows the experimental result of temperature measurement for the case of quartz sand particles. The location of the temperature measurement is in the light tower of two-tower area at middle height of the particle bed. In the figure, the temperature becomes highest at the particle supply rate of 20 g/min and air mass flow rate of 110 L/min. However, the temperature becomes lowest at the same particle supply rate at air mass flow rate of 120 L/min. The temperature measurement was also made in the loop seal with exit port. The temperature was found lower by about 20 - 40℃ in the loop seal than in the two tower. It is suggested that the thermal insulation is not enough for attenuating the heat losses. In the figure, the maximum temperature which occurs at the former condition is nearly 430 °C. Therefore, the newly designed apparatus can absorb the concentrated light at the higher temperatures than criterion of the synthetic oil.

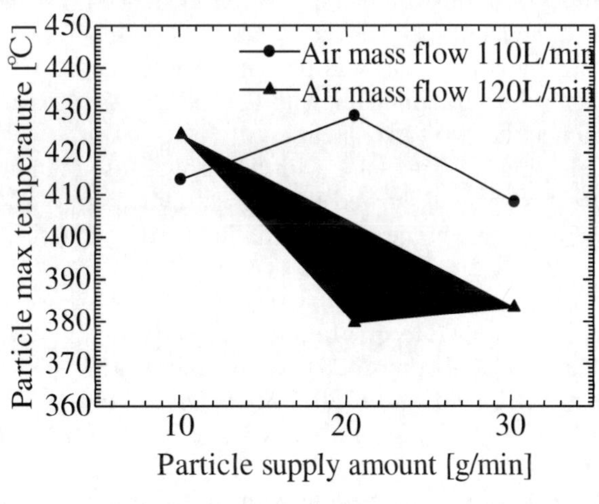

Figure 6 Temperature in two tower of steady-flow-type receiver using quartz sand.

The temperature measurement at the same position using brown fused alumina is shown in figure 7. This type of particle was used because it has preferable absorption characteristics. Quartz sand is white and its absorption coefficient is low. Brown fused alumina is abrasives produced from melting calcined bauxite. Its particles are dark colored and absorption coefficient is high. The experimental data are limited to only five cases. The temperatures tend to decrease at two level of air mass flow rate. The maximum temperature is achieved at particle supply rate of 10 g/min at air mass flow rate of 120 L/min. This maximum value is 490 °C, and this is quite higher than the limitation of synthetic oil usage. In this condition, The thermal power of heated particles is 62 W, and that of air exit port is 192

W. The efficiency is thus calculated as 33.9 %. It is suggested that heat losses occur due to the air flow exiting from the particle outlet and inlet ports. This heat leakage can be fixed by the improvement of the design of receiver.

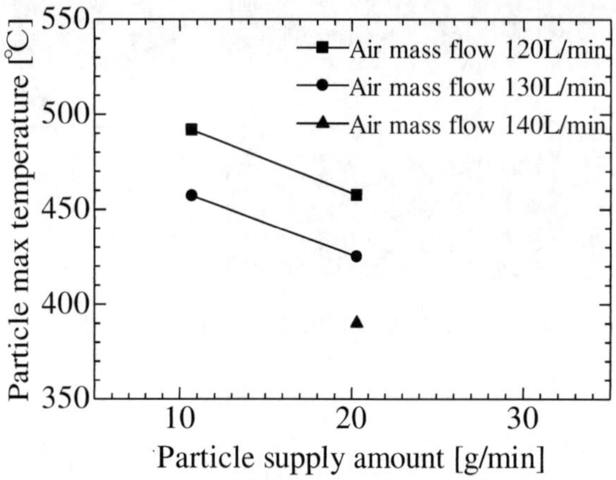

Figure 7 Temperature in two tower of steady-flow-type receiver using brown fused alumina.

5. Conclusions

This paper describes visualization, simulation and irradiation experiment of the steady-flow-type solar receiver. Conclusions can be summarized as follow:

(1) The newly designed steady-flow-type receiver comprises of a two-tower-type inner-circulating fluidized bed placed between two loop seals with inlet and exit ports. Visualization of cold model demonstrated that this apparatus could maintain steady flow of the particle and the inner-circulating of the particle at the same time. Two-dimensional simulation evidenced that the particle of quartz sand can stream in the down-sized model of the receiver.

(2) The receiver for irradiation experiment was designed based on the visualization and numerical results. This was tested using the sun simulator with 0.75 kWth. The experiment using quartz sand achieved 430 °C at maximum at the two-tower area.

(3) The receiver was tested using sun simulator with the brown fused alumina with higher radiation absorption coefficient. This experiment showed the highest temperature at 490 °C at the two-tower area. This highest temperature is quite higher than the limitation of synthesis oil usage (400 °C).

6. References

[1] Stein W H, Buck R, 2017 *Solar Energy* Volume **152**, 91-105

[2] Iaquaniello G, Montanari W, Salladini A 2017 *Solar Energy* **157** 1056-1063

[3] Krenzke P T, Fosheim J R, Davidson J H 2017 *Solar Energy* **156** 48-72

[4] Matsubara K, Sakai H, Kazuma Y, Sakurai A, Kodama T, Gokon N, Cho H S, Yoshida K 2015 *Energy Procedia* **69** 487-496

[5] Kodama T, Gokon N, Cho H S, Matsubara K, Kaneko H, Senuma K, Itoh S, and Yokota S 2017 *AIP Conference Proceedings* **1850**, 100012

IOP Conf. Series: Materials Science and Engineering **556** (2019) 012060 doi:10.1088/1757-899X/556/1/012060

Direct Simulation of Volumetric Solar Receiver with Highly Concentrated Radiation

Mitsuho Nakakura[1], **Koji Matsubara**[2,4], **Selvan Bellan**[3,4] **and Tatsuya Kodama**[2,4]
[1]Graduate School of Science and Technology, Niigata University, Niigata, Japan
[2]Faculty of Engineering, Niigata University, Niigata, Japan
[3]Center for Transdisciplinary Research, Niigata University, Niigata, Japan
[4]Pacific Rim Solar Fuels Research Center, Niigata University, Niigata, Japan

E-mail: f16k001f@mail.cc.niigata-u.ac.jp

Abstract. The volumetric solar receiver is a promised technology for high-temperature solar absorption in the concentrated solar power, industrial solar thermal usage and solar fuel production. Existing numerical simulation separated computations of light and convective flow and two-way coupling was not perfectly integrated between radiation and convective transport. This paper describes conjugate analysis of radiation, convection and conduction heat transfer in the volumetric receiver for perfect three-way coupling between three mechanisms of heat transfer. The numerical scheme was used for the volumetric receiver with plane surface and that with cut-back inlets for optimization of the geometry. The simulation was made for single cell of honeycomb with 2.4 mm pitch and 0.5 mm wall thickness. Two cases of surface geometry was computed: plane surface; cut-back inlet with 10 mm depth. Basic equations are momentum and energy equations with radiation transport equation. The discrete ordinates method was applied for solving the radiation intensity to consider perfect interaction between radiation, convection and conduction. The material of the receivers are assumed as stainless steel. The numerical simulation demonstrated the superiority of the cut-back receiver which increased the exit temperature by 30 K and decreased 2 Pa for the smallest beam angle of 10°. The contour plot of total heat flux on the wall revealed that the shadow effect was attenuated by the cut-back inlet leading to heat transfer enhancement. The novelties of this work are to treat perfect interaction between radiation and convection in volumetric receiver and to demonstrate the excellence of cut-back inlet receivers.

Keywords: conjugate radiation and convection analysis; volumetric solar receiver with cut-back inlets; total heat flux including radiation and convection

1. Introduction

The volumetric air receiver is utilized for concentrated solar power generation, solar fuel production and returning carbon dioxide into fuel [1-3]. Conventional receiver adopting molten nitrate salt collect thermal energy at 560 ºC in the newest solar power plant. However, this thermal medium cannot be used at the higher temperatures than 600 ºC since this is thermally broken up beyond this criterion. To increase the heat source temperatures is essential for enhancing the thermal efficiency of energy conversion and for broadening the applicability of solar heat. Therefore, there are research works on air receivers for high-temperature solar acquisition. Numerical simulation is powerful tool for designing and optimizing solar receivers. Most of the simulations made for volumetric solar receivers utilize the volume-averaged equations for solving flow and temperature fields. Recently, porous structures are considered by using the Monte Carlo approach for ray tracing and the conventional Eulerian method for flow and heat transfer. Most of these simulations lack the consideration on full interaction between condensed light and convective transport.

The present paper describes the simulation for single cell of honeycomb receiver with highly concentrated light irradiation [4, 5]. The simulation employs the discrete ordinates (DO) approach for considering three-way coupling between radiation, convection and conduction. Simulation was made for stainless steel honeycomb with plane surface and with cut-back inlets for optimizing the surface geometry.

2. Numerical method

2.1. Numerical model

Figure 1 shows the numerically treated situation, honeycomb irradiated by concentrated light. The air streams inside the honeycomb cells and absorbs sensible enthalpy created on the honeycomb wall. Simulation was made for a single cell of honeycomb as shown in figure 2. Across the top boundary, irradiation and air stream enters the computational domain. The radiation is absorbed and reflected by diffuse reflection on the honeycomb wall. Re-radiation occurs from the honeycomb wall according to the Stefan-Boltzmann law. The net radiation on the wall is balanced with the convective transport in the fluid side and the conduction in the solid side. All the three mechanisms of heat transport are considered in the simulation.

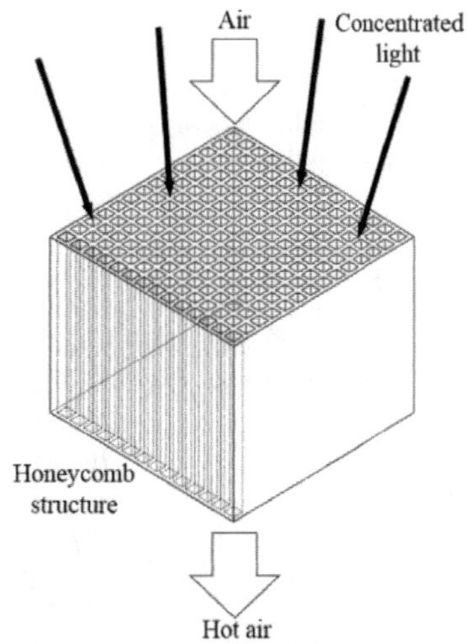

Figure 1. Honeycomb receiver with concentrated light irradiation.

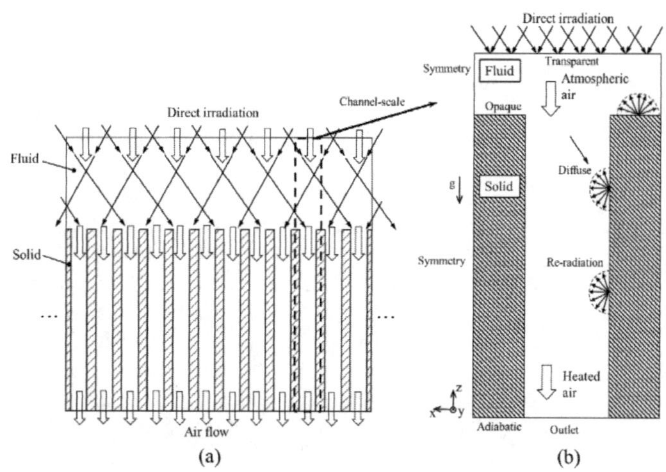

Figure 2. Computational domain.

2.2. Cell geometry

Two kinds of receivers shown in figure 3 are treated in the simulation. One is a simple cell with plane surface. The cell pitch is 2.9 mm, and the cell wall is 0.5 mm. The other is a cut-back inlet cell with 2.4 mm × 2.4 mm cut with 10.0 mm depth. These cell have 50 mm length. The previous study experimentally tested the honeycomb block of ceramics with the same geometry as the simple cell [4]. The experimental data was used for validation of the numerical simulation [5]. This paper simulates the honeycomb cell made of stainless steel. The complex geometry of ceramics requires high level of manufacturing technology. Whereas recently emerging three-dimensional metal printer enabled complex geometry of metal at low level of fabrication skills.

2.3. Computational conditions

Table 1 summarizes computational conditions. In this paper, the heat flux of incidence is fixed at 500 kW/m^2 and air mass flow rate at 6.17×10^{-3} g/s. The maximum incidence angle is changed from 10 to 90° in nine steps for each of two kinds of cells. Grid convergence test was made between two kinds of grid resolution with number of grid cells of 34,560 and 276,480. The discrete ordinates number was changed in the range of 3×3×8 to 18×18×8. There are almost no effects from the grid resolution and discrete ordinates in these conditions. The simulations listed in table 1 was made using the grid number of 34,560 and discrete ordinates of 18×18×8.

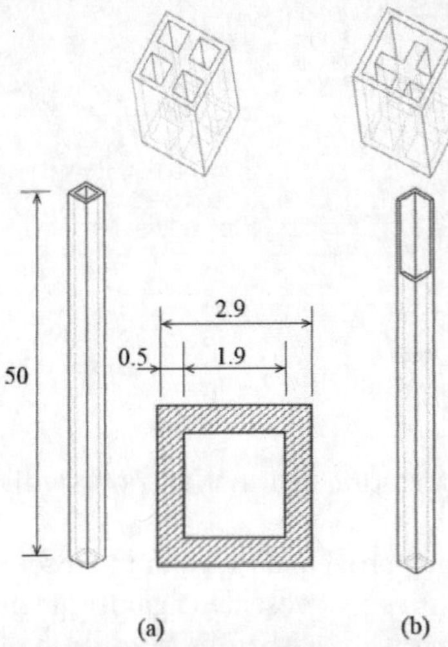

Figure 3. Simple cell and cell with cut-back inlet.

Table 1. Computational conditions.

Case	Channel geometry	Maximum incidence angle of DNI, α (deg.)	No. of DO discretization	DNI, I_0 (kW/m^2)	Air mass flow rate, $\dot{m}$ (g/s)
1	simple / cut-back	10	18×18	500	6.17×10^{-3}
2	simple / cut-back	20	18×18	500	6.17×10^{-3}
3	simple / cut-back	30	18×18	500	6.17×10^{-3}
4	simple / cut-back	40	18×18	500	6.17×10^{-3}
5	simple / cut-back	50	18×18	500	6.17×10^{-3}
6	simple / cut-back	60	18×18	500	6.17×10^{-3}
7	simple / cut-back	70	18×18	500	6.17×10^{-3}
8	simple / cut-back	80	18×18	500	6.17×10^{-3}
9	simple / cut-back	90	18×18	500	6.17×10^{-3}

3. Numerical result

The distribution of solid and fluid mean temperatures along with vertical position are presented in figure 4. The mean temperature of solid is the temperatures averaged across the cross-section. The mean temperature of fluid is bulk mean temperature. In figure 4 (a), the solid temperature increases from the simple cell to cut-back cell by 70 K at the inlet. Two cells are made of the same material with the same absorption coefficient; however the solid temperature differs a lot between two cases. It is thus suggested that the effective absorptivity is changed by the cut-back inlets. The temperature difference between two cases remain in the deep area of each cell. In figure 4 (b), the temperature of cut-back inlet is clearly higher than the simple cell at the cell exit. This demonstrates that the thermal performance is improved by the cut-back inlet and modification of the effective absorptivity thus resulted. Regarding the effects from the maximum incident angle on the cut-back cell, the smaller angle of incident resulted in the lower solid temperatures at the inlet area but increases the solid temperatures at the deep area of cell. This tendency is similarly observed on the fluid temperatures. The fluid temperature of cut-back for the case of lower incident angle becomes higher than that for higher incident angle. These trends are consistent with the fact that the irradiation can intrude more deeply when the incident angle becomes smaller. Therefore, low angle of incident radiation increases the thermal performance of the receiver to achieve the higher heat acquisition by the air flow.

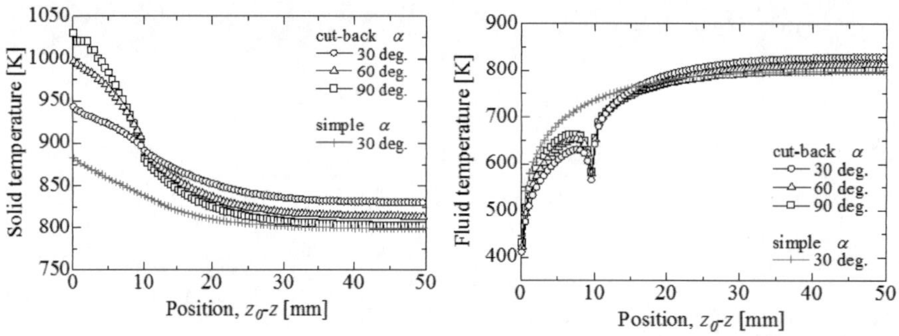

(a) (b)

Figure 4. Mean temperature distribution along with vertical position.

Figure 5 shows the thermal power transported to air flow at the receiver exit and the pressure drop at the same position. In the figure, the quantities are presented against the maximum incidence angle. In figure 5, clear advantage of the cut-back receiver can be observed in the thermal power and pressure drop. Namely, the thermal power increases and the pressure drop decreases by changing the simple cell to cut-back cell. There are relatively high dependency of thermal power on the maximum incidence angle. The thermal power increases as the incidence angle decreases as implied from the temperature distribution earlier discussed.

In figure 5 (a), the difference between simple and cut-back cell becomes larger as the maximum incidence angle decreases. In figure 5 (b), pressure drop increases as the maximum incidence angle decreases to 30° or 20° and this decreases for smaller angles. This dissimilarity of thermal and flow performance may come from the change of the temperature distribution. When the incidence angle reduces, the temperature decreases around the inlet area to reduce the pressure loss but increases in the deep area to increase it. The pressure drop characteristics are resultant of the two opposite effects to decrease and increase it. The decrease of the pressure loss by the decreasing the maximum incident angle from 20 - 30° means that decreasing temperature near the receiver inlet and resulted reducing pressure loss effects excels increasing temperature in deep area and resulted increasing pressure loss.

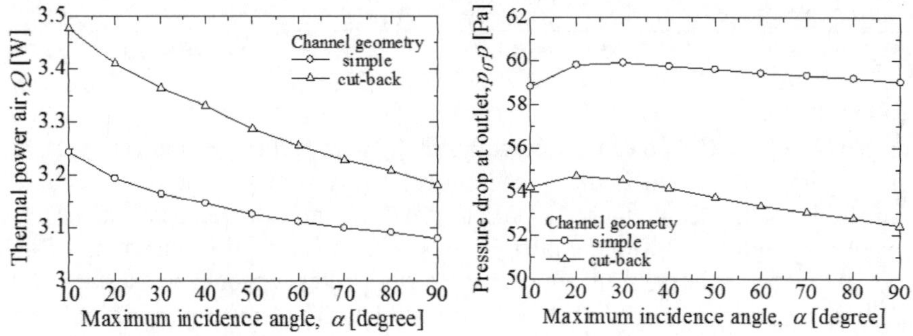

(a) (b)

Figure 5. (a) Thermal power air and (b) pressure drop at the receiver exit.

Wall variables are visualized using contour map for discussion of the heat transfer mechanisms. Cell walls A, B, C, D are defined as shown in figure 6. The temperature distribution on each walls of two types of cells are presented in figure 7.

In figure 7 (a), temperature on simple cell inner walls increases across the whole area as the maximum incidence angle decreases. In figure 7 (b), temperature on cut-back cell decreases around the leading edge and increases in the deep area. Therefore, incidence angle effects to the temperature differ between two cases regarding trends near the inlet area. There are conspicuous temperature decrease on the cut-back cell near the leading edge compared with other area. When α = 90°, a lot of light beam reaches inner walls near the leading edge due to low shadow effect. When α = 30°, light beam tends to come to deep inside the cell and temperature decreases near the leading edge. The cut-back receiver thus is affected by the incidence angle more greatly than the simple receiver.

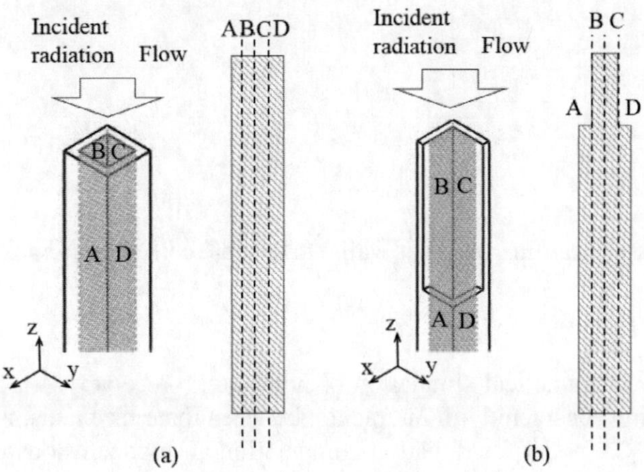

Figure 6. Definition of inner walls: (a) simple cell; (b) cut-back inlet cell.

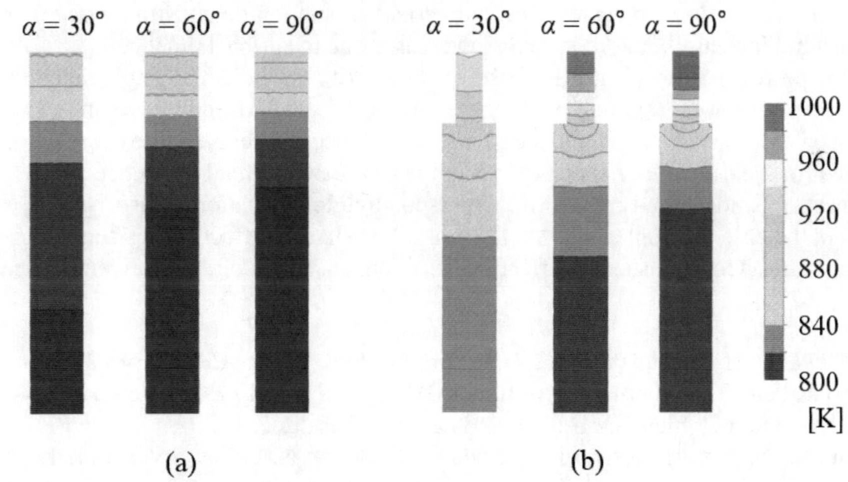

Figure 7. Temperature distribution on inner walls: (a) simple cell; (b) cut-back inlet cell.

Total heat flux is presented in figure 8 by the same manner. In this figure, positive heat flux aims from the wall to fluid side and negative heat flux from the wall to solid side. In the case of simple cell, there are negative values of heat flux in a short distance from the leading edge, which correspond to the direct irradiation for the maximum incidence angle α = 90°. This negative value decreases with decrease of the incidence angle. In the case of cut-back cell, negative value little apart from the leading edge is conspicuous, which corresponds to very large direct irradiation. This negative value decreases with decrease of the incidence angle similarly to the case of simple cell. There are remarkable difference between two types of receiver in the negative values corresponding to the direct irradiation. The improved efficiency of the cut-back cell is attributed to the attenuation of shadow effect which take in the light more efficiently on the inner walls.

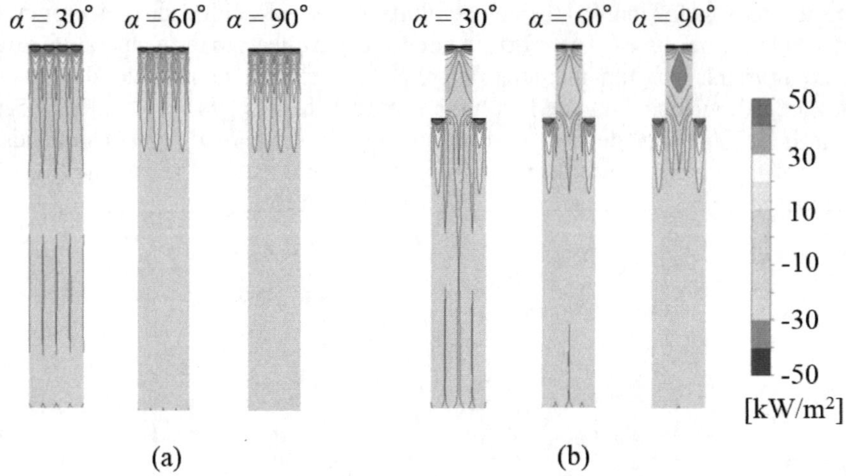

Figure 8. Total heat flux on inner walls: (a) simple cell; (b) cut-back inlet cell.

4. Conclusions

This paper describes direct numerical simulation of honeycomb receiver with plane surface and cut-back inlets. The simulation considered full interaction between three mechanisms of heat transfer using the discrete ordinates (DO) model for radiation computation. The conclusions can be summarized as follow:

(1) The honeycomb receiver with cut-back inlet cell is superior to the receiver with plane surface cell in heat transfer and pressure loss characteristics. Numerical simulation demonstrated that thermal power (receiver efficiency) increases and pressure loss decreases due to cut-back inlet cell.

(2) The thermal power increases with decrease of the maximum incidence angle since the radiation intrudes more deeply. However, pressure loss increases with decrease of incidence angle up to a certain criterion and decreases at very small incidence angle. This is because temperature decrease near the inlet area contributed for the decrease of pressure loss in the case of very small incidence angle.

(3) On the inner walls, the negative heat flux corresponding to irradiation is larger in the case of cut-back cell than in the case of simple cell. Attenuation of the shadow effect is thus suggested in the case of cut-back cell. The attenuated shadow effect comes to enhancement of the receiver efficiency.

5. References

[1] Sano Y, Iwase S, Nakayama A 2012 *J. Sol. Energy Eng. Trans.-ASME* **134** 021006

[2] Capuano R, Fend T, Stadler H, Hoffschmidt B, Pitz-Paal R 2017 *Renew. Energy* **114** 556-66

[3] Du S, Li M J, Ren Q, Liang Q, He Y L 2017 *Energy* **140** 1267-75

[4] Nakakura M, Matsubara K, Cho H S, Kodama T, Gokon N, Bellan S, Yoshida K 2015 *Energy* **141** 2337-50

[5] Nakakura M, Bellan S, Matsubara K, Kodama T 2018 *Sol. Energy* **170** 606-17

A method for a cradle-to-cradle life cycle assessment of integrated collector-storage solar water heaters

Ruth M Saint[1], Francesco Pomponi[2] and John I Currie[3]
[1]School of Engineering & the Built Environment, Edinburgh Napier University, Edinburgh, UK
[2]Resource Efficient Built Environment Lab (REBEL), Edinburgh Napier University, Edinburgh, UK
[3]Scottish Energy Centre, Edinburgh Napier University, Edinburgh, UK
E-mail: r.saint@napier.ac.uk

Abstract. Solar thermal technologies are a potentially free and limitless source of energy. Due to the high demand for domestic hot water and the longer insolation periods during summer in Scotland, solar water heating (SWH) systems are a promising renewable energy technology. One of the systems within this technology is the integrated collector-storage solar water heater (ICSSWH). Existing research has predominantly focussed on the technical and economic aspects, with aims to reduce costs whilst maintaining operational performance. This begs the question, what are the true environmental costs of ICSSWH systems? Life cycle assessment (LCA) is an established methodology to evaluate the environmental performance of a product or a system across its life cycle. The aim of this paper is therefore to present an ICSSWH system from an LCA standpoint, also considering the current shift towards a more circular economy. Preliminary field data from a novel ICSSWH designed at Edinburgh Napier University will be presented. The paper will detail the data sources, impact assessment methods and software tools that will be used. Key outcomes from this research show that a holistic ICSSWH design, which considers whole-life implications, does not compromise efficiency and economic viability.

Keywords: Integrated collector-storage solar water heater; Life cycle assessment; Circular economy; Design for disassembly and reuse

1. Introduction

Given the global commitments to mitigate climate change and to limit global temperature rise to below 2°C from preindustrial levels [1], many countries are migrating to renewable energy sources with increasing interest in their research and development. Amid this zeal for zero/low carbon technologies there is a danger that large scale production of supposedly 'green' solutions will lead to their unsustainable growth, from a holistic life cycle viewpoint. Every product has associated embodied impacts that are necessary to produce, maintain, and dispose of it. For this purpose, a life cycle assessment (LCA) should be carried out to ensure that a product is environmentally sustainable, as well as technologically and economically feasible. This paper presents an LCA evaluation method of a unique integrated collector-storage solar water heater (ICSSWH) system under Scottish weather conditions, designed with the circular economy and reuse potential in mind.

2. Background

The consumption of energy in the UK is increasing across all sectors with the biggest change occurring in the domestic sector, which showed a 3.1% rise from 2015 to 2016 [2]. Scotland aims to provide 50% of total power output – in the forms of electricity, heat, and transport – by 2030 using renewable sources which would require the renewable energy contribution to triple compared to 2014 [3]. 53% of the energy consumed in Scotland is in the form of heat, 13% of which is dedicated to domestic water heating [4]. This need for hot water, alongside the Scottish climate targets [5], calls for greater efforts to be made in renewable heat production. A potential contributing solution to this demand is the passive ICSSWH.

2.1. The ICSSWH system

Several studies have outlined the potential of ICSSWHs in climates similar to Scotland [6–8] and indeed the Scottish climate has been shown to be favourable for solar water heating due to its long heating season [9]. However, the literature surrounding these systems is largely focussed on improving performance and efficiency and reducing cost [10–13]. Relatively few studies focus on the environmental impact throughout the whole life cycle of the product, despite the importance of comparing the operational energy savings against the embodied energy to truly evaluate the sustainability of such systems.

The system under evaluation is a newly developed ICSSWH which offers improvements over its predecessors [9,14–16] and advantages over its competitors. Also, it can be disassembled to avoid any materials in the system being processed as waste at the end of its useful life thus reducing its life-cycle impacts and offering greater potential in terms of a circular economy. Figure 1 illustrates the basic collector-storage unit. During experimental testing, two variations of these collectors were mounted side-by-side. Each collector has the same dimensions, volume, absorber area, and collector material. They differ in that one collector has an internal baffle plate ("Baffled") while the other has three elongated heat transfer fins ("Finned"). These configurations were chosen based on an extensive literature review surrounding ICS systems and the system is being compared against itself, under different design conditions, as opposed to commercially available solar systems. Three different design conditions have been considered for "Baffled" vs. "Finned": basic system set-up; additional insulation applied to the top third of the absorber plate and; a night cover.

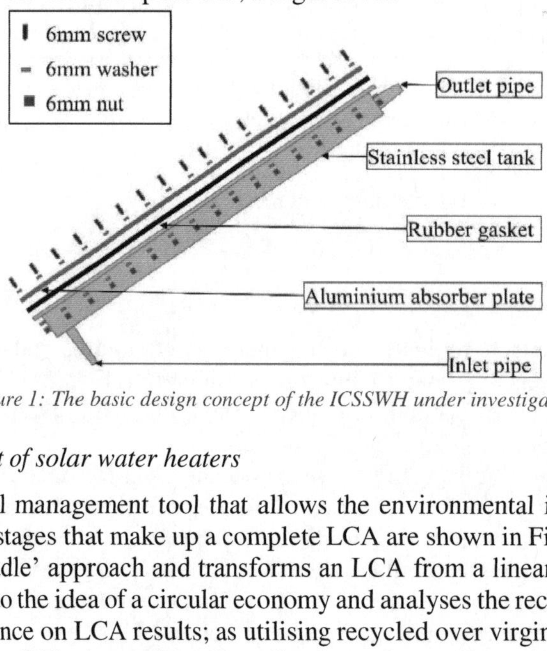

Figure 1: The basic design concept of the ICSSWH under investigation.

2.2. Life cycle assessment of solar water heaters

LCA is an environmental management tool that allows the environmental impact of a product to be quantified. The different stages that make up a complete LCA are shown in Figure 2. The final stage, D, considers a 'cradle-to-cradle' approach and transforms an LCA from a linear analysis to circular. This end-of-life stage appeals to the idea of a circular economy and analyses the recycle/reuse potential which can have a positive influence on LCA results; as utilising recycled over virgin materials greatly reduces the environmental impact [17]. A number of studies have focussed on the LCA of SWH systems, however, only one focusses on ICSSWH in the UK [18]. Therefore, there is a gap in the literature concerning the LCA of ICSSWH in the UK.

Following a systematic review of the literature, it is clear that the same methodology metric is rarely used making direct comparison, as well as repeatability, difficult. This is a common problem for LCAs in the built environment [19] and although many studies follow the ISO 14040 standard [20], they use different databases and analytical software. This inconsistency across the literature makes it hard to identify the steps within the LCA that have the greatest impact and, therefore, the greatest improvement potential [21]. The majority of studies also claim to conduct a cradle-to-grave LCA, which covers Stages A-C (Fig. 2). However, results are often presented as aggregated indexes, i.e. bulk calculations, not taking into account the truncation error inherent in many material databases [22]. The cradle-to-cradle LCA methodology applied in this paper is discussed in Section 4, where the materials and assembly of the ICSSWH under investigation are designed with the reuse/recycle potential as a strong consideration.

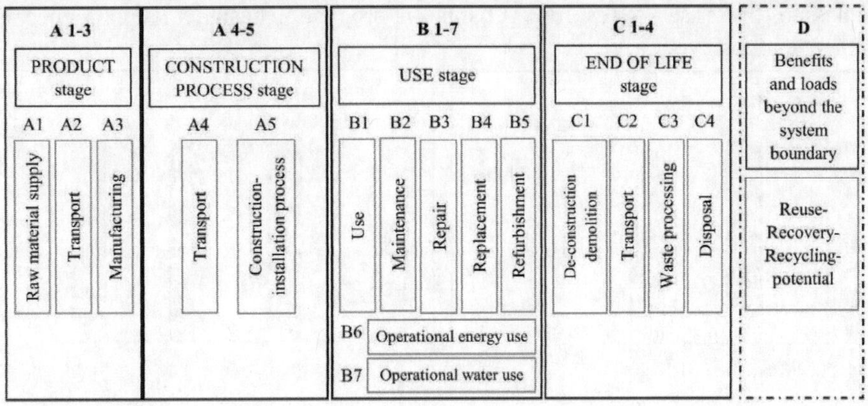

Figure 2: Different stages of the building life-cycle, from cradle-to-cradle. Adapted from BSI [23].

3. Experimental results

Testing of the novel ICSSWH, under different design configurations, is ongoing and presented here is preliminary data. As such, a comprehensive LCA on this iteration of the collector has yet to be completed. A full annual test is desired to ensure as transparent and accurate an analysis as possible – once the energy contribution over a full year is known the carbon and energy savings can be calculated. Preliminary data will be presented to allow for a conservative estimation of annual contribution.

3.1. Energy calculations

The energy each ICSSWH can provide is derived from: $Q = mC_p\Delta T$ (1), where, Q is the energy produced by the system, m is the mass of water in the system, C_p is the specific heat capacity of water, and ΔT is the maximum bulk water temperature difference. Equation (1) enables calculation of the energy contribution from each system. Figures 3 and 4 give a snapshot of the energy that can be generated by the 'Finned' and 'Baffled' collector designs, respectively, under the different design conditions. This data is for a week period in autumn when insolation levels are lower than summer.

3.2. Estimated annual contribution

Based on a study monitoring the hot water consumption of 32 households in Edinburgh [24], a hot water demand of 50 l/person/day, at 55°C, is assumed. The average temperature of mains water is assumed to be 10°C, therefore, the temperature difference between the inlet water and required domestic hot water is 45°C. Under these assumptions and using equation (1), the energy required for a single occupancy household is approximately 2.6 kWh/day, 954 kWh/year. This is a rough estimation as hot water usage is heavily dependent on the lifestyle of the occupant, the time of year, and other external factors. Therefore, this value is being used as a conservative approach. Figure 5 shows the preliminary results of ongoing data collection. Partial experimental data collected from in-situ tests, with the systems being subject to a standardised domestic hot water profile [25], was used and the monthly energy contribution was extrapolated to give conservative estimations for annual performance. Once a full year of data has been collected, an in-depth LCA will be carried out using the following methodology and software tools.

4. LCA methodology

Given that a number of studies utilise the ISO 14040 methodology [20], that framework will also be adhered to in this work. An LCA should follow four main steps and the first is to identify the goal and scope of the current study. This includes defining the system boundary, the functional unit (FU; to allow comparability and reproducibility), and the depth and breadth of the assessment. Secondly, the life cycle inventory (LCI) stage requires the necessary data collection. Thirdly, the life cycle impact assessment (LCIA) allows the quantification of potential environmental impacts based on the data collected for the LCI stage. Finally, an improvement analysis can be done to assess any possible solutions to the environmental issues that arise – for example, changes in production design, materials, energy use, waste management, reuse and recycling, etc.

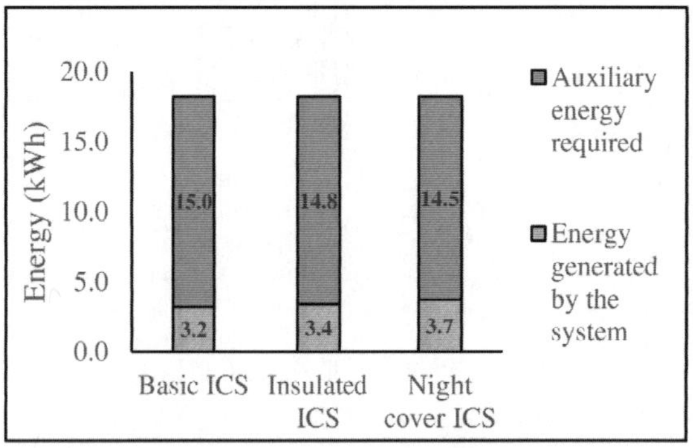

Figure 3: Energy contribution from FINNED ICSSWH system for 7 days in September/October.

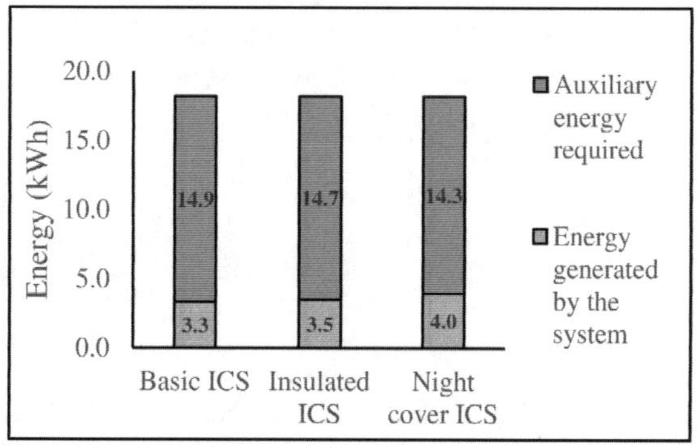

Figure 4: Energy contribution from BAFFLED ICSSWH system for 7 days in September/October.

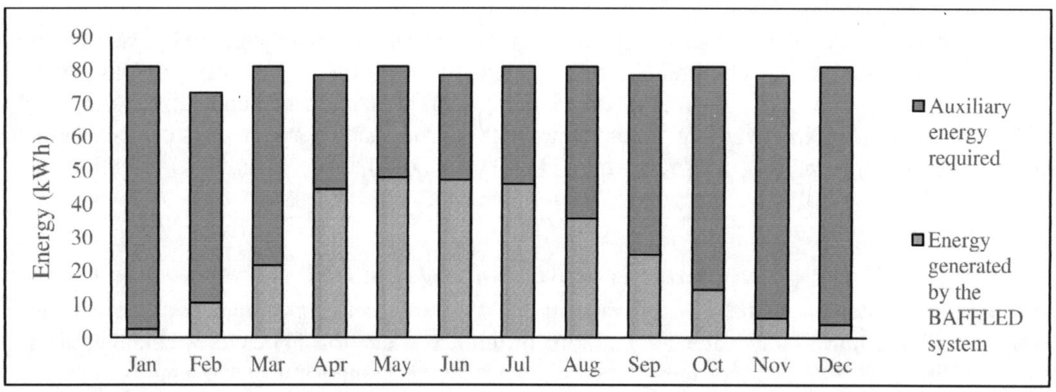

Figure 5: Estimated annual energy contribution for the BAFFLED system. Estimations extrapolated from partial data.

4.1. Goal and scope

The goal of the present research is to determine the environmental impact of an ICSSWH under different design configurations (three testing phases as per Section 2.1). To make the system more attractive from an LCA point of view the absorber plate can be detached from the storage tank. The absorber plate is made of aluminium for its high thermal conductivity but it also has a much higher environmental impact. Therefore, the storage tank is made of stainless steel and the two are sealed together using a rubber gasket. This greatly improves the collectors recycling potential as the steel and aluminium can be cleanly separated and reused. This design aspect was chosen with a circular economy ethos in mind.

4.1.1. Functional unit and system boundaries. The FU is defined as one ICSSWH unit designed to provide domestic hot water for a single occupancy dwelling, based on a consumption of 50 l/day, over a service life of 20 years. One FU includes the collector and associated fittings, detailed in Figure 2. Pipework to and from the collector and the surrounding insulation is excluded as it is assumed that this is already in place before the installation of the unit. This collector was designed for integration into the structural insulated panels (SIPs) used in warm roof construction. Therefore, the SIP itself provides the supporting frame and insulation for the back and sides of the collector.

The difference between the "Baffled" and "Finned" systems will have an impact across the whole life-cycle – Stage A will depend on the embodied carbon and energy associated with materials; Stage B will depend on the operational performance throughout the useful life of the collectors; Stage C and D will depend on the recycle potential of the different materials. In terms of the useful life of the collector, a lifespan of 20 years is assumed. Transport was set at 50km as it is assumed that any travel is within a 50km radius, for example, manufacturer to site.

4.2. Life cycle inventory (LCI)

The LCI brings together all the materials used in the construction of the product and the creation of the FU, i.e. one ICSSWH unit providing hot water for one person. The data should show the quantity and total mass of each component, from which the percentage of the overall product weight can be calculated. The ISO 14040 standard [20] states that any component whose mass is less than 1% of the overall weight has a negligible effect in the analysis and can thus be neglected. However, to disaggregate the results as fully as possible and generate a more transparent and comprehensive LCA, a cut-off criteria of 0.5% on the mass will be applied in the analysis.

4.3. Life cycle impact assessment (LCIA)

Embodied energy is the energy expended throughout the systems useful life (i.e. from sourcing raw materials to the deconstruction of the FU) and is added to give a positive impact on the total energy, i.e. the energy needed. Operational energy, which is the useful energy generated by the FU during its useful life, is added to give a negative impact on total energy, i.e. the energy generated by the system. An LCIA of the ICSSWH FUs will be done for both operational and embodied energy, considering both a linear and circular economy to determine the positive (or negative) impact of the final supplementary stage of the LCA, Stage D.

Three main LCA methods exist in the built environment; process, input-output, and hybrid. Each has its own advantages and disadvantages with the hybrid analysis appearing to be the most accurate method [26]. However, as a hybrid LCA is a combination of process and input-output data it is a highly manual and time-consuming process [27]. Therefore, a process LCA is still a valid and viable method which is also recommended by the European and International Standards that are specifically developed for the construction sector [28].

The LCA tool that will be utilised in the analysis is OpenLCA v1.6.2 equipped with the Ecoinvent database v3.1, using global data. The impact assessment method that will be used to determine the embodied carbon is the CML Baseline (v4.4, January 2015) and for the embodied energy, the Cumulative Energy Demand (CED, v1.0.1, January 2015) will be used. These tools were adopted based on the types of analytical software and databases that are commonly used throughout European studies. The accuracy of the LCA relies on the integrity and applicability of the database used and they are often specific to geographic regions. Ecoinvent is considered to be a high quality database for European studies [29] and the CML tool provides robust and accurate results [30].

4.4. Improvement analysis

This final stage of the LCA process involves assessing the possibilities of improving any aspect of a product's life-cycle in order to improve its environmental impact, by reducing the embodied carbon and energy. By disaggregating the LCI and LCIA and ensuring the LCA boundaries are well defined and comprehensive, areas of improvement can be pinpointed, such as inefficiencies in production design, materials and energy use or waste management/recycling [20]. In the current, ongoing work, potential contributions will be targeted as well as assessing the impact of the chosen heat retention methods on the operational carbon and energy savings.

5. Discussion and Conclusions

A process LCA approach to assess the newly developed ICSSWH designs was presented in this paper based on experimentally validated numerical data. By adopting this approach the authors found [31] that by taking a circular LCA approach and considering the recycle/reuse potential of the system, the newly designed ICSSWH had an embodied carbon saving of 47%. Using past data, operational carbon savings equated to a carbon payback time of just 7 months, when replacing an electric system. The work presented here contributes a frame of reference to estimate the potential energy contribution for the current iteration of the system and outlines how LCA methodology can and should be used in this as well as similar projects.

6. References

[1] United Nations. Paris Agreement. 21st Conf Parties 2015:3. doi:FCCC/CP/2015/L.9.
[2] Department for Business Energy & Industrial Strategy. Energy consumption in UK. vol. 112. 2017. doi:10.1073/pnas.1423686112.
[3] Stuart N. Can a sustainable Scotland be achieved? Renew Energy Focus 2016;17:58–9. doi:10.1016/j.ref.2016.02.003.
[4] Wheelhouse P. Draft Scottish Energy Strategy: The Future of Energy in Scotland 2017.
[5] Scottish Government. Draft Climate Change Plan: The draft third report on policies and proposals 2017-2032. 2017.
[6] Kaushik SC, Kumar R, Garg HP, Prakash J. Transient analysis of a triangular built-in-storage solar water heater under winter conditions. Heat Recover Syst CHP 1994;14:337–41. doi:10.1016/0890-4332(94)90037-X.
[7] Macgregor K, Grassie T. Comparative Testing of a Novel Freeze-Tolerant Solar Water Heating System 1995.
[8] Smyth M, Eames PC, Norton B. Evaluation of a freeze resistant integrated collector/storage solar water-heater for northern Europe. Appl Energy 2001;68:265–74. doi:10.1016/S0306-2619(00)00049-0.
[9] Junaidi HA. Optimized Solar Water Heater for Scottish Weather Conditions 2007.
[10] Singh R, Lazarus IJ, Souliotis M. Recent developments in integrated collector storage (ICS) solar water heaters: A review. Renew Sustain Energy Rev 2016;54:270–98. doi:10.1016/j.rser.2015.10.006.
[11] Smyth M, Eames PC, Norton B. Integrated collector storage solar water heaters. Renew Sustain Energy Rev 2006;10:503–38. doi:10.1016/j.rser.2004.11.001.
[12] Kumar R, Rosen MA. Review of solar water heaters with integrated collector-storage units. Int J Energy, Environ Econ 2013;21:343–82.
[13] Jamar A, Majid ZAA, Azmi WH, Norhafana M, Razak AA. A review of water heating system for solar energy applications. Int Commun Heat Mass Transf 2016;76:178–87. doi:10.1016/j.icheatmasstransfer.2016.05.028.
[14] Currie J, Garnier C, Muneer T, Grassie T, Henderson D. Modelling bulk water temperature in integrated collector storage systems. Build Serv Eng 2008;29:203–18. doi:10.1177/0143624408094277.
[15] Garnier C. Performance measurement and mathematical modelling of integrated solar water heaters. PhD Thesis 2009.
[16] Saint R, Currie J, Garnier C. Technological development of integrated collector storage solar water heaters: A review. 8th Int. Conf. SOLARIS 2017, 2017.
[17] Allen SR, Hammond GP, Harajli HA, Mcmanus MC, Winnett AB. Integrated appraisal of a Solar Hot Water system. Energy 2010;35:1351–62. doi:10.1016/j.energy.2009.11.018.
[18] Smyth M, Eames PC, Norton B. Life Cycle Assessment of a Heat Retaining Integrated Collector/Storage Solar Water Heater (ICSSWH). Energy 21st Century World Renew. Energy Congr. VI, Brighton.: 2000, p. 1036–40.
[19] Pomponi F, Moncaster A. Embodied carbon mitigation and reduction in the built environment – What does the evidence say? J Environ Manage 2016;181:687–700. doi:10.1016/j.jenvman.2016.08.036.
[20] ISO14040. Environmental management — Life cycle assessment — Principles and framework 2006.
[21] Ardente F, Beccali G, Cellura M, Brano V Lo. Life cycle assessment of a solar thermal collector. Renew Energy 2005;30:1031–54. doi:10.1016/j.renene.2004.09.009.
[22] Lenzen M, Dey C. Truncation error in embodied energy analyses of basic iron and steel products. Energy 2000;25:577–85. doi:10.1016/S0360-5442(99)00088-2.
[23] BSI. BS EN 15978:2011 Standards Publication Sustainability of construction works — Assessment of environmental performance of buildings — Calculation method. London: 2011.
[24] McLennan C. Solar water heating. Edinburgh: 2006.
[25] European Commission. Mandate to CEN and CENELEC for the elaboration and adoption of measurement standards for household appliances: water-heaters, hot water storage appliances and water heating systems. 2002:1–14.
[26] Pomponi F, Lenzen M. Hybrid life cycle assessment (LCA) will likely yield more accurate results than process-based LCA. J Clean Prod 2018;176:210–5. doi:10.1016/j.jclepro.2017.12.119.
[27] Crawford RH, Bontinck PA, Stephan A, Wiedmann T. Towards an Automated Approach for Compiling Hybrid Life Cycle Inventories. Procedia Eng 2017;180:157–66. doi:10.1016/j.proeng.2017.04.175.
[28] Moncaster AM, Song JY. A comparative review of existing data and methodologies for calculating embodied energy and carbon of buildings. Int J Sustain Build Technol Urban Dev 2012;3:26–36. doi:10.1080/2093761X.2012.673915.
[29] Martínez-Rocamora A, Solís-Guzmán J, Marrero M. LCA databases focused on construction materials: A review. Renew Sustain Energy Rev 2016;58:565–73. doi:10.1016/j.rser.2015.12.243.
[30] Martínez E, Blanco J, Jiménez E, Saenz-Díez JC, Sanz F. Comparative evaluation of life cycle impact assessment software tools through a wind turbine case study. Renew Energy 2015;74:237–46. doi:10.1016/j.renene.2014.08.004.
[31] Saint R, Pomponi F, Garnier C, Currie J. Whole life design and resource reuse of an integrated collector-storage solar water heater. Eng Sustain 2018.

The analysis on applicability of hourly horizontal diffuse radiation estimation model of Shanghai Zone

Qimeng Cao , Liu Yang and Ying Yu
Xi'an University of architecture and Technology, Xi'an, china
E-mail: yangliu@xauat.edu.cn

Abstract: For shading design and energy consumption simulation, hourly diffuse radiation data is necessary. In order to estimate the total solar radiation on surfaces with different spatial orientations, the knowledge of all components on the horizontal plane is required. Available measured data is, however, typically less. The 2004-2013 year's measured data of Shanghai were localized revising of three categories models, comparing the measured data and estimation data by the correlation coefficient and error index. The results show that the three types of models have high accuracy at higher solar altitude and the accuracy of multi parameter logistic function is best. For stations with only daily solar radiation data, the third types of models can be used to calculate hourly diffuse solar radiation.

Keywords: hourly global solar radiation; hourly diffuse solar radiation; clearness index; daily diffuse radiation; solar altitude angle

1. Introduction

To analysis buildings' energy performance under the effect of solar radiation and calculate solar radiation on a titled surface, it is necessary to know the direct and diffuse solar radiation. Moreover, design and sizing of solar energy systems (e.g. photovoltaic cells, solar-thermal collectors) needs accurate solar radiation data. The most accurate way to obtain solar radiation data is to establish station to collect in real time using high-precision measurement equipment. However, due to cost and maintenance, concurrent measured data of global and diffuse radiation on horizontal or inclined surface are available only for a limited number of locations in China. Therefore, it is necessary to established mathematical method using radiation-related parameters.

The diffuse fraction model, which is the ratio of the hourly horizontal diffuse solar radiation to the hourly horizontal global solar radiation, was first derived from Origill's study [1]. Origill divided the hourly clearness index (ratio of horizontal global radiation and extra-terrestrial radiation) into three intervals and established the relationship between diffuse fraction and hourly clearness index in each interval. Erbs et al. [2] fitted the fourth degree polynomial functions between them. Then, many scholars studied the regional adaptability of the Origill´ and Erbs´ model, and made revised to this model, in terms of first to fourth degree polynomial functions, first to three paragraph curve and other variables such as air temperature, relative humidity, cloud amount, sunshine hours ratio, air quality, and daily clearness index [3-11]. Other forms [12-14], with hourly clearness index, apparent solar time, solar altitude angle, and daily clearness index, are the Gompertz function and Logistic function. Some researches were about the relationship between the ratio of the diffuse rate[15](the ratio of the hourly diffuse solar radiation to the hourly horizontal extra-terrestrial solar radiation) and clearness index. Daily diffuse radiation[16] were decomposed, according to the changing rules of the sun from sunrise to sunset. This article calls daily diffuse decomposition model.

Many scholars verified the applicability of the diffuse fraction model in different regions[17-19]. The results showed the calculated are closed to the measured values. However, the diffuse fraction model needs hourly global horizontal radiation data as a support. There are more daily radiation data in

China. Compared with this, the development of the daily diffuse decomposition model has more development, but its accuracy remains to be verified.
This paper localizes the diffuse fraction, diffuse rate, and daily diffuse decomposition model and verifies the daily diffuse decomposition model with meteorological elements based on the Baoshan station of Shanghai.

2 Data Sources and Quality Control

2.1 Data Sources
For this research, hourly and daily global, diffuse radiation data were recorded in Baoshan(latitude: 31.24°, longitude: 121.46°, altitude: 8 m) form January 2004 to December 2013. Other parameters such as hourly extra-terrestrial radiation and clearness index can be calculated.

2.2 Quality Control
Measurement technology and equipment accuracy will bring errors, so it is necessary to quality control the measured data. The steps are as follows [20]:
①The first test proceeds by eliminating entries that show a solar altitude angle less than 7°.
②The second test: clearness index and the diffuse ratio are positive and have values between zero and one(inclusive)
③The third test: global and diffuse irradiation are compared with their corresponding Page model upper and lower boundaries.
④A fourth test involves the construction of k_d-k_t quality control envelope. This is a statistical procedure that requires estimation of k_t banded mean, weighted mean and standard deviations of "k_d" values (Δ). Typically the k_t range of data may be divided in, say, ten bands of equal width. For each band the above mentioned statistics is obtained. From this information an envelope may be drawn that connects those points that, respectively, represent the top ($\overline{k}_d+2\Delta$) and bottom ($\overline{k}_d-2\Delta$) curves. The data lying within this envelope represent quality data—free of any measurement related errors.
Though the four steps, the final database consist of about 37413 hourly values.

3 Method

3.1 Statistical Tests and Comparison Indices
To compare the performance of the models, five statistical indicators were employed[6,13], namely the correlation coefficient (R^2), the mean deviation (MBE), the relative mean absolute error (MABE%), the relative root mean square error (RMSE%), and the relative error (RE). The closer R^2 is to 1, the more accurate the model is. The remaining parameters is to zero, the more accurate the model is.

3.2 Used Models
Based on the measured radiation data after the quality control between 2004 to 2009, a total of nine models were fitted and represented by M1-M9, as shown in formula (1) to (16). Meteorological parameters, such as Temperature (T), Relative Humidity (RH), Atmospheric Pressure (P), have a complex relationship with solar radiation. Using SPSS to analysis the correlation between hourly diffuse radiation and temperature, relative humidity and Atmospheric pressure, the correlation coefficients were 0.367, -0.427, and -0.191, respectively. The relative humidity, which is the most relevant factor, was selected to modify M9. The modified model was represented by M10, as shown in formula (17).
1) M1

$$k_d = 0.946k_t^3 + 1.05k_t^2 - 4.93k_t + 3.28 \quad (1)$$

2) M2

$$k_d = 5.816k_t^4 - 6.726k_t^3 + 0.667k_t^2 - 0.058k_t + 1.0 \qquad 0 \le k_t \le 0.85 \quad (2)$$

$k_d = 0.4 \qquad 0.85 < k_t \leq 1 \qquad (3)$

3) M3

$k_d = 1.0 \qquad k_t \leq 0.1 \qquad (4)$

$k_d = 5.801k_t^4 - 6.578k_t^3 + 0.49k_t^2 + 0.008k_t + 0.993 \qquad 0.1 < k_t \leq 0.8 \qquad (5)$

$k_d = 0.35 \qquad 0.8 < k_t \leq 1 \qquad (6)$

4) M4

$k_d = 1.0 \qquad k_t \leq 0.1 \qquad (7)$

$k_d = -1.235k_t + 0.022t + 0.974 \qquad 0.1 < k_t \leq 0.85 \qquad (8)$

$k_d = 0.4 \qquad 0.85 < k_t \leq 1 \qquad (9)$

5) M5

$k_d = 1.0 \qquad k_t \leq 0.1 \qquad (10)$

$k_d = -1.242k_t + 0.037\sin\alpha + 0.023t - 0.946 \qquad 0.1 < k_t \leq 0.85 \qquad (11)$

$k_d = 0.4 \qquad 0.85 < k_t \leq 1 \qquad (12)$

6) M6

$$k_d = \frac{1}{1+\exp(6.554k_t - 4.127)} \qquad (13)$$

7) M7

$$k_d = \frac{1}{1+\exp(8.174k_t - 0.195t - 2.572)} \qquad (14)$$

8) M8

$$k_D = 1.187k_t^3 - 2.683k_t^2 + 1.761k_t - 0.044 \qquad (15)$$

9) M9

$$\frac{I_{dh}}{I_d} = \frac{\pi}{24} \cdot \frac{\cos\omega - \cos\omega_s}{\sin\omega_s - \omega_s\cos\omega_s} \qquad (16)$$

10) M10

$$\frac{I_{dh}}{I_d} = 0.021 + 0.986 \cdot \frac{\pi}{24} \cdot \frac{\cos\omega - \cos\omega_s}{\sin\omega_s - \omega_s\cos\omega_s} - 0.032RH \qquad (17)$$

4 Results

The above-mentioned 10 models were validated using data from the 2010 to 2013 period in Shanghai, and the error indicators of each model were calculated. Fig. 1 shows the R^2, MBE, MABE%, and RMSE% of M1-M9. The cumulative distribution function of RE (in %) for M1 to M9 are showed in Fig.2. Figure 3 show the relationship between model accuracy and solar altitude. Table 1 shows, for both original(M9) and modified by the RH(M10), the R^2, MBE, MABE% and RMSE%. In the evaluation of the merits of the model, the method of reference [21], that is, the MBE<10%, RMSE <30% means good model; the 10% <MBE <20%, 30% <RMSE <40% means good enough model; the MBE> 10%, RMSE> 40% means bad model.

IOP Conf. Series: Materials Science and Engineering **556** (2019) 012062 doi:10.1088/1757-899X/556/1/012062

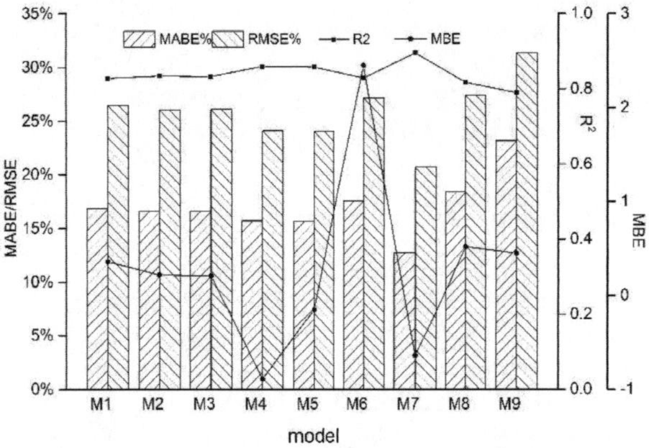

Fig.1 The R^2、MBE、MABE% and RMSE% of M1-M9

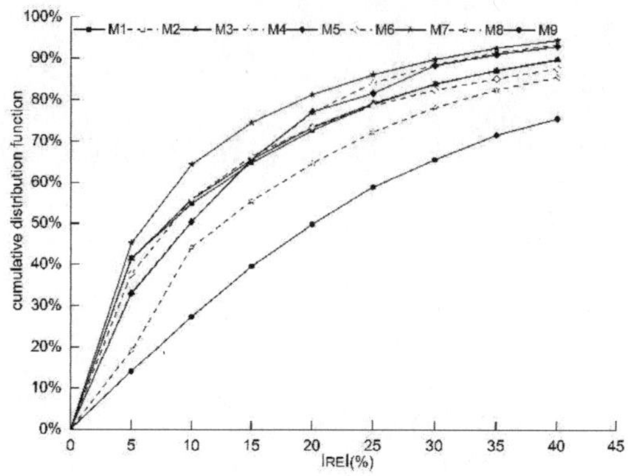

Fig.2 The cumulative distribution function (CDF) of the relative errors(in ±%) for the eight models with the original coefficients.

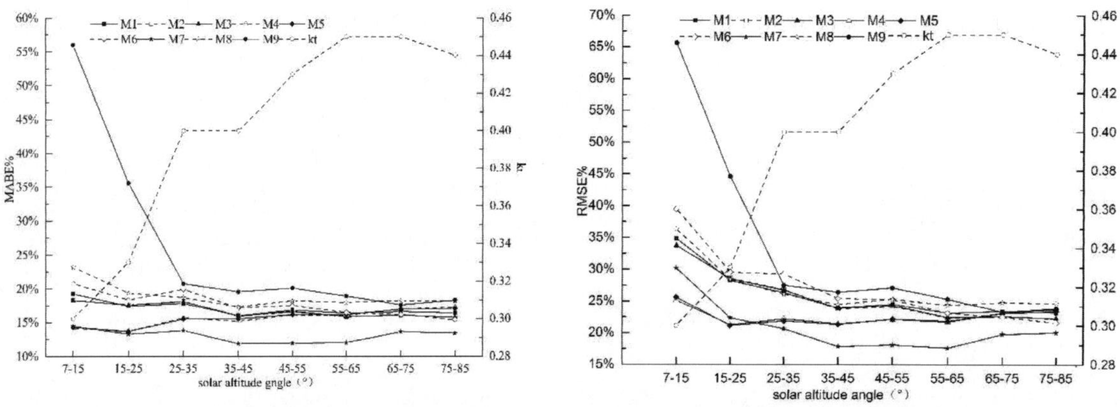

a MABE% b RMSE%

Fig.3 MABE%、RMSE% and k_t of M1-M9 with solar altitude

Table 1 The R^2、MBE、MABE% and RMSE% of M9 and M10

model	R2	MBE	MABE%	RMSE%
M9	0.79	0.45	23.17%	31.36%
M10	0.83	0.79	22.82%	31.16%

5 Discussion

(1) The R^2 of the diffuse fraction models' value is the highest, the diffuse rate models' is next and the daily diffuse decomposition models' is the smallest in Fig.1. The MABE% and RMSE% have the opposite changing tendency. This shows that the diffuse fraction model has the highest accuracy, followed by the diffuse rate model and the daily diffuse decomposition model.

(2) The R^2, MABE% and RMSE% of the univariate model (M1-M3, M6) with the clearness index as variable are 0.82 to 0.85, 16% to 18% and 26% to 28%, respectively; while, the multivariate model (M4, M5, M7) is greater than 0.85, 12% to 16.3%, and 20% to 24.2%, respectively. This indicates that the accuracy of the multivariate model is higher than that of the univariate model.

(3) The values of R^2, MABE% and RMSE% of M6 is smallest and biggest in the univariate model, which indicates that the accuracy of the piecewise polynomial model is higher than the logistic function model. The values of R^2, MABE% and RMSE% of M7 is biggest and smallest in the multivariate model, which indicates that logistic function is better than the linear function. The error

indexes of M4 and M5 are basically the same, indicating that the solar altitude angle has no obviously effect on its accuracy. Based on the principle of simple calculation, M4 is priority than M5.
(4) Figure 2 shows that the results of M1 to M9 with a relative error less than 10% are about 50%, 55%, 56%, 56%, 56%, 50%, 64%, 44%, 44%, respectively, implying the accuracy of M7 is higher than others.
(5) The MABE% and RMSE% of M1 to M9 decrease with the increase of the solar elevation angle, while, the change of the clearness index is opposite, indicating that the accuracy of these models is increasing with the solar altitude angle and clearness index increasing. The trend of M9 is more obvious. When the solar altitude angle is ≥25°, the MABE%, RMSE% are 17% to 21%, 23% to 28%, respectively. According to the literature [21], the solar elevation angle is bigger, the accuracy is better.
(6) In the revised model (Table 1), R^2 increased by 0.4, MABE% decreased by 0.35%, and RMSE% decreased by 0.2%, indicating that the model´ accuracy can be improved by adding the correction of the relative humidity.

6 Conclusion

(1) The accuracy of the three types of models increases with the increase of the solar altitude.
(2) For the hourly diffuse radiation of Shanghai and its similar areas, the multi-parameter logistic function model (M7) is recommended, followed by the multi-parameter diffuse fraction model (M4) added with apparent solar time correction.
(3) The diffuse rate model (M8) could not improve the accuracy of the model. In case of the hourly global solar radiation data, this model is not recommended.
(4) Although the accuracy of M9 is slightly lower when the solar elevation angle is low, it can be used to calculate the hourly diffuse radiation to compensate the lack of hourly radiation data. The correction of relative humidity improves the accuracy of the model, but it is not obvious and more stations are needed to be verified.

Acknowledgments

This work was supported by “the 13th Five-Year” National Science and Technology Major Project of China (NO. 2018YFC0704500) and National Natural Science Foundation of China (51608423, 51508444).

Nomenclature

I_{Gh} : hourly horizontal global solar radiation, MJ/m^2;

$I_{\mathrm{Gh,o}}$: hourly horizontal extra-terrestrial solar radiation, MJ/m^2;

I_{dh} : hourly horizontal diffuse solar radiation, MJ/m^2;

I_{d} : hourly horizontal daily solar radiation, MJ/m^2;

k_t: hourly clearness index, $k_t = I_{\mathrm{Gh}}/I_{\mathrm{Gh,o}}$;

k_d: diffuse fraction, $k_{\mathrm{d}} = I_{\mathrm{dh}}/I_{\mathrm{Gh}}$;

k_D: diffuse rate, $k_D = I_{\mathrm{dh}}/I_{\mathrm{Gh,o}}$

t : apparent solar time, h;

α : solar altitude angle, deg;

ω : hour angle, deg;

ω_s : sunset angle, deg;

RH: relative humidity, %;

T: temperature, ℃;

P: atmospheric pressure, Pa;

Reference

[1] Orgill JF and Hollands KGT Correlation equation for hourly diffuse radiation on a horizontal surface 1997 *J. Solar Energy* **19** 357–9.

[2] Erbs DG, Klein SA and Duffie JA Estimation of the diffuse radiation fraction for hourly, daily and monthly-average global radiation 1982 *J. Solar Energy* **28** 293–302.

[3] Clarke P, Munawwar S, Davidson A, Muneer T and Kubie J Technical note: an investigation of possible improvements in accuracy of regressions between diffuse and global solar radiation 2007 *J. Building Services Engineering Research and Technology* **28** 189–97.

[4] Karatasou S, Santamouris M and Geros V Analysis of experimental data on diffuse solar radiation in Athens, Greece, for building applications 2003 *J. International Journal of Sustainable Energy* **23** 1–11.

[5] Reindl DT, Beckman WA and Duffie JA. Diffuse fraction correlations 1990 *J. Solar Energy* **45** 1–7.

[6] Miguel A, Bilbao J, Aguiar R, Kambezidis H and Negro E Diffuse solar irradiation model evaluation in the north Mediterranean belt area 2001 *J. Solar Energy* **70** 143–53.

[7] Jacovides CP, Tymvios FF, Assimakopoulos VD and Kaltsounides NA Comparative study of various correlations in estimating hourly diffuse fraction of global solar radiation 2006 *J. Renewable Energy* **31** 2492–504.

[8] Soares J, Oliveira AP, Boznar MZ, Mlakar P, Escobedo JF and Machado AJ Modeling hourly diffuse solar-radiation in the city of Sao Paulo using a neural-network technique 2004 *J. Applied Energy* **79** 201–14.

[9] Mondol J D, Yohanis Y G and Norton B Solar radiation modelling for the simulation of photovoltaic systems 2008 *J. Renewable Energy* **33** 1109-20.

[10] Muneer T. and Munawwar S Improved accuracy models for hourly diffuse solar radiation 2006 *J. Journal of Solar Energy Engineering* **128** 104-17.

[11] Paulescu E and Blaga R Regression models for hourly diffuse solar radiation 2016 *J.* Solar Energy **125** 111-24.

[12] Boland J, Ridley B and Brown B Models of diffuse solar radiation 2008 *J. Renewable Energy* **33** 575-84.

[13] Zhang Qingyuan and Joe Huang Standard meteorological database for Chinese Architecture 2004 *M. Machinery Industry Press* p9

[14] Ridley B, Boland J and Lauret P Modelling of diffuse solar fraction with multiple predictors 2010 *J. Renewable Energy* **35** 478-83.

[15] Maxwell, E L A quasi-physical model for converting hourly global horizontal to direct normal insolation 1987 *J. Solar Energy Institute*, Golden, CO. Report SERI/TR-215-3087.

[16] Jain P C, Jain S and Ratto C F A new model for obtaining horizontal instantaneous global and diffuse radiation from the daily values 1988 *J. Solar Energy* **41** 397-404.

[17] Li Zhengrong, Yao Wanxiang and Zhao qun Study on the comparison of models for daily diffuse solar radiation on a horizontal surface 2013 *J. Acta Energiae Solaris Sinica* **34** 794-99.

[18] Su Hua, Tian Shengyuan and Su Fenxian Models to separate daily diffuse radiation from daily total radiation for energy consumption analysis of air-conditioning system 2002 *J. Journal of Chongqing University: Natural Science Edition* **25** 73-6.

[19] Feng Wei, Dong Hong, Yang Liu and Liu Jiaping Comparison of the fraction model of direct and diffuse solar radiation-Taking Beijing area as an example 2015 *J. Journal of Civil, Architectural Environmental Engineering* **37** 12-7.

[20] Younes S, Claywell R and Muneer T Quality control of solar radiation data: Present status and proposed new approaches 2005 *J. Energy* **30** 1533-49.

[21] Badescu V, Gueymard C A and Cheval S et al Accuracy and sensitivity analysis for 54 models of computing hourly diffuse solar irradiation on clear sky 2013 *J. Theoretical and Applied Climatology* **111** 379-99.

Improvement of Airtightness for Lightweight Prefabricated Building Envelope through Optimized Design of Panel Joints

Zhenghao Lin[1,2], **Yehao Song**[1,2,*] **and Dongchen Han**[1,2]
[1]School of Architecture, Tsinghua University, Beijing, China
[2]Key Laboratory of Urban-Rural Eco Planning and Green Building, Ministry of Education, Beijing, China
E-mail: zhenghao.lin@foxmail.com

Abstract. Airtightness of building envelope is a key factor affecting the indoor comfort, air quality, structural durability and energy efficiency, especially for the lightweight prefabricated building systems aiming to achieve the nearly zero energy goals. For panelised systems, the air mainly leaks through the openings and panel joints. However, most current studies concentrate on the airtightness of the former ones, such as windows and doors, studies on joint design are relatively rare. As a result, this study firstly proposes two joint prototypes, the straight joints and the special-shaped joints, with various design parameters. Secondly, the air infiltration volume for different joints are simulated and calculated by the CFD method. Besides, the volume was further compared to find out the optimized joint design. Finally, the airtightness requirements of Passivhaus are used to evaluate the performance of the optimized tongue and groove joint with the conventional straight ones. The results prove that the well-designed joints deliver better airtightness especially for the high-performance standard buildings.

Keywords: Air tightness; CFD; lightweight prefabricated buildings; panel joints; energy efficiency; nearly zero energy buildings

1. Introduction

Airtightness refers to the ability of building envelope to resist air infiltration. Not only can unorganized air infiltration gravely affects indoor comfort, air quality and building durability, but also causes huge extra energy consumption. Therefore, improving the overall airtightness of buildings has been the common ground for high-performance and low-energy consumption architectural research in recent years. For example, nearly zero energy building systems (NZEBs) like Germany's Passivhaus, Italy's Casaclima, and Switzerland's Minergie-P have all introduced unequivocal airtightness indicators [1].

In China, how to realize the nearly zero energy consumption of prefabricated buildings has evolved into one of the hot spots in architecture. However, because the prefabricated buildings, especially the lightweight ones, are usually assembled with composite sandwich panels and the like, whose joints tend to be the Achilles heel of the overall airtightness [2]. On the one hand, the "narrow and deep" joints in lightweight prefabricated buildings (LPBs) inflict difficulties in sealing on the construction site; on the other hand, the aging of the seal fillings will result in the exposure and infiltration. Thus, the airtightness of LPBs cannot be resolved solely by the construction accuracy and material performance. Instead, from the perspective of tectonics, this paper discusses how to improve the foundational airtightness through the optimized design of the panel joints.

Currently, the research into airtightness primarily focuses on three facets. First, as for the air leakage of door and window gaps, relevant research has established a predictive model based on experiments and experience, such as the "gap method" [3]. The α and b in the formula $Q = \alpha(\Delta P)^b$ are characteristic variables of infiltrated wind, which are subject to the influences of such factors as gap widths, geometric

* Corresponding author. Tel.: +86 010 62785691. E-mail address: ieohsong@tsinghua.edu.cn (Y.Song).

IOP Conf. Series: Materials Science and Engineering **556** (2019) 012063 doi:10.1088/1757-899X/556/1/012063

shapes, and actuating pressure differences. Yet, as the size, shape, connection mode and service state of the panel joints of LPBs still quite differ from those of doors and windows, their applicability remains to be tested. Second, some research focuses on the influencing mechanism of air infiltration on energy consumption, which sees the indoor air exchange frequency as the airtightness variable to simulate the change of indoor energy consumption under specific climate conditions, thus determining the airtight grade based on the energy conservation laws and regulations [4]. Nevertheless, such research rarely compares influences of different structure designs of joints on the airtight grade. Ultimately, the current research into building airtightness with nearly zero energy buildings is chiefly based on field measurement of "blower-door-test" upon completion [5]. Yet, its experimental difficulties and costs are higher, making it hard to systematically compare and analyse varying joints and structures
Therefore, this paper leverages Airpak 3.0, a Computational Fluid Dynamics (CFD) simulation software, to compare the influences of different structural design parameters of the panel joints of LPBs on air infiltration, taking account of the construction efficiency.

2. Methodologies

2.1. About Airpak 3.0
Airpak is a CFD software designed for building environment, which takes Fluent as the solution core and adopts a finite volume method based on unstructured grids. It can simulate the physical phenomena such as air flow, heat transfer and pollution in the research object. In this study, Airpak 3.0 is adopted to simulate the panel joints of LPBs.

2.2. Modelling and parameter setting
A 2m×2m×0.2m vertical wall is set up as CFD model in this study, which is formed by two horizontal panels, with a vertical joint at the centre axis of the wall. As this study focuses on the air flow phenomenon at the joint, the wall panel model is placed in a 2m×2m×8m tubular flow field (Figure 1). The wall divides the tubular flow field into two small ones, representing the outdoor and indoor environment. As per the "blower-door-test", a stable pressure difference ΔP perpendicular to the wall is applied to both ends of the flow field (usually 50pa), forcing outdoor air to infiltrate into indoor spaces via joints. By simulating the median cross section of the flow field as the representative plane (Figure 2), the average wind speed of the outlet side of the joint chamber is obtained, and the total volume of air penetrating through the joint per second is calculated and used as the evaluation indicator of joint airtightness. The calculation method is described in formula 1.

$$Q = F \cdot v_j = w \cdot l \cdot v_j \qquad (1)$$

In this formula, Q represents the air infiltration volume (m^3/s), F stands for the area (m^2) of the indoor air outlet of the joint chamber, v_j refers to the average wind speed (m/s) of air outlet, w signifies the width (m) of the air outlet, and l represents the unit length (m) of the air outlet (1m by default).

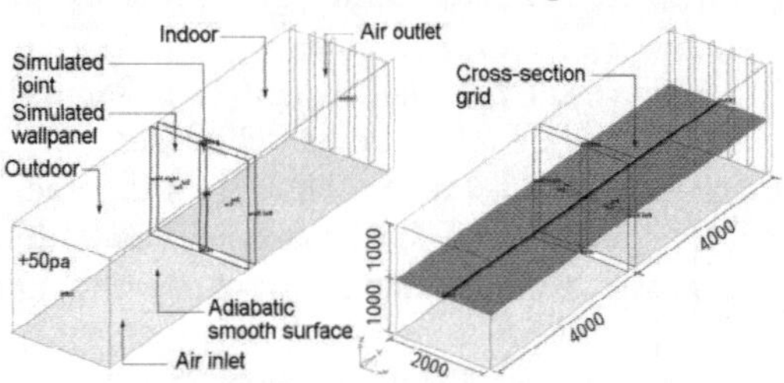

Figure 1. CFD wall panel model (exemplified by the straight joint).

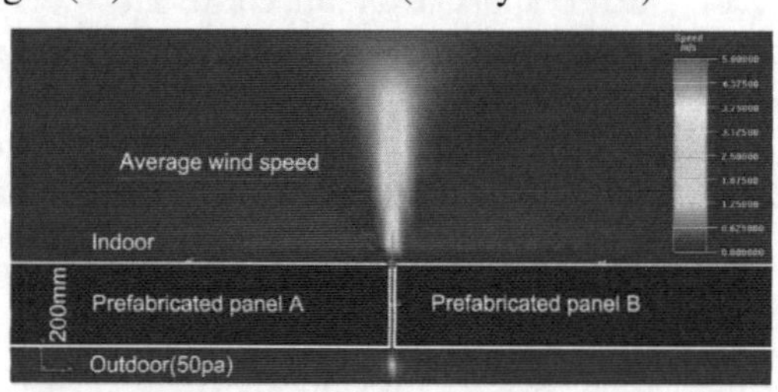

Figure 2. A cloud atlas of the wind speed of the Airpak's wall panel cross section.

2.3. Prototype classifications of joints and simulation arrangements
The panel joint prototypes of LPBs can be divided into straight joints and special-shaped joints, of which the latter ones are mainly categorized into staggered joints and tongue and groove (T&G) joints (Figure 3). For the convenience of simulation and discussion, this paper simplifies the prototypes of these joints

and simulates the influences of different structural design parameters on their airtight performance by control variables.

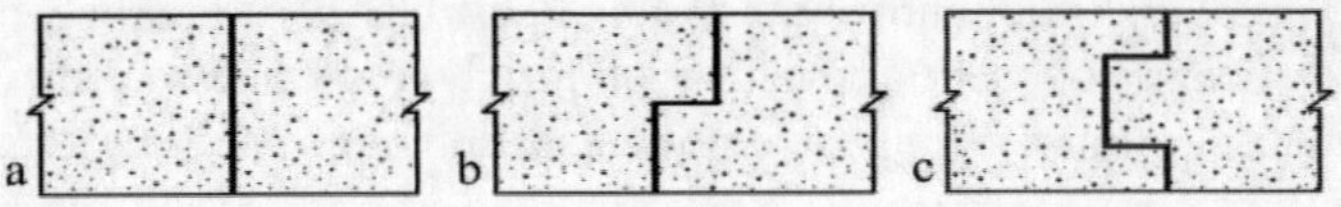

Figure 3. Three types of joint prototypes: a. straight joint, b. staggered joint, c. tongue and groove joint.

3. Results and discussions

3.1. Simulation of the straight joints

The straight joints are simple in structure, low in costs, hard to be damaged and good for the universality of construction. However, they easily cause air leakage. Therefore, this paper discusses the effects of two basic design parameters of the depth and width of joints on airtightness (Figure 4).

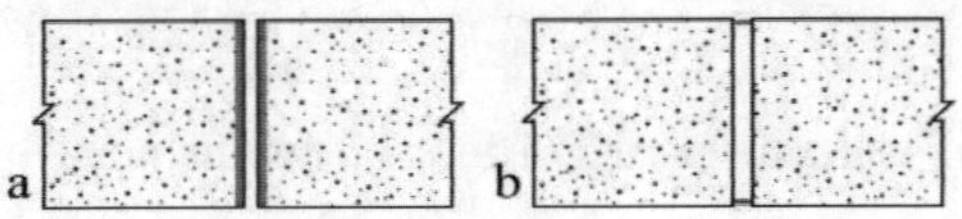

Figure 4. Geometric parameters of straight joints: a. depth of joint and b. width of joint.

3.1.1. Depth of joint. For straight joint, the depth equals the thickness of the envelope panel. A 2mm-width straight joint is selected to simulate and compare the effects of various joint depths ranging from 50mm to 400mm on airtightness. As shown in Figure 5, with the increase of panel thickness, the volume of infiltrated air through the joint continuously drops. When the panel is thicker than 100mm, the air tightness at the joint is greatly improved; when the panel is thicker than 200mm, the downward curve of air infiltration gets gentle, indicating a limited increase in airtightness. Therefore, for straight joints, the panel thickness should be around 200 mm, too thin an insulation (< 200 mm) does not help achieve the insulation performance of NZEB. On the other hand, too thick plates (⩾ 200 mm) will undermine the overall construction efficiency of LPBs.

3.1.2. Width of joint. The width of joints in this simulation does not refer to the "designed width", it instead should be interpreted as the "tightness" between panels, characterizing the fineness of the assembly construction or the performance of the sealing material. 17 sets of data are included in this group of simulations, covering the width from 0.5 mm to 10 mm. As shown in Figure 6, with the increase of the width, the air infiltration will also increase. The Origin 9.0 statistical software is used to regress the data. When the width is within 5mm, it is exponentially related to the air infiltration (R^2=0.998) (Figure 7); when it is over 5mm, there is a significant linear relationship (R^2=0.999) (Figure 8), suggesting different mechanisms of air infiltration. However, both manifest strong positive correlation, namely the narrower joint, the better. That not only requires better manufacturing and assembly quality, but also calls for the architects' awareness of gap control at the design stage.

As such, this paper chooses 2mm as the default joint width for subsequent simulation: first, the 2mm can characterize the air infiltration mechanism within the width range of 0-5 mm; second, adopting the 2mm joint width helps more to improve the quality of Airpak grid division.

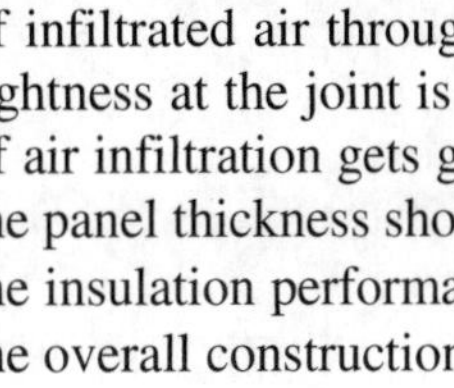

Figure 5. Relation between air leakage and depth of joint.

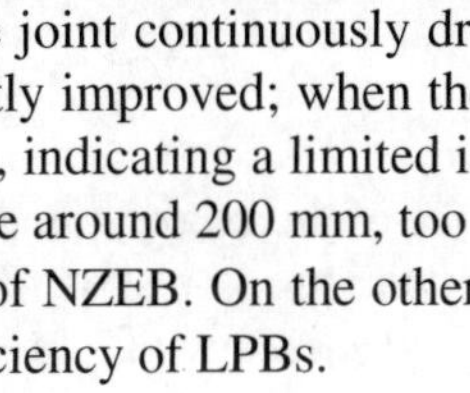

Figure 6. Relation between air leakage and width of joint.

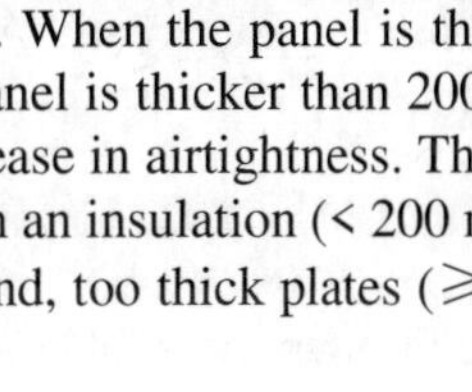

Figure 7. Fitting curve of airtightness (width within 5mm).

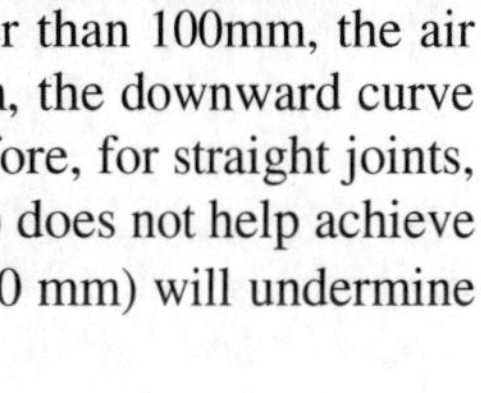

Figure 8. Fitting curve of airtightness (width over 5mm).

3.2. Simulation of special-shaped joints

Compared with straight joints, the special-shaped joints can provide a longer air flow path, and their "mutually embedded" panel-end can improve the assembly efficiency of building envelope. Therefore, this section of the study focuses on the improvement of airtightness while considering the construction efficiency, so as to find the optimized joint by comparing different geometric parameters (Figure 9).

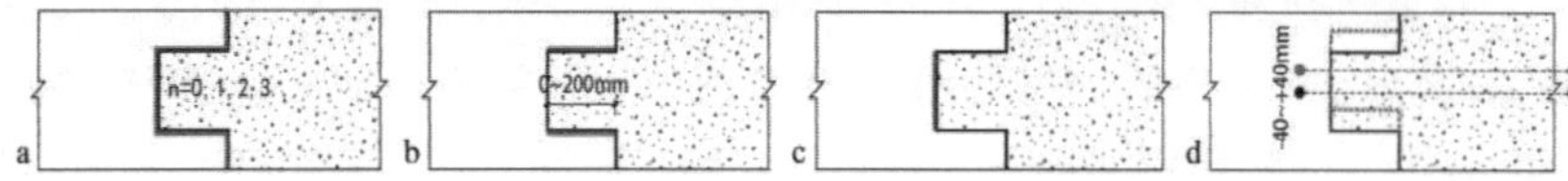

Figure 9. Geometric structural parameters of T&G joints: a. number b. length c. width d. position.

3.2.1. Tongue and groove number. The T&G number represents the complexity of the joint structure. The simulation sets five groups of joints for comparison with different T&G numbers (Figure 10).

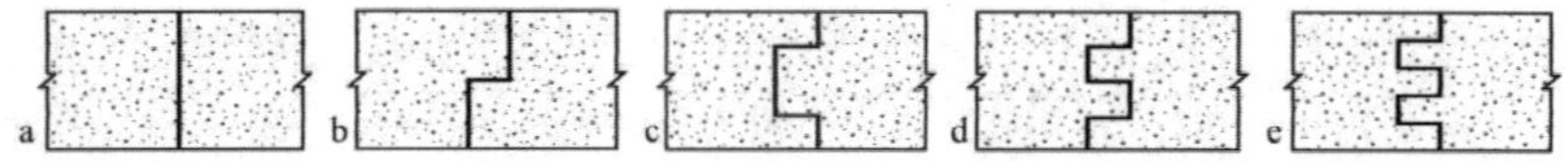

Figure 10. Five groups of joint structures with different T&G numbers for simulation.

As shown in Figure 12, the air infiltration volume of the special-shaped joints is lower than that of the straight joints, but the airtightness improvement of the staggered joint is insignificant, indicating that T&G joints are better. On the other hand, with the increase of T&G number, the joint airtightness is also improved. Nonetheless, complicated panel-end structures are bound to cause higher production costs, processing difficulties and transportation restrictions, and hence more mature LPBs in the market often adopt staggered joints or single T&G joints. After all-round deliberations, this paper chooses the prototype of single T&G joint for further simulation.

3.2.2. Tongue and groove length. The T&G length is one of the core parameters of the T&G joints. This simulation set nine T&G lengths (0~200mm) for comparison. As shown in Figure 13, as the length increases, the air infiltrating through the joint gradually declines with a linear trend, indicating that the longer the T&G is, the better the airtightness is.

From the perspective of construction, over-length T&G will not only exacerbate the difficulty of panel prefabrication, but also inflict damages on transportation and installation; additionally, the joint with large depth is hard to seal on site, thus inducing higher risks of air leakage. According to the author's study, most manufacturers suggest that the T&G length should not exceed 1/2 of the panel thickness. Thus, this paper chooses a 100 mm T&G length as the basic condition for subsequent simulation.

3.2.3. Tongue and groove width. This paper compares the airtight performance of joints with different T&G widths. Eight groups of data (40~180 mm) are given in the simulation. As shown in Figure 14, as the width grows, the air infiltration increases, with its amplitude being very small and basically negligible. It suggests that the T&G width is not a key factor affecting the airtightness of joints. Meanwhile, like the T&G length, the width should not be less than 1/2 of the panel thickness for the sake of structural rationality. Therefore, a 100mm T&G width is used in the following study.

3.2.4. Tongue and groove position. The relative position of the T&G towards the panel depth is another design parameter for single T&G joint. This simulation sets a total of 5 types of T&G positions for comparison (Figure 11). As can be seen from Figure 15, when the T&G position is closer to air outlet (low pressure area), the smaller the air infiltration through the joint, indicating a quite significant linear relationship. It shows that the positional relationship of the T&G structure is one of the key factors affecting the airtightness.

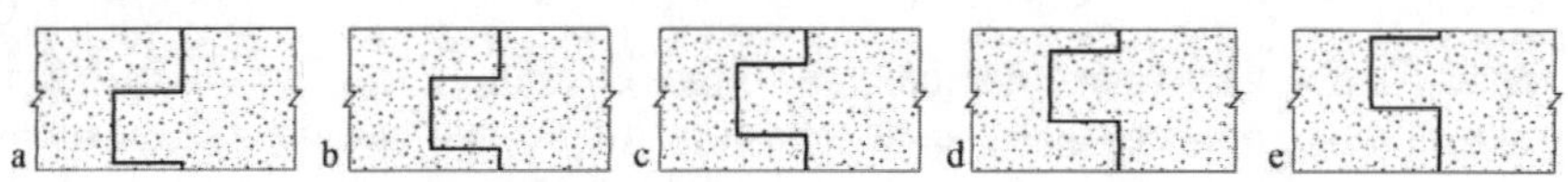

Figure 11. Five tongue and groove positions.

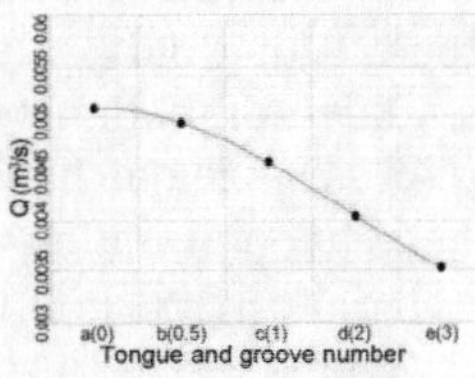

Figure 12. Relation between air leakage and T&G number.

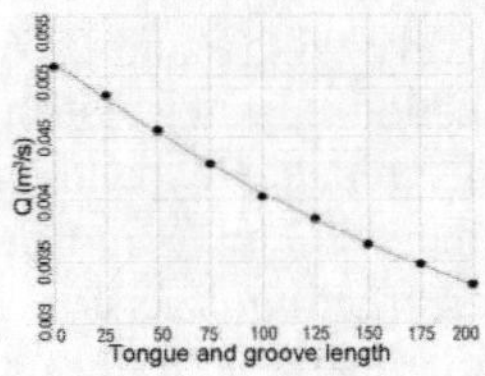

Figure 13. Relation between air leakage and T&G length.

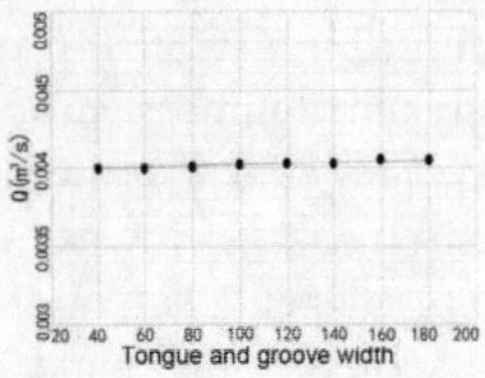

Figure 14. Relation between air leakage and T&G width.

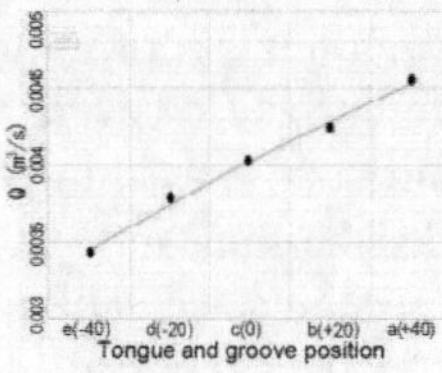

Figure 15. Relation between air leakage and T&G position.

3.3. A comparison of the optimized tongue and groove joint and the straight joint

Based on the above simulation, this paper extracts the “innermost” single T&G joint obtained from Section 3.2.4 as the "optimized joint" and takes the straight joint as the "conventional joint", so as to further compare their airtight performance under different ΔP ranging from 5pa to 50pa.

As shown in Figure 16, the larger ΔP is, the greater air infiltration volume of two types of joints is; besides, the airtightness of the optimized T&G joint is continuously significantly better than that of the straight one. However, if the difference between two joints is compared, the smaller ΔP, the more obvious the airtight improvement of the optimized one. For example, under the ΔP of 5pa, the air infiltration of the optimized T&G joint can be reduced by 57.3 % compared with that of the straight one; yet under the ΔP of 50pa, the figure is only 32.6%. As ΔP mostly ranges from 5~20pa in actual operating conditions, thus the optimized T&G joint is more practical. Besides, to verify the feasibility of applying the “gap method” to LPBs joints, the author has regressed and fitted the airtight data of the two in the form of power exponent equation $Q = \alpha(\Delta P)^b$, discovering the R^2 indicator could reach 0.998 (Figure 17) and 0.997 (Figure 18) respectively. That is, the fitting is very ideal, which shows that the "gap method" was applicable to the estimation of the air infiltration of the panel joints of LPBs.

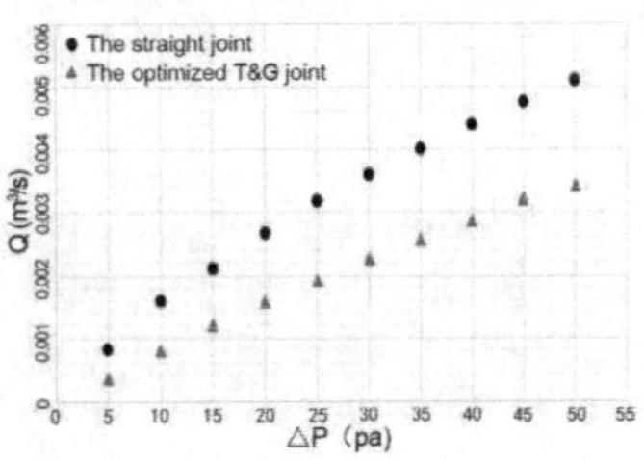

Figure 16. Relation between airtightness and ΔP.

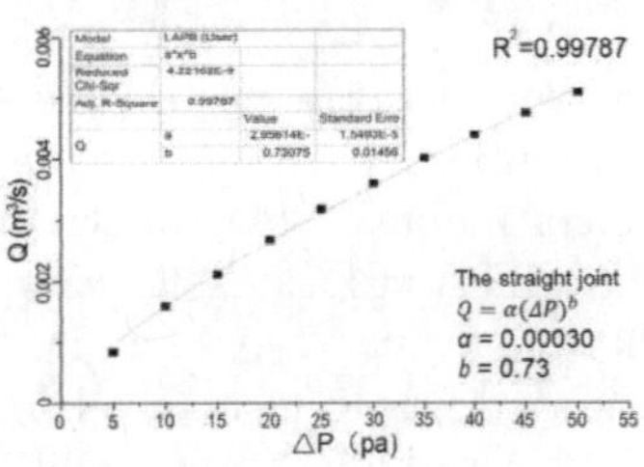

Figure 17. Fitting curve of airtightness under different ΔP (the straight joint).

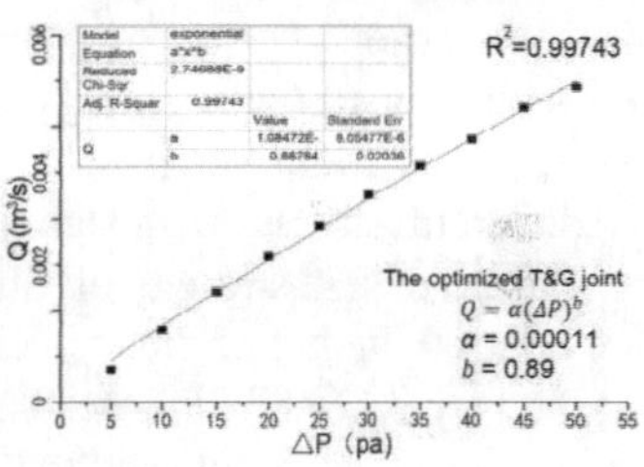

Figure 18. Fitting curve of airtightness under different ΔP (the optimized T&G joint).

Since the goal of the optimized design of joints is to realize nearly zero energy consumption of LPBs, the airtight requirements of Passivhaus (n50 $\leqslant$ 0.6/h) are introduced to gauge the performance of the above two types of joints. This paper pre-sets a building with a net indoor size of 3m×3m×3m, with its each facade composed of three prefab panel, with a total of 12 vertical joints, and the total length is 36m (Figure 19). According to Passivhaus, the upper limit indoor air infiltration per hour (V) is 16.2m³ under 50pa pressure difference. Meanwhile, two sets of joint widths (1mm and 2mm) indicating panel assembly tightness are respectively applied to the optimized T&G joint and straight joint, then the actual air infiltration per hour (v_i) of them in the pre-set building is simulated and calculated (Table 1). According to formula 2, the "lowest sealing ratio (K_i)" of the four groups of joints can be calculated, and K_i represents the minimum ratio of the “perfectly sealed joints” in each room to meet the same Passivhaus airtight requirement.

$$K_i = \left(1 - \frac{V}{v_i}\right) \times 100\% \tag{2}$$

As shown in Table 1, the K_i of the optimized T&G joint is very close to that of the straight one at the lower tightness of connection (2mm width), respectively 96.4% and 97.6%, representing only 1.2% difference. However, at higher tightness of connection (1mm width), the values of both drop significantly, reaching 80.5% and 89.9%, with their discrepancy expanding to 9.4%. On the one hand, under the same airtight requirement, the lower K_i of the T&G joints indicates the joint optimization in the design stage can reduce the dependence on the construction precision and material performance to ensure the overall airtightness; on the other hand, comparing the K_i values under two joint widths, it can be seen that for building systems with higher connection tightness, the airtightness benefits brought by the optimized design of joints are more obvious. Therefore, it is of practical significance for LPBs aiming at nearly zero energy consumption.

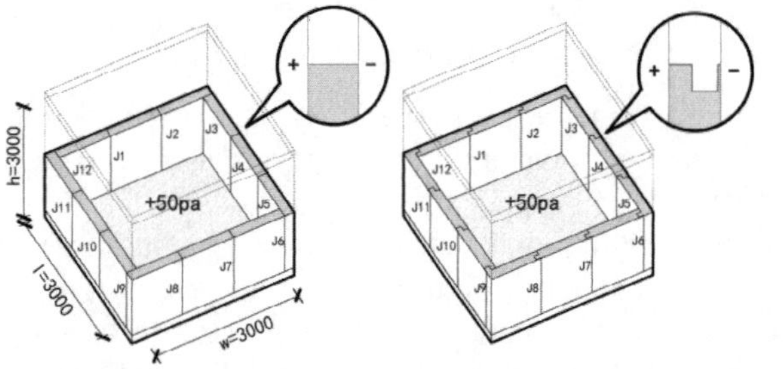

Figure 19. The pre-set building for the airtightness simulation of optimized T&G joints and straight joints

Table 1. Relevant airtightness indicators of two types of joints.

Joint Type	Optimized T&G Joint 1mm	Optimized T&G Joint 2mm	Straight Joint 1mm	Straight Joint 2mm
Infiltration rate Q [m^3 /(s·m)]	0.64×10^{-3}	3.43×10^{-3}	1.24×10^{-3}	5.09×10^{-3}
Actual air infiltration v_i (m^3)	82.9	444.5	160.7	659.7
Lowest sealing ratio K_i (%)	80.5	96.4	89.9	97.6

4. Conclusion

To realize nearly zero energy consumption of LPBs, this paper proceeds from the airtightness of the panel joints, compares the changes of air infiltration volume of straight joints and special-shaped joints in different structural design parameters through CFD simulation and calculation, and obtains the optimized design strategy of joints. The results suggest that the air tightness of the optimized T&G joints are better than that of the conventional straight ones, and the air infiltration volume of the former can be slashed by 32.6~57.3%. Besides, if the goal was to meet the airtightness requirements of Passivhaus, the lower K_i values proved that a well-designed joint system would effectively cut the dependency on construction accuracy and material properties, especially for the building systems with high-performance standards.

References

[1] Mlecnik E, Kaan H and Hodgson G. 2008. “Certification of passive houses: A Western European overview”. PLEA (Dublin). Issue 2008: 1061-1066.
[2] Ahmad K, Hua G, Andreas A and Jiwu R. 2017. “Experimental study of thermal and airtightness performance of structural insulated panel joints in cold climates”. Building and Environment 115: 345-357.
[3] She MH. 1987. “Estimation of infiltration from leakage and climate indicators”. Energy and Buildings. 10(1): 81-86.
[4] Lisa N, Andrew P, Steven E. 2014. “Consideration of Envelope Airtightness in Modelling Commercial Building Energy Consumption”. International Journal of Ventilation 12(4): 369-378
[5] Targo K. 2007. “Air tightness and air leakages of new lightweight single-family detached houses in Estonia”. Building and Environment 42(6): 2369-2377.

Acknowledgments

This work is supported by the National Natural Science Foundation of China (number 51678324).

BIM-based real time building energy simulation and optimization in early design stage

Hao Gao[1], Liwenbo Zhang[1], Christian Koch[2] and Yupeng Wu [1]
[1] Department of Architecture and Built Environment, Faculty of Engineering, The University of Nottingham, University Park, Nottingham, NG7 2RD, UK
[2] Intelligent Technical Design, Bauhaus-Universität Weimar, Marienstr. 13a, 99423 Weimar, Germany
E-mail: Yupeng.Wu@nottingham.ac.uk

Abstract: Increasing requirements of sustainability has prompted architects to pay more attention to the energy performance of their design. The design decisions, such as building form, orientation, window size, and construction materials, made in the early design stages have the significant impact on the building energy performance. However, architects who has limited knowledge of energy modelling usually suffer from selecting and using the common building energy simulation tools. Consequently, building energy simulations are rarely used in the early design process. It is imperative to develop a tool integrated with the deisgn tool to help architect to understand the effects of various design stratigies on building energy performance. Fortunately, the simplified building energy calculation method and Building Information Modelling (BIM) provide an innovative approach for architect. In this paper, a BIM-based real time building energy simulation and optimization tool is proposed and developed based on quasi-steady state method, Revit API and Dynamo, which can assist architects to optimize the building form, building orientation and window to wall ratio (WWR) based on building energy performance in the early design stage to provide better stratigeis for buildings energy efficient design.

Keywords: Early design stage, BIM-based building energy simulation, Design optimization, Revit.

1. Introduction

The concept design stage, also called early design stage, is an especially crucial period for entire building design, when the project goals, design requirements, site development, initial architecture concepts, etc. are analysed and determined [1]. Building energy performance is largely impacted by the early design decisions. For instance, the building energy conception could be reduced by 30-40% without any additional cost if the building shape and orientation are well designed[2]. Many researchers claim that the building performance simulation tools should be integrated into the early building design stage to assist the designers to obtain high performance buildings [2-4]. The application of building energy simulation tools for concept design is a typical example. To achieve the greatest possible energy saving for buildings, building energy simulation should be imbedded into the early design stage. Design decisions on the building form, orientation, window to wall ratio, and construction materials are generally made during this stage and can have a profound impact on how the building will perform [2]. However, building energy simulation tools are rarely used in the early design stage in practice. The reasons include that the conventional building energy simulation is mainly used to perform compliance check for green building rating and to predict the building energy conception after building design [4]; building energy simulation is a costly and time-consuming process [5]; most of the current building energy simulation tools are inadequate, user hostile and too incomplete to be used by the architect (who is the main performer of the concept design) during the early design stage [2]. In addition, the architects prefer to explore "what if" scenarios and evaluate a number of design alternatives in quick way, but the common building energy simulation tools may take between 20s and 5min for one time simulation depending on the building size on a standard computer [6]. Usually, the concept design decisions are mainly determined by the architects based on their prior successfully experience in design energy efficient buildings, rule of thumb and the regulation compliance, rather than energy performance feedback from the building energy simulation tools [5]. Consequently, building design are usually far from the optimized design. Many researchers were conscious of the absent of the building energy simulation in the early design stage, and they developed some building energy simulation methods for the early design stage. Ma and Fukuda [7], Ladybug [8], and Jakubiec and Reinhart [9] provided methods to evaluate the building energy performance in the early design stage with the purpose of minimizing the total energy consumption by using Rhino conjunction with Grasshopper. Negendahl and Nielsen [3], and Ahuja, et al. [10] evaluated the energy flows in a building at the macro level by using the quasi steady state method for the building concept design. Mengana and Mousiadis [11] utilized the visual programming tool of Dynamo together with design tool Autodesk Revit and Green Building Studio to investigate the different energy possibilities and outcomes during the early design stage. Carlos and Nepomuceno [12] presented a simpler methodology in the form of a spreadsheet for the energy analysis in early design stage. It This project is going to develop a BIM-based real time building energy simulation method for architect using in the early design stage for better energy efficient design in buildings.

2. Methodology

The proposed BIM-based building energy simulation and optimization system adopts the monthly quasi-steady state calculation method, according to BS EN ISO 52016. The quasi-steady state model applies aggregated transmission, ventilation, internal and solar heat flow based on average internal and external temperatures, to estimate the monthly heating and cooling demands. The calculation of the monthly energy need for heating and cooling is assumed without the influence of a specific choice of technical building systems by using standard indoor environment condition. This is the consideration for that the actual systems are absent or undersized in early design stage. The proposed BIM-based building energy simulation and optimization system is developed based on Revit Platform API and Dynamo BIM. In this paper, the customized Revit API is used to extract project data for energy calculation and analysis, and generate reports within Revit User Interface (UI) by Visual C#. The visual programming of Dynamo enables the users to create the customized program by manipulating the Dynamo's nodes graphically rather than by specifying them textually. The Dynamo program is designed to assist the user to modify the building model quickly and efficiently in this paper. The general working process of the proposed BIM-based building energy simulation and optimization system is presented in the Figure 1.

Energy demand for heating and cooling a building is given by:

$$Q_{H,n,an} = \sum_{m=1}^{12} Q_{H,n,m} \quad (1)$$

if $\gamma_{H,m} \le 0$ and $Q_{H,hg,m} > 0$:

$$Q_{H,n,m} = 0$$

if $\gamma_{H,m} \le 0$:

$$Q_{H,n,m} = 0$$

else:

$$Q_{H,n,m} = Q_{H,ht,m} + \eta_{H,hg,m} Q_{H,hg,m} \quad (2)$$

$$Q_{C,n,an} = \sum_{m=1}^{12} Q_{C,n,m} \quad (3)$$

If $(1/\gamma_{C,m}) > 2$:

$$Q_{C,n,m} = 0$$

else:

$$Q_{C,n,m} = \alpha(Q_{C,hg,m} + \eta_{C,ht,m} Q_{C,ht,m}) \quad (4)$$

where $Q_{H,n,an}$ is the annual heating demand (kWh), $Q_{H,n,m}$ is the monthly heating demand (kWh), $Q_{C,n,an}$ is the annual cooling demand (kWh), $Q_{C,n,m}$ is the monthly heating demand (kWh), $Q_{H,ht,m}$

and $Q_{C,ht,m}$ are the monthly heat transfer for the heating and cooling mode (kWh), $Q_{H,hg,m}$ and $Q_{C,hg,m}$ are the monthly heat gain for the heating and cooling mode, $\gamma_{H,m}$ and $\gamma_{C,m}$ are the dimensionless heat-balance ratio for the heating and cooling mode, $\eta_{C,ht,m}$ is the dimensionless gain utilization factor, $\eta_{C,ht,m}$ is the dimensionless heat transfer utilization factor, and α is the dimensionless reduction factor for intermittent cooling which is set as 1 in the proposed system.

The input for the monthly heating and cooling demands calculation consists of building geometry, climate data, materials of building fabric, and the basic building operation data. In the proposed BIM-based building energy simulation system, the building geometry is extracted automatically from the Revit architecture model by customized Revit API, mainly including the total exterior wall area, total exterior window area, the total area of different direction of exterior walls (e.g. south facing walls), the total area of different direction of exterior windows (e.g. south facing windows), total floor gross area, ground floor gross area, the perimeter of ground floor, total building volume, building orientation, and window to wall ratio (WWR). The output of the proposed BIM-based building energy simulation and optimization system consists of 3 parts, including general results, optimization results, and detailed results. The general results include monthly heating demand, monthly cooling demand, annual heating demand, annual cooling demand, and annual total energy demand. The optimization results are used to indicate the float of the heat and cooling energy consumption due to the change of window to wall ratio (WWR) and the building orientation. The proposed system is able to provide suggestions of optimized WWR, optimized building orientation, and optimized combination of average WWR and building optimization. For the users, who has relevant energy knowledge background, the system could also provide some results from the calculation process, such as heat transfer by ventilation and transmission respectively, internal gains, solar gains, heat balance ratio, etc..

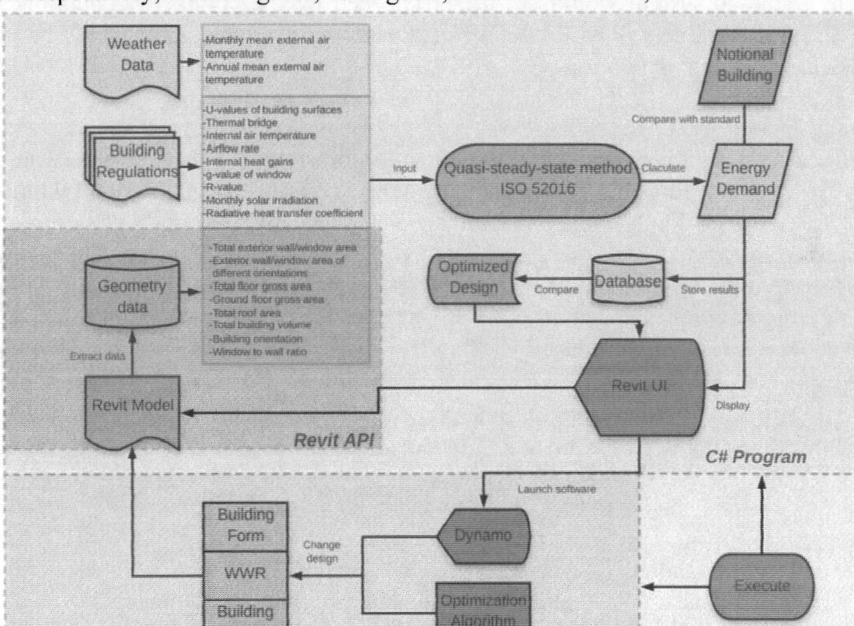

Figure 1. The general work process of the BIM-based building energy simulation and optimization system

3. Verification

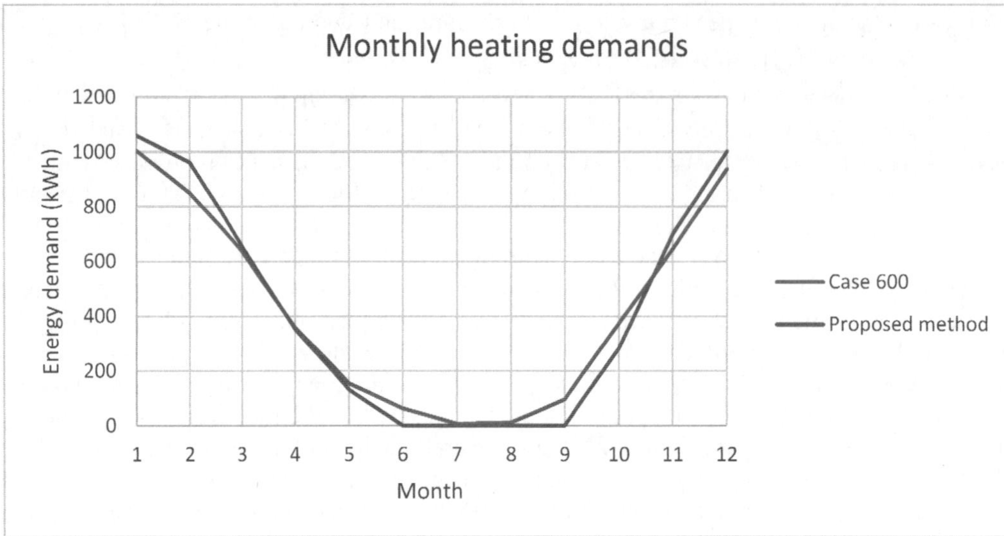

Figure 2. Verification of the heating demands

This is just an initial step of verification of the proposed system in terms of the heating demands. The results of the proposed tool are initially verified based on the BSETEST 600 case as described in ANSI/ASHRAE, which shows that the heating demands estimated by the proposed tool are acceptable. The proposed system would be further checked by the commercial building simulation software, e.g. IES-VE and EnergyPlus. More information about the verification would be provided by the future research.

4. Conclusion

A BIM-based real time building energy simulation and optimization tool is developed in this paper, which can assist architects to optimize the building form, building orientation and window to wall ratio (WWR) based on building energy performance in the early design stage. This approach is developed based on monthly quasi-steady state method, Revit API and Dynamo. The principle of the monthly quasi-steady state method based on BS EN 52016, the workflow of the Revit API developed by Visual C# programming, and the framework of the Dynamo program developed by visual programming are presented and introduced in this paper. The results of the proposed tool are initially verified based on the BSETEST 600 case as described in ANSI/ASHRAE, which shows that the heating demands estimated by the proposed tool are as excepted. In the future work, the verification process would be conducted in detail, including the verification on heat transmission, ventilation heat transfer, internal heat gain, and the solar heat gain.

5. Reference

[1] *Integrated Whole Building Design Guidelines*. Available: http://www.mfe.govt.nz/sites/default/files/integrated-building-guidelines.pdf
[2] E. Elbeltagi, H. Wefki, S. Abdrabou, M. Dawood, and A. Ramzy, "Visualized strategy for predicting buildings energy consumption during early design stage using parametric analysis," *Journal of Building Engineering,* vol. 13, pp. 127-136, 2017.
[3] K. Negendahl and T. R. Nielsen, "Building energy optimization in the early design stages: A simplified method," *Energy and Buildings,* vol. 105, pp. 88-99, 2015.
[4] T. L. Hemsath, "Conceptual energy modeling for architecture, planning and design: Impact of using building performance simulation in early design stages," in *13th Conference of International Building Performance Simulation Association*, 2013.
[5] S. Bambardekar and U. Poerschke, "The architect as performer of energy simulation in the early design stage," in *Building Simulation*, 2009, pp. 1306-1313.

IOP Conf. Series: Materials Science and Engineering **556** (2019) 012064 doi:10.1088/1757-899X/556/1/012064

[6] A. Hollberg, T. Lichtenheld, N. Klüber, and J. Ruth, "Parametric real-time energy analysis in early design stages: a method for residential buildings in Germany," *Energy, Ecology and Environment,* vol. 3, pp. 13-23, 2018.

[7] M. Qingsong and H. Fukuda, "Parametric office building for daylight and energy analysis in the early design stages," *Procedia-Social and Behavioral Sciences,* vol. 216, pp. 818-828, 2016.

[8] M. S. Roudsari, M. Pak, and A. Smith, "Ladybug: a parametric environmental plugin for grasshopper to help designers create an environmentally-conscious design," in *Proceedings of the 13th international IBPSA conference held in Lyon, France Aug*, 2013.

[9] J. A. Jakubiec and C. F. Reinhart, "DIVA 2.0: Integrating daylight and thermal simulations using Rhinoceros 3D, Daysim and EnergyPlus," in *Proceedings of building simulation*, 2011, pp. 2202-2209.

[10] S. Ahuja, P. Chopson, J. Haymaker, and G. Augenbroe, "Practical energy and cost optimization methods for selecting massing, materials, and technologies," *of Architectural Research,* p. 280, 2015.

[11] S. Mengana and T. Mousiadis, "Parametric BIM: Energy Performance Analysis Using Dynamo for Revit," ed, 2016.

[12] J. S. Carlos and M. C. Nepomuceno, "A simple methodology to predict heating load at an early design stage of dwellings," *Energy and Buildings,* vol. 55, pp. 198-207, 2012.